Fluid Mechanics

(for Rajasthan Technical University)

Dr. Kamlesh Purohit
Professor
Dept. of Mechanical Engineering,
JNV University, Jodhpur

Dr. S.P. Harsha
Assistant Professor
Dept. of Mechanical & Industrial Engineering,
IIT, Roorkee

Dr. R.K. Purohit
Retd. Associate Professor
Dept. of Mechanical Engineering,
JNV University, Jodhpur

SCIENTIFIC
PUBLISHERS (INDIA)

Published by:
Scientific Publishers (India)
5-A, New Pali Road, P.O. Box 91,
Jodhpur – 342 001 (India)

E-mail: info@scientificpub.com
Website: www.scientificpub.com

ISBN: 978-81-7233-816-9

Lasertype set: Rajesh Ojha
Printed in India

4ME3: FLUID MECHANICS

Syllabus

UNIT -1 — Basic Definitions and Fluid Properties; Definition of Fluid, Incompressible and compressible fluids, Fluid as a continuum, Mass, Density, specific weight, relative density, specific volume, Bulk modulus, velocity of sound Ideal fluid Viscosity. Newtonian and Non Newtonian fluid, Kinematic viscosity, Effect of temperature and pressure on viscosity, surface tension capillarity, vapour pressure and cavitation.

UNIT -2 — Kinematics and conservation of Mass : Flow classifications. Fluid velocity and acceleration, streamlines and the stream function. Pathlines and streak lines. Deformation of a fluid element, verticity and circulation. Irrotational and Rotational flow. Flownet, Laplace equation. Conservation of mass and the continuity equation for three dimensions. Fluid Momentum: The Momentum theorem Applications of the momentum theorem Equation of motion, Euler's equation of motion Integration of Euler's equation of motion. Bernoulli's equation. Applications of Bernoulli's Pitot tube, Equation of motion for Viscous fluid, Navier Stoke's equation.

UNIT -3 — Orifice discharging free, Jet, vena contracts, co-efficient of contraction, velocity and discharge, coefficient of resistance. Orifices and mouthpieces Nozzles and weires. Flow Through Pipes: Reynold's experiment Darcy's Weisback equation. Loss of head due to sudden enlargements, contraction, entrance, exit obstruction, bend, pipe fittings. Total and Hydraulic grandient lines, Flow through pipe line. Pipes in series, parallel Transmission of power through pipes.

UNIT -4 — Laminar Flow: Simple solution of Navier Stokes equations. Hagen – Poiseuille flow. Plans Poiseuille flow and coutte flow. Turbulent Flow; Variation of friction factor with Reynold's number. The Prandt Mixing length hypothesis applied to pipe flow, velocity distribution in smooth pipes, sough pipes. The Universal pipe friction laws, Colebrook. White formula. Dimensional Analysis: Buckingham variables, Model Similitude, Force ratio, Reynolds, Froude's Mach, Weber and Euler numbers and their applications. Undistorted model distorted model scale effect.

UNIT -5 — The Boundary Layer: Description of the boundary layer. Boundary Layer thickness boundary layer separation and control. The Prandtl boundary layer equation. Solution for cominar boundary layer. The momentum equation for the boundary layer. The flat plate in uniform free stream with no pressures gradients. Approximate momentum analysis laminar boundary Aerofoils Theory. Flow round a body; Drag skin friction drag, pressure drag, combined skin friction & pressure drag (Profile drag) wave drag, lift induced drag. Flow past sphere & Cylinder.

Preface

It gives me immense pleasure in placing this book `Fluid Mechanics for RTU' in the hands of B.Tech IV Semester students of Mechanical Engineering of Rajasthan Technical University (RTU).

The subject matter has been evolved from first principles and the treatment is then extended to considerable depth so that this single book covers the complete course of IV Semester of RTU.

The importance of illustrative examples cannot be under estimated, as they help in understanding the concepts of the subject clearly. Besides, they also follow the style of question set in RTU.

We are thankful to colleagues, friends and students who encouraged us to write this book.

We express our appreciation and gratefulness to our publisher Tanay Sharma for his most cooperative, painstaking attitude and untiring effort in bringing out this book in a very short period.

We have used several references and we are grateful to those learned authors.

Jodhpur

Dr. Kamlesh Purohit
Dr. S.P. Harsha
Dr. R.K. Purohit

Contents

1

Basic Concept Relating to Fluids

1.1 INTRODUCTION

Fluid mechanics is the science which deals with the application of the fundamental principles of general mechanics to fluid at rest and in motion. In the development of the principles of fluid mechanics, some fluid properties play principal roles. Other only minor roles or no roles at all. In fluid statics, weight is the important property, whereas in fluid flow, density and viscosity are predominant properties. Where appreciable compressibility occurs, principles of thermodynamics must be considered. Vapour pressure becomes important when negative pressure gauge are involved, and surface tension affects static and flow conditions in small passages. The introductory text will provide a sound foundation in the fundamental aspects of fluid mechanics.

1.2 DEFINITION OF A FLUID

A fluid is a substance that deforms continuously under the action of externally applied forces.

Another property that defines a fluid is the fact that fluids do not have a predetermined shape but, rather, assume the shape of their container. For this reason, a liquid can be easily transferred from one container to another. These fluid properties are in contrast to the behaviour of solids that have predefined shapes (i.e. they cannot be easily ``poured" into shapes other than their original shape) and that deform by a finite amount (not continuously) under the influence of externally applied forces.

1.3 FLUID AS A CONTINUUM

There are two possible mathematical approaches to the treatment of problems in fluid mechanics. In the **microscopic approach**, fluid behaviour is described by specifying motion of the individual molecules. In the **macroscopic approach,** fluid behaviour is described by specifying motion of small fluid volumes containing many molecules. The macroscopic approach uses the average effect of many molecules and means that we are treating the fluid as a continuous medium or a **continuum**. In this text, we will use the continuum approach throughout.

1.4 INCOMPRESSIBLE AND COMPRESSIBLE FLOW

A flow is classified as being incompressible or compressible, depending on the level of variation of density during flow. A flow is said to be incompressible if the density remains nearly constant throughout. Compressible flow is defined as variable density flow.

Liquids are usually referred as incompressible. A pressure of 210 atm, for example, causes the density of liquid water at 1 atm to change by just 1 percent. Gases on the other hand, are highly compressible. A pressure change of just 0.01 atm, for example causes a change of 1 per cent in the density of atmospheric air.

1.5 BASIC DEFINITIONS

The following properties of fluids are of general importance to the study of fluid mechanics.

1.5.1 Mass

Mass denoted by M, is the quantity of matter, it is a function of the internal structure, and the dimensions of a substance. In the SI system the unit used to define mass is the kilogram, kg.

1.5.2 Density

The density of a fluid, designated by the Greek symbol ρ (rho), is defined as its mass per unit volume. Density is typically used to characterize the mass of a fluid system. In SI unit it is expressed as kg/m^3.

The value of density can vary widely between different fluids, but for liquids, variation in pressure and temperature generally have only a small effects on the value of ρ. The density of water at 4°C is 1000 kg/m^3. The density of a gas is strongly influenced by both pressure and temperature. The density of air at 21.2°C is 1.2 kg/m^3.

1.5.3 Specific Volume

The specific volume, v is the volume per unit mass and is therefore the reciprocal of the density-that is,

$$v = \frac{1}{\rho} \tag{1.1}$$

This property is not commonly used in fluid mechanics but is used in thermodynamics.

1.5.4 Specific Weight

The specific weight of a fluid, designated by the symbol w, is defined as its weight per unit volume. The weight, W, of an object is the force applied to that object by the action of gravity. Then the weight of a mass is equal to the product of the mass times acceleration due to gravity (W = Mg = ρVg)

Thus mathematically

$$w = \frac{W}{V} = \frac{Mg}{V} = \frac{\rho V g}{V}$$

$$w = \rho g \tag{1.2}$$

The specific weight of water is 9810 N/m³. The specific weight of air is 11.8 N/m³.

1.5.5 Relative density

The relative density of a fluid, designated as S, is defined as the ratio of the density of the fluid to the density of water at the specified temperature. Usually the specified temperature is taken as 4°C, and at this temperature the density of water is 1000 kg/m³. Relative density is also called specific gravity. In equation form relative density is expressed as

$$w = \frac{\rho_{\text{fluid}}}{\rho_{\text{water at 4°C}}} = \frac{\rho_f}{1000} \tag{1.3}$$

and since it is the ratio of densities, the value of S does not depend on the system of units used. It is a dimensionless quantity. For example, the relative density of water at 4°C, is 1 and for mercury at 20°C is 13.55.

The specific gravity of a gas is the ratio of its density to that of either hydrogen or air at some specified temperature and pressure, but there is no general agreement on these standards, and so they must be stated in any given case.

Since the density of a fluid varies with temperature, specific gravities must be determined and specified at particular temperatures.

1.6 VISCOSITY

Viscosity is that property of a fluid which, by virtue of cohesion and interaction between fluid molecules offers resistance to shear deformation. Different fluids deform at different rates under the action of some shear stress. Viscosity of liquids vary inversely with temperature while of gases vary directly with temperature. Gases as well as liquids have viscosity, although the viscosity of gases is less evident in every day life.

Consider the motion of a fluid between two very long parallel plates at a small distance h apart. The upper plate is moved at a constant velocity U in its own plane and the lower plate remains stationary. The fluid between the two plates moves with velocity U = u (y) that would be found to vary linearly as shown in Fig 1.1. Thus, a velocity gradient, du/dy, is developed in the fluid between the plates.

Newton postulated that the shear stress within a fluid is proportional to the rate of shearing strain or velocity gradient in a direction perpendicular to the layers. That is

$$\tau \alpha \frac{du}{dy}$$

$$\tau = \mu \frac{du}{dy} \tag{1.4}$$

Where τ = shear stress

$\frac{du}{dy}$ = velocity gradient (Fig. 1.1)

μ = a proportionality constant

(τ is the Greek letter "tau" and μ is the Greek letter "mu"). The proportionality constant (μ) in equation (1.4) is called the coefficient of viscosity or, more simply, the viscosity of the fluid. It is also termed `absolute' or (since it involves force) `dynamic' viscosity to distinguish it from kinematic viscosity. The symbols η and (Greek `eta') are both widely used for viscosity, in this book μ will be used.

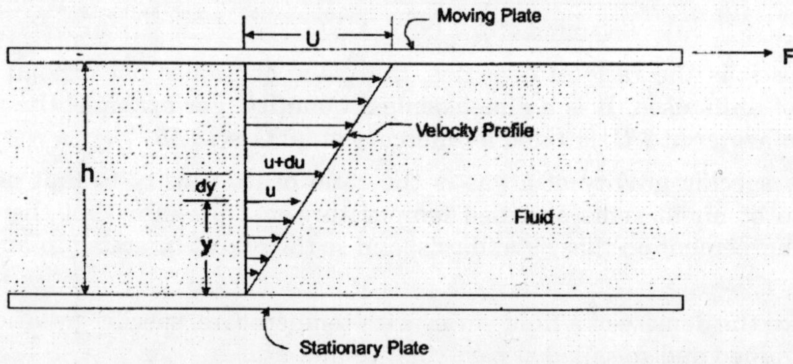

Fig. 1.1. Velocity distribution in a viscous fluid between two parallel plates

1.6.1 Units of Viscosity

Because viscosity is a derived quantity, its units must be derived from equation (1.4) Solving for μ gives

$$\mu = \frac{\tau}{du/dy}$$

Substituting units only gives

$$\mu = \frac{N/m^2}{m/s/m} = \frac{N\,s}{m^2}$$

Since 1 Pa = 1N/m², we can also express μ as

$$\mu = \text{Pa s}$$

Sometimes, when units for μ are being combined with other terms-especially density - it is convenient to express μ in terms of kg rather than Newton (N).

$$\text{Since } 1\text{ N} = \frac{1\text{ kg m}}{\text{s}^2}$$

μ can be expressed as

$$\mu = \frac{\text{N s}}{\text{m}^2} = \frac{\text{kg m}}{\text{s}^2} \times \frac{\text{s}}{\text{m}^2} = \frac{\text{kg}}{\text{ms}}$$

Thus,

$$\mu = \frac{\text{N s}}{\text{m}^2} = \text{Pa s} = \frac{\text{kg}}{\text{ms}}$$

In C.G.S. units, viscosity is measured in poise. The unit is named after Poiseuille, who did some of the first work on viscosity, In S.I. units

$$1\text{ poise} = \frac{1}{10}\frac{\text{N s}}{\text{m}^2} \quad (1.5)$$

or

$$\frac{1\text{ N s}}{\text{m}^2} = 10\text{ poise} \quad (1.6)$$

One hundred centipoise equal a poise.

$$\text{Accordingly} \quad 1\text{ cp} = \frac{1}{1000}\text{Ns/m}^2$$

Typical values of viscosity

For water $\mu = 1.14 \times 10^{-3}$ kg/ms

For air $\mu = 1.78 \times 10^{-5}$ kg/ms

1.6.2 Dimensional Formula of Viscosity

Viscosity is defined as the ratio of a shear stress to a velocity gradient.

$$\tau = \frac{\text{Force}}{\text{Area}} = \frac{F}{L^2}$$

$$\frac{du}{dy} = \frac{LT^{-1}}{L} = T^{-1} = \frac{1}{T}$$

$$\mu = \frac{\tau}{du/dy} = \left[\frac{FT}{L^2}\right] = [\,FT\,L^{-2}]$$

Since $[F] = [M L T^{-2}]$

Therefore,

$$\mu = [M L T^{-2} T L^{-2}] = [M L^{-1} T^{-1}]$$

Although this last form is usually more suitable in dimensional analysis, the version involving [F] is more useful in defining units of viscosity. Therefore,

$$\mu = [F T L^{-2}] = [M L^{-1} T^{-1}]$$

1.6.3 Kinematic Viscosity

In many problems involving viscosity, we are concerned with the magnitude of the viscous forces compared with the magnitude of the inertia forces, that is, those forces causing acceleration of the fluid. Since the viscous forces are proportional to the viscosity μ, and the inertia forces to the density ρ, the ratio of viscosity divided by density is frequently involved, and in fact has come to be regarded almost as a property of the fluid in its own right.

The ratio of dynamic viscosity to density is known as the kinematic viscosity and is denoted by the symbol υ (Greek "nu") so that

$$\nu = \frac{\mu}{\rho} \tag{1.7}$$

Since μ and ρ are both properties of the fluid, ν is also a property.

1.6.4 Units of Kinematic viscosity

We can derive the units for kinematic viscosity by substituting the previously developed units for μ and ρ :

$$\nu = \frac{\mu}{\rho} = \frac{\text{kg/ms}}{\text{kg/m}^3} = \text{m}^2/\text{s}$$

The basic unit m^2/s is too large for most purposes and so the mm^2/s is generally employed.

In C.G.S. units, kinematic viscosity is measured in Stokes, thus honouring the Sir George Stokes, who contributed much to the theory of viscous fluids. This unit too was rather large, and the centistokes (cs), i.e., 10^{-2} stokes = mm^2/s was in wide use.

$$1 \text{ stoke} = 10^2 \text{ mm}^2/s = 1 \text{ cm}^2/\text{s} = 10^{-4} \text{ m}^2/s \tag{1.8}$$

$$1 \text{ cs} = 1 \text{ mm}^2/\text{ s} = 10^{-6} \text{ m}^2/\text{ s} \tag{1.9}$$

Water has a kinematic viscosity of exactly 1 mm^2/s at 20.2°C.

Typical values of kinematic viscosity at 20°C and 1 atm

For water $\nu = 1.01 \times 10^{-6}$ m^2/s

For air $\nu = 1.51 \times 10^{-5}$ m^2/s

1.6.5 Dimensional formula of kinematic viscosity

$$\nu = \frac{\mu}{\rho}$$

Since $\mu = [M L^{-1} T^{-1}]$

$\rho = [M L^{-3}]$

$$\text{Therefore, } \nu = \left[\frac{M L^{-1} T^{-1}}{M L^{-3}}\right] = [L^2 T^{-1}]$$

Kinematics is defined as the study of motion without regard to the causes of the motion, and so is concerned with lengths and time intervals only, not with masses. That is why the name kinematic viscosity, now in universal use, has been given to the ratio μ/ρ.

1.6.6 Newtonian and non Newtonian fluids

Fluids for which the shear stress is proportional to velocity gradient (rate of deformation or shear strain rate) are called **Newtonian fluids.** Most common fluids such as water, air, alcohol, gasoline and oils are Newtonian fluids.

The plot of Fig. 1.2 is a straight line that passes through the origin. This result is a characteristic of a Newtonian fluid, but there are other types of fluids called non-Newtonian fluids. A graph of shear stress versus velocity gradient, called a **rheological diagram**, is shown in Fig. 1.2 for several types of fluids. Newtonian fluids follow Newtons law of viscosity (equation 1.4).

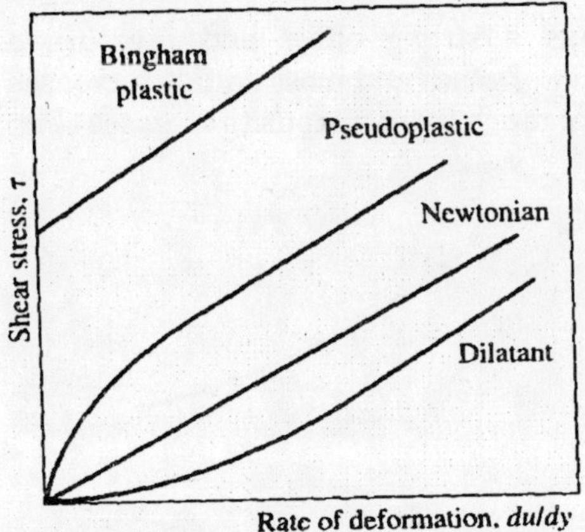

Fig. 1.2. Rehelogical diagram

Non-Newtonian fluids are considered shear thinning, or *pseudoplastic,* when their viscosity is decreasing with increasing velocity gradient. Moltan plastics and paint exhibit such behaviour. Other fluids show shear thickening, or dilatant, when the viscosity increases with increasing velocity gradient. This behaviour is usually observed in mixtures of liquid and solid particles. Fig. 1.3 shows different types of fluid behaviour.

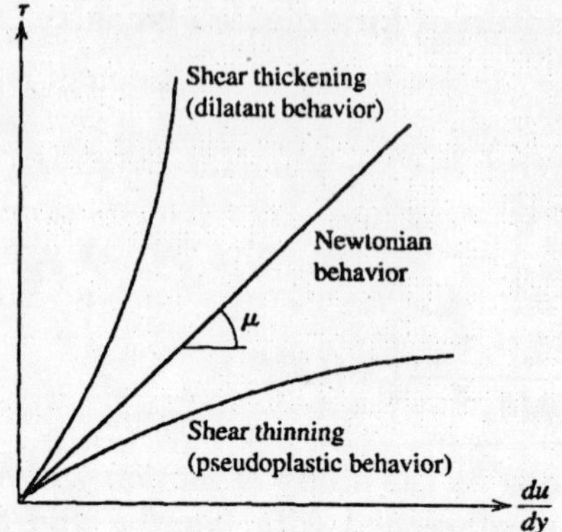

Fig. 1.3. Different types of fluid behaviour

A power law equation called the Ostwald deWaele equation usually gives an adequate description

$$\tau = k\left(\frac{du}{dy}\right)^n$$

where, k is called a consistency index with dimensions of Ns^n/m^2

n is a flow behaviour index.

(a) pseudoplastic fluids n < 1

(b) dilatant fluids n > 1

1.6.7 Effects of temperature and pressure on viscosity

Temperature has a strong effect and pressure a moderate effect on viscosity. The viscosity of gases and most liquids increases slowly with pressure. Liquids in general are much more viscous than gases (Fig. 1.4).

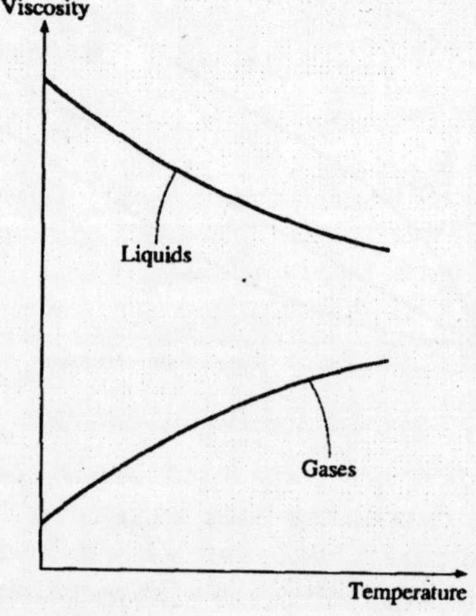

Fig. 1.4. Variation of viscosity for liquid and gases

The kinematic viscosity of liquids is largely a function of temperature since this is the case for both dynamic viscosity and density. Thus values of both dynamic viscosity and kinematic viscosity may be given in graphical or tabular form as a function of temperature. The kinematic viscosity of gases, however, is a function of pressure as well as temperature. The dynamic viscosity of gases is essentially dependent on temperature only, but the density of a gas is a function of both pressure and temperature and is calculated from a gas equation of state.

The viscosities of same fluids are plotted against temperature in Fig. 1.5. The kinematic viscosities are plotted against temperature in Fig. 1.6.

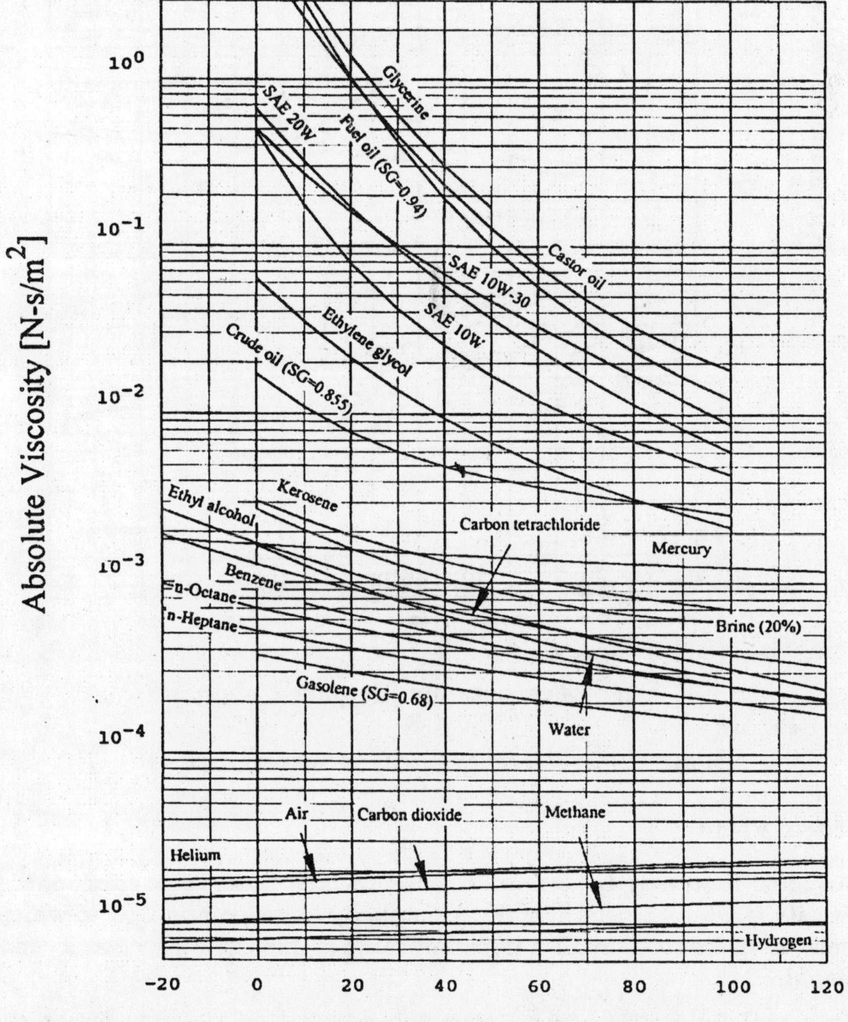

Fig. 1.5. Plot of dynamic viscosity versus temperature at 1 atm.

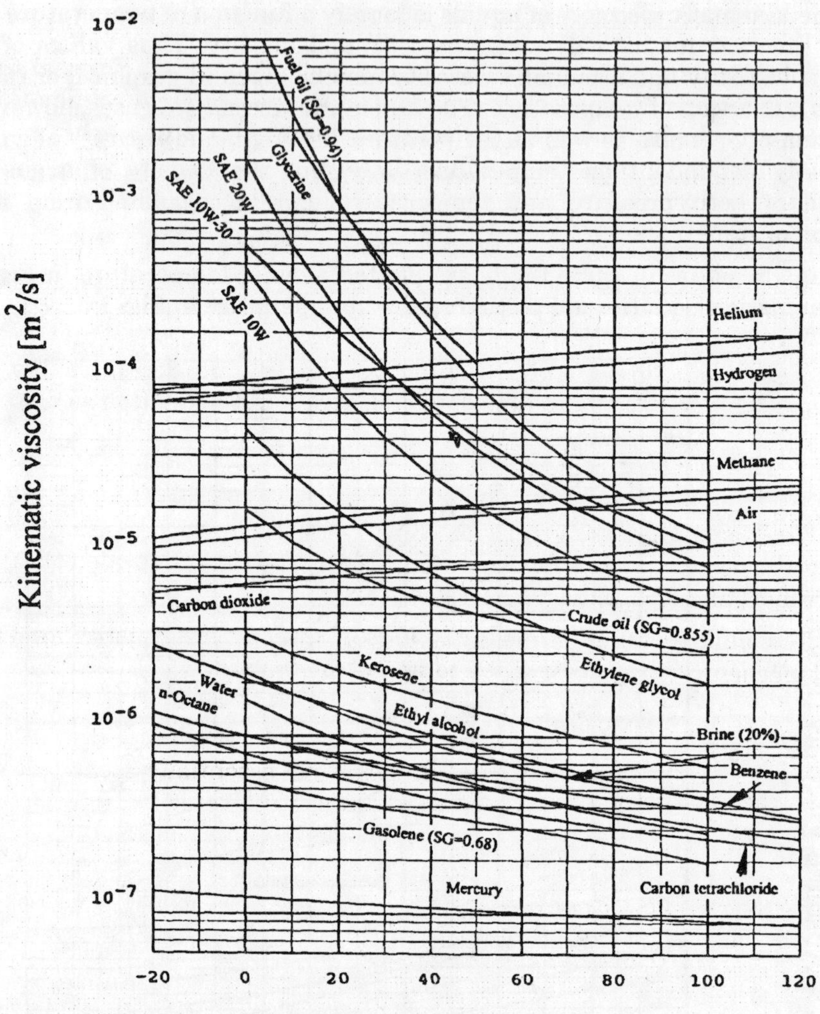

Fig. 1.6. Plot of kinetic viscosity versus temperature

1.6.8 Ideal Fluid

An ideal fluid may be defined as one in which there is no friction; that is, its viscosity is zero. For mathematical simplicity in describing flow in which the influence of viscosity is small a hypothetical fluid having zero viscosity may be postulated.

In a real fluid either liquid or gas, tangential or shearing forces always come into being whenever motion takes place, thus giving rise to fluid friction, because these forces oppose to the movement of one particle past another. These friction forces are due to viscosity.

1.7 COMPRESSIBILITY AND ELASTICITY OF FLUIDS

All fluids are compressible under the application of an external force and when the force is removed they expand back to their original volume exhibiting the property that stress is proportional to volumetric strain.

The compressibility is defined in terms of an average bulk modulus of elasticity.

K = pressure change/volumetric strain.

$$K = -\frac{dp}{dV/V} \tag{1.10}$$

Since for a given mass of fluid as the volume decreases the density increases then $-dV = d\rho$ and the volumetric strain can be written as $d\rho/\rho$.

$$\text{Hence} \quad K = \frac{dp}{d\rho/\rho} \tag{1.11}$$

The denominators of equations (1.10) and (1.11) are dimensionless, so that K has the dimensions of pressure, or force per unit area. Equation (1.10) is used for liquids, and equation (1.11) for gases.

The bulk modulus of elasticity K is of interest in acoustics as well as in fluid mechanics. The speed of sound in any medium is

$$a = \sqrt{K/\rho} \tag{1.12}$$

and for a gas, sound waves are transmitted essentially isentropically (See chapter of compressible flow), so that the velocity of sound in perfect gas is

$$a = \sqrt{k\,R\,T} \tag{1.13}$$

Where $k = c_p/c_v = 1.4$ for air

$R = 0.287$ kJ/kg K for air

Typical values of K for water and air are given below

At atmospheric pressure

For water $K = 2.18 \times 10^9$ Pa

For air $K = 1.013 \times 10^5$ Pa

Which means that air is approximately 20000 times more compressible than water.

1.8 SURFACE TENSION

Surface tension is the surface force that develops at the interface between two immiscible liquids or between liquid and gas or at the interface between a liquid and a solid surface. Because of surface tension, small water droplets, gas bubbles, drops of mercury, and the like to tend to maintain spherical shapes (Fig. 1.7).

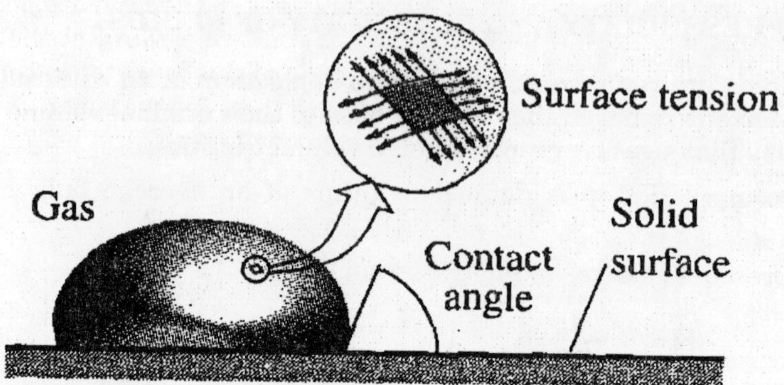

Fig. 1.7. Schematic of surface tension physics

At the interface between a liquid and another liquid or a gas, the difference in the molecular structure of the two substances result in an imbalance of the molecular forces. These forces can be quite strong, but they decay very rapidly with distance from the interface. Therefore, intermolecular force interactions result in a zero net force. However, molecules at interfaces interact with molecules of the same fluid on only one side. Consequently, such molecules experience a net force that puts the interface under tension. The ultimate magnitude and direction of this tension force is determined not only by what happens on either side of the interface; but by the way molecules of the two fluids interact with each other. Therefore, surface tension is a property specific to the ``participating" fluids. For instance, surface tension between water and air is different from that between water and another gas. Surface tension forces are also sensitive to the physical and chemical condition of the solid surface in contact, such as its roughness, cleanliness, or temperature.

The amount of force per unit length necessary to deform this interfacial surface is called the surface tension and is denoted by the Greek letter σ (sigma), and is usually expressed in the unit N/m.

$$\sigma = \frac{\text{Force}}{\text{Length}} = \frac{\text{MLT}^{-2}}{\text{L}} = [\text{MT}^{-2}]$$

1.9 CAPILLARITY

Capillarity is defined as a phenomenon of rise or fall of a liquid surface in a small tube relative to the adjacent general level of liquid when the tube is held vertically in the liquid. The rise of liquid surface is known as capillary rise while the fall of the liquid surface is known as capillary depression (Fig. 1.8). The curved free surface of a liquid in a capillary tube is called the meniscus.

When a liquid-gas interface ends at a solid wall a, the contact angle that the interface makes with the wall (Fig. 1.9) is a function of what the two fluids are, and also can depend on the composition of the wall. **Contact angle** is defined as the angle between the wall and the surface, measured in the liquid from the wall. A contact angle of zero is referred to as **perfect wetting**. A contact angle of 180° is referred to as **complete nonwetting**. If the wall attractive forces (**adhesive forces**) are stronger than the fluid attractive forces (**cohesive forces**), the surface of the fluid is concave upward at the wall, and we say that the fluid wets the solid. An example of this is Water in contact with glass. If the fluid attractive forces are stronger than the wall attractive forces, the surfaces of the fluid is concave downward at the wall, and we say that the fluid does not wet the solid. An example of this is Mercury in contact with glass.

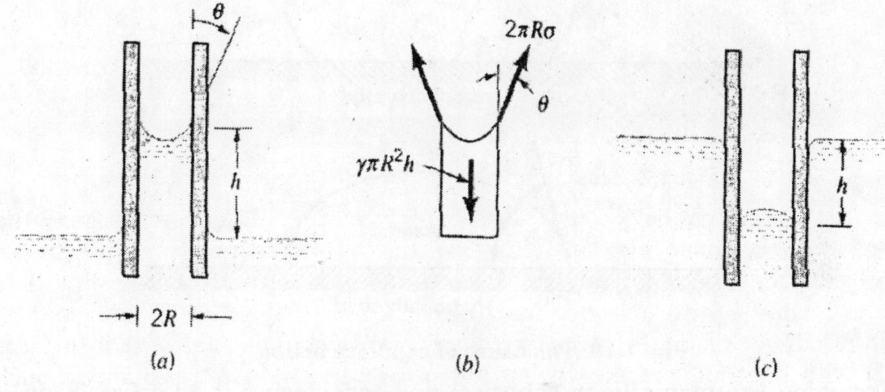

Fig. 1.8. Capillary action

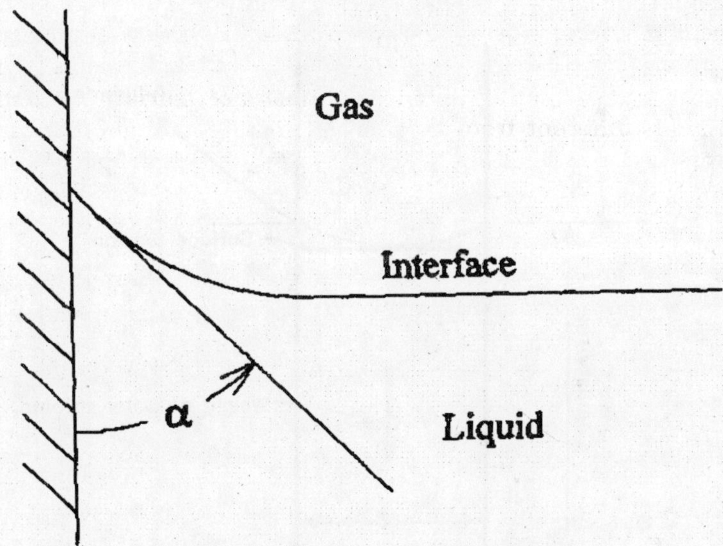

Fig. 1.9. Definition of contact angle

Another important contributor to the physical effect of capillary rise is to be the adhesion of water to most solid materials. Solids that have positive adhesion to water are called **hydrophilic** (Water liking), and that those that repel water are **hydrophobic.** The latter have negative adhesion to water (Fig. 1.10).

Hydrophobic materials have a contact angle with water that is larger than 90°. For example, the contact angle between water and paraffin is 107°. Silver, on the other hand, is neutral to pure water; its contact angle is nearly 90°. Quartz and other materials found in porous soils have a contact angle with water that is less than 90°; this means that they wet well by water.

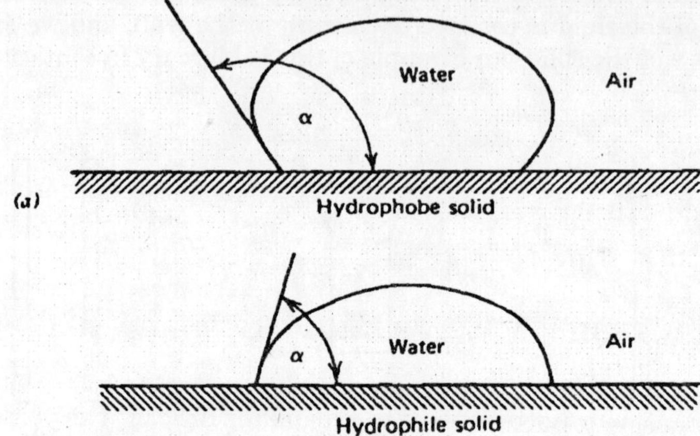

Fig. 1.10. The cause of capillary action

Assuming the meniscus is spherical and considering it as a free body as shown in Fig. 1.11.

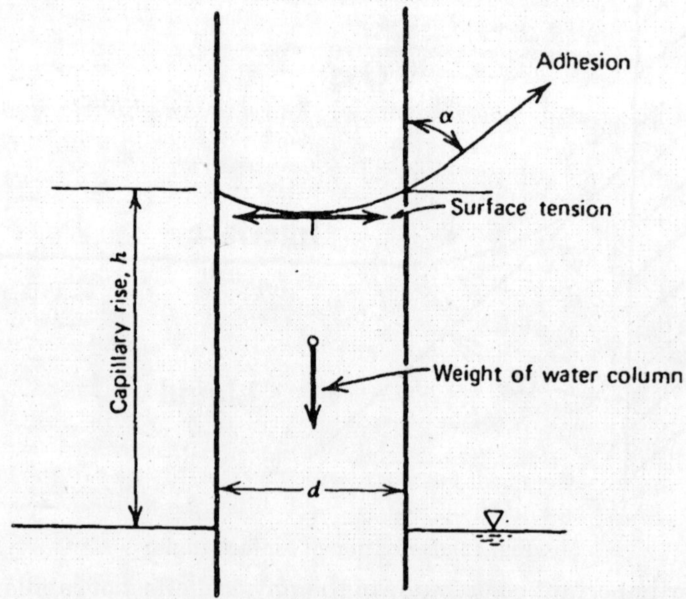

Fig. 1.11. Free body diagram

Weight of the column of liquid

$$W = \rho \; g \, V = \rho g \pi \; R^2 \, h \tag{1.14}$$

The force due to the surface tension acts at the circumference where the liquid touches the glass wall.

$$T = \sigma 2 \pi R \quad (1.15)$$

Summing forces in the vertical direction shows that the vertical component of the surface tension force must equal the weight of the liquid column.

$$W = T \cos \theta \quad (1.16)$$

$$\rho g \pi R^2 h = \sigma 2 \pi R \cos \theta$$

$$h = \frac{2 \sigma \cos \theta}{\rho g R} = \frac{4 \sigma \cos \theta}{\rho g D} \quad (1.17)$$

When liquid wets the wall

$$h = 4 \sigma / \rho g D \quad (1.18)$$

1.10 PRESSURE INSIDE A WATER DROPLET, SOAP AND BUBBLE

A sketch of a droplet in equilibrium is shown in Fig. 1.12. The pressure difference from inside to outside tends to expand the droplet. The pressure force is constrained by the surface tension. The bursting force is pressure difference times area, and this force is balanced by the surface tension force. A force balance on the droplet of Fig. 1.12 gives.

$$(p_1 - p_o) \pi R^2 = \sigma (2 \pi R)$$

$$p_1 - p_0 = \frac{2 \sigma}{R} \quad (1.19)$$

However, the factor 2 in the force balance for the bubble is due to the bubble consisting of a film with two surfaces (inner and outer surfaces) and thus two circumferences in the cross section.

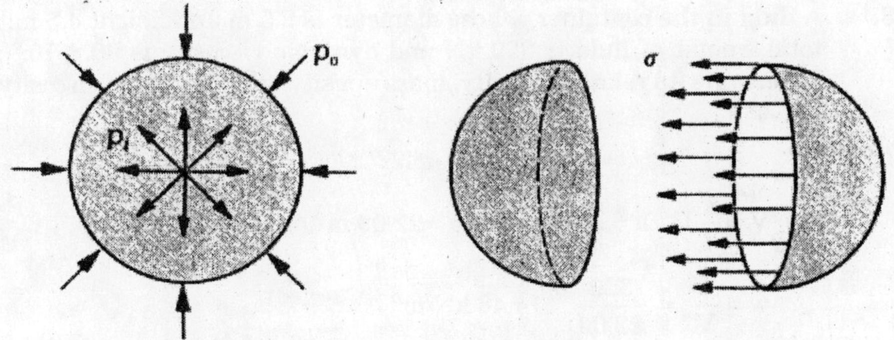

Fig. 1.12. Pressure inside a droplet of a liquid and soap bubble

Pressure force = Surface tension force

$$(p_1 - p_o) \pi R^2 = 2 \sigma (2 \pi R)$$

$$p_1 - p_0 = \frac{4 \sigma}{R} \quad (1.20)$$

ILLUSTRATIVE EXAMPLES

1.1. Determine the dynamic viscosity of an oil with a kinematic viscosity of 0.035 stoke and relative density 0.9.

Solution

$$S = \frac{\rho_{oil}}{\rho_{water}} \quad \text{or} \quad 0.9 = \frac{\rho_{oil}}{1000}$$

$$\therefore \rho_{oil} = 900 \text{ kg/m}^3$$

$$\nu = \frac{\mu}{\rho} \quad \text{or} \quad 0.035 \times 10^{-4} = \frac{\mu}{900}$$

$$\mu = 3.15 \times 10^{-3} \frac{\text{N s}}{\text{m}^2} = 3.15 \times 10^{-3} \text{ Pa s} \qquad \text{Ans}$$

1.2. A fluid has a specific weight 9500 N/m³ and a dynamic viscosity of 0.3 poise. Determine its relative density, kinematic viscosity, mass density.

Solution

$$S = \frac{w_f}{w_{water}} = \frac{9500}{9810} = 0.968 \qquad \text{Ans}$$

$$\rho_f = S\,\rho_{water} = 0.9684 \times 1000 = 968.4 \text{ kg/m}^3 \qquad \text{Ans}$$

$$\nu = \frac{\mu}{\rho} = \frac{0.3/10}{968.4} = 3.0979 \times 10^{-5} \text{ m}^2/s = 0.309 \text{ stoke} \qquad \text{Ans}$$

1.3. A fluid in the container, whose diameter is 2.5 m and height 4.5 m. The total weight of fluid is 320 kN and dynamic viscosity is 90 x 10^{-5} Pas. Determine its relative density, mass density and kinematic viscosity.

Solution

$$V = \frac{\pi}{4} D^2 h = \frac{\pi}{4}(2.5)^2 \times 4.5 = 22.09 \text{ m}^3$$

$$w = \frac{W}{V} = \frac{320}{22.09} = 14.48 \text{ kN/m}^3$$

$$S = \frac{w_f}{w_{water}} = \frac{14.48}{9.81} = 1.476 \qquad \text{Ans}$$

$$\rho_f = S\rho_{water} = 1.476 \times 1000 = 1476 \text{ kg/m}^3 \qquad \text{Ans}$$

$$\nu = \frac{\mu}{\rho} = 90 \times \frac{10^{-5}}{1476} = 6.0975 \times 10^{-7} \text{ m}^2/\text{s} \qquad \text{Ans}$$

1.4. If a Bromine has viscosity 1.02 x 10^{-3} kg/ms and density 3120 kg/m³, what are its kinematic viscosity and specific gravity.

Solution

$$\nu = \frac{\mu}{\rho} = \frac{1.02 \times 10^{-3}}{3120} = 3.27 \times 10^{-7}\ \text{m}^2/\text{s} \qquad \text{Ans}$$

$$S = \frac{\rho_l}{\rho_w} = \frac{3120}{1000} = 3.12 \qquad \text{Ans}$$

1.5. The kinematic viscosity and specific gravity of a liquid are 8 stokes and 1.6 respectively. Calculate the viscosity of this liquid.

Solution

$$S = \frac{\rho_{liquid}}{\rho_{water}} \quad \text{or,} \quad 1.6 = \frac{\rho_l}{1000} \quad \therefore \rho_t = 1600\ \text{kg/m}^3$$

$$\nu = \frac{\mu}{\rho} \quad \text{or} \quad 8 \times 10^{-4} = \frac{\mu}{1600}$$

$$\therefore \mu = 8 \times 10^{-4} \times 1600 = 1.28\ \text{Pas} \qquad \text{Ans}$$

1.6. If the absolute pressure of methane at 40°C is 8.3 bar, find its specific volume, density, relative density and specific weight. Take R = 518 J/kgK.

Solution

$$p = 8.3\ \text{bar} = 8.3 \times 10^5\ \text{N/m}^2$$

$$T = 273 + 40 = 313\ \text{K}$$

$$\nu = \frac{RT}{p} = \frac{518 \times 313}{8.3 \times 10^5} = 0.1953\ \text{m}^3/\text{kg} \qquad \text{Ans}$$

$$\rho = \frac{1}{v} = \frac{1}{0.1953} = 5.12\ \text{kg/m}^3 \qquad \text{Ans}$$

$$S = \frac{\rho_l}{\rho_w} = \frac{5.12}{1000} = 5.12 \times 10^{-3} \qquad \text{Ans}$$

$$w = \rho g = 5.12 \times 9810 = 50.2272\ \text{N/m}^3 \qquad \text{Ans}$$

1.7. A shaft of diameter 74.90 mm rotates in a bearing of diameter 75.03 mm and of length 75 mm. The annular space between the shaft and the bearing is filled with oil having a coefficient of viscosity 0.096 kg/ms. Determine the power used in overcoming viscous resistance in this bearing at 1400 revolution per min.

Solution

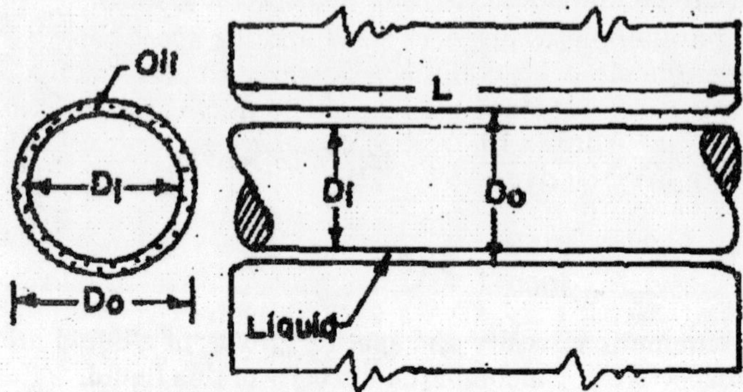

Fig. 1.14. Shaft and bearing P. 1.7.

$$u = du = \frac{\pi \, Di \, N}{60} = \frac{\pi \times 0.749 \times 1400}{60} = 5.49 \text{ m/s}$$

$$y = dy = \frac{Do - Di}{2} = \frac{0.07503 - 0.7490}{2} = 6.5 \times 10^{-5} \text{ m}$$

$$A = \pi D_i L = \pi \times 0.0749 \times 0.075 = 0.01764 \text{ m}^2$$

$$\therefore F = A \mu \frac{du}{dy} = 0.01764 \times 0.096 \times \frac{5.49}{6.5 \times 10^{-5}} = 143.0945 \text{ m}$$

$$P = F \times u = 143.0945 \times 5.49 = 785.589 \text{ W}$$ Ans

1.8. A cylinder of 120 mm radius rotates concentrically inside of a fixed cylinder of 130 mm radius. Both cylinders are 300 mm long. Determine the viscosity of the liquid which fills the space between the cylinders if a torque of 0.88 Nm is required to maintain an angular velocity of 2 π radian/s.

Solution : Given :

$r_i = 120 \text{ mm} = 0.12 \text{ m}$, $r_0 = 130 \text{ mm} = 0.13 \text{ m}$

$l = 300 \text{ mm} = 0.3 \text{ m}$, $w = 2\pi \text{ rad/s}$, $T = 0.88 \text{ Nm}$

$u = du = w\, r_i = 2\pi \times 0.12 = 0.754 \text{ m/s}$

$d_y = r_0 - r_1 = 0.13 - 0.12 = 0.01 \text{ m}$

$A = \pi D_1 l = \pi \times 2 \times 0.12 \times 0.3 = 0.226 \text{ m}^2$

$T = F r_i$ or $0.88 = F \times 0.12$

$\therefore F = 7.34 \text{ N}$

$$F = A \mu \frac{du}{dy} \quad \text{or} \quad 7.34 = 0.226 \times \mu \times \frac{0.754}{0.01}$$

$\mu = 0.43 \text{ Ns/m}^2$ Ans.

1.9. Two parallel plates are held apart and the space between them is filled with crude oil. A force of 2 N is required to drag the upper plate of 900 cm² area at a constant velocity of 800 mm/s. The viscosity of the oil is 1.36 poise. What is the size of the space between the plates? Assume straight line velocity distribution.

Solution

$$\tau = \mu \frac{du}{dy} \quad \text{or} \quad \frac{F}{A} = \mu \frac{du}{dy} \quad \text{or} \quad F = A\mu \frac{du}{dy}$$

Here $F = 2\text{ N}$ or $A = 900 \times 10^{-4}\text{ m}^2$, $u = du = 800\text{ mm/s}$

$$\mu = \frac{1.36}{10}\text{ N s/m}^2 \quad \text{or} \quad 2 = 900 \times 10^{-4} \times \frac{1.36}{10} \times \frac{0.8}{dy}$$

$$dy = \frac{900 \times 1.36 \times 0.8}{2 \times 10^5} = 4.896 \times 10^{-3}\text{ m} = 4.896\text{ mm} \qquad \text{Ans}$$

1.10. The dynamic viscosity of an oil, used for lubricating between a shaft and sleeve is 10 poise. The shaft rotates at 200 rpm. The power lost in the bearing for a sleeve length of 100 mm is 250 Watts. The thickness of the oil film is 1.5 mm. Calculate the diameter of the shaft and sleeve.

Solution:

Power absorbed = T x w

$$250 = T \times \frac{2\pi N}{60} \qquad \therefore\ T = 11.93\text{ Nm}$$

$$T = F \times \frac{Di}{2} = A\mu \frac{du}{dy} \times \frac{Di}{2}$$

$$11.93 = \pi Di\, l \times \mu \times \frac{du}{dy} \times \frac{Di}{2} = \pi \times l \times \mu \times \frac{\pi Di N}{60 \times 1.5 \times 10^{-3}} \times \frac{Di}{2} \times Di$$

$$= \frac{\pi^2 D_i^3 \times l \times \mu \times N}{60 \times 2 \times 1.5 \times 10^{-3}}$$

$$D_i^3 = \frac{11.93 \times 120 \times 1.5}{\pi^2 \times 0.1 \times 1 \times 200 \times 1000} = 0.0108$$

$$D_1 = 0.2215\text{ m} = 221.5\text{ mm} \qquad \text{Ans}$$

$$D_o = 2y + D_i = 2 \times \frac{15}{10} + 221.5 = 3 + 221.5 = 223.5\text{ mm} \qquad \text{Ans}$$

1.11. The dynamic viscosity of an oil, used for lubricating between a shaft and sleeve is 15 Poise. The shaft rotates at 200 rpm. The power lost in the bearing for a sleeve length of one meter is 500 Watts. The diameter of the sleeve is 250.15 mm. Calculate the diameter of the shaft and the thickness of the oil film.

Solution:

Power absorbed = T x w

$$500 = \frac{T \times 2\pi N}{60} \quad \text{or} \quad T = \frac{60 \times 500}{2\pi \times 200} = 23.83 \text{ Nm}$$

$$T = F \times \frac{D_i}{2} = A \times \mu \times \frac{du}{dy} \times \frac{D_i}{2}$$

$$23.93 = \pi D_i\, l \times \frac{15}{10} \times \frac{\pi D_i N}{60} \times \frac{1}{\frac{D_o - D_i}{2}} \times \frac{D_i}{2}$$

$$= \frac{D_i^3}{\left(\frac{25.015}{100} - D_i\right)} \times \left[\pi^2 \times 1 \times 2 \times 200 \times \frac{15}{10 \times 60}\right]$$

or, $(23.83)(0.25015 - D_i) = D_i^3 \times 49.348$

or, $0.25015 - D_i = D_i^3 \times 49.348$

or, $D_i^3 + 0.4849\ D_i - 0.1213 = 0$

By trial and error, $D_i = 0.25$ m **Ans**

$$\therefore\ dy = t = \frac{D_o - D_i}{2} = \frac{0.25015 - 0.25}{2} = 0.0075 \text{ m}$$ **Ans**

1.12. A flywheel weighing 600 N has a radius of gyration of 300 mm. When it is rotating at 600 rpm, its speed reduces 1 rpm/s owing to fluid viscosity between sleeve and shaft. The sleeve length is 50 mm; shaft diameter is 20 mm; and radial clearance is 0.05 mm. Determine the fluid viscosity.

Solution

$$I = mK^2, \quad \alpha = 1 \text{ rpm/s} = \frac{2\pi}{60} \text{ rad/s}^2$$

$$= \frac{600}{9.81} \times (0.3)^2 = 5.5045 \text{ kg m}^2$$

Torque $= I\alpha$

$$= 5.5045 \times \frac{2\pi}{60} = 0.576 \text{ Nm}$$

$$A = \pi D_i l = \pi \times \frac{20}{1000} \times \frac{50}{1000} = 3.14 \times 10^{-3} \text{ m}^2$$

$$du = u = \frac{\pi D_i N}{60} = \frac{\pi \times \frac{20}{1000} \times \frac{1}{2}}{60} = 0.6283 \text{ m/s}$$

$$dy = t = 0.05 \text{ mm}$$

Torque $T = F \times \frac{D_t}{2}$

$$0.576 = F \times \frac{20}{1000} \times \frac{1}{2}$$

$$\therefore F = 57.6 \text{ N}$$

$$\therefore F = A\mu \frac{du}{dy} = 3.14 \times 10^{-3} \times \mu \times \frac{0.6283}{0.05/1000}$$

$$57.6 = \mu \times 39.477$$

$$\mu = 1.46 \text{ N s/m}^2$$ Ans.

1.13. **A hydraulic lift used for automobile, has a 250 mm diameter ram which slides in a 250.18 mm diameter cylinder, the annular space being filled with oil having a kinematic viscosity 3.7 stokes and specific gravity of 0.85. If the rate of travel of the ram is 150 mm /s, find the frictional resistance when 3.3 m of ram is engaged in the cylinder**

Solution

$$du = 150 \text{ mm/s} = 0.15 \text{ m/s}$$

$$dy = \frac{D_o - D_i}{2} = \frac{25.018 - 25}{2 \times 1000} = 9 \times 10^{-5} \text{ m}$$

$$S = \frac{\rho_l}{1000}$$

$$\therefore \rho_l = 0.85 \times 1000 = 850 \text{ kg/m}^3$$

$$v = \frac{\mu}{\rho}$$

$$3.7 \times 10^{-4} = \frac{\mu}{850}$$

$$\mu = 3.7 \times 10^{-4} \times 850 = 0.3145 \text{ N s/m}^2$$

$$F = A\,\mu \frac{du}{dy} = (\pi \times 0.25 \times 3.3) \times 0.3145 \times \frac{0.15}{9 \times 10^{-5}}$$

$$= 1358.54 \text{ N}$$ Ans

1.14. **(a) A shaft 80 mm in diameter is being pushed through a bearing sleeve 80.2 mm diameter and 300 mm long. The clearance, assumed uniform, is filed with oil at 20°C with kinematic viscosity 50 stokes and relative density 0.892. If the shaft moves axially at 500 mm/s, estimate the resistance force exerted by the oil on the shaft.**

(b) If the shaft is now fixed axially and rotated inside the sleeve at 1800 rpm, estimate the resisting torque exerted by the oil and the power required to rotate the shaft.

Solution

(a) $$S = \frac{\rho_l}{\rho_w}$$

$$0.892 = \frac{\rho_l}{1000}$$

$$\rho_l = 0.892 \times 1000 = 892 \text{ kg/m}^3$$

$$\mu = \rho\,\nu = (0.892 \times 1000) \times 0.005 = 4.46 \text{ N s/m}^2$$

$$dy = \frac{80.2 - 80}{2 \times 1000} = 0.0001 \text{ m}$$

$$A = \pi\, D_i\, L = \pi \times \frac{80}{1000} \times \frac{300}{1000} = 0.0754 \text{ m}^2$$

$$F = A\,\mu \frac{dv}{dy} = \frac{0.0754 \times 4.46 \times 0.5}{0.0001} = 1683 \text{ N}$$ Ans.

(b) $$du = wR$$

$$= \frac{2\pi \times 1800}{60} \times \frac{8}{1000 \times 2} = 7.54 \text{ m/s}$$

$$T = F \times R = \tau\, A\, R = \mu \frac{dv}{dr} A \times R$$

$$= 4.46 \times \frac{7.54}{0.0001} \times 0.0754 \times \frac{8}{100 \times 2} = 1014 \text{ N/mm}$$

$$P = T\,w = 1014 \times \frac{2\pi \times 1800}{60} = 191.178 \text{ kW}$$ Ans

1.15. **A block weighing 300 N and having an area of 0.04 m² is sliding down the inclined plane at 20° to the horizontal with a viscosity of 7 x 10⁻³ Ns/m² and oil film thickness of 0.005 mm. If we use a linear velocity profile in the oil, what is the terminal speed of the block?**

Solution

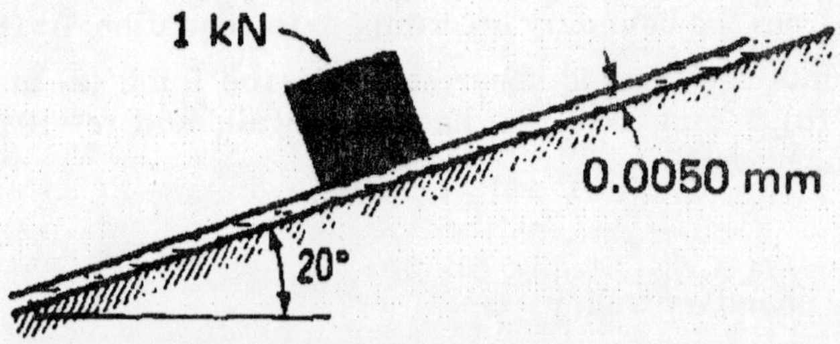

Fig. 1.15. Inclined plane P. 1.15.

$$\tau = \mu \frac{du}{dy} = 7 \times 10^{-3} \times \frac{u_t}{\left(\frac{0.005}{1000}\right)} = 1400\ u_t$$

$$F = \tau A = 1400\ u_t \times \left(\frac{200}{1000}\right)^2$$

$$1000 \sin 20° = 56\ u_t$$

$$\therefore u_t = \frac{1000 \sin 20°}{56} = 6.11 \text{ m/s}$$ Ans

1.16. A journal bearing consists of an 80 mm shaft in an 80.3 mm sleeve 100 mm long, the clearance space (assumed to be uniform) being filled with lubricating oil at 40°C (μ = 0.11 Ns/m²). Calculate the rate at which heat is generated at the bearing when the shaft turns at 120 rpm.

Solution

$$du = \frac{\pi D_i N}{60} = \pi \times \frac{80}{1000} \times \frac{120}{60} = 0.5027 \text{ m/s}$$

$$dy = \frac{80.3 - 80}{2 \times 1000} = 0.00015 \text{ m}$$

$$\tau = \mu \frac{du}{dy} = 0.11 \times \frac{0.5027}{0.00015} = 368.6 \text{ N/m}^2$$

$$F = \tau A = 368.6 \times \pi D_i l$$

$$= 368.6 \times \pi \times \frac{80}{1000} \times \frac{100}{1000} = 9.264 \text{ N}$$

Rate of energy loss = F x v = 9.264 x 0.5027 = 4.657 Nm/s

Rate of heat generated = 4.657 J/s Ans

1.17. A Viscous fluid of dynamic viscosity 0.25 kg/ms flows past a fixed boundary with laminar flow. The velocity of flow, u, varies with the distance y from the boundary according to the equation. u=18y (20–1000 y)

Calculate the viscous shear stress in the fluid: (a) at the boundary wall, (b) 5 mm from the boundary wall, and (c) 10 mm from the boundary wall.

Solution: (a)

At the boundary wall, y = o

$$\frac{du}{dy} = 18(20 - 2000\,y)$$

$$\therefore \left(\frac{du}{dy}\right)_{y=0} = 18 \times 20 = 360 \text{ per second}$$

$$\therefore \tau = u\frac{du}{dy} = 0.25 \times 360 = 90 \text{ N/m}^2$$ Ans

(b)

$$\frac{du}{dy} = (360 - 36000\,y)$$

$$\left(\frac{du}{dy}\right)_{y=5mm} = 360 - 36000 \times 5 \times 10^{-3}$$

$$= 360 - 180 = 180 \text{ per second}$$

$$\tau = \mu\frac{du}{dy} = 0.25 \times 180 = 45 \text{ N/m}^2$$

(c)

$$\frac{du}{dy} = (360 - 36000\,y)$$

$$\left(\frac{du}{dy}\right)_{y=10mm} = 360 - 36000 \times 10 \times 10^{-3} = 0$$

$$\tau = \mu\frac{du}{dy} = 0.25 \times 0 = 0$$ Ans

1.18. Calculate the viscous shear stress on a flat plate when air of viscosity 17.8×10^{-6} Ns/m^2 flows parallel to the plate with a velocity distribution given by the relation $u = 4 \times 10^{+4} (y - 100\,y^2)$ in which v is in m/s and y is the distance from the plate in meters.

Solution

$$u = 4 \times 10^4 \left(y - 100\,y^2\right)$$

$$\frac{du}{dy} = 4 \times 10^4 \,(1 - 200\,y)$$

Hence, on the plate where $y = 0$

$$\therefore \frac{du}{dy} = 4 \times 10^4 \text{ per second}$$

Therefore the viscous shear stress

$$\tau = u\frac{du}{dy} = 17.8 \times 10^{-6} \times 4 \times 10^4 = 0.712 \text{ N/m}^2$$ Ans

1.19. Parabolic velocity distribution for laminar flow is given by

$$u = u_{max}\left(1 - \frac{r^2}{R^2}\right)$$

Assuming a parabolic distribution having its vertex 100 mm from the boundary, calculate the shear stresses for y = 0, 25 mm, 50 mm, 75 mm, 100 mm. Take dynamic viscosity. 4 cp

Solution

Parabolic velocity distribution.

$$u = u_{max}\left(1 - \frac{r^2}{R^2}\right)$$

$$u = 2400\left[1 - \frac{(100 - y^2)}{100^2}\right] = 2400 - \frac{2400}{10000}\,(100 - y)^2$$

$$u = 2400 - 0.24\,(100 - y^2)\text{ mm/s}$$

$$\frac{du}{dy} = -\,0.24 \times 2\,(100 - y)$$

$$= -0.48\,(100 - y)$$

$$\tau = \mu\,\frac{du}{dy}$$

$$= 4 \times \frac{1}{100} \text{ x } \frac{1}{10} \times [-0.48\,(100 - y)]$$

$$= 4 \times 10^{-6} \times 10^{-3}\ [-0.48\,(100 - y)]$$

$$= -\,4 \text{ x } 10^{-9} \text{ x } 0.48\,[\,100 - y]$$

$$= -\,1.92 \text{ x } 10^{-9}\,(100 - y)\text{ N/mm}^2$$ Ans

$$\tau_{y=0} = 1.92 \times 10^{-9} \times 100 = -1.92 \times 10^{-7}\text{ N/mm}^2$$ Ans

$$\tau_{y=25mm} = -1.92\times10^{-9} \times 75 = -144\times10^{-9}\ N/mm^2$$ Ans

$$\tau_{y=50\,mm} = -1.92\times10^{-9} \times 50 = -96\times10^{-9}\ N/mm^2$$ Ans

$$\tau_{y=75\,mm} = -1.92\times10^{-9} \times 25 = -48\times10^{-9}\ N/mm^2$$ Ans

$$\tau_{y=100\,mm} = 0$$ Ans

1.20. An infinite plate is moved over a fixed plate on a layer of liquid. For small gap width 0.3 mm, we assume a linear velocity distribution in the liquid. If the fluid properties are, dynamic viscosity is 0.65 centipoise, and specific gravity 0.88. Calculate

(a) the absolute viscosity of fluid in kg/ms.

(b) The kinematic viscosity of the fluid in stokes.

(c) The shear stress on the upper plate,

(d) The shear stress on the lower plate, and

(e) Indicate the direction of each shear stress calculated in parts (c) & (d)

Solution

(a) $$\mu = 0.65\ c\,P = \frac{0.65}{100}\ \text{poise} \qquad = \frac{0.65}{100} \times \frac{1}{10}$$ Ans

(b) $$\nu = \frac{\mu}{\rho} = \frac{0.65\times10^{-3}}{0.88\times100} = 7.38\times10^{-7}\ m^2/s = 7.38\times10^{-3}\ \text{stokes}$$ Ans

(c) $$\tau_{upper} = \mu\left(\frac{du}{dy}\right)_{y=0.33} = \frac{0.65}{1000} \times \frac{0.3}{0.3/1000} = 0.65\ N/m^2$$ Ans

(d) $$\tau_{lower} = \mu\frac{u}{h} = \frac{0.65}{1000} \times 1000 = 0.65\ N/m^2$$ Ans

(e) Direction of shear stress an upper and lower plates

1.21. The velocity distribution for laminar flow between parallel plates is given by

$$u = u_{max}\left[1-\left(\frac{2y}{h}\right)^2\right]$$

Where h is the distance separating the plates, and the origin is placed midway between the plates. Consider a flow of water at 10°C, with U_{max} = 0.3 m/s and h = 0.5 mm. Calculate the shear stress on the upper plate and give its direction. Take μ at 10°C = 1.311 x 10^{-3} kg/ms.

Solution

$$\tau = \mu \frac{du}{dy}$$

Here $u = u_{max}\left[1 - \frac{4y^2}{h^2}\right]$ or, $\frac{du}{dy} = -U_{max}\frac{8y}{h^2}$

$$= -\frac{0.3 \times 8 \times y}{(0.5\times10^{-3})^2} = -9.6\,y \times 10^6\ \mu$$

$$\therefore \tau = (-9.6\,y)\times10^6\ \mu = -9.6\times10^6 \times 0.25 \times 10^{-3}\ \mu$$

$$= -2.4 \times 10^3\ \mu = -2.4 \times 10^3 \times 1.311 \times 10^{-3} = 3.1464\ \text{N/m}^2$$ Ans

1.22. The space between two flat and parallel walls 25 mm apart is filled with water of dynamic viscosity 1.12 x 10^{-3} kg/ms. A flat plate 250 mm square and 2mm thick is pulled through the space in such a manner that one surface remain parallel at 9 mm distance from the wall. Assuming that the velocity profiles between plate and walls are linear, determine the force and power required to maintain a plate velocity 200 mm/s. Neglect the resistance to motion on the edges of the plate.

Solution

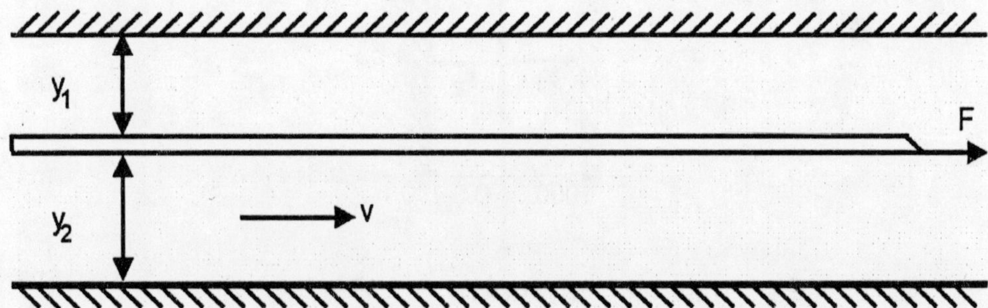

Fig. 1.16. Flat and parallel plates P. 1.22

$$F_1 = \tau_1 A = \mu \frac{u}{y_1} A, \qquad F_2 = \tau_2 A = \mu \frac{u}{y_2} A$$

Total force on the plate

$$F = F_1 + F_2 = \mu \frac{u}{y_1} A + \mu \frac{u}{y_2} A = \mu v A\left[\frac{1}{y_1} + \frac{1}{y_2}\right]$$

$$F = 1.12 \times 10^{-3} \times 0.2 \times 0.25 \times 0.25 \left[\frac{1}{9/1000} + \frac{1}{14/1000}\right]$$

$$= 1.4 \times 10^{-5}\left[\frac{1}{9\times10^{-3}} + \frac{1}{0.014}\right]$$

$$= 1.4\times10^{-5}\left[\frac{0.023}{9\times10^{-3}\times0.014}\right] = 2.55\times10^{-3} = 0.0025\text{ N}$$

Power to maintain motion

$$P = F\times v = 0.00255 \times 0.2 = 0.000511 \text{ Watts}$$ Ans

1.23. A vertical gap of 24 mm of infinite extent contains oil of specific gravity 0.8 and kinematic viscosity 40 stokes. A square metal plate of 1.5 m side and negligible thickness weighing 100 N is to be lifted through the gap and at a constant speed of 100 mm/s. Assuming that the velocity profiles between plate and walls are linear, determine the force and power required to maintain a constant velocity. Neglect the resistance to motion on the edges of the plate.

Solution

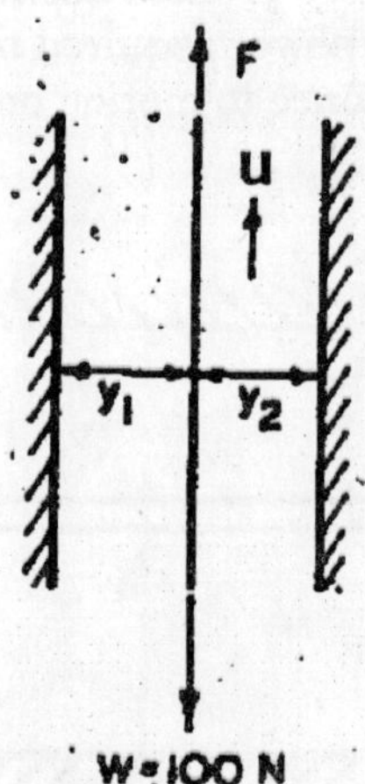

Fig. 1.17. Square metal plate P. 1.23.

$$y_1 = y_2 = \frac{24}{2\times1000} = 0.012 \qquad S = 0.8 = \frac{\rho_t}{1000}$$

$$\therefore \rho_t = 800 \text{ kg/m}^3 \qquad \nu = \frac{\mu}{\rho_t}$$

$$4\times10^{-3} = \frac{\mu}{800} \qquad \therefore \mu = 3.2 \text{ kg/ms}$$

$$\tau = \mu\frac{u}{y} = 3.2 \times \frac{0.1}{0.012} = 26.67 \text{ N/m}^2$$ Ans

1.24. A high pressure steel container is partially full of the liquid at a pressure of 10 atm. The volume of the liquid is 1.232 liters. At a pressure of 25 atm, the volume of the liquid equal 1.231 liters. What is the average bulk modulus of elasticity of the liquid over the given range of pressure if the temperature after compression is allowed to return to the original temperature ? What is the coefficient of compressibility (β)?

Solution

$$K = \frac{\Delta P}{\frac{\Delta V}{V}} = \frac{(25-10)\times 101.3}{\frac{1.231-1.232}{1.232}} = 1.872 \times 10^6 \text{ kN/m}^2 = 1872 \text{ MN/m}^2$$

$$\beta = \frac{1}{K} = \text{Coefficient of Compressibility}$$

$$= \frac{1}{1872} = 0.000534 \text{ m}^2\text{/MN}$$ Ans

1.25. A heavy tank contains oil (A) and water (B) and the air pressure is varied. The dimensions shown in Fig 1.20 corresponding to the atmosphere pressure of the air. If air is slowly added from a pump to bring pressure P upto a 1 M Pa gauge, what will be the total downward movement of the free surface of oil and air? Take average values of bulk moduli of elasticity of the liquids to be, for the pressure range, 2050 MN / m² for oil and 2075 MN/m² for water. Assume the container does not change volume. Neglect hydrostatic pressures.

Solution:

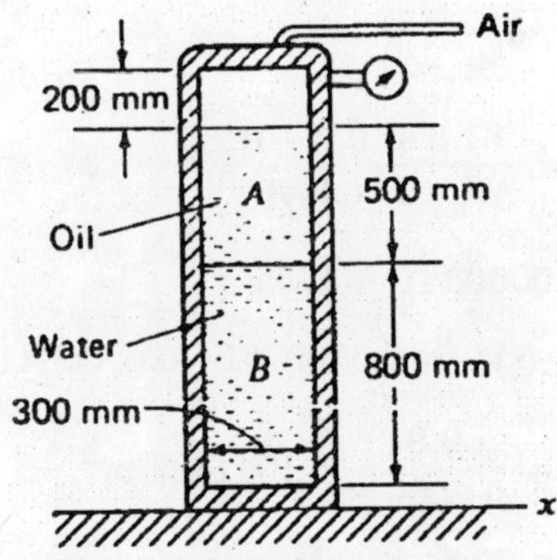

Fig. 1.18. Heavy tank P. 1.25.

For oil

$$K = -\frac{\Delta P}{\frac{\Delta V}{V}} \quad \text{or,} \quad 2050 = \frac{1-0}{\frac{\Delta \text{ Voll}}{\frac{\pi}{4}(0.3)^2 \times 0.5}}$$

$$\therefore \Delta V_{oil} = -0.00001724 \text{ m}^3$$

and for water

$$2075 = \frac{1-0}{\dfrac{\Delta V_{water}}{\frac{\pi}{4}(0.3)^2 \times 0.8}}$$

$$\therefore \quad \Delta V_{water} = -0.00002725 \text{ m}^3$$

$$\Delta V_{total} = \Delta V_{oil} + \Delta V_{water} = -0.0004449 \text{ m}^3$$

Let x = the distance the upper surface moves

$$\therefore \Delta V_{total} = -\frac{\pi}{4}(0.3)^2 \times x \qquad -0.00004449 = -\frac{\pi}{4}(0.3)^2 \times x$$

$x = 0.000629 \text{ m} = 0.629 \text{ mm}$ Ans

1.26. At a depth of 8 km in the ocean, the pressure is 81.8 MPa. Assume specific weight at the surface is 10050 N/m³ and the average bulk modulus of elasticity is 2.34 x 10⁹ N/m² for that pressure range (a) What will be the change in specific volume between that at the surface and at that depth?

(b) What will be the specific volume at that depth ?

(c) What will be the specific weight at that depth ?

Solution

$$\text{(a)} \quad v_{s_1} = \frac{1}{\rho_1} = \frac{g}{w_1} = \frac{9.81}{10050} = 0.0009761 \text{ m}^3/\text{kg}$$

$$K = -\frac{\Delta P}{\Delta v_s / v_{s_1}}$$

$$2.34 \times 10^9 = \frac{81.8 \times 10^{+6} - 0}{\Delta v_s / 0.0009761}$$

$\therefore \Delta v_s = -0.0000341 \text{ m}^3/\text{kg}$ Ans

(b) $v_{s_2} = v_{s_1} + \Delta v_s = 0.0009761 - 0.0000341 = 0.000942 \text{ m}^3/\text{kg}$ Ans

(C) $w_2 = g/v_{s_2} = \dfrac{9.81}{0.000942} = 10414 \text{ N/m}^2$ Ans

1.27. Distilled water at 10°C stands in a glass tube of 8 mm diameter at a height of 25 mm. What is the true static height. Take σ = 0.0742 N/m

Solution

$$h = \frac{2\sigma\cos\theta}{\rho g r} = \frac{2 \times 0.0742 \times \cos 0^\circ}{9810 \times 4/1000}$$

$= 0.0038 \text{ m} = 3.8 \text{ mm}$ Ans

1.28. **At 30°C what diameter glass tube is necessary to keep the capillary height change of water less than 1 mm? Take σ = 0.0712 N/m, ρ = 996 kg/m³**

Solution

$$h = \frac{2\sigma\cos\theta}{\rho g r} \quad \text{or,} \quad \frac{1}{1000} = \frac{2 \times 0.0712 \times \cos 0^\circ}{996 \times 9.81 \times r}$$

$$r = 0.0146 \text{ m} = 14.6 \text{ mm}$$

$$d = 29.2 \text{ mm or greater}$$ Ans

1.29. **Find the height through which water rises by capillary action in a glass tube of 2 mm bore if the surface tension at the prevailing temperature is 75 μ N/mm.**

Solution

$$h = \frac{4\sigma\cos\theta}{\rho g d}$$

$$= \frac{4 \times 75 \times 10^{-3} \times 1}{9810 \times 2 \times 10^{-3}} = 0.01529 \text{ m} = 15.29 \text{ mm}$$ Ans

1.30. **At 60°C the surface tension of mercury and water is 0.47 N/m and 0.0662 N/m, respectively. What capillary height changes will occur in these two fluids when they are in contact with air in glass tube of radius 0.25 mm? Use θ = 0° for water and θ = 130° for mercury and ρg = 132300 N/m³ for mercury and 9650 N/m³ for water.**

Solution

For water

$$h = \frac{2\sigma\cos\theta}{\rho g r} = \frac{2 \times 0.0662 \times \cos 0^\circ}{9650 \times 0.25 \times 10^{-3}} = 54.9 \text{ mm}$$

For mercury

$$h = \frac{2\sigma\cos\theta}{\rho g r} = \frac{2 \times 0.47 \times \cos 130^\circ}{132300 \times 0.25 \times 10^{-3}} - 0.0183 \text{ m} = -18.3 \text{ mm}$$ Ans

1.31. **What will be the capillary rise at 20°C in a clean glass tube of 1 mm bore containing (a) water, (b) mercury and (c) what will be the correspo-nding values in a tube 10 mm bore? Density of water is 1 Mg/m³ and of mercury 13.6 Mg/m³ and g = 9.807 m/s². Take surface tension for water 74 μ N/mm and for mercury 465 μ N/mm.**

Solution

(a) For water, average rise

$$h = \frac{4\,\sigma \cos\theta}{\rho g\ d} = \frac{4 \times 75 \times 1}{1 \times 9.807 \times 1} = 30 \text{ mm} \qquad \text{Ans}$$

(b) For mercury, average rise

$$h = \frac{4\,\sigma \cos\theta}{\rho g\ d} = \frac{4 \times 465 \times \cos \times 130^\circ}{13.6 \times 9.807 \times 1} = -8.5 \text{ mm} \qquad \text{Ans}$$

(c) In a tube 10 mm bore, which is about the size used in gauges in many engineering recorders, the figures would be 3 mm and - 0.85 mm. Ans

1.32. Air is introduced through a nozzle into a tank of water to form a stream of bubbles. If the bubbles are intended to have a diameter of 2 mm, calculate by how much the pressure of the air at the nozzle must exceed that of the surrounding water. Assume that $\sigma = 72.7\ \mu$ N/mm.

Solution

$$p = \frac{2\,\sigma}{r} = \frac{2 \times 72.7 \times 10^{-3}}{1 \times 10^{-3}} = 143.4 \text{ N/m}^2 \qquad \text{Ans}$$

1.33. A U tube is made up of two capillaries of bore 1mm and 2mm respectively. The tube is held vertically and is partially filled with a liquid of surface tension 0.05 N/m and zero contact angle. Calculate the mass density of the liquid if the estimated difference in the level of two menisci is 12.5 mm.

Solution

Let h_1 and h_2 be the heights of liquid columns in two limbs of bore d_1 and d_2 respectively.

Then, $h_1 = \dfrac{4\,\sigma \cos\theta}{\rho g\ d_1}$ and $h_2 = \dfrac{4\,\sigma \cos\theta}{\rho g\ d_2}$

$$\therefore\ h_1 - h_2 = \frac{4\,\sigma\ \cos\theta}{\rho g}\left[\frac{1}{d_1} - \frac{1}{d_2}\right]$$

$$0.0125 = \frac{4\,\sigma \times 1}{\rho \times 9.81}\left[\frac{1}{1 \times 10^{-3}} - \frac{1}{2 \times 10^{-3}}\right]$$

$$\therefore\ \rho = \frac{8000}{9.81} = 815.49 \text{ kg/m}^3 \qquad \text{Ans}$$

1.34. Calculate the gauge pressure and the absolute pressure within (a) a droplet of water 4 mm in diameter and (b) a jet of water 4 mm in diameter. Assume surface tension of water as 73 μN/mm and the atmospheric pressure as 1.01325 bar.

Solution

Assuming that the droplet is to be spherical and the jet is to be cylindrical.

For droplet $h = \frac{P}{\rho g} = \frac{4\sigma}{\rho g d}$ [cos o° = 1]

$$P = \frac{4\ \sigma}{d} = \frac{4 \times 0.073}{0.004} = 73\ N/m^2 = 73\ Pas$$

$$P_{absolute} = P_{atmospheric} + P_{gauge}$$

$$= 1.01325 + \frac{73}{10^5} = 1.01398\ bar$$ Ans

For jet $P = \frac{2\sigma}{d} = \frac{2 \times 0.073}{0.004} = 36.5\ N/m^2 = \frac{36.5}{10^5}$ bar

$$P_{absolute} = P_{atmospheric} + P_{gauge}$$

$$= 1.01325 + \frac{36.5}{10^5} = 1.013615\ bar$$ Ans

1.35. The glass tube in Figure 1.22 is used to measure pressure p_1 in the water tank. The tube diameter is 1 mm and the water is at 30°C, After correcting for surface tension what is the true water height in the tube? What percent error is made if no correction is computed?

Take $\sigma = 0.0712$ N/m, $\rho g = 9770\ N/m^3$

Solution

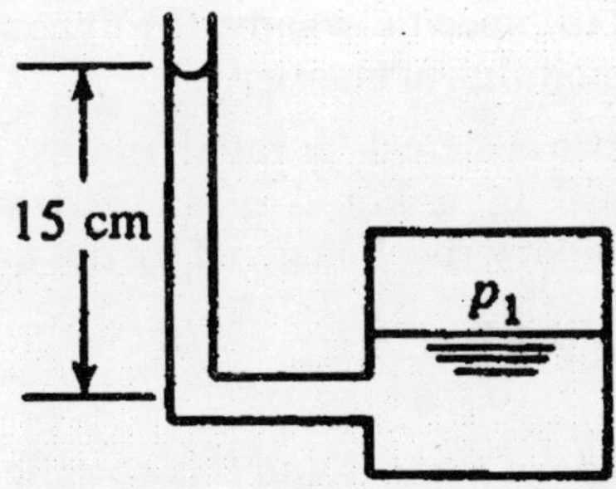

Fig. 1.19. Water tank P. 1.35

$$h = \frac{2\sigma \cos\theta}{\rho\ g r} = \frac{2 \times 0.0712 \times \cos\theta}{9770 \times\ 0.5 \times 10^{-3}} = 0.029\ m = 29\ mm$$

True water height in the tube = 150 – 29 = 121 mm

Neglecting Capillary correction causes $\frac{29}{150}$ = 19.33% error. Ans

EXERCISES

A. THEORY

1.1. Differentiate between solid, liquid and gas.

1.2. Define a fluid. How are the fluids classified ?

1.3. Define specific gravity, specific weight, and mass density. What are their units?

1.4. What do you understand by absolute viscosity and kinematic viscosity of a liquid? What is their relationship and units ?

1.5. What do you understand by a Newtonian and Non-Newtonian fluid ?

1.6. Explain the terms : Bulk modulus, surface tension, capillary rise.

1.7. What is the difference between a fluid and solid? Difference between compressible fluids and incompressible fluids.

1.8. What is the difference between an ideal fluid and a real fluid ?

1.9. What is vapour pressure? What is its significance in flow problems ?

1.10. Obtain an expression for the bulk modulus of elasticity K of a fluid in terms of the pressure p, and the mass density ρ.

1.11. Develop the expression for the relation between gauge pressure p inside a droplet of liquid and the surface tension.

1.12. Determine the approximate height to which a liquid will rise in a capillary tube exposed to atmosphere.

1.13. Define the following fluid properties and mention their values in respect of water (a) specific weight (b) dynamic viscosity (c) surface tension (d) bulk modulus of elasticity.

1.14. Show that for two vertical parallel plates distance t apart, held partially immersed in a liquid of specific weight ρ g and surface tension σ, the capillary rise h is given by the formula

$$h = \frac{2\sigma\cos\theta}{S\rho g t}$$

where θ is the angle of contact and S is specific gravity.

1.15. Define dynamic viscosity and kinematic viscosity. What are their units ? What is the effect of temperature and pressure on viscosity? How is fluid classified on the basis of Newton's law of viscosity ?

1.16. Write short notes on compressibility and elasticity of a fluid.

1.17. Is there any anology of Hookes law in fluids ? If so, state the parallel law in fluids.

1.18. Explain with the aid of sketches the following fluid properties (a) viscosity (b) surface tension and (c) capillarity.

1.19. Explain the following properties of fluids.

(a) Newtonian and Non Newtonian fluids (b) Vapour pressure (c) Compressibility.

1.20. Determine the dimensions (M, L, T) of the following quantities and suggest appropriate S.I. units. (a) kinetic energy (b) power (c) kinematic viscosity (d) surface tension (e) bulk modulus of elasticity.

B. UNSOLVED PROBLEMS

1.1. Convert 15 poises to kinematic viscosity in stokes if the liquid has relative density 0.9 [16.7 stokes]

1.2. Convert 1.796 x 10^{-3} Pa s to kinematic viscosity in m^2/s if the liquid has the relative density 0.998. [1.799 μ m^2/s]

1.3. The specific gravity and kinematic viscosity of a mercury are 13.6 and 1.16 x 10^{-3} stokes respectively. Calculate the viscosity of this liquid in poises. [1.578 – 10^{-2} poises]

1.4. The kinematic viscosity and dynamic viscosity of a liquid are 8 x 10^{-3} stokes and 5 x 10^{-3} poise respectively. Calculate the density and relative density of this liquid. [625 kg/m^3, 0.625]

1.5. A rectangular plate 1 m x 0.5 m weighing 1 kN slides down a 30° inclined surface at a uniform velocity of 2 m/s. If the 2 mm gap between the plate and the inclined surface is filled with castor oil, determine the viscosity of the oil. [1 N s/m^2]

1.6. A shaft of 100 mm diameter rotates at 60 rpm in a 200 mm long bearing. Taking that the two surfaces are uniformly separated by a distance of 0.5 mm and taking the linear velocity distribution in lubricating oil having dynamic viscosity 4 cp, find the power absorbed in the bearing. [0.0495 Watt]

1.7. The space between two square flat parallel plates is filled with oil. Each side of the plate is 600 mm. The thickness of the oil film is 12.5 mm. The upper plate, which moves at 2.5 m/s requires a force of 100 N to maintain the speed. Determine (a) the dynamic viscosity of the oil in poise, (b) the kinematic viscosity of the oil in stokes if the specific gravity of the oil is 0.95. [13.89 poise, 14.62 stokes]

1.8. Two horizontal plates are placed 12.5 cm apart, the space between them being filled with oil of viscosity 14 poises. Calculate the shear stress in the oil, if the upper plate is moved with a velocity of 2.5 m/s. [280 N/m^2]

1.9. A dashpot 100 mm diameter and 125 mm long moves down in 100.5 mm diameter cylinder. The oil filling the annular space has a viscosity of 1 poise. Find the speed with which the piston slides down if load on the piston is 10 N. [636 mm/s]

1.10. A Newtonian fluid is filled in the clearance between a shaft and a concentric sleeve. The sleeve attains a speed of 0.5 m/s, when a force of

400 N is applied to the sleeve parallel to the shaft. Determine the speed if a force of 2 kN is applied. [2.5 m/s].

1.11. Find the kinematic viscosity of an oil having density 1250 kg/m^3. The shear stress at a point in oil is 1N/m^2 and the velocity gradient at the point is 0.5 per second [16 stokes]

1.12. A 150 mm diameter vertical cylinder rotates concentrically inside another cylinder of diameter 151 mm. Both cylinders are 250 mm high. The space between the cylinders is filled with a liquid whose viscosity is unknown. If a torque of 12 Nm is required to rotate the inner cylinder at 100 rpm, determine the viscosity of the liquid. [8.646 poise]

1.13. If the velocity distribution for flow over a flat plate is given by $u = \frac{3}{2} y - y^{3/2}$ where u is the point velocity in m/s at a distance y m above the plate. Determine the shear stress at y = 90 mm. Take μ = 8 poise. [0.0840 N/m^2]

1.14. If the velocity distribution of a fluid over a plate is given by $v = \frac{3}{4} y - y^2$ where v is the velocity in m/s at a distance y meters above the plate, determine the shear stress at y = 150 mm. Take dynamic viscosity of the fluid as 8.5 x 10^{-4} N s/m^2. [3.825 x10^{-4} N/m^2]

1.15. A viscous fluid of dynamic viscosity 0.25 Ns/m^2 flows past a fixed boundary with laminar flow. The velocity of flow, u, varies with the distance y from the boundary according to the equation.

$$u = 20\, y\, (25 - 1000\, y)$$

Calculate the viscous shear stress in the fluid.

(a) 5 mm from the boundary wall, (b) at the boundary wall, and (c) at the 12.5 mm from the boundary wall. [75 N/m^2, 125 N/m^2 and O]

1.16. Calculate the viscous shear stress on a flat plate when air of viscosity 17.83 x 10^{-6} kg/ms flows parallel to the plate with a velocity distribution given by the numerical formula u = 40000 (y – 164 y^2) in which u is in m/s and y is the distance form the plate in meters [0.7132 N/m^2].

1.17. Find the torque and power required to overcome the friction in a bearing in which 50 mm diameter shaft rotates at 500 rpm. The length of bearing is 80 mm and the clearance between the bearing and the shaft is 0.5 mm which is filled with lubricating oil having a viscosity of 0.1 poise. [0.0083 Nm, 0.435 Watts]

1.18. A cylinder of 300 mm diameter rotates concentrically inside a fixed cylinder of 310 mm diameter. Both the cylinders are 300 mm long. Determine the viscosity of the liquid which fills the space between the cylinders if a torque of 0.98 Nm is required to maintain an angular velocity of 60 rpm. [0.123 Ns/m^2]

1.19. The space between two square flat parallel plates is filled with oil. Each side of the plate is 600 mm. The thickness of oil film is 12.5 mm. The upper plate, which moves at 2.5 m/s requires a force of 100 N to maintain speed. Determine (a) the dynamic viscosity of oil in stokes, if the specific gravity of oil is 0.95. [13.89 poise, 14.62 stokes]

1.20. A flat disc of diameter 100 mm rotates on a table separated by an oil film of 1.5 mm thickness. If the torque required to rotate the disc at 50 rpm is 3×10^{-4} N m, find the viscosity of oil. It may be assumed that the velocity gradient in the oil film is linear. Derive the formula which you have used. [0.00877 Ns/m^2]

1.21. A concentric cylinder viscometer may be formed by rotating the outer member of a pair of closely fitting cylinders. For small clearances a linear velocity profile may be assumed in the liquid filling the clearance gap. A viscometer has an inner cylinder 75 mm diameter and 150 mm height, with a clearance gap width of 0.02 mm. A torque of 0.012 Nm is required to turn the outer cylinder at 100 rpm. Determine the viscosity of the liquid in the clearance gap of the viscometer. [8.07×10^{-4} Ns/m^2]

1.22. A drop of water has a diameter of 3 mm. What is the excess pressure inside, if the surface tension of water is 0.074 N/m? [98.66 N/m^2].

2

Static Pressure and its Measurement

2.1 INTRODUCTION

In this chapter the topics are principally concerned with an analysis of the conditions within a fluid that is at rest. When a body of fluid is motionless, there are no relative motions within it, so that Newton's law of viscosity, would require an absence of shear stresses within the fluid body. Consequently, the only stresses that could occur within the fluid are normal stresses, which must be equal to the pressure if no relative motion exist. Thus, our principal concern is to investigate pressure and its variation throughout a fluid. The absence of shearing greatly simplifies the analysis, and as we will see, allows us to obtain relative simple solutions to many important practical problems.

2.2 PRESSURE AT A POINT

The term pressure is used to indicate the normal force per unit area at a given point acting on a given plane within the fluid mass of interest. The pressure at a point is the limit of the ratio of normal force to area as the area approaches zero size at a point.

$$P = Lt_{\delta A \to o} \frac{\delta F}{\delta A} = \frac{dF}{dA}$$

It is important to note that if an imaginary surface is assumed within a fluid body, the fluid pressure and pressure force on the imaginary surface are exactly the same as those acting on any real surface by virtue of Newton's third law of motion.

2.3 PASCAL'S LAW

At any point in a fluid at rest, the pressure is transmitted equally in all directions. This means it is independent of the orientation of the surface around the point.

Let us Consider the free body diagram, illustrated in Fig 2.1, that was obtained by removing a small triangular wedge of fluid from some arbitrary location within a fluid mass. Since we are considering the situation in which there are no shearing stresses, the only external force acting on the wedge are due to the pressure and the weight.

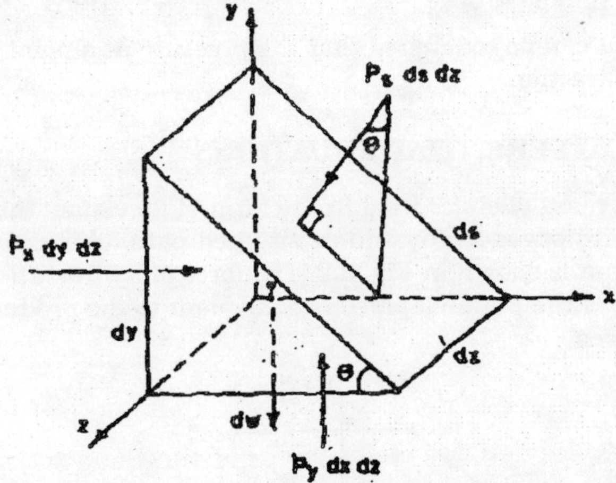

Fig. 2.1. Force on an arbitrary wedged shaped element of fluid.

Suppose that the pressure in three directions of the prism are p_x, p_y and p_z. Under the action of the pressure forces the prism is in equilibrium, therefore, the forces in horizontal and vertical directions (x and y directions) must be balanced.

Resolving horizontallly (x direction)

$$\Sigma F_x = p_x\, dy\, dz - p_s\, ds\, dz \sin\theta = 0$$

$$p_x\, dy\, dz = p_s\, ds\, dz \sin\theta$$

Here, $\sin\theta = \dfrac{dy}{ds}$

Therefore, $p_x\, dy\, dz = p_s\, ds\, dz \dfrac{dy}{ds}$

$$p_x\, dy\, dz = p_s\, dy\, dz$$

$$\therefore\ p_x = p_s \tag{2.1}$$

Resolving vertically (y direction)

$$\Sigma F_y = p_y\, dx\, dz - p_s\, ds\, dz\, \cos\theta = 0$$

$$p_y\, dx\, dz = P_s\, ds\, dz\, \cos\theta$$

Here, $\cos\theta = \dfrac{dx}{ds}$

Therefore, $p_y\, dx\, dz = p_s\, ds\, dz\, \dfrac{dx}{ds}$

$$p_y\, dx\, dz = P_s\, dx\, dz$$

$$\therefore p_y = p_s \tag{2.2}$$

From equations (2.1) and (2.2)

$$p_x = p_y = p_s \tag{2.3}$$

From this it can be concluded that the pressure at a point in a fluid at rest is same in any direction.

2.4 EULER'S DIFFERENTIAL EQUATIONS

Consider a mass of static fluid in the form of an elementary parallelepiped of volume dV with edges dx, dy and dz situated parallel to coordinate axes x, y and z. The system is shown in Fig. 2.2. The force of hydrostatic pressure acting against anyside of the parallelepiped is equivalent to the product of pressure, p, over the sides area.

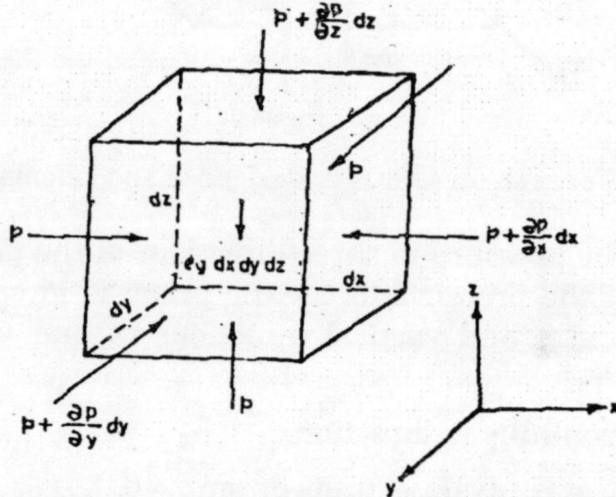

Fig. 2.2. Balance of forces on an element.

To develop on expression for pressure, we must assume p to be a function of all three coordinates, p = f (x, y, z). The gravity force acting on this volume is the product of its mass times acceleration due to gravity.

Select the z axis in the direction of gravity as shown in Fig 2.2 with no relative motion between fluid particle, there are no shear forces acting on the element, only normal forces (due to pressure) and the gravity force.

The force of gravity acting in the downward direction is parallel to the z axis.

$$-g\,d\,(\text{mass}) = -g\,\rho dV = -\rho\,g\,dx\,dy\,dz$$

The pressure force acting normal to the lower side of the parallelepiped of the z axis is pdxdy if the pressure changes at a given point in the z-direction is $\frac{\partial p}{\partial z}$, then along edge dz it is $\left(\frac{\partial p}{\partial z}\right)$ dz.

The pressure acting on the opposite (upper) side is $\left[p+\frac{\partial p}{\partial z}dz\right]$, and the projection of this pressure force onto the z-axis is

$$-\left(p+\frac{\partial p}{\partial z}dz\right)dx\ dy$$

Hence, the resulting pressure force on the z-axis is

$$p\,dx\,dy-\left(p+\frac{\partial p}{\partial z}dz\right)dx\ dy\ =\ -\frac{\partial p}{\partial z}dx\,dy\,dz$$

and the sum of the force projected on to the z-axis is equal to zero, That is,

$$-\rho g\,dx\,dy\,dz-\frac{\partial p}{\partial z}dx\,dy\,dz=0$$

and because, dx dy dz = dV ≠ 0, we obtain

$$-\rho g-\frac{\partial p}{\partial z}=0$$

The Projection of gravity forces on the x - axis equals zero. Thus, we may write

$$p\,dy\,dz-\left(P+\frac{\partial p}{\partial x}dx\right)dy\ dz\ =0$$

After simplification, we obtain $-\frac{\partial p}{\partial x}=0$

And, corresponding, for the y – axis

$$-\frac{\partial p}{\partial y}=0$$

Thus, the equilibrium conditions of the elementary parallelepiped may be expressed by the following system of equations :

$$-\frac{\partial p}{\partial x}=0\quad \text{and}\quad -\frac{\partial p}{\partial y}=0$$

$$\therefore -\rho g-\frac{\partial p}{\partial z}=0 \tag{2.4}$$

These expression are known as Eulers differential equations.

2.4.1 The basic equation of hydrostatics

Equation (2.4) shows that the pressure of a static fluid varies vertically only.

In otherwords, the pressure at any point on a horizontal plane of the fluid parallelepiped is the same. For the system of equations in equation (2.4), the partial derivatives $\frac{\partial p}{\partial x}$ and $\frac{\partial p}{\partial y}$ are equal to zero. Consequently, the partial derivative $\frac{\partial p}{\partial z}$ may be substituted by $\frac{dp}{dz}$ and, therefore,

$$-\rho g - \frac{dp}{dz} = 0$$

$$-dp - \rho g\, dz = 0$$

$$dz + d\left(\frac{p}{\rho g}\right) = 0$$

For a homogeneous, imcompressible fluid, density is constant. Therefore,

$$d\left(z + \frac{p}{\rho g}\right) = 0$$

After integration, we obtain

$$z + \frac{p}{\rho g} = \text{Constant} \qquad (2.5)$$

For any two horizontal planes 1 and 2, equation (2.5) may be written in the following form

$$z_1 + \frac{p_1}{\rho g} = z_2 + \frac{p_2}{\rho g} \qquad (2.6)$$

Equation (2.6) expresses the condition of hydrostatic equibibrium.

2.4.2 Pressure head

From equation (2.6)

$$\frac{p_2}{\rho g} - \frac{p_1}{\rho g} = z_1 - z_2$$

or, $\frac{p_2 - p_1}{\rho g} = h$ where h is difference in elevation.

$$\text{or,} \quad p_2 - p_1 = \rho g h \qquad (2.7)$$

$$p_2 = p_1 + \rho g h$$

Where h is the difference in elevation. In fluid mechanins the term $\frac{p}{\rho g}$ is usually referred to as the pressure head and z as the position head (Levelling head). In S.I. units both may be expressed in m of liquid.

Engineers prefer to use pressure head or head measurements because, difference in ground level can affect the pressure in a pipeline. It is then an easy matter to add (or subtract) changes in ground level to pressure values because they both have the same units.

When head is used to measure water pressure it is important to say that it is water, i.e., head of water. If another liquid is used, e.g. mercury, then the term `head of mercury' must be used. Even though the pressure may be the same , the head will be quite different to different liquids.

Equation $p = \rho gh = wh$ can be used to obtain a relationship between the heights of columns of different liquids which would develop the same pressure at any point. Thus, if h_1 and h_2 are the heights of columns of liquids of specific weight w_1 and w_2 required to develop the same pressure p, at any point; then from the relation $p = \rho gh = wh$

$$p = w_1 h_1 = w_2 h_2 \tag{2.8}$$

If S_1 and S_2 are the relative densities of the two liquids and w is the specific weight of water then since $w_1 = S_1 w$ and $w_2 = S_2 w$, then

$$h_1 S_1 = h_2 S_2 \tag{2.9}$$

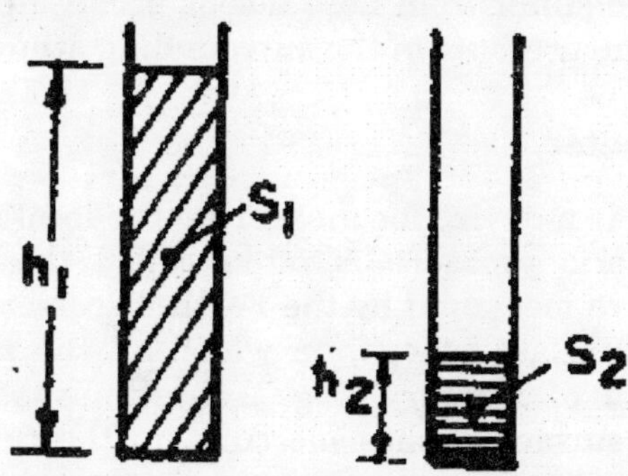

Fig. 2.3. Equal pressure at any point.

2.4.3 The hydrostatic paradox

The area of the base of the tank has no effect an the pressure nor does the weight of the water in tank (Fig. 2.4). It does not matter if the water is in a large reservoir or in a narrow tube. When the depth is the same then the pressure is the same.

It is to be noted that the equation $p = \rho qh$ works for all fluids and not just water, provided of course that the correct value of density is used for the fluid concerned.

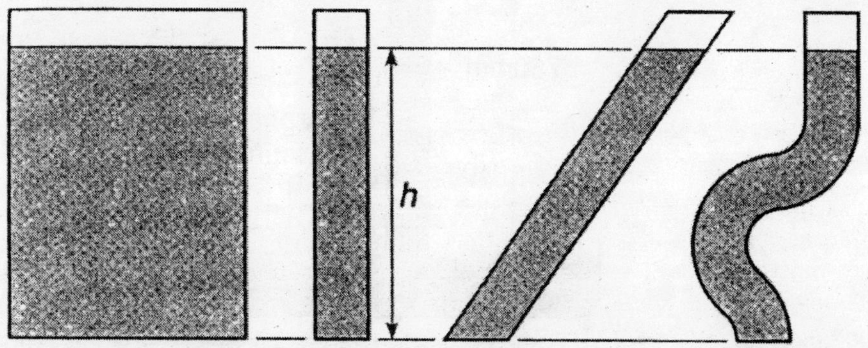

pressure is the same at the base of all these containers

Fig. 2.4. The hydrostatic paradox

2.5 ATMOSPHERIC PRESSURE

The pressure of the atmospheric is all around us pressing on our bodies. Fortunately there is an equal and opposite pressure from within our bodies that balances this pressure and so we feel no effect (Newtons third law). If for some reason one side of this pressure system was removed then the result would be catastrophic (like stepping out from a pressurized space rocket into space) would probably explode because the outside pressure is removed but the internal pressure remains.

At high altitudes, where atmospheric pressure is less than at the Earth's surface, some people suffer from nose bleeds because their blood pressure is much higher than the pressure of the surrounding atmosphere.

2.5.1 Fortin barometer

The barometer is a device for measuring the local atmospheric pressure, thus the atmos-pheric pressure is often called the barometric pressure. Atmospheric pressure measured by the Fortin barometer using a mercury. It consists of a vertical glass tube, from which all the air has been removed, immersed in a bath of mercury (Fig. 2.5). Atmospheric pressure pushing down on the open surface pushes mercury up the tube. The height of the mercury column is a measure of atmospheric pressure. It is normally measured in mm and the long term average value at sea level is 760 mm.

Atmospheric pressure can be measured using any liquid and not just mercury. If water is used the height of the water column would be quite different.

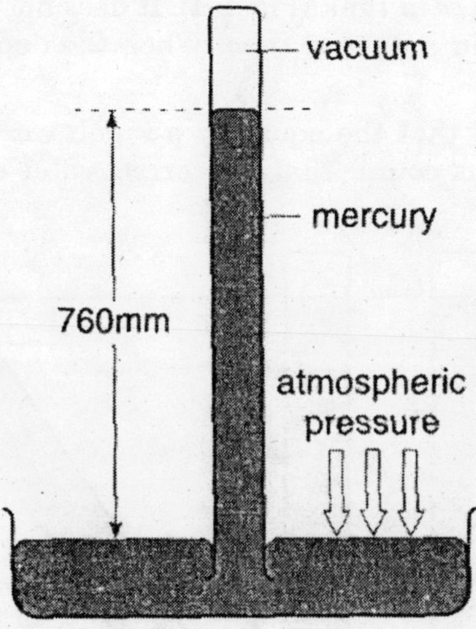

Fig. 2.5. Fortin barometer

It is better to convert 1 bar into height of water column.

Using the relation

$$p = \rho gh$$

$$1 \times 10^5 = 1000 \times 9.81 \times h$$

$$h = \frac{10^5}{10^3 \times 9.81} = \frac{100}{9.81} \text{ m of water}$$

$$\therefore \quad h = 10.19367992 = 10.2 \text{ m of water}$$

Therefore,

1 m of water = 9.81 kPa

1 mm of water = 9.81 Pa

In terms of mercury column.

Using the relation

$$h_1 S_1 = h_2 S_2$$

$$h_m \times 13.6 = h_w \times 1$$

$$h_m = \frac{100}{9.8} \times \frac{1}{3.8} = 0.75 \text{ m}$$

$$= 750 \text{ mm of mercury}$$

Standard value of mercury for atmospheric pressure is 760 mm.

Therefore,

$$p = S \rho gh$$

$$= 13.6 \times 1000 \times 9.81 \times 0.76$$

$$= 1.01325 \text{ bar}$$

For conversion purpose

1 bar = 10.2 m of water = 750 mm of mercury

1 mm of water = 9.81 Pa

1 mm of mercury = 133.4 Pa

Therefore, for standard atmosphere

1.01325 bar = 10.34 m of water = 760 mm of mercury.

2.6 APPLICATION OF THE BASIC EQUATION OF FLUID STATICS

The fundamental problems of fluid statics is the determination of the static pressure in a fluid. Pressure is usually expressed with reference to either absolute zero pressure (or complete vacuum) or local atmospheric pressure. However, any other arbitrary datum may also be used as reference. Most pressure measuring devices, however, are calibrated to read zero in the atmosphere, and so they indicate the difference between the absolute pressure and the local atmospheric pressure. This difference is called the gauge pressure. Pressures below atmospheric pressure are called vacuum pressure (negative gauge pressure) and are measured by vacuum gauge which indicate the difference between the atmospheric pressure and the absolute pressure.

∴ Absolute pressure = Atmospheric pressure ± Gauge pressure.

It can be written as

Absolute pressure = Atmospheric pressure + Gauge pressure

or,

Absolute pressure = Atmospheric pressure – vacuum pressure

Gauge pressure – The pressure which is above the atmosphere pressure is known as gauge pressure. The atmospheric pressure is a zero gauge pressure (Fig 2.6)

Vacuum Pressure – The pressure which is below the atmospheric pressure is known as vacuum pressure or negative gauge pressure (Fig 2.6).

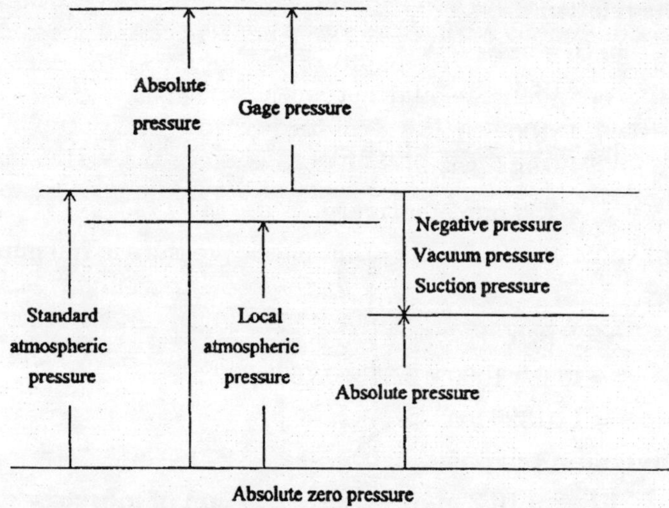

Fig. 2.6. Units and scales for pressure measurement

2.7 MANOMETERS

Manometers are devices that employ liquid columns for determining differences in pressure.

A general procedure should be followed in working all manometers problems.

(1) Start at one end (or any meniscus if the circuit is continuous) and write the pressure there in an appropriate unit (say pascals) or in an appropriate symbol if it is unknown.

(2) Add or subtract to this the change in pressure, in the same unit, from one meniscus to the next (plus if the next meniscus is lower, minus if higher).

(3) Continue until the other end of the gauge (or the starting meniscus) is reached and equate the expression to the pressure at that point, known or unknown.

The expression will contain one unknown for a simple manometer or will give a difference in pressure for the differential manometers.

No formulas for particular manometer should be memorized. It is much more satisfactory to work them out from the general procedure for each case as needed.

2.7.1 Piezometer

The most elementary manometer, usually called a piezometer; is shown in Fig 2.7, it measures the pressure in a liquid when it is above zero gauge (above atmospheric pressure). A glass tube is mounted vertically so that it is connected to the space with in the container. Liquid rises in the tube until equilibrim is reached. The pressure is then given by the vertical distance h from the meniscus (liquid surface) to the point where the pressure is to be measured, expressed in units of length of the liquid in the container. It is obvious that the piezometer would not work for negative gauge pressures, because air would flow into the container through the tube. It is also impractical for measuring large pressures at A, since the vertical tube would need to be very long. If the relative density of the liquid is S, the pressure at A is hS units of length of water.

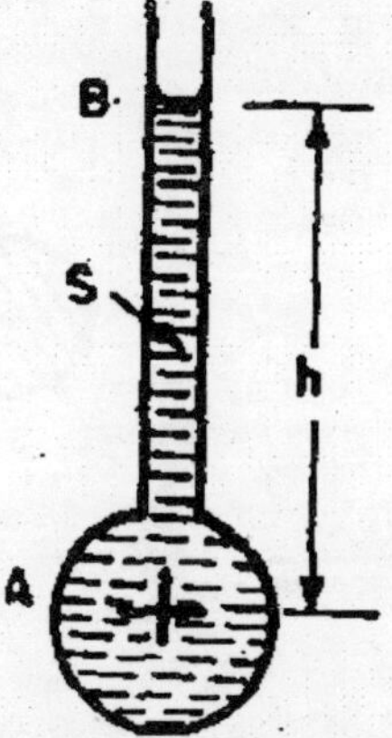

Fig. 2.7. Piezometer

The general procedure

$$h_A - hS = h_B = \text{Zero gauge pressure}$$

$$= \text{Atmospheric pressure}$$

$$= 0$$

$$\therefore \quad h_A = hS$$

2.7.2 Simple manometers

It is not always convenient to use a piezometer particularly when measuring high pressures. An alternative is to use a U-tube manometer. The bottom of the U tube is filled with a different liquid which does not mix with that in the pipe. When measuring pressures in a water system, oil or mercury is used. Mercury is a very useful because high pressures can be measured with a relatively small tube.

The basic assumption is that as the mercury in the two limbs of the manometer is not moving then the pressures in the two limbs must be the same. If they were not then the mercury would move. So if a horizontal line is drawn through the liquid surface in the left limb and extended to the right limb then it can be assumed that :

Pressure at left limb = Pressure at right limb.

This is the fundamental assumption on which all manometer calculations are based.

For measurement of small negative or positive gauge pressures in a liquid the tube may take the form shown in Fig 2.8. With this arrangement the meniscus may come to rest below A, as shown. Since the pressure at the meniscus is zero gauge and since pressure decreases with elevation.

For greater negative (Fig 2.9) or positive gauge pressures a second liquid of greater specific gravity is employed. It must be immiscible in the first fluid, which may now be a gas. If the specific gravity of the fluid at A is S_1 and the specific gravity of the manometer liquid is S_2 the equation for pressure at A may be written thus starting at either A or the upper menisucs and proceeding through the manometers.

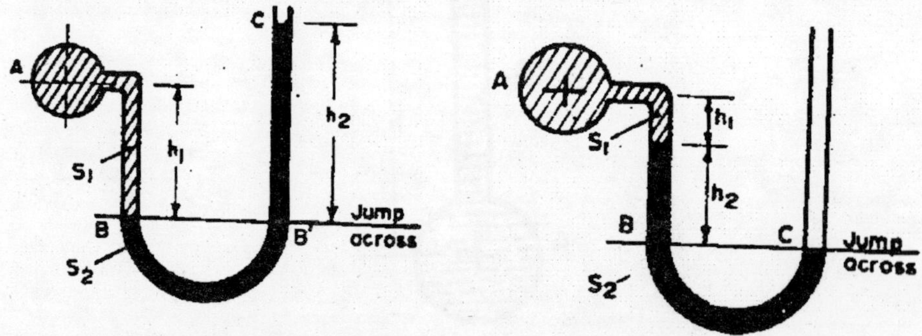

Fig. 2.8. Simple manometer

Fig. 2.9. Simple manometer

Procedure for Fig. 2.8

$$h_A + h_1 S_1 - h_2 S_2 = h_C$$

= open to atmospheric pressure

= zero gauge pressure = 0

$$\therefore h_A = h_2 S_2 - h_1 S_1$$

Procedure for Fig. 2.9

$$h_A + h_1 S_1 + h_2 S_2 = h_C = \text{open to atmospheric pressure}$$

= zero gauge pressure = 0

$$\therefore h_A = -h_1 S_1 - h_2 S_2 \quad \text{[Gives the vacuum pressure]}$$

It is to be noted that the pressure at B – B' in Fig 2.8 and B – C in Fig 2.9. This is because a continuous length of the same fluid connects the two equal levels. The hydrostatic relation requires this as a form a Pascals law.

Any two point at the same elevation in a continuous mass of the same static fluid will be at the same pressure.

This rule can be used to facilitate calculations in multiple-U tube problems. We can jump across the U tube if we stay within the same fluid.

2.7.3 Differential Manometers

A differential U tube manometer (Fig 2.10) and differential inverted U tube manometer (Fig 2.11) determines the difference in pressures at two points A and B when the actual pressure at any point in the system cannot be determined.

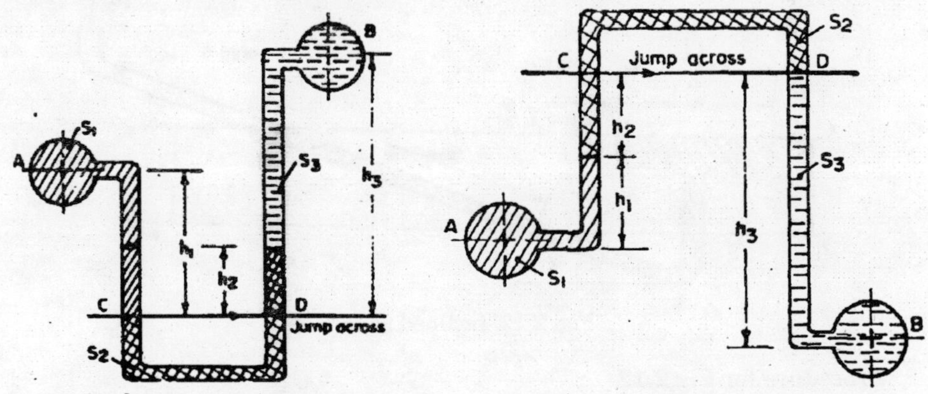

Fig. 2.10. Differential U tube manometer

Fig. 2.11. Differential inverted U tube manometer

Procedure for Fig. 2.10

$$h_A + h_1 S_1 - h_2 S_2 - h_3 S_3 = h_B$$

$$\therefore h_A - h_B = h_2 S_2 + h_3 S_3 - h_1 S_1$$

Procedure for Fig 2.11

$$h_A - h_1 S_1 - h_2 S_2 + h_3 S_3 = h_B$$

$$\therefore h_B - h_A = h_3 S_3 - h_1 S_1 - h_2 S_2$$

2.8 MEASUREMENT OF SMALL PRESSURE DIFFERENCE

In many flow systems, particularly with gas flow, the pressure difference to be measured are small and the range of pressure differences involved is small. It is therefore, necessary to modify the basic U tube, so that they may record the relatively small range of pressure differences accurately. The following methods by which this may be accomplished will be considered here and they are

(1) Inclined gauge

(2) U tube with enlarged ends or micromanometer and

(3) Micrometer type manometer

2.8.1 Inclined gauge

In this type of gauge the vertical leg of the U tube has an enlarged reservoir of area A, and the other leg of the U tube is inclined at an angle α to the horizontal as shown in Fig 2.12.

When no pressure difference is applied to the gauge, the level of the manometeric liquid having a relative density S_m is at x – x

Applying a pressure difference ($P_C - P_D$) where $P_C > P_D$ to the gauge causes the surface of separation to fall to y – y.

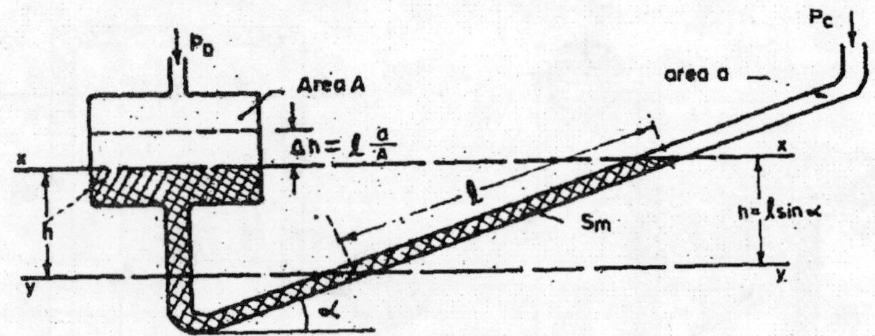

Fig. 2.12. Inclined gauge

Procedure for Fig 2.12.

$$h_D + \Delta h\, S_m + h\, S_m = h_C$$

or,
$$h_C - h_D = h\, S_m + l \frac{a}{A} S_m$$

$$h_C - h_D = h\, S_m + \frac{h}{\sin\alpha} \cdot \frac{a}{A} S_m = h\, S_m \left[1 + \frac{a}{A \sin\alpha}\right]$$

$$h_C - h_D = l\, S_m \left[\sin\alpha + \frac{a}{A}\right]$$

$$\therefore (p_C - p_D) = \rho g\ (h_C - h_D)\ P_a$$

This assumption that the density of the fluid flowing through the system is negligible compared to the manometric liquid and is therefore sensibly true if the fluid flowing through the system is a gas.

The sensitivity of this type of gauge depends essentially on the angle of inclination. The scale on this instrument is placed along the inclined leg, thus the reading actually taken is `l' and the pressure of difference may be read accurately to 1 mm and estimated to 1/10 mm.

2.8.2 Micromanometers

For the measurement of very small pressure differences a wide variety of special manometers have been developed. A number of devices may be used to increase the accuracy of a reading. For example a meniscus may be observed through a small telescope containing a horizontal cross wire, and the assembly may be raised or lowered by a slow motion screw with a micrometer scale. Or a scale floating on the surface of a liquid may be optically magnified. When an additional gauge liquid is used in a U-tube a large difference of meniscus level may be produced by a small pressure difference. One arrangement is illustrated in Fig 2.13.

The ends of this manometer consists of larger (and equal) areas containing two fluids that will not mix. Before any pressure is applied, the common surface of the two liquids is at x x and the heights of the two liquids above this level are h_1 and h_2. Since the fluids must be in equilibrium,

$$h_1 S_1 = h_2 S_2 \tag{2.10}$$

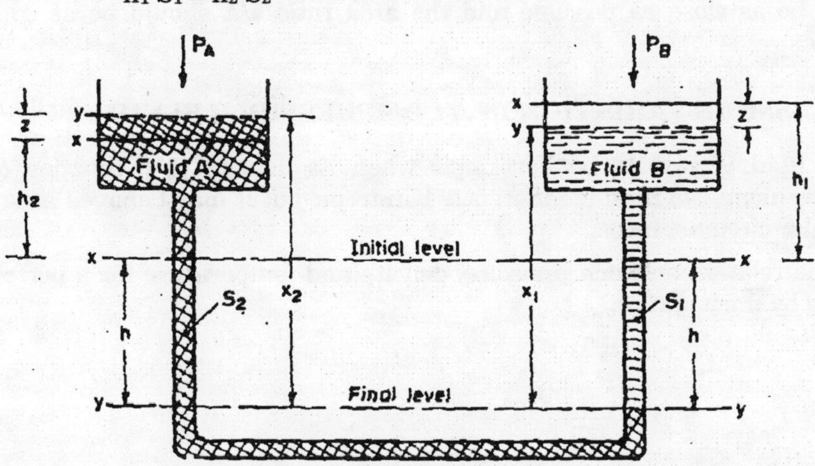

Fig. 2.13. Micromanometer

Where S_1 and S_2 are the relative densities of the fluids. A pressure is applied to the right arm of the tube, which causes the fluids to be displaced, and the common surface descends an amount h to the new level y y. The levels of the two liquids above the common surface are now X_1 and X_2

Since mass must be conserved, the mass of fluid displaced in one end must equal the mass of fluid displaced in the other end, or

$$Az = a\,h$$

$$\therefore z = \left(\frac{a}{A}\right)h$$

Where A is the area of cross section of large ends- both taken to be equal and `a' the area of cross section of the tube. From geometry,

$$X_1 = h_1 + h - z \quad \text{and} \quad X_2 = h_2 + h + z \tag{2.12}$$

Using the general procedure for the given micromanometer.

$$h_B + X_1 S_1 - X_2 S_2 = h_A$$

or, $$h_B - h_A = X_2 S_2 - X_1 S_1$$

$$= (h_2 + h + z) S_2 - (h_1 + h - z) S_1 \quad \text{[From (2.12)]}$$

$$= h_2 S_2 + h S_2 + z S_2 - h_1 S_1 - h S_1 + z S_1 \quad \text{[From (2.10)]}$$

$$= h S_2 + \frac{a}{A} h S_2 - h S_1 + \frac{a}{A} h S_1 \quad \text{[From (2.11)]}$$

$$= h \left[S_2 \left(1 + \frac{a}{A} \right) - S_1 \left(1 - \frac{a}{A} \right) \right] \quad (2.13)$$

$$= h \left[(S_2 - S_1) + (S_2 + S_1) \frac{a}{A} \right] \quad (2.14)$$

From equation (2.14) we conclude that for greatest sensitivity S_1 and S_2 should be as close as possible and the area ratio a/A should be as small as possible.

2.9 PRESSURE VARIATION IN A COMPRESSIBLE FLUID

A fluid is said to be barotropic when its density is a function of the pressure alone. No fluid is inherently barotropic but it may behaved as such in particular circumstances.

The relation between pressure, density and temperature for a perfect gas is given by the equation

$$p = \rho R T.$$

$$\therefore \rho = \frac{p}{RT} \quad (2.15)$$

According to equation

$$\frac{dp}{dz} = -\rho g \quad (2.16)$$

From Eq (2.15), $\frac{dp}{dz} = -\frac{p}{RT} g$

or, $$\frac{dp}{p} = -\frac{g\,dz}{RT} \quad (2.17)$$

2.9.1 Variation under Isothermal conditions

The equation (2.17) can be integrated from $p = p_1$ when $z = z_1$, to $p = p_2$ when $z = z_2$.

$$\therefore \int \frac{dp}{p} = -\int \frac{g dz}{RT}$$

$$\ln p_2/p_1 = -\left(\frac{g}{RT}\right)(z_2 - z_1)$$

$$\frac{p_2}{p_1} = \exp.\left\{-\left(\frac{g}{RT}\right)(z_2 - z_1)\right\} \tag{2.18}$$

2.9.2 Variation under adiabatic conditions

If conditions are adaibatic the relationship between pressure and density is given by

$$\frac{p}{\rho^k} = \text{Constant} = C$$

$$\therefore \rho = \left(\frac{p}{C}\right)^{1/k} \quad \text{Here, } C = \frac{p_1}{p_1^k} = \frac{p_2}{\rho_2^k}$$

From equation (2.10)

$$\frac{dp}{dz} = -\rho g = -\left(\frac{p}{C}\right)^{1/k} g$$

or,
$$dz = -\frac{dp}{(p/C)^{1/k}\, g}$$

Intergrating from $p = p_1$ when $z = z_1$ to $p = p_2$ when $z = z_2$

$$z_2 - z_1 = -\frac{C^{1/k}}{g}\left[\frac{p^{-\frac{1}{k}+1}}{-\frac{1}{k}+1}\right]_{p_1}^{p_2}$$

$$= -\frac{C^{1/k}}{g}\left(\frac{k}{k-1}\right)\left[p_2^{\frac{k-1}{k}} - p_1^{\frac{k-1}{k}}\right]$$

$$= -\left(\frac{k}{k-1}\right) p_1^{\frac{k-1}{k}} \frac{c^{1/k}}{g}\left[\left(\frac{p_2}{p_1}\right)^{\frac{k-1}{k}} - 1\right]$$

$$= -\left(\frac{k}{k-1}\right) \frac{p_1^{1/k}}{\rho_1 g} \times p_1^{\frac{k-1}{k}}\left[\left(\frac{p_2}{p_1}\right)^{\frac{k-1}{k}} - 1\right]$$

$$= -\left(\frac{k}{k-1}\right) \frac{p_1}{\rho_1 g}\left[\left(\frac{p_2}{p_1}\right)^{\frac{k-1}{k}} - 1\right]$$

$$z_2 - z_1 = -\left(\frac{k}{k-1}\right)\frac{RT_1}{g}\left[\left(\frac{p_2}{p_1}\right)^{\frac{k-1}{k}} - 1\right]$$

$$\text{or,}\quad \left(\frac{p_2}{p_1}\right)^{\frac{k-1}{k}} = \frac{-g(z_2 - z_1)}{RT_1}\left(\frac{k-1}{k}\right) + 1$$

$$\frac{p_2}{p_1} = \left[1 - \frac{g(z_2 - z_1)}{RT_1} \times \left(\frac{k-1}{k}\right)\right]^{\frac{k}{k-1}} \tag{2.19}$$

For Polytropic process

$$\frac{p_2}{p_1} = \left[1 - \frac{g(z_2 - z_1)}{RT_1} \times \left(\frac{n-1}{n}\right)\right]^{\frac{n}{n-1}} \tag{2.20}$$

Thus rate of change of temperature with altitude - the temperature lapse rate-can also be found for adiabatic conditions. From the characteristic equation $\rho = p/RT$ and, since, from equation (2.16).

$$dz = -\frac{dp}{\rho g}$$

$$dz = -\left(\frac{RT}{gp}\right)dp \tag{2.21}$$

For adiabatic conditions

$$\frac{p}{\rho^k} = \frac{p_1}{\rho_1^k} \quad \text{and} \quad \frac{p}{p_1} = \left(\frac{T_1}{T}\right)^{\frac{k}{k-1}}$$

$$p = p_1\left(\frac{T_1}{T}\right)^{\frac{k}{k-1}}$$

Differentiating

$$dp = -\left[\frac{k}{k-1}\right] p_1 T_1^{\frac{k}{k-1}} T^{\frac{-1}{1-k}} dT$$

Substituting in equation (2.21) and solving we get

$$dz = \left(\frac{k}{k-1}\right)\left(\frac{R}{g}\right)dT$$

$$\text{or,}\quad \frac{dT}{dz} = \text{Temperature gradient}$$

$$= \left(\frac{k-1}{k}\right)\left(\frac{g}{R}\right) = -\lambda \ = \text{Temperature lapse rate} \qquad (2.22)$$

2.9.3 Variation of pressure and density with altitude for a constant temperature gradient

An important special problem is that of a perfect gas with a constant temperature lapse rate, for this is a good approximation to average condition in the troposhere, i.e. that part of the earth's atmosphere lying between the earths surface and the stratosphere, in which the temperature is nearly constant. Let the absolute temperature at height z above sea level be

$$T = T_o - \lambda z \qquad (2.23)$$

Where λ is the constant lapse rate

we know that

$$\rho = \frac{p}{RT}$$

From (2.23), $$\rho = \frac{p}{R(T_o - \lambda z)} \qquad (2.24)$$

We also know that

$$dp = -\rho g\, dz$$

$\therefore$ From equation (2.24)

$$\frac{dp}{p} = \frac{-gdz}{R\,(T_o - \lambda z)}$$

Since p_o is the pressure when z is zero we get

$$ln\ \frac{p}{p_o} = \frac{g}{\lambda R}\ln\left(1-\frac{\lambda z}{T_o}\right)$$

$$p = p_o\left(1-\frac{\lambda z}{T_o}\right)^{g/\lambda R} = P_o\left(\frac{T_o - \lambda z}{T_o}\right)^{g/\lambda R}$$

$$p = P_o\left(\frac{T}{T_o}\right)^{\left(\frac{g}{\lambda R}-1\right)} \qquad (2.25)$$

Hence also by the equation of state

$$\rho = \rho_o\left(\frac{T}{T_o}\right)^{\left(\frac{g}{\lambda R}-1\right)} = \rho_o\left(1-\frac{\lambda z}{T_o}\right)^{\frac{g}{\lambda R}-1} \qquad (2.26)$$

2.9.4 Variation of temperature and pressure in the atmosphere

In the atmosphere n varies with altitude. For the first 9 km (approx.) above the sea level the temperature decreases uniformly, that is

$$\frac{dT}{dz} = \text{Constant} = -\lambda \tag{2.27}$$

Combining with the equation p = ρRT and the equation $\frac{p}{\rho n} = C$ We have

$$\frac{T}{T_o} = 1 - \frac{gz}{RT_o}\left(\frac{n-1}{n}\right) \tag{2.28}$$

Differentiating with respect to z we get

$$\frac{1}{T_o}\frac{dT}{dz} = -\frac{g}{RT_o}\left(\frac{n-1}{n}\right) - \frac{dT}{dz} = \frac{g}{R}\left(\frac{n-1}{n}\right)$$

$$\lambda = \frac{g}{R}\left(\frac{n-1}{n}\right)$$

$$n\lambda R = gn - g$$

$$\therefore n = \frac{1}{1 - \frac{\lambda R}{g}} \tag{2.29}$$

The temperature lapse rate observed in this region is approximately 0.0065 k/m.

$$\therefore \frac{\lambda R}{g} = 0.1902 \text{ and } n = 1.238$$

Above 11000 m lies the stratosphere, in which conditions are assumed to be isothermal, with the temperature constant at –56.5°C. Fig 2.14 Shows the variation of temperature with altitude in the I.S.A. The atmospheric pressure at sea level is assumed to be equivalent to 760 mm of mercury, the temperature 15°C and the density 1.225 kg/m³.

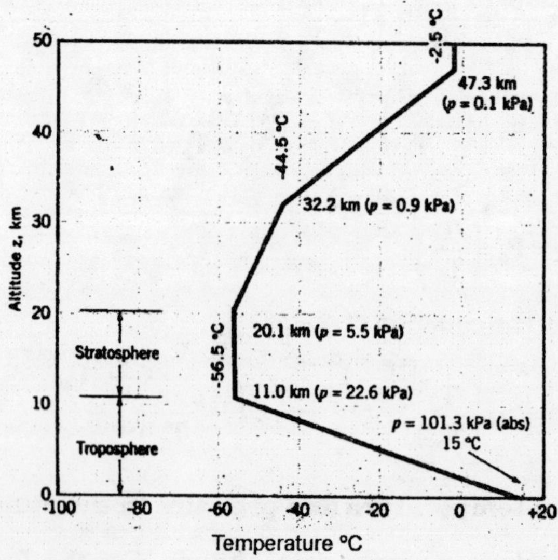

Fig. 2.14. Variation of pressure and temperature with altitude

In the real atmosphere, the troposphere extends to an average of 11000 m, but can vary from 7600 m at the poles to 18000 m at the equator. While the temperature, in general, falls steadily with altitude meteorological conditions can arise in the lower layers which produce temperature inversion-the temperature increasing with altitude. In the stratosphere, the temperature remains substantially constant up to approximately 32000 m; it then rises to about 70°C before falling again. Fig 2.15 shows typical values for pressure and temperature in the real atmosphere.

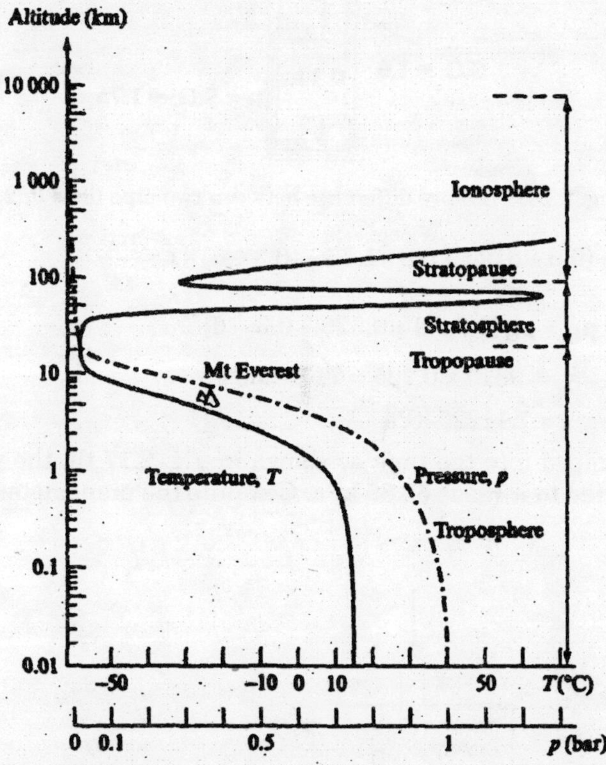

Fig. 2.15. Typical values of pressure and temperature.

ILLUSTRATIVE EXAMPLES

2.1 The manometer shown in the Fig. 2.16 is used to measure the pressure difference between two pipe lines A and B, each containing water. The manometer fluid has a relative density of 1.75, and the elevation difference between the two pipes is 0.5 m. The difference between the two meniscuses is 0.3 m. What is the pressure difference p_A-p_B?

Solution

Let the distance from B to the first interface be d. Start at point B and go to the next interface, adding (going down) or subtracting (going up) the distance times the relative density of the fluid. Continue until point A is reached.

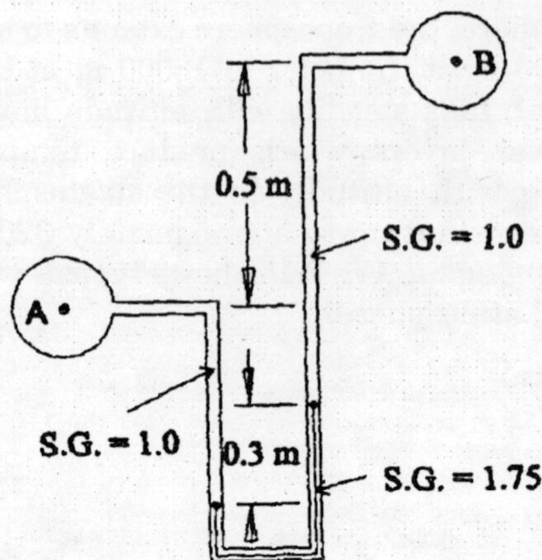

Fig. 2.16. Pressure difference between two pipe lines P. 2.1

$$\frac{p_B}{\rho g} + (0.5 + d) \times 1 + 0.3 \times 1.75 - (0.3 + d) \times 1 = \frac{p_A}{\rho g}$$

$$p_A - p_B = \rho g\,[0.5 + d + 0.525 - 0.3 - d]$$

$$= 9810 \times 0.725 = 7112.25 \text{ N/m}^2$$

$$= 7.11225 \text{ kPa}$$

Ans

2.2 Air pumped into the tank as shown in Fig. 2.17 till the pressure gauge connected to it reads 58.86 kPa. Compute the manometer deflection y.

Solution

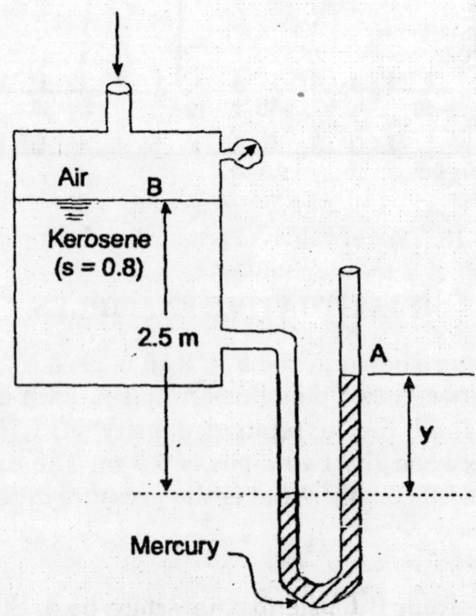

Fig. 2.17. Tank P. 2.2

Air pressure head

$$\frac{p_B}{\rho g} = \frac{58.86 \times 1000}{9810} = 6 \text{ m of water}$$

The general procedure for the given problem.

$$\frac{p_A}{\rho g} = \text{open to atmosphere} = \text{zero gauge pressure}$$

Therefore,

$$\frac{p_B}{\rho g} = 6 = y \times 13.6 - 2.5 \times 0.8$$

$$y \times 13.6 = 6 + 2 = 8$$

$$y = \frac{8}{13.6} = 0.588 \text{ m}$$ Ans

2.3 Find the difference of pressure ($p_B - p_A$), if x = 3m, y = 2m, and z = 1m and fluids 1, 2 and 3 are kerosene, water and petrol respectively. The relative densities of kerosene and petrol are 0.8 and 0.78 respectively (Fig. 2.18).

Solution

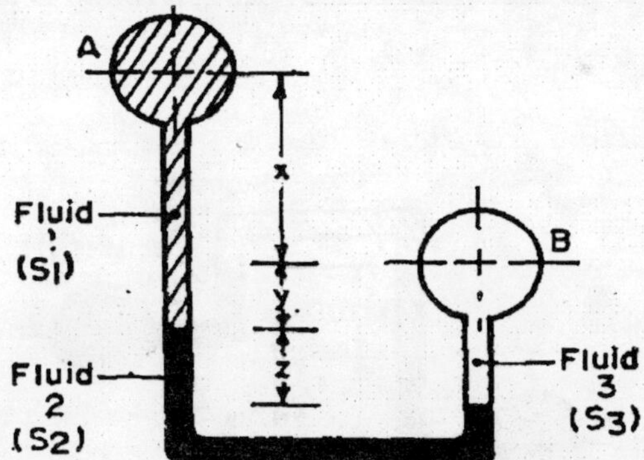

Fig. 2.18. Differential U tube manometer P. 2.3

$$h_A \times x S_1 + y S_1 + z S_2 - (y + z) S_3 = h_B$$

$$h_A + (5) \times 0.8 + 1 \times 1 - (2+1) \times 0.78 = h_B$$

$$h_A - h_B = 3 \times 0.78 - 4 - 1 = 2.34 - 5 = -2.66 \text{ m of water}$$

$$\therefore P_A - P_B = 9810 \times (-2.66) = -26.094 \text{ kN/m}^2$$

or, $P_B - P_A = +26.094$ kPa Ans

2.4 Find z (Fig 2.19) if $p_B - p_A$ = 10 kPa, fluid 1 is kerosene, fluid 2, is mercury and fluid 3 is water. Also, x = o and y = 1 m.

Solution

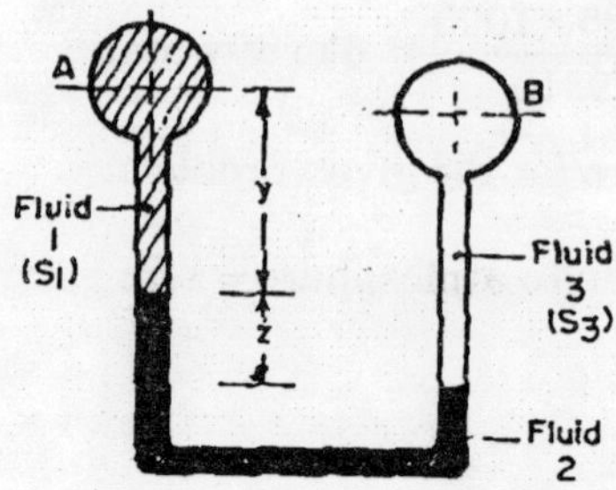

Fig. 2.19. Differential U tube manometer P. 2.4

The general solution for the given manometer is as follows.

$$h_A + y \times S_1 + z \times S_2 - (y+z) \times S_3 = h_B$$

or, $h_B - h_A = 1 \times 0.8 + z \times 13.6 - (1+z) \times 1$

$1.019 = 0.8 + 13.6\,z - 1 - z$

$1.019 = -0.2 + 12.6\,z$ [Here 1 kPa = 0.1019 m of water]

$$z = \frac{1.219}{12.6} = 0.09677 = 96.77 \text{ mm}$$ Ans.

2.5 Calculate the pressure difference between A and B for the set up shown in Fig 2.20.

Solution

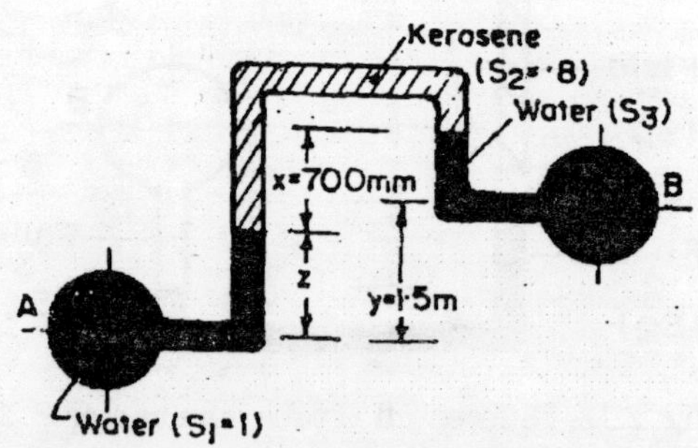

Fig. 2.20. Inverted U tube differential manometer P. 2.5

The general procedure for the given manometer is as follows.

$$h_A - z \times S_1 - 0.7 \times S_2 + (z - 0.8) \times S_3 = h_B$$

$$h_A - h_B = z\,S_1 + 0.7 \times 0.8 - z\,S_3 + 0.8\,S_3$$

$$= z \times 1 + 0.56 - z \times 1 + 0.8 \times 1 = 0.56 + 0.8 = 1.36$$

$\therefore\ p_A - p_B = 9.81 \times 1.36 = 13.34$ kPa Ans

2.6 In Figure 2.21 the liquids at A and B are water and the manometer liquid is oil of relative density 0.8

(a) Determine the difference in pressue in Pascals.

(b) If pressure at B is equal to 50 kPa and the barometer reading is 730 mm of mercury, find the pressure at A, in m of water absolute.

Solution: We know that 1 bar = 750 mm of mercury

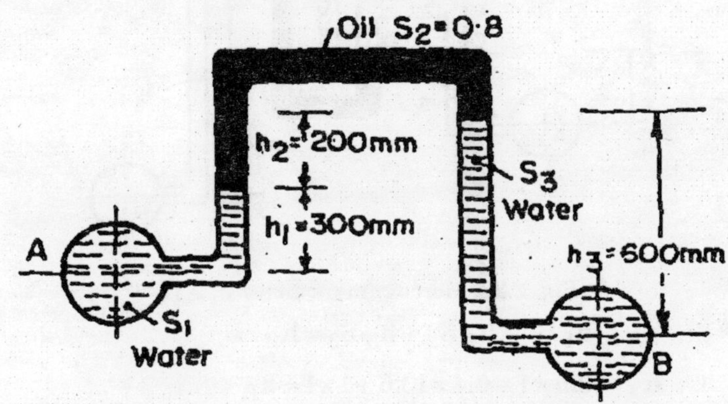

Fig. 2.21. Differential inverted U tube manometer P. 2.6

∴ 730 mm of mercury is equal to 0.9733 bar

Using the general procedure starting from A

(a) $$h_A - h_1 S_1 - h_2 S_2 + h_3 S_3 = h_B$$

$$h_A - 0.3 \times 1 - 0.2 \times 0.8 + 0.6 \times 1 = h_B$$

$$h_A - h_B = -0.6 + 0.3 + 0.16 = -0.14 \text{ m of water}$$

$$p_A - p_B = \rho h (h_A - h_B)$$

$$= 9810 (-0.14) = -1373.4 \text{ Pascal}$$ Ans

(b) $$P_B = \rho g h_B$$

$$50 \times 1000 = 9810 \times h_B \quad \therefore h_B = 5.096$$

$$\therefore h_{B\,absolute} = h_{B\,gauge} + \text{Atmospheric pressure}$$

$$= 5.096 + 0.9733 \times \frac{100}{9.81}$$

$$= 5.096 + 9.921 = 15.017 \text{ m of water}$$

$$\therefore h_{B\,absolute} = h_{B\,gauge} - 0.14$$

$$= 15.017 - 0.14 = 14.877 \text{ m of water}$$ Ans

2.7 A mercury manometer is used to measure the pressure difference between two pressure vessels. For the arrangement and readings shown in Figure 2.22, determine the piezometric pressure difference when the vessels A and B contain (a) water and (b) air of density 1.2 kg/m^3.

Solution

Using the general procedure starting from A. Taking jump across from C to D and E to F. Taking S_1 and S_3 is water.

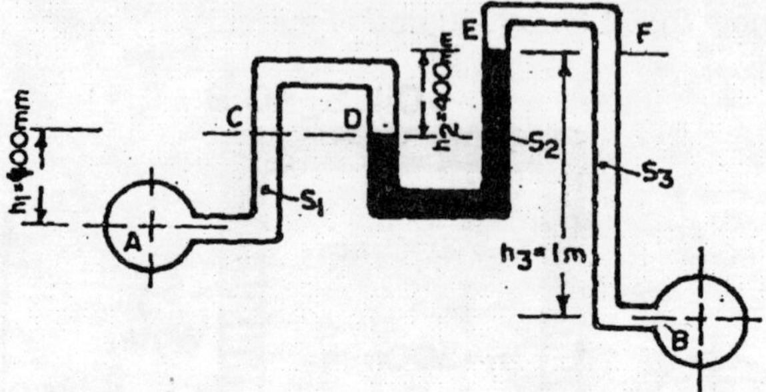

Fig. 2.22. Mercury manometer P. 2.7.

$$h_A - h_1 S_1 - h_2 S_2 + h_3 S_3 = h_B$$

$$h_A - 0.5 \times 1 - 0.4 \times 13.6 + 1 \times 1 = h_B$$

$$h_A - h_B = 0.5 + 0.4 \times 13.6 - 1 = 4.94 \text{ m of water}$$

$$\therefore\ p_A - p_B = \rho g (h_A - h_B) = 9.81 \times 4.94 = 48.46 \text{ kPa} \qquad \text{Ans}$$

(b) Taking S_1 and S_3 is air

$$h_A - h_1 S_1 - h_2 S_2 + h_3 S_3 = h_B$$

$$h_A - 0.5 \times \frac{1.2}{1000} - 0.4 \times 13.6 + 1 \times \frac{1.2}{1000} = h_B$$

$$h_A - h_B = 6 \times 10^{-4} + 5.44 - 1.2 \times 10^{-3} = 5.4394 \text{ m of water}$$

$$p_A - p_B = 53.36 \text{ kPa} \qquad \text{Ans}$$

2.8 For the set up shown in Figure 2.23, what is the absolute pressure in drum A? Assume an atmospheric pressure of 760 mm of mercury.

Solution

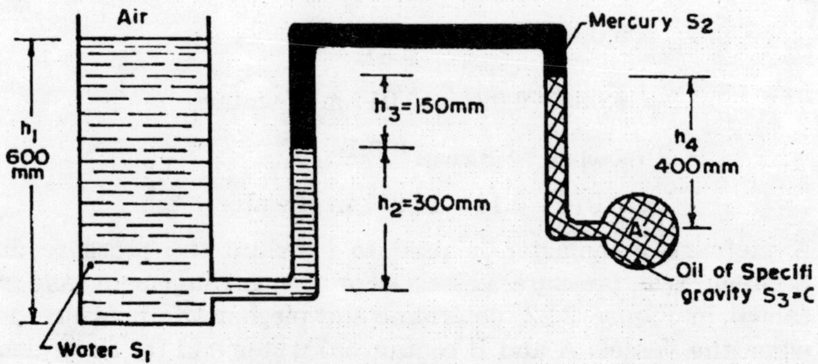

Fig. 2.23. Manometer P. 2.8.

Using the general procedure, starting from A.

$$h_{A_{abs}} - h_4 S_3 + h_3 S_2 + h_2 S_1 - h_1 S_1 = 0.76 \times 13.6$$

$$h_{A_{abs}} - 0.4 \times 0.8 + 0.150 \times 13.6 + 0.3 \times 1 - 0.6 \times 1 = 10.336$$

$$h_{A_{abs}} = 0.32 - 0.3 + 0.6 - 2.04 = -1.44 + 10.336 = 8.896 \text{ m of water}$$

$$P_{A_{absolute}} = 9.81 \times 8.896 = 87.27 \text{ kN/m}^2$$ Ans

2.9 The pressure at point B is 0.2 bar. Find the pressure in kN/m^2 in the closed air space at the top of the tank A as shown in Figure 2.24.

Solution

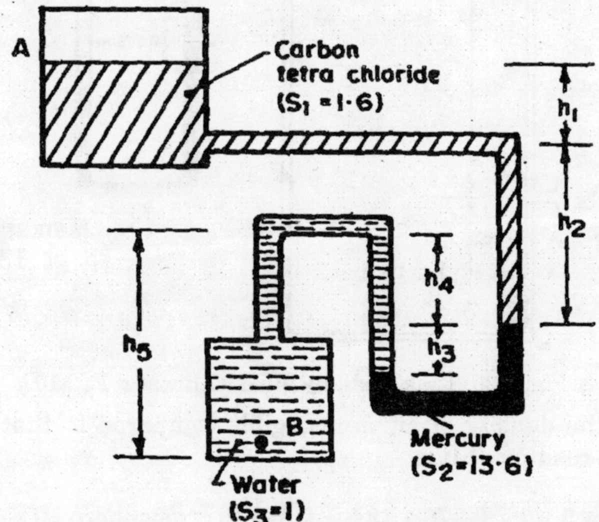

Fig. 2.24. U tube manometer P. 2.9

Using the general procedure starting from A.

$$h_A + h_1 S_1 + h_2 S_1 + h_3 S_2 - (h_3 + h_4) S_3 + h_5 \times S_3 = h_B \qquad (1)$$

Here

$h_1 = 700$ mm, $h_2 = 1$ m, $h_3 = 500$ mm, $h_4 = 300$ mm, $h_5 = 1$ m

$S_1 = 1.6$, $S_2 = 13.6$, $S_3 = 1$

$$p_B = 0.2 \text{ bar} \qquad \therefore\ h_B = \frac{0.2 \times 10^5}{9810} = 2.038$$

Putting all the values in equation (1)

$$h_A + 0.7 \times 1.6 + 1 \times 1.6 + 0.5 \times 13.6 - 0.5 \times 1 - 0.3 \times 1 + 1 \times 1 = 2.0385$$

$$h_A = 2.038 - 1.12 - 1.6 - 6.8 + 0.5 + 0.3 - 1 = -10.52 + 2.838 - 1$$

$$p_A = \rho g h_A = 9.81 \times (-7.682) = -75.36 \text{ kPa}$$ Ans

2.10 The tank shown in Fig. 2.25 is partially filled with water. If the barometric pressure is 712.5 mm of mercury, determine @ the pressure gauge reading at B and (b) the absolute pressure of the air in the tank.

Solution :

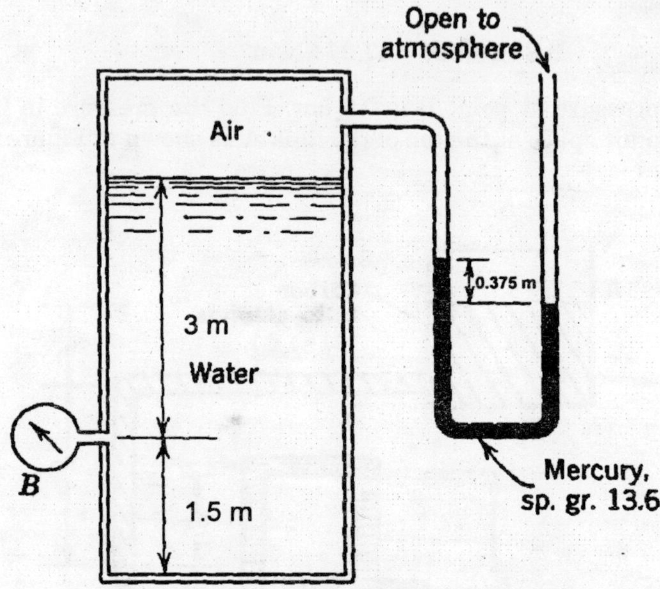

Fig. 2.25. Tank partially filled with water P. 2.10

(a) Since the density of air is negligible compared to that of liquids, the gauge pressure reading at B is.

$$\frac{p_B}{\rho g} - 3 \times 1 + 0.375 \times 13.6 = \text{Open to atmosphere} = O$$

$$p_B = \rho g\,(-5.1 + 3) = -9810 \times 2.1$$

$$= -20601 \text{ Pa} = 20.6 \text{ kPa (Vacuum)}$$

The pressure gauge at B indicates a reading of 20.6 kPa vacuum. Ans.

(B) Barometric pressure = 712.5 mm of mercury

$$\frac{P_{air}}{\rho g} = \frac{712.5}{1000} = 0.375$$

$$= 0.3375 \text{ m of mercury}$$

$$= 13.6 \times 9810 \times 0.3375$$

$$= 45027.9 \text{ Pa} = 45.028 \text{ kPa (Absolute)}$$ Ans

2.11 Two U tube manometer, one upright and the other inverted U tube type, are connected across a water line and oil line as shown in Figure 2.26, If $h_1 = 1.2$ m a = 0.2 m, b = 1 m, what shall h_2 be?

Solution: I Case

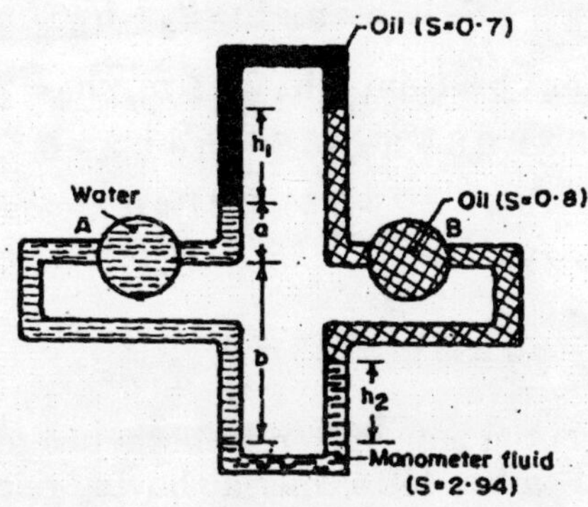

Fig. 2.26. Two U tube manometers P. 2.11

Below the centre line A - B.

$$h_A + b\times 1 - h_2 \times 2.94 - (b - h_2)\times 0.8 = h_B$$

Here $a = 0.2\,m$, $b = 1\,m$, $h_1 = 1.2\,m$

$$h_A + 1 - h_2 \times 2.94 - (1 - h_2)\times 0.8 = h_B$$

$$h_A - h_B = h_2 \times 2.94 - 1 + (1 - h_2)\times 0.8$$

$$= h_2 \times 2.94 - h_2 \times 0.8 - 1 + 0.8 = h_2\,(2.14) - 0.2 \quad (1)$$

II Case Above centre line A – B.

$$h_A - a \times 1 - h_1 \times 0.6 + (h_1 + a) \times 0.7 = h_B$$

$$h_A - h_B = 0.2\times 1 + 1.2\times 0.7 - (1.2 + 0.2)\times 0.8 = 0.2 + 0.84 - 1.12 = -0.08 \quad (2)$$

$$h_2 \times 2.14 - 0.2 = -0.08$$

$$h_2 \times 2.14 = 0.12$$

$$h_2 = \frac{0.12}{2.14} = 0.056\text{ m} = 56\text{ mm} \qquad \text{Ans}$$

2.12 For the U tube manometer connected as shown in the Figure 2.27, If a = 300 mm, b = 400 mm and c = 500 mm, find the difference of mercury column in the two limbs.

Solution

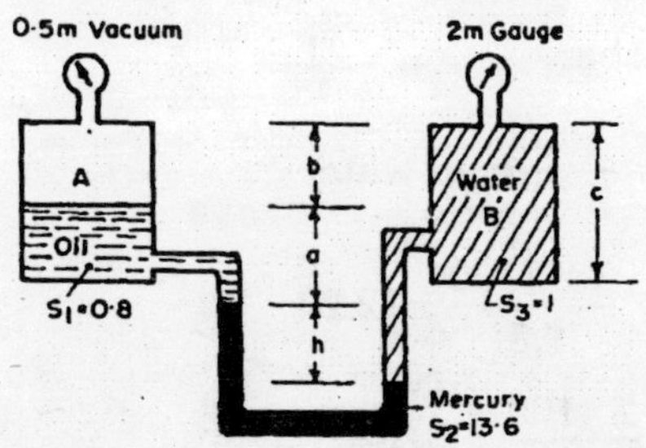

Fig. 2.27. U tube manometer P. 2.12

Using the general procedure starting from A

$$h_A + as_1 + h\,S_2 - (h + a + b)\times S_3 = h_B$$

Here $a = 0.3$ m, $b = 0.4$m $h_A = 0.5$ m, $h_B = 2$ m

$$-0.5 + 0.3 \times 0.8 + h \times 13.6 - (h + 0.3 + 0.4) \times 1 = 2$$

$$h \times 13.6 - h \times 1 = 2 + 0.5 - 0.24 + 0.7$$

$$12.6\,h = 2.96$$

$$h = \frac{2.96}{12.6} = 0.235 \text{ m} = 235 \text{ mm}$$ Ans

2.13 A U tube differential manometer connects two pressure pipes A and B. Pipe A contains carbon tetra chloride having relative density 1.6 under a pressure of 1.2 bar and pipe B contains oil of relative density 0.8 under a pressure of 2 bar. The pipe A lies 2.5 m above the centre line of pipe B, find the difference of pressure measured by mercury as fluid filling U tube.

Solution

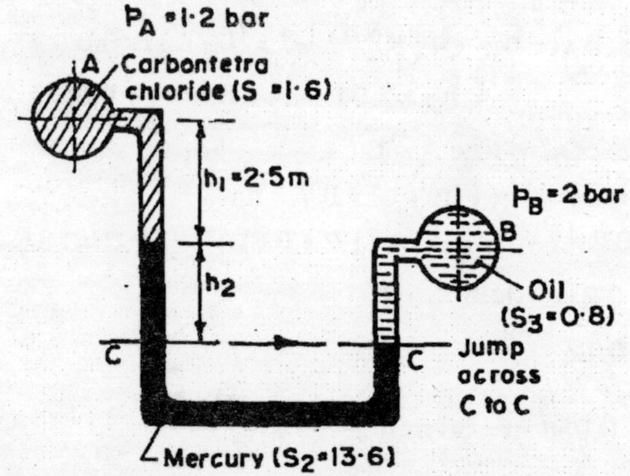

Fig. 2.28. U tube differential manometer P. 2.13.

Using the general procedure starting from A

$$h_A + h_1\,S_1 + h_2\,S_2 - h_2\,S_3 = h_B$$

Here

$$p_A = 1.2 \text{ bar}, \qquad p_B = 2 \text{ bar}$$

$$1 \text{ bar} = \frac{100}{9.81} \text{ m of water}$$

$$\therefore\ h_A = \frac{1.2\times 100}{9.81} = 12.23$$

and $$h_B = \frac{2\times 100}{9.81} = 20.38$$

Therefore,

$$12.23 + 2.5 \times 1.6 + h_2 \times 13.6 - h_2 \times 0.8 = 20.38$$

$$h_2 (12.8) = 20.38 - 12.23 - 4 = 4.15$$

$$h_2 = \frac{4.15}{12.8} = 0.3342 = 324.2 \text{ mm} \qquad \text{Ans}$$

2.14 A U tube containing mercury has its right hand limb open to atmosphere and the left hand limb connected to a pipe conveying water under pressure, the difference in levels of mercury in the two limbs being 200 mm. If the mercury level in the left limb is 400 mm below the centre line of the pipe. Find the absolute pressure in the pipe line. Also find the new difference in levels of the mercury in the U tube, if the pressure in the pipe falls by 2 kPa. Take atmospheric pressure 1.01325 bar.

Solution:

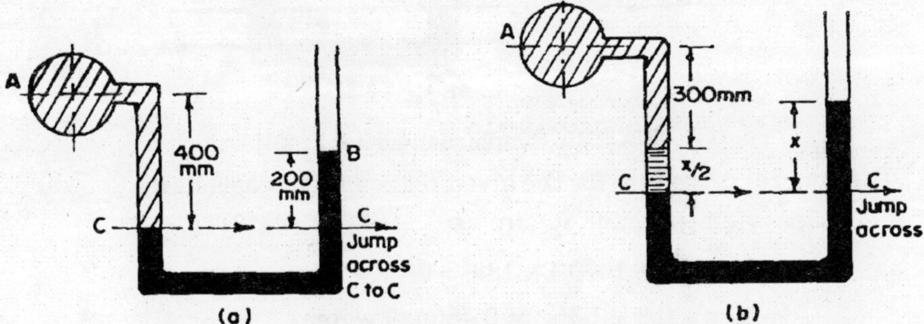

Fig. 2.29. Simple manometer P. 2.14 (a and b)

The general procedure for the given manometers is as follows.

$$h_A + 0.4 \times 1 - 0.2 \times 13.6 = 0$$

$$h_A = 2.32 \text{ m of water}$$

$$p_A = 9.81 \times 2.32 = 22.7592 \text{ kPa}$$

The absolute pressure

$$= 101.325 + 22.7592 = 124.0842 \text{ kPa}$$

When the manometer is not connected to the system the mercury levels in both the limbs equalise and are 300 mm below the centre line of the pipe.

$$2 \text{ k Pa} = 0.2038 \text{ m of water}$$

$\therefore$ The new pressure head at A = 2.32 – 0.2038

$$= 2.116 \text{ m of water}$$

$$\therefore \; h_A + \left(0.3 + \frac{x}{2}\right) \times 1 - x \times 13.6 = 0$$

$$2.116 + 0.3 = x \times 13.6 - \frac{x}{2}$$

$$x = \frac{2.416}{13.1} = 0.18442 \text{ m} = 184.42 \text{ mm}$$ Ans

2.15 In Figure 2.30, originally there is a 100 mm manometer reading. If the absolute pressure at A is doubled, what then would be the manometer reading? The atmospheric pressure is 101.3 kPa.

Solution

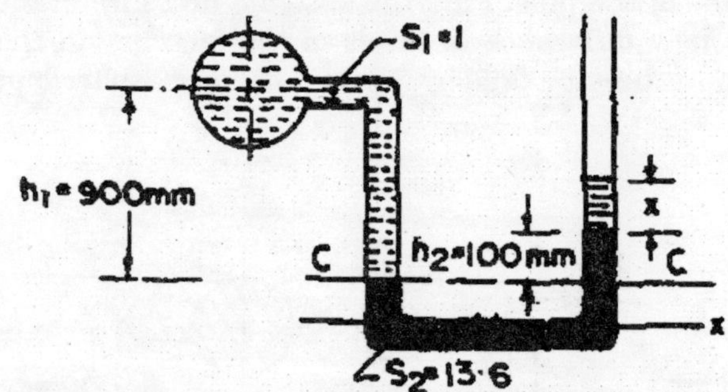

Fig. 2.30. Simple manometer P. 2.15

The general procedure for the given manometer is as follows.

$$h_A + h_1 S_1 - h_2 S_2 = 0$$

$$h_A + 0.9 \times 1 - 0.1 \times 13.6 = 0$$

$$h_A = -0.9 + 1.36 \quad = 0.46 \text{ m of water}$$

$$\therefore h_{A\,absolute} = \text{Atmospheric pressure} + \text{Gauge pressure at A}$$

$$= 10.326 + 0.46 = 10.786 \text{ m of water}$$

As per the question $h_A = 2 \times 10.786 = 21.572$ absolute

$$\therefore h_A + (h_1 + x)\, S_1 - (h_2 + 2x) \times S_2 = h_{atmosphere}$$

$$21.572 + (0.9 + x) \times 1 - (0.1 + 2x) \times 13.6 = 10.326$$

$$21.572 + 0.9 - 0.1 \times 13.6 - 10.326 = 2x \times 13.6 - x$$

$$10.786 = 27.2\,x - x = 26.2\,x$$

$$\therefore x = \frac{10.786}{26.2} = 0.4116$$

New manometer reading $2x + h_2$

$$= 2 \times 0.4116 + 0.1 = 0.8233 + 0.1$$

$$= 0.9233 = 923.3 \text{ mm}$$ Ans.

2.16 In Fig 2.31, A contains water, and the manometer fluid has a relative density of 2.94. When the surface of separation of water and manometer fluid is at C C. When the left menisucs is at C C, $h_A = 100$

mm of water, find the reading of the right of the right meniscus for P_A = 8 kPa.

Solution

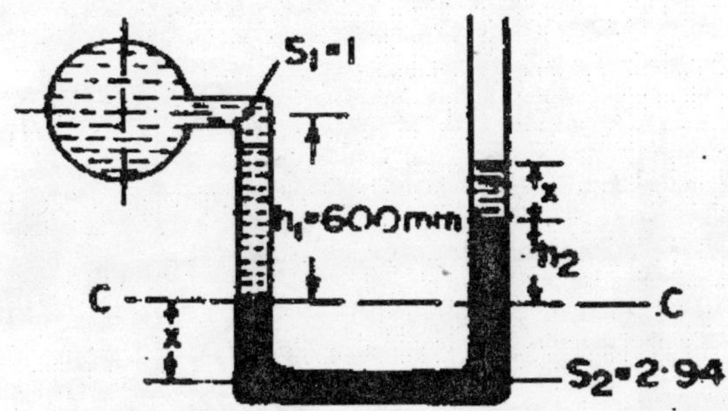

Fig. 2.31. Simple manometer P. 2.16

Manometer reading drops x in left side and rise x in right side.

I Case h_A = 100 mm of water

$$h_A + h_1 S_1 - h_2 S_2 = 0$$

$$100 + 600 \times 1 - h_2 \times 2.94 = 0$$

$$h_2 = \frac{700}{2.94} = 238.095 \text{ mm}$$

II Case, $p_A = 8\ kPa$

$$\therefore h_A = \frac{8000}{9810}$$

$$\therefore h_A + (0.6 + x) \times S_1 - (h_2 + 2x) \times S_2 = 0$$

$$\frac{8000}{9810} + (0.6 + x) \times 1 - (0.238 + 2x) \times 2.94 = 0$$

$$0.815 + 0.6 - 0.700 = 5.88\,x - x$$

$$0.715 = 4.88\text{ x}$$

$$\therefore x = \frac{0.715}{4.88} = 0.1465 \text{ m}$$

The reading of the manometer

$= 2x + h_2 = 2 \times 0.1465 + 0.238095 = 0.293 + 0.238095$

$= 531$ mm Ans

2.17 A U tube manometer is used to measure the pressure of kerosene in pipeline A. The manometer fluid is mercury and recorded head is 500 mm. If the mercury-kerosene interface is 750 mm above the pipe centre line determine the gauge pressure of the kerosene. If condition in the pipe line now change so that mercury- air interface falls 80 mm below the mercury interface with no change in manomtere position determine the absolute pressure in the pipe line. Atmospheric pressure is 101.325 kPa. Relative densities of kerosene and mercury are 0.82 and 13.6 respectively.

Solution

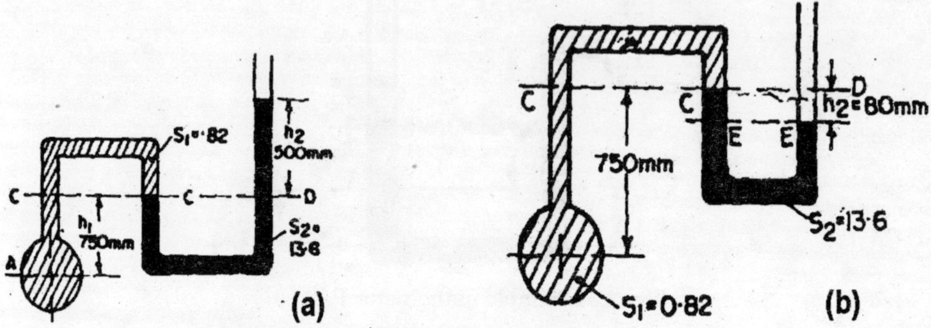

Fig. 2.32 (a). U tube manometer I case P. 2.17
(b) U tube manometer II case P. 2.17.

I Case:

$$h_A - h_1 S_1 - h_2 S_2 = 0$$

$$h_A = h_1 S_1 + h_2 S_2 = 0.75 \times 0.82 + 0.5 \times 13.6$$

$$= 0.615 + 6.8 = 7.415 \text{ m of water}$$

$$p_A = \rho g\, h_A = 9.81 \times 7.415 = 72.74\ kPa$$

II Case: The mercury-kerosen interface is at C-C and mercury-air interface is at C-D. At D the level drops to below 80 mm from C-D interface. Therefore, the new interface is at E-E.

$$h_A - h_1 S_1 + h_2 S_2 = 0$$

$$h_A - 0.75 \times 0.82 + 0.08 \times 13.6 = 0$$

$$h_A = -1.088 + 0.615 = -0.473$$

$$\therefore P_A = \rho g h_A$$

$$= -9.81 \times 0.473 = -4.64\ kN/m^2$$

$$P_{Aabsolute} = Atmospheric\ pressure + Gauge\ pressure\ at\ A.$$

$$= 101.325 + (-4.64) = 96.685\ kN/m^2$$ Ans

2.18 A differential manomter is connected at two points A and B of two pipes. The pipe A contains a liquid of relative density 1.5 while pipe B contains a liquid of relative density 0.9. The pressure at A and B are 1 bar and 1.8 bar respectively. The vertical distance between A and B is 3m. B is below A. The vertical distance between point B and mercury level in the right limb B is (2 + h) mater. Find the value of h. The difference in mercury level in the differential manometer is h.

Solution

Using the general procedure starting from A.

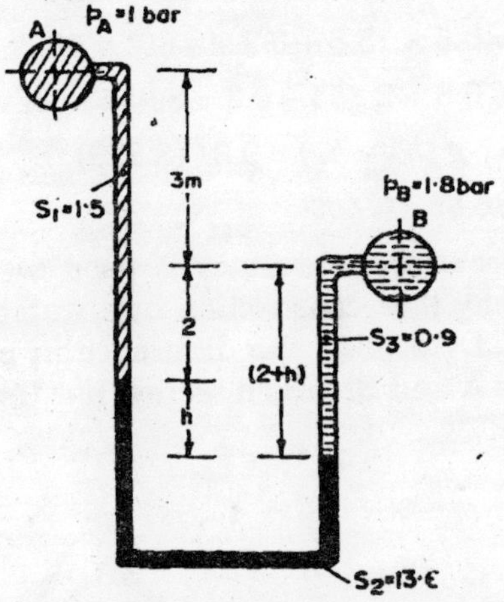

Fig. 2.33. Differential manometer P. 2.18.

$$h_A + 3 \times S_1 + (2) \times S_1 + h \times S_2 - (2+h) \times S_3 = h_B$$

$$h_A = 1 \times \frac{100}{9.81} = 10.19 \quad \text{and} \quad h_B = 1.8 \times \frac{100}{9.81} = 18.348$$

$$\therefore\ 10.19 + 3 \times 1.5 + 2 \times 1.5 + h \times 13.6 - (2 + h) \times 0.9 = 18.348$$

$$10.19 + 4.5 + 3 + 13.6\,h - 1.8 - h \times 0.9 = 18.348$$

$$(13.6 - 0.9)\ h = 2.458$$

$$h = \frac{2.458}{12.7} = 0.1935 = 193.5 \text{ mm}$$ Ans.

2.19 Find the difference in pressure between tanks A and B in Fig 2.45 if h_1 = 300 mm, h_2 = 500 mm and h_3 = 200 mm.

Solution

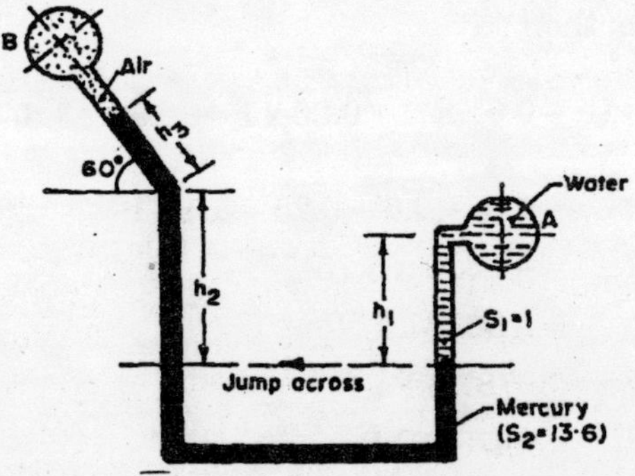

Fig. 2.34. U tube manometer P. 2.19

Using the general procedure starting from A, neglecting air.

$$h_A + h_1 S_1 - h_2 S_2 - h_3 \sin 60°S_2 = h_B$$

$$h_A - h_B = h_2 \; S_2 + h_3 \sin 60°S_2 - h_1 S_1$$

$$= 0.5 \times 13.6 + 0.2 \sin 60° \times 13.6 - 0.3 \times 1$$

$$= 6.8 + 2.355 - 0.3 = 8.855 \text{ m of water}$$

$$p_A - p_B = \rho g \,(h_A - h_B) = 9.81\,(8.855)$$

$$= 86.86 \; kN/m^2$$

2.20 A mercury differential manometer is used to measure the change in pressure caused by flow constriction in a water pipe system, as shown in the Fig. 2.35. Determine the difference in pressure in between the two pipe sections A and B. Which section has the high pressure?

Solution:

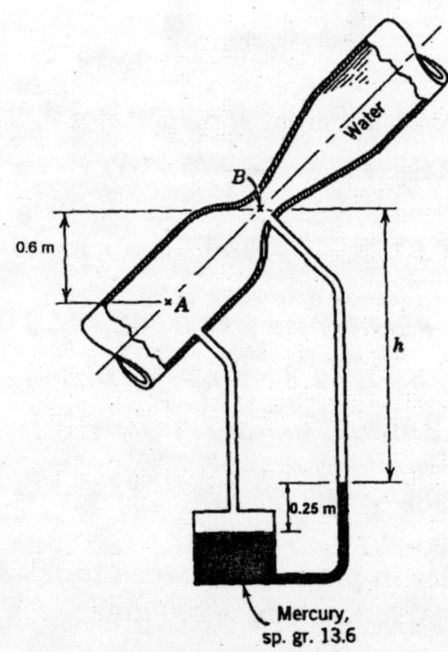

Fig. 2.35. Mercury differential manometer P. 2.20

Let h be the distance from B to the surface of separation between mercury and water in the tube; then

$$\frac{p_A}{\rho g} + (h - 0.6) \times 1 + 0.25 \times 1 - 0.25 \times 13.6 - h \times 1 = \frac{p_B}{\rho g}$$

$$p_B - p_A = \rho g \,(h - 0.6 + 0.25 - 3.4 - h)$$

$$= -9810 \times 3.75$$

$$= -36787.5 \text{ N/m}^2$$

$$p_A - p_B = 36787.5 \text{ Pa}$$

$$= 36.7875 \text{ kPa}$$

Obviously, p_A is greater than p_B.

2.21 Show that the ratio of the pressures p_2/p_1 and densities ρ_2/ρ_1 for altitudes Z_2 and Z_1 in the isothermal atmosphere are given by

$$\frac{p_2}{p_2} = \frac{\rho_2}{\rho_1} = e^{-g(Z_2 - Z_1)/RT} = e^{-g\frac{\Delta Z}{RT}}$$

What increase in altitude is necessary in the stratosphere to halve the pressure. Assume a constant temperature of – 56.5°C and gas constant R = 287 J/kgK.

Solution

$$\frac{p_2}{p_1} = e^{-\frac{g\,\Delta Z}{RT}}$$

$$0.5 = \frac{1}{e^{g\,\Delta Z / RT}}$$

$$e^{g\,\Delta Z/RT} = 2$$

$$\frac{g\,\Delta z}{RT} = \ln 2 = 0.6931$$

$$\Delta z = \frac{0.6931 \text{ x } 287 \text{ x } T}{9.81} = 20.27\ T$$

$$= 20.27\ (273 - 56.5) = 20.27 \text{ x } 216.5 = 4390.32 \text{ m}$$ Ans

2.22 From observation it is found that at a certain altitude in the atmosphere the temperature is –25°C and the pressure is 45.5 kN/m² while at sea level the corresp-onding values are 15°C and 101.5 kN/m². Assuming that the temperature decreases uniformly with increasing altitude, estimate the temperature lapse rate and the presssure and density of the air at an altitude of 300 m.

Solution

We know that

$$\ln\frac{p_2}{p_1} = \frac{g}{\lambda R}\ln\frac{T_2}{T_1}$$

$$\therefore \lambda = \frac{g}{R}\frac{\ln T_2/T_1}{\ln P_2/P_1} = \frac{9.81}{287} \text{ x } \frac{\ln\frac{284}{288}}{\ln\frac{45.5}{101.5}}$$

$$= \frac{9.81}{287} \text{ x } 0.186 = 6.37^\circ \text{ C/1000 m}$$

$$T = T_o - \lambda z = 288 - 0.00637 \times 3000 = 268.89$$

$$\frac{p}{p_o} = \left(1 - \frac{\lambda z}{T_o}\right)^{\frac{g}{\lambda R}} = \left(1 - \frac{0.00637 \times 3000}{288}\right)^{9.81/0.00637x287} = (1 - 0.06635)^{5.365}$$

$$\frac{p}{p_o} = (0.93365)^{5.365} = 0.69189$$

$$\therefore\ p = 0.7010 \text{ bar}$$

$$p = \rho RT$$

$$0.7010 \times 10^5 = \rho \times 287 \times 268.89$$

$$\rho = 0.9084 \text{ kg/m}^3$$

Alternatively Using the relation

$$\frac{p_1}{p_2} = \left(\frac{T_1}{T_2}\right)^{n/n-1}$$

$$\frac{101.5}{45.5} = \left(\frac{288}{248}\right)^{n/n-1}$$

$$2.23 = (1.16)^{n/n-1} \qquad \therefore\ n = 1.23$$

Temperature gradient $\frac{\delta T}{\delta Z} = -\left(\frac{n-1}{n}\right)\frac{g}{R}$

$$\lambda = \frac{0.23}{1.23} \times \frac{9.81}{287} = 0.00639 = 6.39°\text{C/km}$$

$$p = p_1\left(1 - \frac{\lambda}{T_1}\Delta Z\right)^{g/\lambda R} = 101.5\left[1 - \frac{0.00639}{288} \times 3000\right]^{\frac{9.81}{0.00639x287}}$$

$$= 101.5\,[0.9337]^{5.35} = 70.22 \text{ kN/m}^2$$

$$T = T_1\left(\frac{p}{p_1}\right)^{\frac{n-1}{n}} = 288\left(\frac{70.22}{101.5}\right)^{0.1867} = 268.825$$

$$\rho = \frac{P}{RT} = \frac{70.22 \times 1000}{287 \times 268.825} = 0.910 \text{ kg/m}^3$$ Ans

2.23 At the top of the mountain the temperature is –5°C and barometer reads 600 mm of mercury whereas the reading at the foot of the mountain is 1 bar. Calculate the height of the mountain.

Solution

We know the relation

$$\frac{T}{T_o} = \left[1 - \frac{n-1}{n}\frac{g\Delta z}{RT_o}\right]$$

$$\frac{268}{T_o} = \left[1 - \frac{0.4}{1.4} x \frac{9.81}{287} \frac{\Delta Z}{T_o}\right]$$

or, $$T_o = 268 + 9.766 \times 10^{-3} \Delta z \qquad (1)$$

also $$\frac{T}{T_o} = \left(\frac{P}{P_o}\right)^{\frac{n-1}{n}}$$ [1 bar = 750 mm of mercury]

$$\frac{T}{T_o} = \left(\frac{600}{750}\right)^{0.4/1.4}$$

$$\frac{268}{268 + 9.766 \times 10^{-3} \Delta z} = (0.8)^{0.2875} = 0.9382$$

$$251.44 + 9.1624 \times 10^{-3} \Delta z = 268$$

$$\Delta z = \frac{268 - 251.44}{9.1624 \times 10^{-3}} = 109.14 \times 17.44 = 1903.418 \text{ m}$$

2.24 In the lower part of the earth's atmosphere the temperature T may be considered to vary with altitude z according to the linear relation $T = T_o - \lambda z$. Prove that the variation of pressure is given by the relation.

$$\frac{p}{p_o} = \left[1 - \frac{\lambda z}{T_o}\right]^{\frac{g}{\lambda R}}$$

Assume that $dp = -\rho g\, dz$; determine the density of air at the bottom of a mine shaft 800 m below sea level. Assume sea level conditions of p_o = 1.01325 bar and To = 288.15 K. Take λ = 6.5°C/km, R = 287/kgk., z = – 800 m

Solution:

$$\frac{p}{p_o} = \left(1 - \frac{\lambda z}{T_o}\right)^{g/\lambda R} = \left[1 + \frac{800 \lambda}{288.15}\right]^{\frac{9.81}{287 x 0.0065}}$$

$$= \left[1 + \frac{800 \times 0.0065}{288.15}\right]^{5.258} = [1 + 0.18]^{5.258} = 1.098$$

$$p = p_o \times 1.098 = 1.11289 \text{ bar}$$

$$T = T_o - \lambda z = T_o - \lambda (-z) = T_o + \lambda Z$$

$$= 288.15 + 0.0065 \times 800 = 288.15 + 5.2 = 293.35$$

$$p = \rho RT$$

$$1.11289 \times 10^5 = \rho \times 287 \times 293.35$$

$$\rho = 1.3218 \text{ kg/m}^3$$ Ans

2.25 A barometer pressure of the atmosphere sea level is equivalent to 750 mm of mercury and its temperature is 288 K. The temperature decreases with increasing altitude at the rate 6.5 K per 1000 m until the stratosphere is reached in which the temperature remains constant at 216.5 K. Calculate the pressure in mm of mercury and density in kg/m³ at an altitude of 14500 m. Assume R = 287 J/kg K.

Solution

$$\frac{p_1}{p_o} = \left[1 - \frac{\lambda \Delta z}{T_o}\right]^{g/\lambda R} = 1\left[1 - \frac{0.0065 \times 11000}{288}\right]^{\frac{9.81}{287 \times 0.0065}}$$

$$= 1\,[1 - 0.24826]^{5.258} = 1\,[0.7517]^{5.258} = 0.225 \text{ bar}$$

$$\frac{P_2}{P_1} = e^{-(z_2 - z_1)g/RT_2}$$

$$= e^{-(14500 - 11000) \times 9.81/287 \times 216.5}$$

$$p_2 = p_1 \; e^{-0.5525}$$

$$p_2 = 0.225 \times 0.575 = 0.1299 \text{ bar}$$

$$p_2 = \rho RT$$

$$0.1299 \times 10^5 = \rho \times 287 \times 216.5$$

$$\rho = 0.209 \text{ kg/m}^3$$

$$p_2 = \rho g h$$

$$0.1299 \times 100 = 13.6 \times 9.81 \times h$$

$$h = 97.36 \text{ mm of mercury}$$ Ans

2.26 In the standard atmosphere the temperature falls linearly with increase in altitude at the rate of 6.6°C/km at sea level the temperature is 15°C and the pressure is 760 mm of mercury. Calculate the pressure and density of air at a height of 6000 m. Derive any formula used. Gas constant R = 0.287 KJ/kgK.

Solution:

$$\frac{p}{p_o} = \left[1 - \frac{\lambda}{T_o}\,\Delta z\right]^{g/\lambda R}$$

$$p = p_o\left[1 - \frac{0.0066 \times 6000}{288}\right]^{\frac{9.81}{287 \times 0.0066}} = 1.01325\,[0.8626]^{5.18}$$

$$= 1.01325 \times 0.465 = 0.4712 \text{ bar}$$ Ans

$$T = T_o - \lambda \Delta z$$

$$= 288 - 0.0066\ (6000) = 248.4\ K$$

$$p = \rho RT$$

$$0.4712 \times 10^5 = \rho \times 287 \times 248.4$$

$$\therefore \rho = 0.66\ kg/m^3$$

Ans

2.27 Prove that in a fluid of density ρ at rest with zero acceleration, the vertical gradient of pressure is given $dp/dz = -\rho g$ where z is measured upwards. The International Civil Aviation Organisation specifies that for certain texts the atmosphere shall be deemed to have properties which for moderate altitudes are found from the following data.

Thermodynamic temperature $T = T_o - z/\beta$

Altitude above sea level = z

A Constant β = 153.85 m/K.

Sea level temperature T_o = 288.15 K

Sea level pressure p_o = 1 atmosphere

Gas constant R = 0.287 kJ/kg K

Gravitational acceleration g = 9.81 m /kJ/kgK.

Determine the three significant figures at an altitude of 10000 m.

Solution

$$T = T_o - \frac{z}{\beta} = 288.15 - \frac{10000}{153.85}$$

$$= 288.15 - 64.998 = 223.15\ K$$

$$\frac{p}{p_o} = \left[\frac{1 - \lambda z}{T_o \beta}\right]^{\frac{g\beta}{R}} = \left[1 - \frac{\lambda z}{T_o}\right]^{\frac{g}{\lambda R}}$$

$$\frac{p}{p_o} = \left[1 - \frac{10000}{153.85 \times 288.15}\right]^{\frac{9.81 \times 153.85}{287}}$$

$$= [1 - 0.22557]^{5.258} = 0.26$$

$$p = p_o \times 0.26 = 1.01325 \times 0.26 = 0.2641\ bar$$

$$p = \rho RT$$

$$0.2641 \times 10^5 = \rho \times 287 \times 223.15$$

$$\rho = 0.4125\ kg/m^3$$

2.28 An atmosphere on a planet has a temperature of 15°C at sea level and drops 1°C per 500 m of elevation. The gas constant R for this atmosphere is 220J/KgK. At what elevation is the pressure 30 percent that of sea level?

Solution

$$\lambda = \frac{1}{500} = 2 \times 10^{-3}\ 0^{o}C/m$$

$$\frac{p}{p_o} = 0.3\left[1 - \frac{\lambda z}{T}\right]^{\frac{g}{\lambda R}}$$

$$0.3 = \left[1 - \frac{0.002\,z}{288}\right]^{\frac{9.81}{0.002x220}} \quad \text{or,} \quad 0.3 = \left[1 - \frac{0.002 \times z}{288}\right]^{22.295}$$

$$(0.3)^{\frac{1}{22.295}} = 1 - \frac{0.002\,z}{288} \quad \text{or,} \quad 0.9474 = 1 - \frac{0.002\,z}{288}$$

$$\frac{0.0002z}{288} = 0.05256 \qquad \therefore z = 7569.88\ m \qquad \text{Ans}$$

EXERCISES

A. THEORY

2.1. Explain the terms, intensity of pressure and pressure head. Why is the pressure expressed in terms of height of liquid column used for all engineering problems?

2.2. Prove that the pressure head is indepedent of the shape and size of container.

2.3 State Pascal's law and give some examples where this principle is used.

2.4 Define the atmospheric pressure. Give its values in different units.

2.5 How does the atmospheric pressure vary with altitude ?

2.6 How is the pressure measured? What type of liquids are used for different manometers measuring law as well as high pressures ?

2.7 State the difference between simple and differential manometers.

2.8 State the advantages of U tube manometers over the piezometer tube.

2.9 How is the negative pressure measured with the help of U tube ?

2.10 State the procedure for formation of manometer equation

2.11 State the advantages of mechanical gauges over the manometers.

2.12 What are sensitive manometers? Name same of them. Explain the modification employed for converting the ordinary manometer into a sensitive one.

2.13 A U tube manometer shows a deffection of h mm of mercury. Prove that the difference of pressure between the two points is given by

$\Delta P = \frac{S\rho gh}{1000}\left(\frac{Sm}{S}-1\right)$ Where Sm = specific gravity of the liquid in the manometer, S = specific gravity of the liquid in the pipe to which the manometer is connected and w = ρg = specific weight of water in N/m³.

2.14 A inverted U tube manometer shows a deflection of h mm. Prove that the difference betweeen the two points is given by $\frac{S\rho gh}{1000}\left(1-\frac{S_o}{S}\right)$ where So = specific gravity of the liquid in the manometer, and S= specific gravity of the liquid in the pipe to which the manometer is connected.

2.15 A U tube manometer shows a deflection of h mm of water. Prove that the difference of pressure between the two points is given by

$$H = \frac{h}{1000}\left(\frac{\rho_{water}}{\rho_{air}}-1\right)$$

2.16 Derive an expression for the measurement of difference of pressure using micromanometer.

2.17 Prove that the static fluid pressure always acts perpendicular to the surface of contact.

2.18 Differentiate clearly betweeen the absolute, gauge on vacuum pressure.

2.19 Explain absolute, gauge and vacuum pressures with the help of a sketch.

2.20 What is meant by Adaibatic, isothermal and International standard Atmosphere?

2.21 Establish the following relation between temperature T and altitude z

$$g(z - z_o) = n\,R/(n-1)\,(T_o - T)$$

where T_o is the temperature at zero altitude n is the polytrotic index and R is the gas constant.

2.22 Show that the ratios of pressure p_2 / p_1 and densities $\frac{\rho_2}{\rho_1}$ for altitudes z_2 and z_1 in an isothermal atmosphere are given by

$$\frac{p_2}{p_1} = \frac{\rho_2}{\rho_1} = e^{\frac{-g(z_2-z_1)}{RT}}$$

2.23 Establish the relation between temperature T and altitude z

$$\frac{T}{T_o} = \left[1 - \frac{g(z-z_o)}{RT_o}\left(\frac{n-1}{n}\right)\right]$$

Where T_o is the temperature at zero altitude h is the polytorpic index and R is the gas constant.

2.24 Show that, for an adiabatic atmosphere $\frac{\rho}{\rho^{\gamma}}$ = constant

$$g(z - z_o) = RT\,\frac{\gamma}{\gamma - 1}\left[1 - \left(\frac{p}{p_o}\right)^{\frac{\gamma - 1}{\gamma}}\right]$$

where subscript o refers to the reference state.

2.25 Prove that the variation of pressure is given by the relation

$$\frac{p_2}{p_1} = \left[1 - \frac{\lambda Z}{T_o}\right]^{\frac{g}{\lambda R}}$$

where λ is the temperature lapse rate.

B. UNSOLVED PROBLEMS

2.1 The system in Fig. 2.46 is at 20°C. If atmospheric pressure is 101.33 kPa and the absolute pressure at the bottom of the tank is 237 kPa, what is the specific gravity of fluid X? Take g = 9.79 m/s² [X = 1.39]

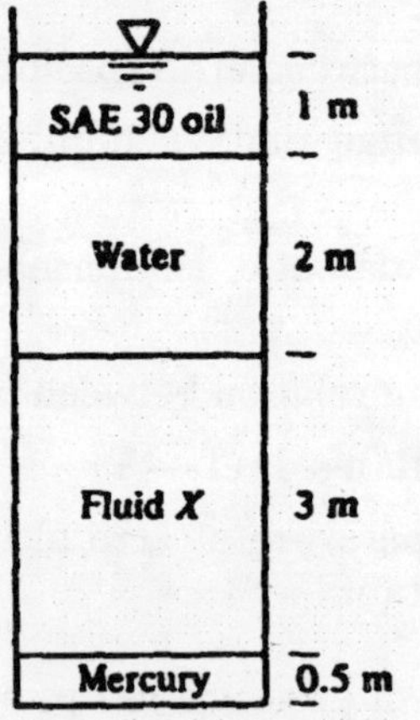

Fig. 2.46. P. 2.1

2.2 Calculate the pressures at A, B, C and D in Fig. 2.47. Take g = 9.79 m/s² [P_A = – 5874 Pa, P_B = 5874 Pa, P_C = 5874 Pa, P_D = 22615 Pa]

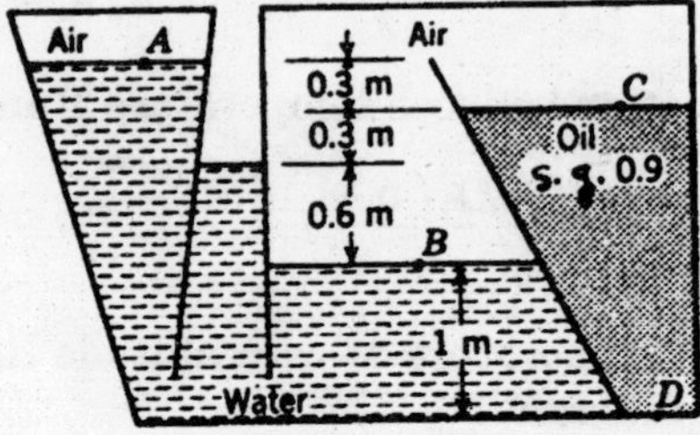

Fig. 2.47. P. 2.2

2.3 The tube shown in Fig. 2.48 is filled with oil. Determine the pressure head at A and at B in meters of water. [h_A = – 2.125 m, h_B = – 0.425 m]

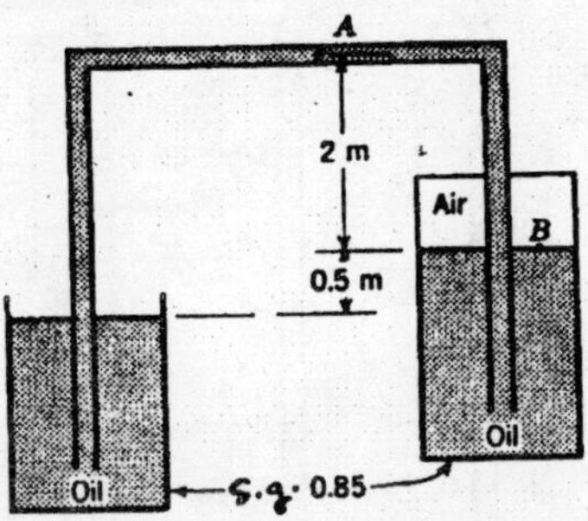

Fig. 2.48. P. 2.3

2.4 A manometer is attached to a tank containing three different fluids, as shown in Fig. 2.49. What will be the difference in elevation of the mercury column in the manometer (i.e. y in Fig. 2.49). Take g = 9.79 m/s^2 [y = 0.627 m]

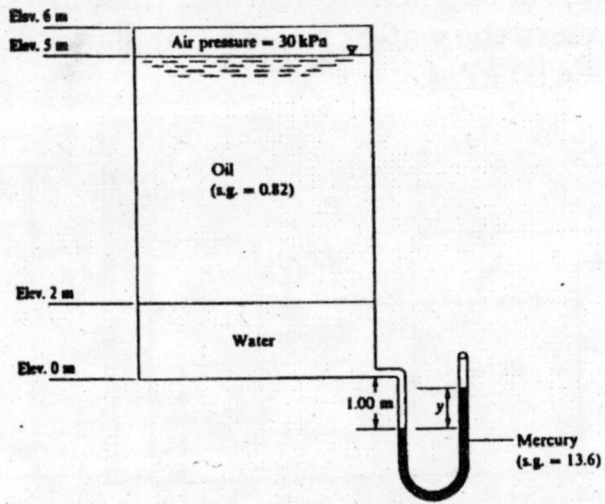

Fig. 2.49. P. 2.4

2.5 In Fig. 2.50, A contains water, and the manometer fluid has a relative density of 2.94. When the left meniscus is at zero on the scale, P_A = 100 mm of water. Find the reading of the right meniscus for P_A = 8 kPa with no adjustment of the U tube or scale. [(a) h = 238.1 mm, (b) h = 385.1 mm]

Fig. 2.50. P. 2.5 Fig. 2.51. P. 2.5

2.6 Determine the gauge pressure at point A in Fig. 2.52 [$P_A = -12.24$ kPa]

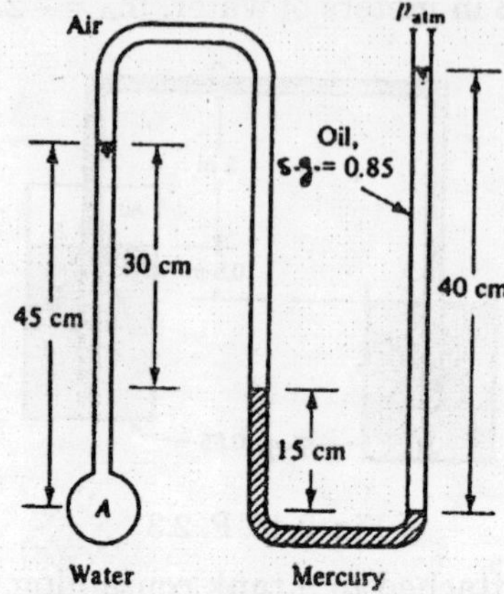

Fig. 2.52. P. 2.6

2.7 An open tube is attached to a tank, as shown in Fig. 2.53. If the water rises to a height of 900 mm in the tube, what are the pressures P_A and P_B of the air above the water? Neglect Capillary effects in the tube [P_A = 2.94 kPa, P_B = 4.90 kPa].

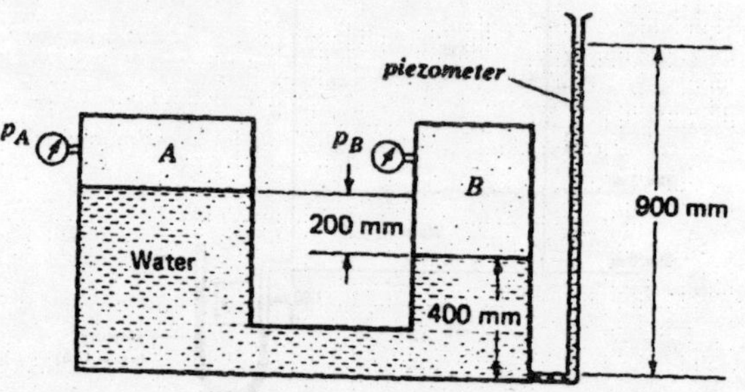

Fig. 2.53. P. 2.7

2.8 A manometer is attached to a water tank, as shown in Fig. 2.54. Find the height of the free water surface above the bottom of the tank. [H = 2.87 m]

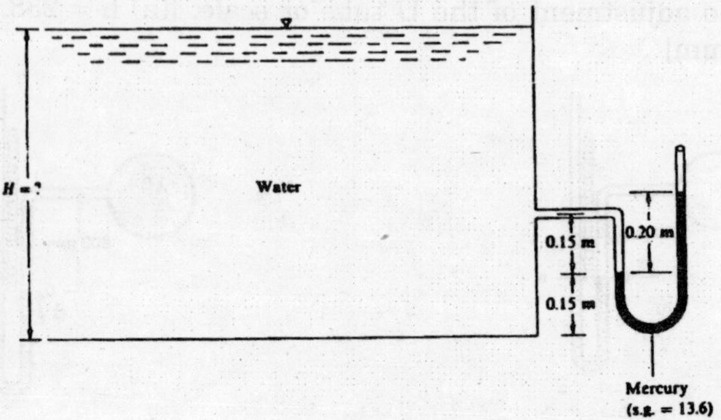

Fig. 2.54. P. 2.8

2.9 In Fig. 2.55, if $P_B - P_A$ = 99 kPa, what must the height H be? [H = 242mm]

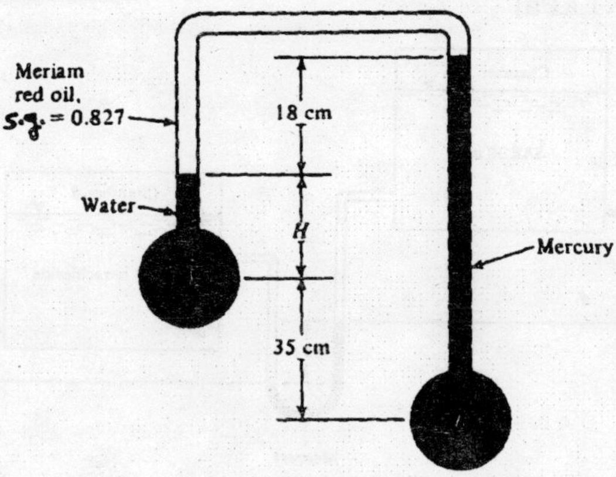

Fig. 2.55. P. 2.9

2.10 Determine the pressure difference between points A and B in Fig. 2.56 [8.95 kPa]

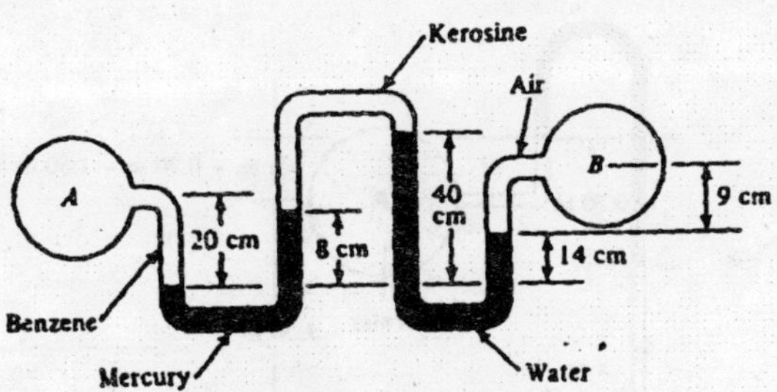

Fig. 2.56. P. 2.10

2.11 A manometer is attached to a pipe containing oil, as shown in Fig. 2.57. Calculate the pressure at point A. [13.02 kPa]

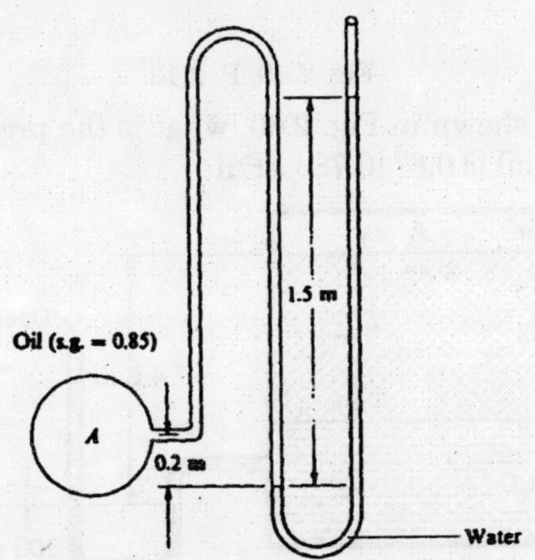

Fig. 2.57. P. 2.11

2.12 A differential manometer is attached to two tanks, as shown in Fig. 2.58. Calculate the pressure difference between chambers A and B. [$P_A - P_B = -37.1$ kPa]

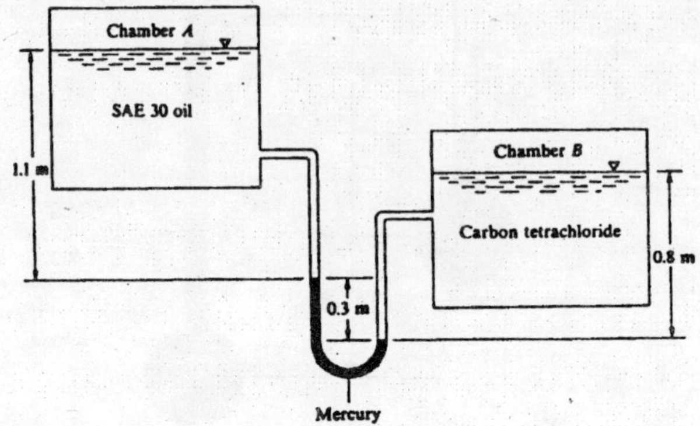

Fig. 2.58. P. 2.12

2.13 Calculate the pressure difference between A and B for the setup shown in Fig. 2.59 [$P_A - P_B = 13.3$ kPa]

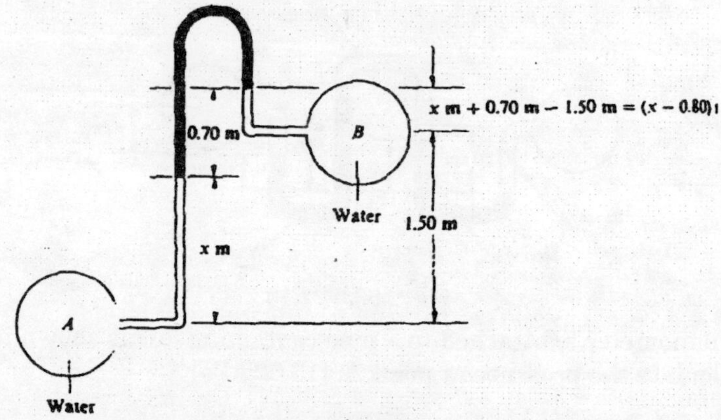

Fig. 2.59. P. 2.13

2.14 For the setup shown in Fig. 2.60, what is the pressure P_A if the specific gravity of the oil is 0.8? [0.783 kPa]

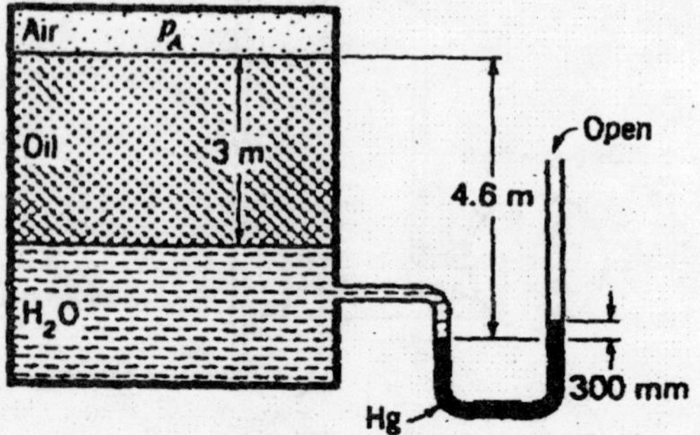

Fig. 2.60. P. 2.14

2.15 For the setup shown in Fig. 2.61, what is the absolute pressure in drum A at position `a'? Assume an atmospheric pressure of 101.3 kPa [86.22 kPa]

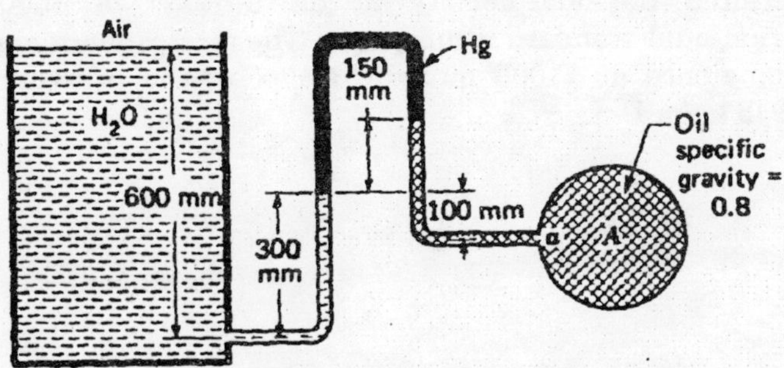

Fig. 2.61. P. 2.15

2.16 At an altitude of 11000 m, the atmospheric temperature is –56.6°C and the pressure is 22.4 kN/m². Assuming that the temperature remains the same at higher altitudes, calculate the density of the air at an altitude of 15000 m. Take R = 287 J/KgK [11.908 kPa, 0.1917 kg/m³]

2.17 Assuming that the temperature of the atmosphere diminishes with increasing altitude at the rate of 6.5°C/km, find the pressure and density at a height of 7000 m if the corresponding values at sea level are 1.01 bar and 1.235 kg/m³ When the temperature is 15°C. Take R = 0.287 kJ/kgK [40.89 kPa, 0.588 kg/m³].

2.18 Calculate the pressure, temperature and density of the atmosphere at an altitude and of 1200 m, if at zero altitude the temperature is 15°C and pressure 1.01 bar. Assume that conditons are adiabatic and R = 0.287 kJ/kgK. [87.33 kPa, 276.28K, 1.1013 kg/m³]

2.19 At sea level air has the following properties, pressure = 101 kN/m², temperature = 288 K and density = 1.22 kg /m³. Make calculations for density and temperature at an altitude of 3500 m. Take n = 1.25. [264.11K, 0.8628 kg/m³]

2.20 Assuming adiabatic conditions, find the temperature drop for an altitude of 3000 m above the sea level. Assume the sea level's temperature as 15°C, the gas constant R for air 0.287 kJ/kg K and n for standard polytropic atmosphere = 1.238. Also find the atmospheric pressure at that altitude if it is given that is 1.01325 bar at sea level. [T = 268.286, Temperature drop $T_o - T$ = 19.714, P = 0.7 bar].

2.21 Find the ratio of the pressures and the densities at 10700 m and at sea level taking the standard atmosphere as having a sea level temperature of 15°C and a lapse rate of 6.5°C/km to a minimum of –56.6°C [0.237, 0.312]

2.22 In the standard atmosphere the temperature at sea level is taken as 15°C and it is assumed that the lapse rate is constant at 6.5°C/km until the stratosphere is reached in which the temperature remains constant at –56.5°C. Calculate the relative pressure and relative density of the atmosphere at a height of 13700 m taking the sea level values as unity and R = 287 J/kgK [0.146, 0.194].

2.23 Determine the air density at an altitude of 15000 m in the International standard atmosphere. The stratosphere may be assumed to commence at 11000 m at which point the pressure is 0.227 bar [0.1943 kg/m^3].

3

Fluid Statics

3.1 INTRODUCTION

In Chapter 2, the concept of pressure was introduced. The equation derived in Chapter 2 can be used to determine the pressure at any point in a fluid. If there is a solid surface in the fluid then the pressure exerted on each element of area on the surface produces a normal pressure force. The total force on the surface, may then be determined from knowledge of statics and mechanics. Only liquids of constant density are dealt with in the following analysis because the changes in pressure in a gas are normally negligible over surfaces of practical interest.

3.2 TOTAL PRESSURE

When a static mass of fluid comes in contact with a surface either plane or curved, a force is exerted by the fluid on the surfaces. This force is known as total pressure or total force. Since for a fluid at rest no tangential force exist, the total pressure or total force acts in the direction normal to the surface.

3.3 CENTRE OF PRESSURE

The point of application of the total pressure on the surface is known as centre of pressure.

3.4 TOTAL PRESSURE AND CENTRE OF PRESSURE ON A VERTICAL PLANE SURFACE

Fig. 3.1 shows a plane surface of arbitrary shape and total area A, wholly submerged in a static mass of liquid of specific weight w. The surface is held in a vertical position, such that the centroid of the surface is at a vertical depth x below the free surface of the liquid.

Let $\overline{x}$ = distance of C.G. of the area from free surface of fluid.

$\overline{h}$ = distance of C.P. from free surface of fluid.

x = distance from the free surface to horizontal strip.

dx = thickness of the horizontal strip

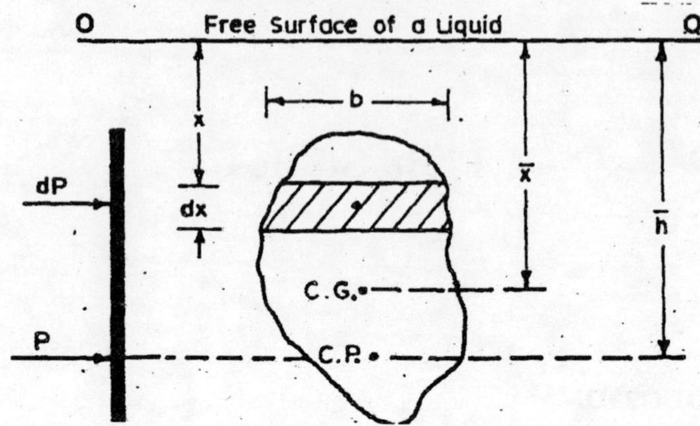

Fig. 3.1. Total pressure and centre of pressure on vertical plane surface.

b = Width of horizontal strip.

A = Area of strip = b dx

p = Pressure on the strip = $\rho g x$

dP = Total pressure on the strip = pA

= $\rho g x$ bdx

P = total pressure on the whole area

= $\int dp = \int \rho g x\,(b\,dx)$;

$P = \rho g \int b\,x\,dx$

We know that

$\int b\,x\,dx$ = First moment of area

$= A\,\bar{x}$

Therefore,

$$p = \rho\, g\, A\,\bar{x} \tag{3.1}$$

The equation (3.1) will hold for all surfaces, whether flat or curved.

The point at which the resultant pressure P acts is called the centre of pressure.

To determine the vertical location of the centre of pressure, consider the moments of the resultant pressure P acting at the centre of pressure, and the summation of the moments of the pressures on the elementary areas, dA, about the free surface, O-O.

Moment of the total pressure (force) about free surface of the fluid

$$= P\,\bar{h} \tag{3.2}$$

Moment of total pressure on a strip about free surface of fluid

$= dP\, x$

$= \rho g x\, (b\, dx)\, x$

$= \rho g x^2\, (b\, dx)$

Sum of moments of all such forces about free surface of fluid

$= \int \rho\, g\, x^2 (b\, dx)$

$= \rho\, g \int b\, x^2\, dx$

We know that

$\int x^2\, b\, dx = I_{OO} =$ Second moment of area

∴ Sum of moments about free surface

$$= \rho\, g\, I_{OO} \qquad (3.3)$$

Equating equations (3.2) and (3.3)

$$P\, \bar{h} = \rho\, g\, I_{OO}$$

We know from the theorem of parallel axis

$$I_{GG} = I_{GG} + A\bar{x}^2$$

where I_{GG} = Moment of Inrtia of area about axis passing through the CG.

$$\therefore p\, \bar{h} = \rho\, g\, \left(I_{GG} + A\bar{x}^2\right)$$

$$= \rho\, g\, I_{GG} + \rho\, g\, A\bar{x}^2$$

$$\bar{h} = \frac{\rho g I_{GG}}{P} + \frac{\rho g\, A\, \bar{x}^2}{P} = \frac{\rho\, g\, I_{GG}}{\rho\, g\, A\bar{x}} + \frac{\rho g A \bar{x}^2}{\rho g A \bar{x}}$$

$$= \frac{I_{GG}}{A\, \bar{x}} + \bar{x} \qquad (3.4)$$

Equation (3.4) gives the position of centre of pressure on a plane surface immersed vertically in a static mass of liquid. Since for any plane surface the factor $\left(\frac{I_{GG}}{A\, \bar{x}}\right)$ is always positive, Eq. (3.4) indicates that $\bar{h} > \bar{x}$, i.e. the centre of pressure is always below the centroid of the area.

Further it is seen that deeper the surface is submerged, i.e. the greater is the value of $\bar{x}$, the factor $\left(\frac{I_{GG}}{A\bar{x}}\right)$ becomes smaller and the centre of pressure comes closer to the centroid of the plane surface. This is so because, as the pressure becomes greater with increasing depth, its variation over a given area becomes smaller in proportion, thereby making the distribution of pressure more uniform. Thus where the variation of pressure is negligible the centre of pressure may be taken as approximately at the centroid. This is justifiable in liquids only if the depth is very large and the area is small and in gases the pressure changes very little with depth.

The lateral location of the centre of pressure can also be readily determined by taking moments about any convenient axis in the same manner.

3.5 TOTAL PRESSURE AND CENTRE OF PRESSURE FOR INCLINED SURFACE

Consider a plane surface of arbitrary shape and total area A wholly submerged in a static mass of liquid of specific weight w. The surface is held inclined such that the plane of the surface makes an angle θ with the horizontal as shown in Fig. 3.2. The intersection of this plane with the free surface of the liquid is represented by axis O-O, which is normal to the plane of the paper.

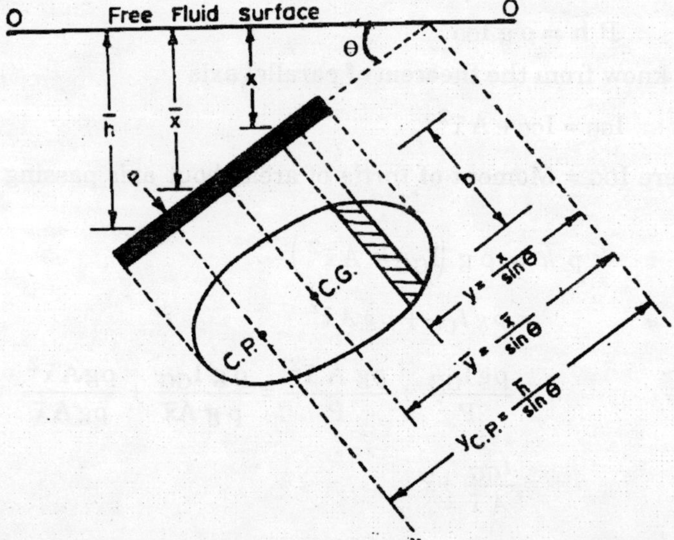

Fig. 3.2. Total pressure and centre of pressure on inclined plane surface.

Let $\bar{x}$ = Vertical depth of C.G. of inclined area from free surface.

$\bar{h}$ = Vertical depth of C.P. of inclined area from free surface.

$\bar{y}$ = Distance of C.G. of the inclined surface from free surface.

y_{cp} = Distance of C.P. of the inclined surface from free surface.

$$dP = \rho\, g \times b\, d\, x$$

From Fig. 3.2

$$\frac{x}{y} = \frac{\bar{x}}{\bar{y}} = \frac{\bar{h}}{y_{cp}} = \sin\theta$$

$$\therefore x = y \sin\theta \text{ and } \bar{x} = \bar{y} \sin\theta$$

Therefore,

$$dP = \rho g\ (y\ \sin\theta)\ bd\ x = \rho g \sin\theta\ y\ dA$$

$$P = \int \rho g \sin\theta\ y\ dA$$

$$= \rho\ g\ \sin\theta \int y\ dA$$

We know that

$$y\ d\ A = A\ \bar{y}$$

$$\therefore P = \rho\ g\ \sin\theta\ A\ \bar{y}$$

$$= \rho g \sin\theta\ A \frac{\bar{x}}{\sin\theta}$$

$$= \rho\ g\ A\ \bar{x} \tag{3.5}$$

The above equation (3.5) is same as equation (3.1), it shows that there is no change in equation for inclined surface.

Centre of pressure can be calculated by taking moments from the free surface at OO.

Moment of total pressure (force) on strip.

$$dP_y = \rho\ g\ (y \sin\theta)\ (bdx)\ y$$

$$dP_y = \rho\ g\ y^2 \sin\theta\ dA$$

$$\int dP_y = \rho\ g \sin\theta \int y^2\ dA$$

Moment of the total pressure (force) about 0-0 is also given by

$$= P\ y_{cp}$$

$$\therefore P\ y_{cp} = \rho\ g\ \sin\theta\ I_{oo}$$

$$P\ y_{cp} = \rho\ g\ \sin\theta\ (I_{GG} + A\ \bar{y}^2)$$

$$P \frac{\bar{h}}{\sin\theta} = \rho\ g\ \sin\theta\ (I_{GG} + A\ \bar{y}^2)$$

$$\bar{h} = I_{GG} \frac{\rho\ g\ \sin^2\theta}{\rho\ g\ A\ \bar{x}} + \frac{\rho\ g\ \sin^2\theta\ A \bar{x}^2}{\rho\ g\ A\ \bar{x}\ \sin^2\theta}$$

$$= \frac{I_{GG}}{A\ \bar{x}} \sin^2\theta + \bar{x}$$

$$\bar{h} = \frac{I_{GG}\ \sin^2\theta}{A\ \bar{x}} + \bar{x} \tag{3.6}$$

If $\theta = 90^\circ$, equation (3.6) become same as equation (3.4) which is applicable for vertical immersed plane surfaces.

Table 3.1 gives the moments of inertia and other geometric properties of different plane surface which are commonly met in actual practice.

TABLE 3.1 Moment of inertia and other geometric properties of plane surfaces

Surface		I_{CG}
Rectangle or square	(diagram: A, h, $\frac{h}{2}$)	$\frac{1}{12}Ah^2$
Triangle	(diagram: A, h, $\frac{h}{3}$)	$\frac{1}{18}Ah^2$
Quadrant of circle (or semicircle)	(diagram: A, r, $\frac{4r}{3\pi}$)	$\left(\frac{1}{4}-\frac{16}{9\pi^2}\right)Ar^2 = 0.0699\ Ar^2$
Quadrant of ellipse (or semiellipse)	(diagram: A, a, $\frac{4a}{3\pi}$)	$\left(\frac{1}{4}-\frac{16}{9\pi^2}\right)Aa^2 = 0.0699\ Aa^2$
Parabola	(diagram: A, h, b, $\frac{3}{8}b$, $\frac{3}{5}h$)	$\left(\frac{3}{7}-\frac{9}{25}\right)Ah^2 = 0.0686\ Ah^2$
Circle	(diagram: A, d, $\frac{d}{2}$)	$\frac{1}{16}Ad^2$
Ellipse	(diagram: A, h, $\frac{h}{2}$)	$\frac{1}{16}Ah^2$

3.6 PRESSURE DIAGRAMS

Total pressure (force) as well as centre of pressure for a plane surface wholly sub-merged in a static mass of liquid, either vertically or inclined, may also be determined by drawing a pressure diagram. A pressure diagram is a graphical representation of the variation in the intensity of pressure over a surface. Such diagrams are shown in Fig 3.3 for horizontal, vertical and inclined surfaces.

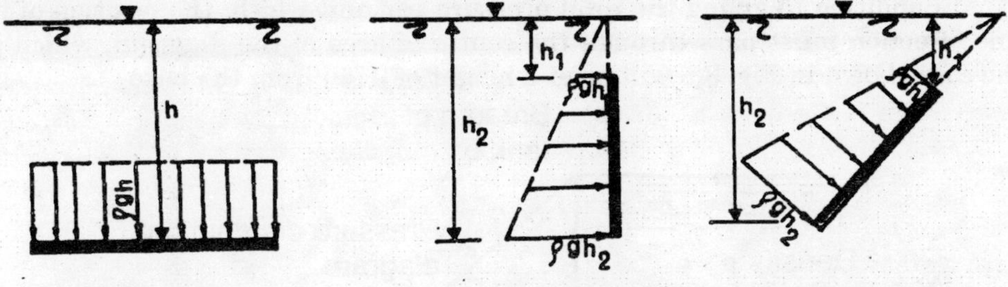

Fig. 3.3. Pressure diagrams

Total thrust on a rectangular vertical surface to water pressure on one side (Fig. 3.4) by pressure diagram.

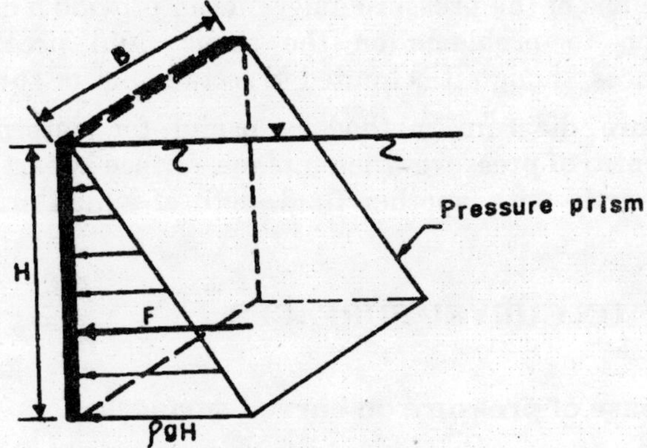

Fig. 3.4. Pressure prism

Average pressure on the surface

$$= \frac{\rho g H}{2}$$

∴ Total force, P = Average Pressure × Area of surface

$$= \left(\frac{\rho g H}{2}\right) H \times B$$

$$= \frac{\rho g H^2}{2} \times B$$

= Volume of the pressure prism. (3.7)

Total force per unit width $= \dfrac{\rho g H^2}{2}$

= Area of the pressure diagram. (3.8)

and the centre of pressure is the centroid of the pressure diagram.

In addition to giving the total pressure per unit width, the position of its line of action must pass through the centre of area of the diagram, which in the case shown in Fig. 3.5 will be at a height of [H/3] from the base.

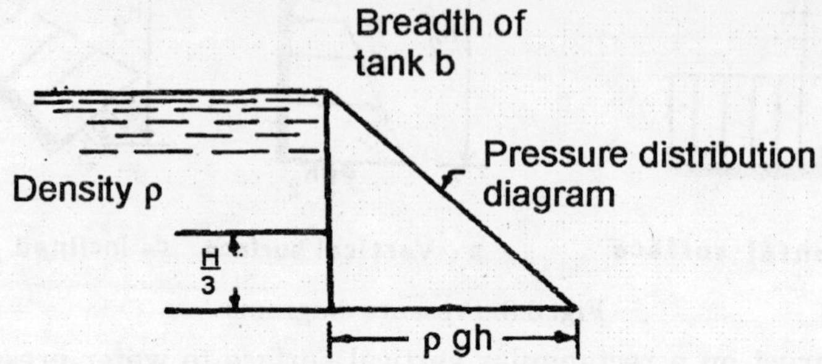

Fig. 3.5. Pressure distribution diagram

Clearly the use of the pressure diagram can provide a quick and straight forward solution to problems on the static fluid pressure exerted on containing surfaces, though it is limited to rectangular or square surfaces.

The pressure diagram method is useful for determining the total pressure and centre of pressure when a plane surface is partly in contact with one fluid and partly with another fluid, both of which are immiscible and static.

3.7 PRESSURE ON CURVED SURFACES

3.7.1 General case of pressure on curved surfaces

Resultant forces on submerged curved surfaces cannot be calculated by the foregoing methods but may be readily determined through computation of their horizontal and vertical components.

Hydrostatic pressure on curved surfaces occur in civil engineering in such applications as gates on dams or sides of culverts, among others. In most instances these surfaces will be of a regular configuration, such as circular, parabolic, elliptical and hyperbolic sections.

For the sake of generality, however, the discussion of pressures on curved surfaces here in is based on a surface with any curvature. For example in Fig. 3.6 consider the prism of fluid ABC A' B' C', having a width equal to one unit and bounded by the curved surface AC, the water surface AB, and the imaginary plane BC.

One force, acting on the surface AC is the pressure exerted by the fluid. This pressure is zero at the fluid surface and equal to ρgh at the point C. This total pressure will act horizontally and can be represented by the pressure diagram in Fig. 3.6. The resultant horizontal force P_H due to hydrostatic pressure acts at a point $\left[\frac{2h}{3}\right]$ below the fluid surface since the projection of the surface is rectangular.

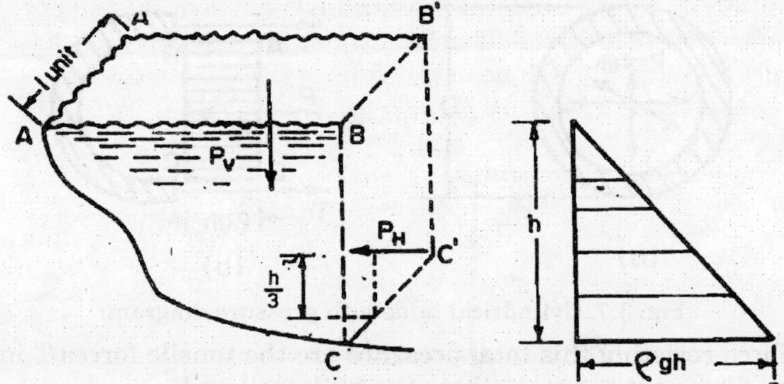

Fig. 3.6. Pressure on curved surface.

The only other force acting on the surface AC, is the force of gravity on the prism of fluid. This force is equal to the volume of liquid in the prism, multiplied by the specific weight of the fluid and will act vertically downwards. This force will be referred to as P_V.

The resultant of pressure P_H and P_V is the total pressure (force) P exerted by the fluid on the surface AC and

$$P = \sqrt{P_H^2 + P_V^2} \tag{3.9}$$

The direction of the resultant force P is given by

$$\tan\theta = \left(\frac{P_V}{P_H}\right)$$

$$\theta = \tan^{-1}\left(\frac{P_V}{P_H}\right) \tag{3.10}$$

3.7.2 Pressure on cylindrical surfaces

Cylindrical surfaces are a common occurrence in civil engineering and other applications. Storage tanks and pipes are almost always circular and are in everyday use.

Fig 3.7(a) represents a cylindrical tank with a vertical longitudinal axis. At any point along this axis the unit pressure on the tank wall will be P = ρgh.

This pressure will be the same at any point in the horizontal plane at the depth h. Consequently, this pressure will also be exerted on the tank walls.

For these reason the total horizontal pressure per unit length on one-half of the tank wall can be represented by the pressure diagram ABCD, as shown in Fig 3.7(b). This total pressure is

$$P = \rho gh \times D \times 1$$

where, D is the diameter of the tank.

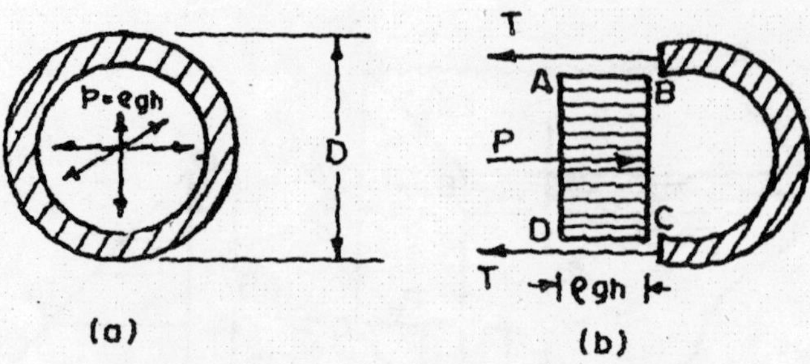

Fig. 3.7. Cylindrical tank with pressure diagram.

The force resisting this total pressure are the tensile forces T in the tank wall. These forces act tangentially to the tank wall and

$$T = \frac{\rho gh \times D}{2}$$

If the cylinder or pipe has a horizontal longitudinal centerline, the above reasoning regarding wall pressures and tensile forces does not apply exactly. It is obvious that in this instance the pressure at the bottom of the tank or pipe would be larger than the pressure at the top by $\Delta P = \rho gh$.

Practical applications of this situation arise in the case of pipes carrying fluids under pressure. In almost all these types of installations, however, the pipe diameter D is only a small fraction of the total head h under which the pipe will operate. Therefore, unless exceptional circumstances prevail, it is generally assumed that the pressure in the horizontal pipe is uniform throughout its cross section and equal to that produced by the head on the pipe's centerline.

ILLUSTRATIVE EXAMPLES

3.1. A rectangular plane area, immersed in water, is 1.5 m by 1.8 m with the 1.5 m side horizontal and the 1.8 m side vertical. Determine the magnitude of the force on one side and the depth of its centre of pressure if the top edge is (a) in the water surface, (b) 1 m below the water surface.

Solution

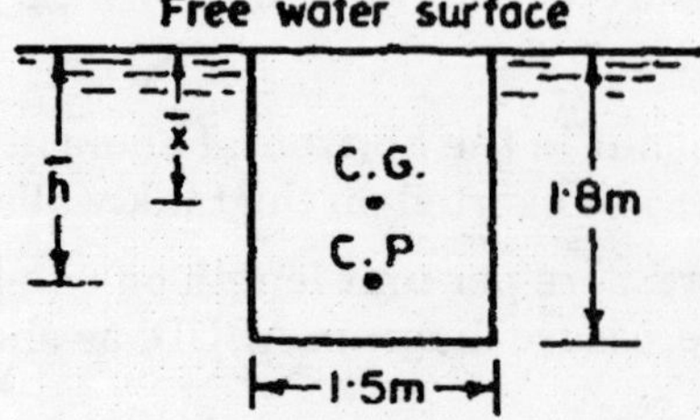

Fig. 3.11. Rectangular plane area P. 3.1(a)

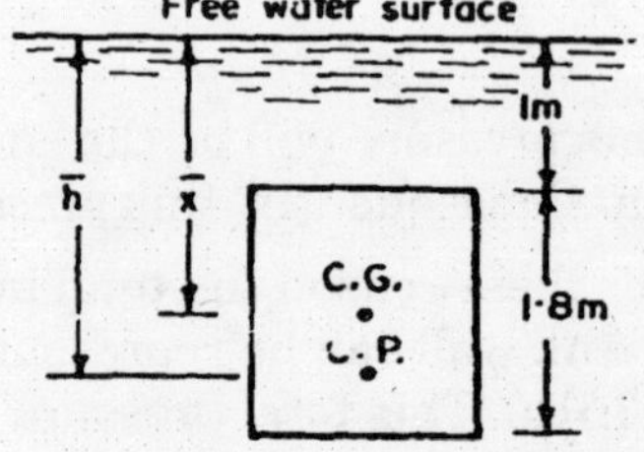

Fig. 3.12. Rectangular plane area P. 3.1 (b)

(a) Top edge is in the water surface.

$$A = 1.8 \times 1.5 = 2.7 m^2$$

$$\bar{x} = \frac{d}{2} = \frac{1.8}{2} = 0.9 m$$

$$\therefore \quad P = wA\bar{x} = 9.810 \times 2.7 \times 0.9 = 23.838\ kN$$ Ans

$$\bar{h} = \bar{x} + \frac{I_{GG}}{A\bar{x}}$$

$$\text{where } I_{GG} = \frac{1}{12} bd^3 = \frac{1}{12} \times 1.5 \times 1.8^3 = 0.729\ m^4$$

$$\therefore \bar{h} = 0.9 + \frac{0.729}{2.7 \times 0.9} = 0.9 + 0.3 = 1.2\ m$$ Ans

(b) Top edge 1 m below the water surface.

$$A = 1.8 \times 1.5 = 2.7\ m^2$$

$\bar{x}$ = Distance from the free surface of water

$$= 1 + \frac{d}{2} = 1 + \frac{1.8}{2} = 1 + 0.9 = 1.9\ m$$

$$P = wA\bar{x} = 9.81 \times 2.7 \times 1.9 = 50.325\ kN$$ Ans

$$\bar{h} = \bar{x} + \frac{I_{GG}}{Ax} = 1.9 + \frac{0.729}{2.7 \times 1.9} = 1.9 + 0.142 = 2.042 m$$ Ans

3.2 **Find the total pressure and the position of the centre of pressure on one face of a 1 m diameter plate which is immersed in such a way that its least and greatest depths of immersion below the free surface of water are 1.5 m and 2.5 m respectively.**

Solution

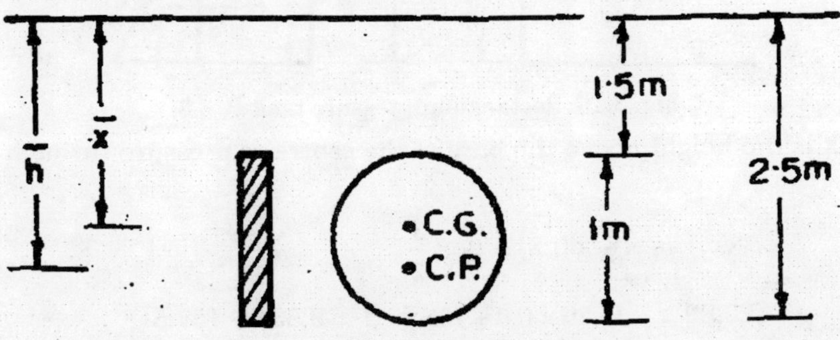

Fig. 3.13. Circular plane P. 3.2.

$$A = \frac{\pi}{4}(D)^2 = \frac{\pi}{4}(1)^2 = 0.785\ m^2$$

$$I_{GG} = \frac{\pi}{64} D^4 = \frac{\pi}{64}(1)^4 = 0.0490 m^4$$

$$\bar{x} = 1.5 + \frac{D}{2} = 1.5 + \frac{1}{2} = 2\ \text{m}$$

$$\omega = 9.81\ \text{kN/m}^3$$

$$\therefore\ P = w A \bar{x} = 9.81 \times 0.785 \times 2 = 15.4\ \text{kN}$$

$$\bar{h} = \bar{x} + \frac{I_{GG}}{A\bar{x}} = 2 + \frac{0.0490}{0.785 \times 2} = 2 + 0.0312$$

= 2.0312 m below the free surface of water **Ans**

3.3. **A rectangular pressure tank of height 3 m contains water to a depth of 1.8 m above which there is air under a pressure of 35 kN/m². Determine the total pressure per meter width of the side of the tank and also find the depth of the centre of pressure.**

Solution

Total pressure due to water

$$P_1 = w A_1 \bar{x}_1 = 9.81 \times 1.8 \times 1 \times 0.9 = 15.89\ \text{kN}$$

Total pressure due to air

$$P_2 = Pressure \times (H \times 1) = PH = 35 \times 3 = 105\ kN$$

Total pressure = $P_1 + P_2$ = 15.89 + 105 = 120.89 kN

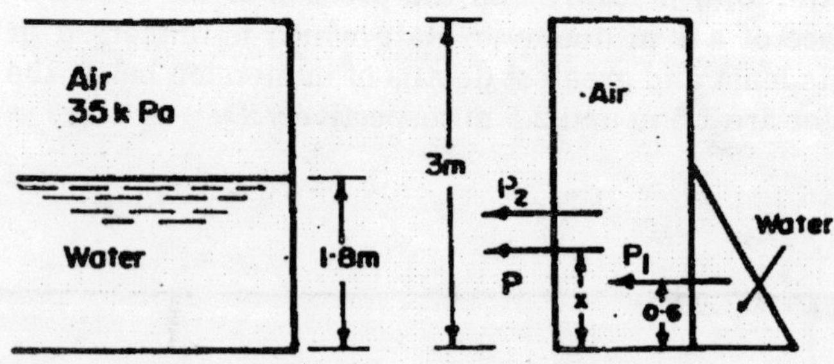

Fig. 3.15. Rectangular pressure tank P. 3.3.

If x is the height above the base of the centre of pressure through which P acts.

$$P \times x = p_1 \times \frac{1}{3} \times (1.8) + P_2 \times \frac{3}{2}$$

$$120.89\ x = 15.89 \times 0.6 + 105 \times 1.5 = 9.534 + 157.5$$

$$120.89\ \text{x} = 167.034$$

$\therefore$ x = 1.3817 m above the base of tank.

or, From water surface

$$= 1.8 - x = 1.8 - 1.3817 = 0.4183 \text{ m}$$

3.4. A circular drum, 1.8 m diameter and 1.2 m long, is submerged with its axis vertical and its upper end at a depth of 1.8 m below water level. Determine (a) the total pressure on the surface of the drum (b) the resultant pressure on the whole surface (c) the depth of centre of pressure of the curved surface.

Solution

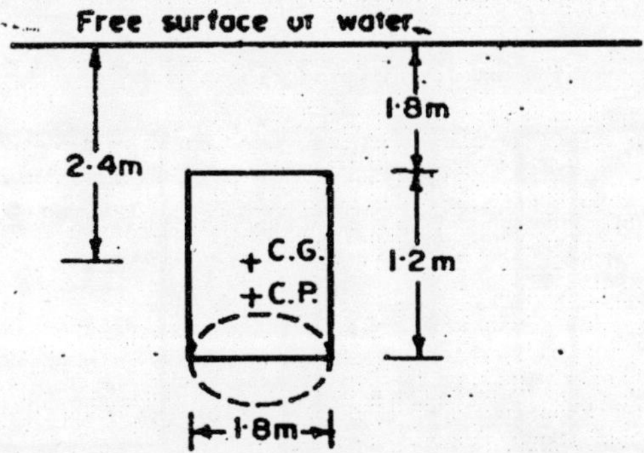

Fig. 3.16. Circular drum P. 3.4.

Total pressure P_1 on top circular area

$$= w\, A_1\, \bar{x}_1 = 9.81 \times \frac{\pi}{4}(1.8)^2 \times 1.8 = 44.93 \text{ kN}$$

Total pressure P_2 on bottom circular area

$$= w\, A_2\, \bar{x}_2 = 9.81 \times \frac{\pi}{4}(1.8) \times 3 = 74.89 \text{ kN}$$

Total pressure P_3 on vertical curved surface

$$= w\, A_3\, \bar{x}_3 = 9.81 \times (\pi \times 1.8 \times 1.2) \times 2.4 = 159.76 \text{ kN}$$

Total pressure P on all surface

$$P_1 + P_2 + P_3 = 44.93 + 74.89 + 159.76 = 279.\ 58 \text{ kN}$$ Ans

(b) Resultant pressure = weight of liquid displaced

$$= 9.81 \times \frac{\pi}{4}(1.8)^2 \times 1.2 = 29.\ 95 \text{ kN}$$

(c) $I_{GG} = \frac{1}{12} bd^3 = \frac{1}{12} \times (1.8) \times (1.2)^3 = 0.2592 \text{ m}^4$

$A = \pi \times 1.8 \times 1.2 = 6.785 \text{ m}^2$

$$\bar{x} = 1.8 + \frac{1.2}{2} = 1.8 + 0.6 = 2.4 \text{ m}$$ Ans

$$\therefore\ \bar{h} = 2.4 + \frac{0.2592}{6.785 \times 2.4} = 2.4 + 0.0159$$

$$= 2.415 \text{ m below free surface of water}$$ Ans

3.5. A vertical rectangular gate 2 m wide and 2.5 m high is subjected to water pressure on one side, the water surface being at the top of the gate. The gate is hinged at the bottom and is held by a horizontal chain at the top. Calculate the tension in the chain.

Solution

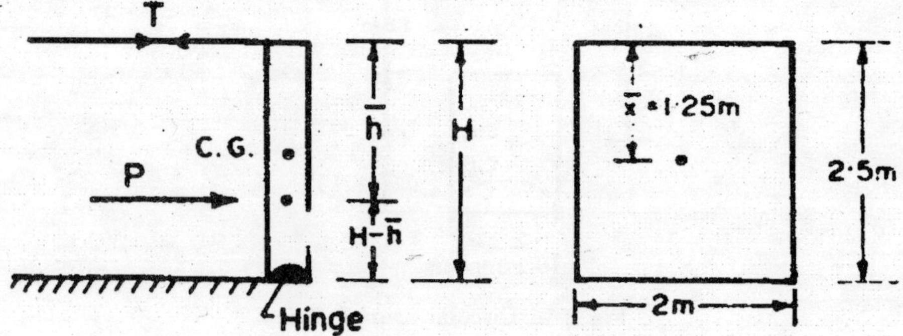

Fig. 3.18. Rectangular gate P. 3.5.

$$A = 2 \times 2.5 = 5 \text{ m}^2$$

$$\bar{x} = \frac{2.5}{2} = 1.25 \text{ m}$$

$$I_{GG} = \frac{1}{12} bd^3 = \frac{1}{12} \times 2 \times (2.5)^3 = 2.604$$

$$P = wA\bar{x} = 9.81 \times 5 \times 1.25 = 61.31 \text{ kN}$$

$$\bar{h} = \bar{x} + \frac{I_{GG}}{A\bar{x}}$$

$$= 1.25 + \frac{2.604}{5 \times 1.25} = 1.6667 \text{ m}$$

Taking moments about the hinge

$$T \times H = P\,(H - \bar{h})$$

$$T = \frac{61.31 \times (2.5 - 1.6667)}{2.5} = 20.43 \text{ kN}$$ Ans

3.6. Calculate the forces acting at the two hinges and at the clamp of the vertical rectangular door shown in Fig. 3.19 P. 3.10 if it is loaded on one face by sea water of density 1030 kg/m³. Density of fresh water is 1000 kg/m³.

Solution

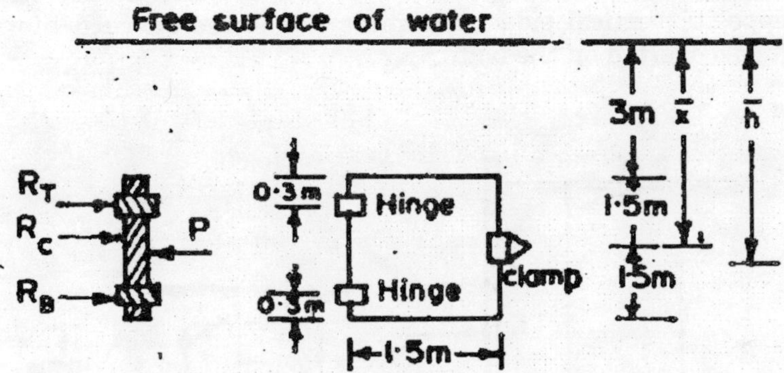

Fig. 3.19. Rectangular door P. 3.6.

$$A = 3\ 1.5 = 4.5\ m^2$$

$$\bar{x} = 3 + 1.5 = 4.5\ m$$

$$P = wA\bar{x} = 9.81 \times 1030 \times 4.5 \times 4.5 = 204.612\ kN$$

By symmetry half of the total pressure will be resisted by the clamp, and other half is resisted by the two hinges.

$$\therefore R_C = \frac{P}{2} = \frac{204.612}{2} = 102.3\,kN$$

$$\bar{h} = \bar{x} + \frac{I_{GG}}{A\bar{x}}$$

$$I_{GG} = \frac{1}{12} \times 1.5 \times 3^3 = 3.375\ m^4$$

$$\bar{h} - \bar{x} = \frac{I_{GG}}{A\bar{x}} = \frac{3.375}{4.5 \times 4.5} = 0.167\ m$$

To find R_T, take moments about R_B

$$R_T \times 2.4 = P \times (1.2 - 0.167) - R_C\,(1.2)$$

$$= 204.612\ (1.033) - 102.3\ (1.2)$$

$$= 211.364 - 122.76 = 88.604$$

$$R_T = \frac{88.604}{2.4} = 36.918\,kN$$ Ans

$$R_C = R_B + R_T$$

$$102.3 = R_B + 36.918$$

$$\therefore R_B = 102.3 - 36.918 = 65.382\ kN$$ Ans

3.7. A rectangular door 1.8 m high and 0.9 m wide closes an opening in the vertical side of a bulkhead which retains water on one side of it to a depth of 1.8 m above the top of the door. The door is supported by two hinges placed 10 cm, from the top and bottom of one of the vertical sides, and is fastened by a bolt fixed at the centre of the opposite vertical side. Determine the forces on each hinge and the forces exerted on the bolt.

Solution

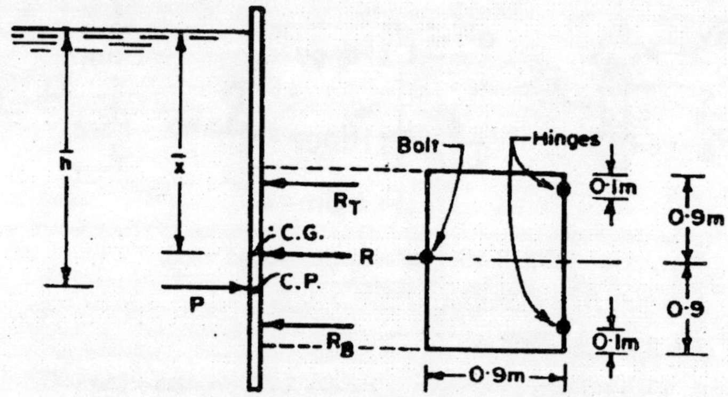

Fig. 3.20. Rectangular door P. 3.7.

$$P = w\,A\,\bar{x} = 9.81 \times (1.8 \times 0.9) \times 2.7 \quad = 42.90 \text{ kN}$$

$$I_{GG} = \frac{1}{12} \times 0.9 \times 1.8^3 \;= 0.4374 \text{ m}^4$$

$$\bar{x} = 1.8 + 0.9 = 2.7 \text{ m}$$

$$A = 1.62 \text{ m}^2$$

$$\bar{h} = \bar{x} + \frac{I_{GG}}{A\bar{x}}$$

$$= 2.7 + \frac{0.4374}{1.62 \, x \, 2.7} = 2.7 + 0.1 = 2.8 \text{ m}$$

and $\bar{h} - \bar{x} = 0.1 \text{ m}$

By symmetry one half of the total pressure P is taken by the hinges and other half by the bolt.

$$\therefore \text{ Force on the bolt R} = \frac{P}{2} = \frac{42.90}{2} = 21.45 \text{ kN}$$

Taking moments about the bottom hinge

$$R_T \times 1.6 = P \times 0.7 - R \, x \, 0.8$$

$$R_T = \frac{42.9 \times 0.7 - 21.45 \times 0.8}{1.6}$$

$$= \frac{30.03 - 17.16}{1.6} = \frac{12.87}{1.6} = 8.043 \text{ kN}$$

$$\therefore \; R_B = R - R_T = 21.45 - 8.043 = 13.407 \text{ kN}$$

3.8. A rectangular tank is filled with kerosene of relative density 0.8 to a depth of 750 mm and carbon disulphide of relative density 1.3 to a depth of 375 mm. Determine the total pressure per metre length of the side wall of the tank and depth of centre of pressure. Assume the two liquids are immiscible.

Solution :

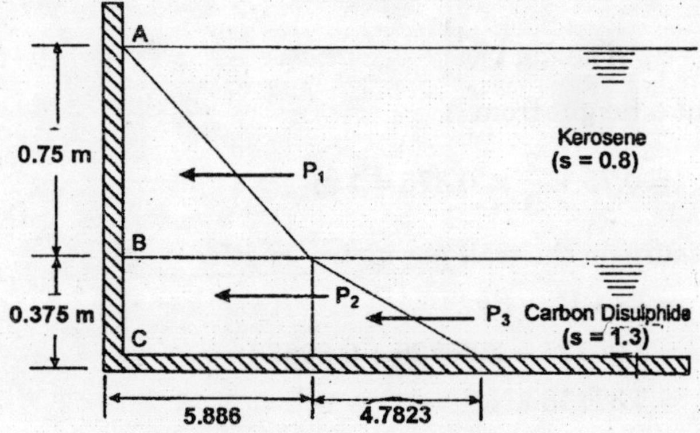

Fig. 3.21. Rectangular tank P. 3.8.

Method I

$$P_B = S \; \rho \, gh = 0.8 \times 9.81 \times 0.75$$

$$= 5.886 \text{ kPa}$$

$$P_C = P_B + S\rho gh$$

$$= 5.886 + 1.3 \times 9.81 \times 0.375$$

$$= 5.886 + 4.7823$$

$$= 10.668 \text{ kPa}$$

Force on AB

$$P_1 = \text{Average pressure} \times \text{Area}$$

$$= \frac{0 + 5.886}{2} \times 0.75 \times 1 = 2.20725 \text{ kN}$$

and acts at a height from A

$$= \frac{2}{3} \times 0.75 = 0.5 \, m$$

Force on BC

P_2 = Pressure × Area

= 5.886 × 0.375 × 1

= 2.20725 kN

and acts at a height from A

$= 0.75 + \frac{0.375}{2}$

= 0.9375 m

P_3 = Average pressure × Area

$= \frac{0 + 4.7823}{2} x\, 0.375 = 1\, m$

= 0.89668 kN

and acts at a height from A

$= 0.75 + \frac{2}{3} \times 0.375 = 1$ m

Total pressure on the wall per metre length

$P = P_1 + P_2 + P_3$

= 2.20725 + 2.20725 + 0.89668

= 5.31118 kN

The centre of pressure is calculated by taking moments about A.

$$P\bar{h} = P_1\bar{h}_1 + P_2\bar{h}_2 + P_3\bar{h}_3$$

= 2.20725 x 0.5 + 2.20725 × 0.9375 + 0.89668 × 1

= 1.103625 + 2.069296 + 0.89668

$5.3118\, \bar{h} = 4.0696$

$$\therefore \bar{h} = \frac{4.0696}{5.3118} = 0.766\, m$$

Therefore, the centre of pressure is 766 mm below the free surface of kerosene i.e. from A. Ans.

Method II.

$$A_1 = \frac{1}{2} \times 5.886 \times 0.75 = 2.20725$$

Pressure force $P_1 = A_1 \times$ width of tank

= 3.20725 × 1 = 2.20725 kN

and this force acts at $\frac{2}{3} \times 0.75 = 0.5\, m$ below the free surface.

A_2 = 5.886 × 0.375

$= 2.20725$

Pressure force $\quad P_2 = A_2 \times$ width of tank

$2.20725 \times 1 = 2.20725$ kN

and this force acts at $0.75 + \frac{0.375}{2} = 0.9375$ m below the free surface.

$$A_3 = \frac{1}{2} \times 4.7823 \times 0.375 = 0.89668$$

Pressure force $P_3 = A_3$ x width of tank

$= 0.89668 \times 1 = 0.89668$ kN

and this force acts at $0.75 + \frac{2}{3} \times 0.375 = 1$ m below the free surface.

Total pressure force on one vertical face of the tank.

$$P_1 = P_1 + P_2 + P_3$$

$$= 2.20725 + 2.20725 + 0.89668 = 5.3118 \text{ kN}$$

The centre of pressure is calculated by taking moments bout free surface i.e. from A.

$$\bar{h} = \frac{2.20725 \times 0.5 + 2.20725 \times 0.9375 + 0.89668 \times 1}{5.3118}$$

$$= \frac{4.0696}{5.3118} = 0.766 \text{ m}$$

$= 766$ mm from A **Ans.**

3.9. A 2m x 2m tank with vertical sides contains oil of density 900 kg/m^3 to a depth of 0.8 m floating on 1.2 m depth of water. Calculate the total thrust and its location on one side of the tank (Fig 3.22)

Solution – Method I

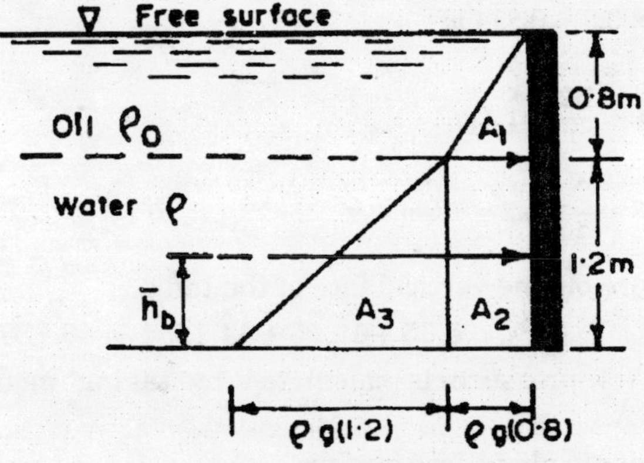

Fig. 3.22. Oil tank P. 3.9.

$$A_1 = \frac{1}{2} \times 0.8 \times 0.8 = 0.32\ m^2$$

$$A_2 = 0.8 \times 1.2 = 0.96\ m^2$$

$$A_3 = \frac{1}{2} \times 1.2 \times 1.2 = 0.72\ m^2$$

Total thrust on one vertical side,

P = volume of the pressure prism

$$= [\rho_o\, g\, A_1 + \rho_o\, g\, A_2 + \rho g\, A_3] \times 2$$

$$= [900 \times 9.81 \times 0.32 \times 900 \times 9.81 \times 0.96 + 1000 \times 9.81 \times 0.72] \times 2$$

$$= 5.65 + 16.95 + 14.126 = 36.7264 \text{ kN}$$ **Ans**

Taking moments about the bottom of the tank

$$36.7264 \times \overline{h_b} = 5.65\left[1.2 + \frac{0.8}{3}\right] + 16.95 \times \frac{1.2}{2} + 14.126 \times \frac{1.2}{3}$$

$$= 8.29 + 10.17 + 5.65 = 24.11$$

$$\overline{h_b} = \frac{24.11}{36.7264} = 0.65647 \text{ m above the base}$$

Method II

Pressure force P_1 = Area of the triangle x width of tank

$$\therefore P_1 = \frac{1}{2}(0.9 \times 9.81 \times 0.8) \times 0.8 \times 2 = 5.65 \text{ kN}$$

$$P_2 = (0.9 \times 9.81 \times 0.8) \times 1.2 \times 2 - 16.95 \text{ kN}$$

$$P_3 = \frac{1}{2}(9.81 \times 1.2) \times 1.2 \times 2 = 14.1264 \text{ kN}$$

From free surface

$$\overline{h_1} = \frac{2}{3} \times 0.8 = 0.534\ m$$

$$\overline{h_2} = 0.8 + \frac{1.2}{2} = 1.4\,m$$

$$\overline{h_3} = 0.8 + \frac{2}{3} \times 1.2 = 1.6\,m$$

Total pressure on one vertical face of the tank.

$$P = P_1 + P_2 + P_3 = 5.65 + 16.95 + 14.1264 = 36.7264 \text{ kN}$$

The centre of pressure is calculated by taking moments about free surface i.e. from A.

Taking moments about free surface.

$$P\bar{h} = P_1\overline{h_1} + P_2\overline{h_2} + P_3\overline{h_3}$$

$$= 5.65 \times 0.534 + 16.95 \times 1.4 + 14.1264 \times 1.6$$

$$= 3.0171 + 23.73 + 22.6 = 49.349$$

$$\therefore \bar{h} = \frac{49.349}{36.7264} = 1.343m \qquad \text{Ans.}$$

Therefore, from the base = 2 – 1.343 = 0.657 m **Ans.**

3.10. **A bulkhead 3 m wide divides a storage tank. On the one side petrol of specific gravity 0.78 is stored to a depth of 1.8 m, while on the other side is oil of specific gravity 0.88 to a depth of 0.9 m. Determine the resultant pressure on the bulkhead and the position at which it acts.**

Solution

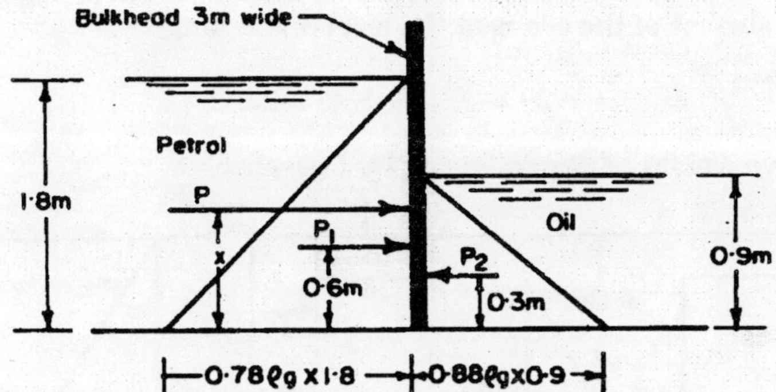

Fig. 3.24. Storage tank P. 3.10.

Total pressure exerted by petrol = P_1

P_1 = area of pressure distribution for petrol × breadth

$$= \frac{0.78\,\rho g \times 1.8}{2} \times 1.8 \times 3$$

$$= 0.78 \times 9.81 \times (1.8)^2 \times \frac{3}{2} = 37.18 \text{ kN}$$

Total pressure exerted by oil

= area of pressure distribution diagram for oil x breadth

$$= \frac{0.88 \times \rho g \times 0.9}{2} \times 0.9 \times 3$$

$$= 0.88 \times 9.81 \times 1.215 = 10.48 \text{ kN}$$

Resultant pressure = $P_1 - P_2$

$$= 37.18 - 10.48 = 26.7 \text{ kN}$$

To find the position at which it acts take moments about the bottom of the bulkhead.

$$P \times x = P_1 \times 0.6 - P_2 \times 0.3$$

$$26.7\,x = 37.18 \times 0.6 - 10.48 \times 0.3$$

$$= 22.308 - 3.144 = 19.164$$

$$x = \frac{19.164}{26.7} = 0.717 \text{ m} = 71.7 \text{ mm} \qquad \text{Ans}$$

3.11. **The fully submerged sea out fall of a 0.6 m square sewer is provided with a vertical top-hung cast steel flap valve, 0.75 m square, with the hinged edge 80 mm above the top of the sewer.**

The vertical line through the centre of gravity of the flap is 15 mm out wards from the centre line of the hinge. The weight of the flap in air is 2.5 kN; the relative density of the cast steel is 7.75 and the density of sewage may be considered as equal to that of sea water i.e. 1028 kg/m³.

Calculate the differential head required just to open the flap when the surface of the sea is at the level of the hinge.

Solution

Effective weight of flap (allowing for buoyance) is

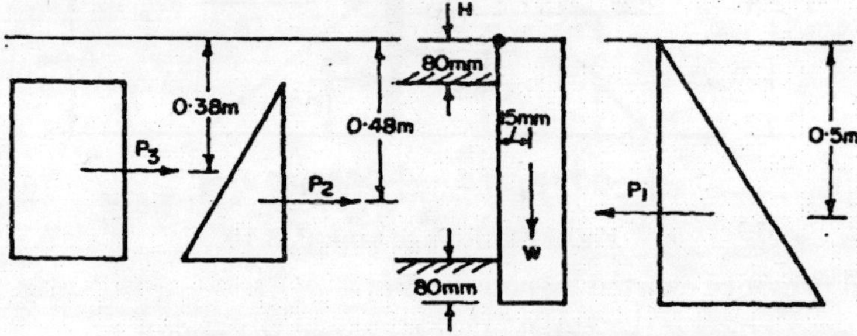

Fig. 3.25. Flap valve P. 3.11.

$$W = 2.5\left[1 - \frac{1.028}{7.75 \times 1}\right] = 2.17 \text{ kN} = 2170 \text{ N}$$

Let x m = head of sewage above top of sewer from Fig. 3.25

$$P_1 = \frac{1}{2} \times (0.75) \times 1028 \times 9.81 \times (0.75)^2 = 2130 \text{ N}$$

$$P_2 = \frac{1}{2} \times (0.6) \times 1028 \times 9.81\ (0.6)^2 = 1090 \text{ N}$$

$$P_3 = 1028 \times 9.81 \times (0.6)^2 \times x = 3630\, x\text{N}$$

Since P_1, P_2 and P_3 act at the centroids of their respective diagrams.

$$\text{Depth of } P_1 \text{ below the hinge} = \frac{1}{3} \times (0.75) = 0.5\ m$$

$$\text{Depth of } P_2 \text{ below the hinge} = \frac{2}{3} \times (0.6) + 0.08 = 0.48m$$

and $$\text{Depth of } P_3 \text{ below the hinge} = \frac{1}{2} \times 0.6 + 0.08 = 0.38m$$

Taking moments about H

$$P_1 \times 0.5 + W \times (0.015) = P_2 \times 0.48 + P_3 \times 0.38$$

$$2130 \times 0.5 + 2170 \times 0.15 = 1090 \times 0.48 + 3630 \text{ x} \times 0.38$$

$$1065 + 32.55 = 523.2 + 1379.4 \text{ x}$$

$$1379.4 \text{ x} = 574.35$$

$$\text{x} = 0.4163 \text{ m}$$

Differential head $= 0.4163 - 0.08 = 0.336$ m $=$ 336 mm Ans

3.12. (a) A circular butterfly gate pivoted about a horizontal axis passing through its centroid is subjected to hydrostatic thrust on one side and counter balanced by a force, F, applied at the bottom as shown in Fig. 3.26. If the diameter of the gate is 4 m and the water depth is 1 m above the gate determine the force F, required to keep the gate in position.

(b) If the gate is to retain water to its top level on the other side also, determine the net hydrostatic thrust on the gate and suggest the new conditions for the gate to be in equilibrium (Fig 3.26b)

Solution

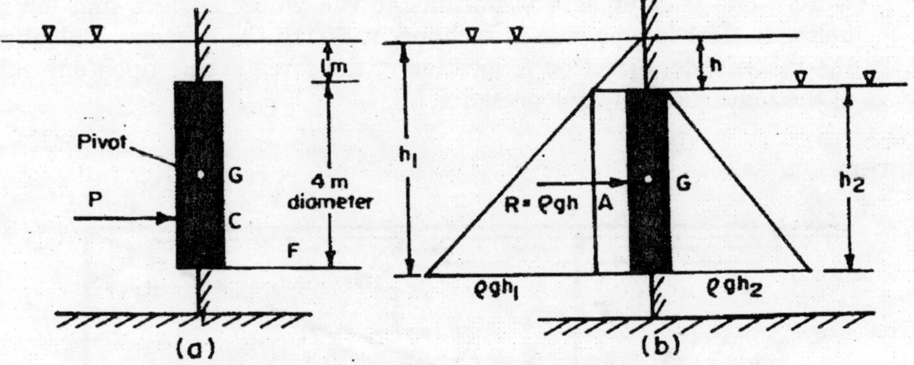

Fig. 3.26. Circular butterfly gate P. 3.12.

Water on one side only

Hydrostatic thrust on the gate

$$P = \rho g \bar{x} A = 9.81 \times 3 \times \frac{\pi}{4} (4)^2 = 369.83 \text{ kN}$$

and the distance, $CG = \frac{I_{GG}}{A\bar{x}}$

$$\bar{h} - \bar{x} = \frac{\pi/64\,(4)^4}{\frac{\pi}{4}(4)^2 \times 3}$$

$$\bar{h} - \bar{x} = 0.333 \text{ m}$$

Taking moment about G,

$$P \times 0.333 = F \times 2$$

$$\frac{369.83 \times 0.333}{2} = F$$

$$\therefore F = 61.64 \text{ kN.}$$

(b) If the gate is retaining water on the other side also, the net hydrostatic thrust is due to the resultant pressure diagram with an uniform pressure distribution of intensity equal to ρgh.

Net hydrostatic thrust R

$$= \rho g h A = 9.81 \times 1 \times \frac{\pi}{4}(4)^2 = 123.28 \; kN$$

This acts through the centroid of the gate, G, and since its moment about G is zero, F = O for the gate to be in equilibrium, for any depth h, of water above the gate on the other side.

3.13. A flat rectangular plate in 1 m wide by 2 m long is immersed in water. The shorter side is parallel to the water surface and 0.5 m below it. The longer side is inclined at 60° to the vertical. Calculate the magnitude, direction & location of the force acting upon one side of the plate due to water pressure.

Solution

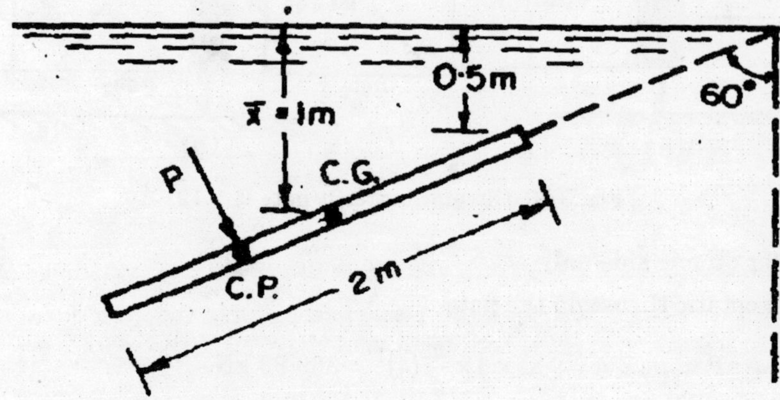

Fig. 3.27. Rectangular tank P. 3.13.

$$A = 2 \times 1 = 2 \text{ m}^2$$

$$\bar{x} = 0.5 + 1 \cos 60^\circ = 0.5 + 0.5 = 1 \text{ m}$$

$$I_{GG} = \frac{1}{12} bd^3 = \frac{1}{12} \times 1 \times 2^3 = 0.67\ m^4$$

Total pressure

$$P = wA\bar{x} = 9.81 \times 2 \times 1 \quad = 19.62 \text{ kN}$$ Ans

Centre of pressure $\bar{h} = \bar{x} + \frac{I_{GG}}{A\bar{x}} \sin^2 \theta = 1 + \frac{0.67}{2 \times 1} \sin^2 30^\circ$

$$= 1 + 0.333 \times 0.25 = 1 + 0.08325 = 1.08325 \text{ m}$$ Ans

3.14. The top edge of a rectangular plane 5 m long by 10 m wide is submerged at an angle of 60°, 5 m below the surface. Compute the head at the centre of gravity, total pressure on the surface of the plane, and the centre of pressure on the plane surface. Finally, how far is the centre of pressure below the surface?

Solution

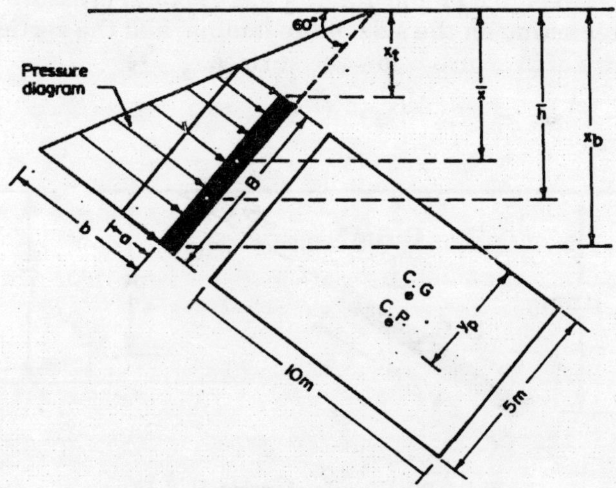

Fig. 3.28. Rectangular plate P. 3.14.

$x_t = 5$ m, $\theta = 60^\circ$

$$\therefore \bar{x} = x + \frac{x}{2} \sin \theta$$

$$= 5 + 2.5 \cos 60^\circ = 5 + 2.5\ (0.866) = 5 + 2.165 = 7.165 \text{ m}$$

The total pressure (force) on the plane is computed from the head at the centroid, the specific weight of the liquid, and the area of the plate.

$$P = wA\bar{x}$$

$$= 9.81 \times (5 \times 10) \times 7.165 = 3514.43 \text{ kN}$$ Ans

The centre of pressure of the plane is computed from the formula of trapezoid. Form the pressure diagram. Since by Pascals law of the pressure heads acting about the top and bottom edges of the inclined plane are the same in all directions.

$$a = x_t = 5 \text{ m}$$

$$b = x_b = 5 + 5 \sin 60° = 5 + 4.33 = 9.33 \text{ m}$$

By pressure diagram.

$$y_P = B\left[1 - \frac{1}{3}\left(\frac{2a+b}{a+b}\right)\right]$$

$$= 3\left[1 - \frac{1}{3}\left(\frac{2 \times 5 + 9.33}{5 + 9.33}\right)\right] = 2.75 \text{ m}$$

Finally, computing the vertical distance to the centre of pressure

$$\bar{h} = 5 + x_p \sin 60° = 5 + 2.75\,(0.866) = 7.38 \text{ m}$$

3.15. A circular lamina 1.25 m in diameter is immersed in water so that the distance of its perimeter measured vertically below the water surface varies between 0.6 m and 1.5 m. Find the total pressure (force) due to the water acting on the side of the lamina, and the vertical distance of the centre of pressure below the surface.

Solution

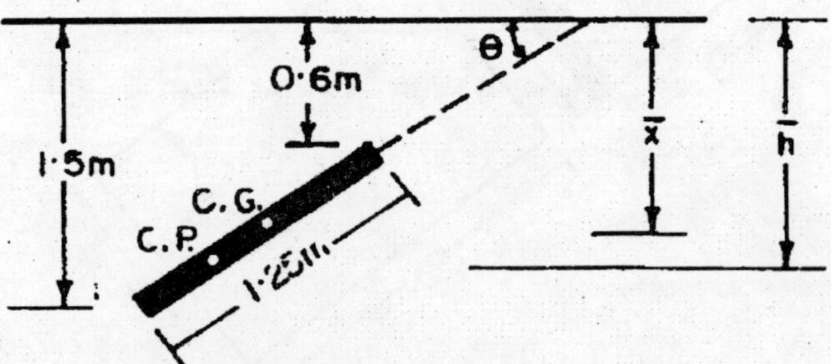

Fig. 3.29. Circular lamina P. 3.15.

$$A = \frac{\pi}{4}(1.25)^2 = 1.228\ m^2$$

$$\bar{x} = \frac{0.6 + 1.5}{2} = 1.05 \text{ m}$$

$$\therefore\ P = wA\bar{x}$$

$$= 9.81 \times 1.228 \times 1.05 = 12.65 \text{ kN}$$

$$I_{GG} = \frac{\pi}{4}(1.25)^4 = 0.1198$$

$$\sin\theta = \frac{1.5-0.6}{1.25} = \frac{0.9}{1.25} = 0.72\ m$$

$$\bar{h} = \bar{x} + \frac{I_{GG}\sin^2\theta}{A\,\bar{x}}$$

$$= 1.05 + \frac{0.1198}{1.228 \times 1.05}(0.72)^2$$

$$= 1.05 + 0.0929 \times 0.5184 = 1.05 + 0.048 = 1.098 \text{ m}$$

Ans

3.16. **An inclined circular gate 1.2 m diameter as shown in Fig 3.30 below is installed to control discharge of water. End A is hinged. Determine the force normal to the gate applied at B to open it.**

Solution

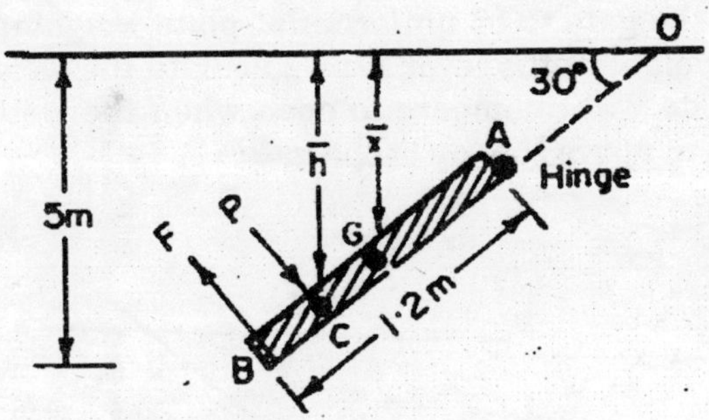

Fig. 3.30. Circular gate P. 3.16.

$$A = \frac{\pi}{4}(1.2)^2 = 1.131\ m^2$$

$$I_{GG} = \frac{\pi}{64}(1.2)^4 = 0.1017 \text{ m}^4$$

$$\bar{x} = 5 - \frac{1.2}{2}\sin 30^\circ = 5 - 0.3 = 4.7 \text{ m}$$

$$P = w\,A\,\bar{x}$$

$$= 9.81 \times 1.131 \times 4.7 = 52.147 \text{ kN}$$

$$\bar{h} = \bar{x} + \frac{I_{GG}}{A\bar{x}}\sin^2\theta$$

$$= 4.7 + \frac{0.1017\sin^2 30^\circ}{1.131 \times 4.7} = 4.7 + 4.78 \times 10^{-3} = 4.704 \text{ m}$$

AB = 1.2 m

AC = AB – BC

= AB – BO + OC

$$= 1.2 - \frac{5}{\sin 30^\circ} + \frac{4.704}{\sin 30^\circ}$$

$$= 1.2 - 10 + 4.704 \times 2 = 0.608$$

Taking moments about hinges

$$F \times AB = P \times AC$$

$$F \times 1.2 = 52.147 \times 0.608$$

$$\therefore F = \frac{31.705}{1.2} = 26.42\,kN$$ Ans

3.17. A rectangular opening in the sloping side of a reservoir containing water is 3 m by 2 m with the 2 m side horizontal. It is closed by means of a gate as shown in Fig. 3.31. The gate is hinged at the top edge and kept closed against the water pressure partly by its own weight and partly by a weight W on a lever arm.

Assuming the gate to be uniform flat plate weighing 5 kN and neglecting the weight of the lever arm, calculate the weight W required so that the gate will commence to open when the water level reaches a height of 1 m above the top of the gate.

Solution

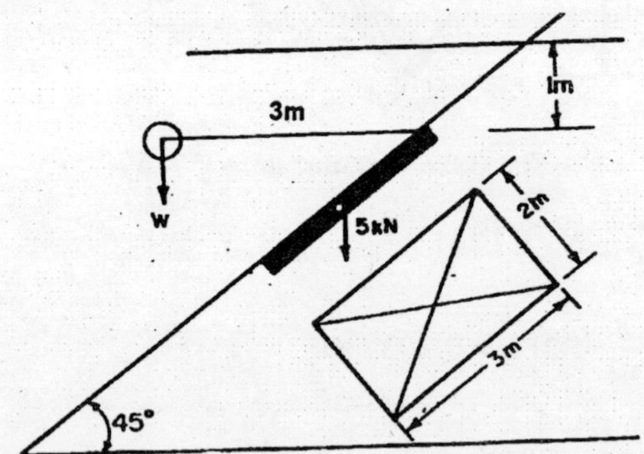

Fig. 3.31. Gate with rectangular opening P. 3.17.

$$\bar{x} = 1 + \frac{3}{2} \sin 45^\circ = 2.06$$

$$P = wA\bar{x}$$

$$9.81 \times (3 \times 2) \times 2.06 = 121.29 \text{ kN}$$

$$I_{GG} = \frac{1}{12} bd^3 = \frac{1}{12} \times 2\,(3)^3 = 4.5 \text{ m}^4$$

$$\therefore \bar{h} = \bar{x} + \frac{I_{GG} \sin^2 45^\circ}{A\bar{x}}$$

$$= 2.06 + \frac{4.5 \times (0.707)^2}{6 \times 2.06} = 2.06 + 0.182 = 2.242 \text{ m}$$

Taking moment about hinge

$$P \ (\bar{h} - 1)/\sin 45° = W \times 3 + 5 \times \frac{3 \sin 45°}{2}$$

$$P \ [2.242 - 1] \times \sqrt{2} = W \times 3 + \frac{15}{2} \times \frac{1}{\sqrt{2}}$$

$$213.04 = W \times 3 + 5.30$$

$$W \times 3 = 207.03$$

$$W = 69.24 \text{ kN}$$ Ans

3.18. An elliptical gate covers the end of a 1.25 m diametere pipe. If the gate is hinged at the top, what normal force F required to open the gate when water is 2.5 m deep above the top of the pipe and the pipe is open to the atmospheric pressure on the other side? Neglect the weight of the gate.

Solution

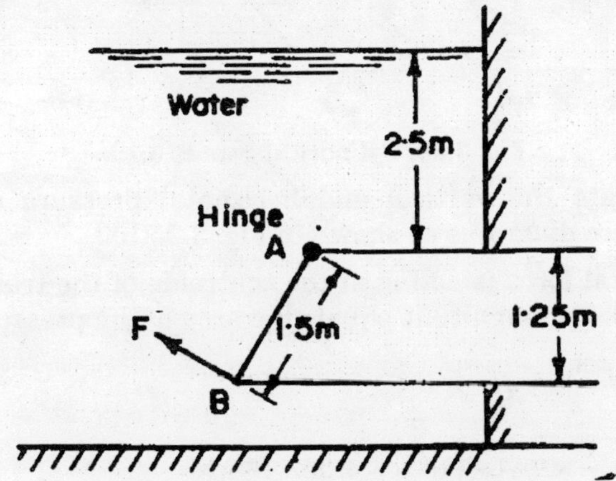

Fig. 3.34. Elliptical gate P. 3.18.

$$A = \frac{\pi}{4} bd = \frac{\pi}{4} \times 1.25 \times 1.5 = 1.472 \text{ m}^2$$

$$\bar{x} = 2.5 + \frac{1.25}{2} = 2.5 + 0.625 = 3.125$$

$$P = wA\bar{x} = 9.81 \times 1.472 \times 3.125 = 45.126 \text{ kN}$$

$$I_{GG} = \frac{\pi}{64} bd^3 = \frac{\pi}{64} (1.25)(1.5)^3 = 0.207$$

$$\sin\theta = \frac{1.25}{1.5} = 0.833$$

$$\bar{h} = \bar{x} + \frac{I_{GG}\sin^2\theta}{A\bar{x}}$$

$$= 3.125 + \frac{0.207 \times 0.694}{1.472 \times 3.125} = 3.125 + 0.03125 = 3.156 \text{ m}$$

Taking moment about hinge A

$$F \times 1.5 \cos\theta = P\left(\bar{h} - 2.5\right)\cot\theta$$

$$F = \frac{P\ (3.156 - 2.5)}{1.5 \times \sin\theta} = \frac{45.126 \times 0.656}{1.25} = 23.68 \text{ kN}$$ Ans

3.19. A cylindrical (drum) gate with a diameter of 3 m is 8 m long. The water is on one side only, reaching the top of the gate. Determine the magnitude and direction of the resultant hydrostatic force acting on the gate.

Solution

Fig. 3.35. Cylindrical gate P. 3.19.

If we separate the vertical and horizontal pressure components, we obtain the pressure diagrams as shown in Fig. 3.35(b).

The horizontal force is acting at the centroide of the triangle, 2 m below the water surface. Its magnitude equals the area of the pressure triangle.

$$P_H = \frac{1}{2}\rho g\ H^2 \times B$$

$$= \frac{1}{2} \times 9.81 \times (3)^2 \times 8 = 353.16 \text{ kN}$$

The vertical force is located a `c' distance away from the vertical centre line of the gate. The value of c is calculated as follows

$$C = \frac{4R}{3\pi}$$

$$= \frac{4 \times 1.5}{3\pi} = 0.636 \text{ m}$$

$$P_v = \frac{\pi}{4} D^2 / 2 \times \rho g \times B$$

$$= \frac{\pi}{8}(3)^2 \times 9.81 \times 8 = 277.37 \text{ KN}$$

The result force R = $\sqrt{P_H^2 + P_V^2} = \sqrt{(353.16)^2 + (277.37)^2} = 449$ kN

The resultant force acts through the intersection of its two components whose co-ordinates are 2 m vertical and 0.636 m horizontal with respect to the top of the water over the gate.

The angle of the resultant force is

$$\tan \infty = \frac{P_V}{P_H} = \frac{277.37}{353.16} = 0.7853$$

$$\therefore \infty = \tan^{-1}(0.7853) = 38°8'\ 45.25''$$ Ans

3.20. A cylindrical drum 2 m diameter and 3 m wide is placed across a stream of the same width. The depths of the water on the upstream and down stream sides of the cylinder are 2 m and 1 m respectively. Determine the magnitude and direction of the resultant water pressure on the cylinder.

Solution

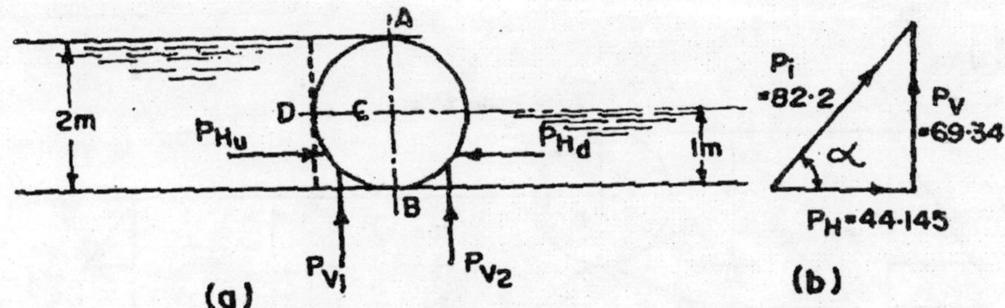

Fig. 3.36. Cylindrical drum P. 3.20.

Let P_{v_1} = Weight of water that would fill the area ACBD

P_{v_2} = Weight of water that would fill the area CBE.

But these areas are, respectively, half and a quarter of the circle of diameter 2 m. The net vertical component $P_v = P_{v_1} + P_{v_2}$

$$P_v = \left[\left[\frac{1}{2}\right]\frac{\pi}{4}D^2 + \left[\frac{1}{4}\right]\frac{\pi}{4}D^2\right] \times 3 \times \rho g$$

$$= \frac{3}{4} \times \frac{\pi}{4} D^2 \times 3 \times \rho\, g$$

$$= \frac{9\pi}{16}(2)^2 \times 9.81 = 69.34 \text{ kN}$$

The horizontal force on the upstream side is

$$P_{H_u} = \rho g \times 2 \times 3 \times 1 = 6\,\rho g$$

$$= 6 \times 9.81 \text{ m} = 58.86 \text{ kN}$$

The horizontal force on the down stream side is

$$P_{H_d} = \rho g \times (1\times3)\times 0.5 = 1.5\ \rho g \quad = 14.715 \text{ kN}$$

Net horizontal component P_H = $P_{H_u} - P_{H_d}$

$$P_H = 58.86 - 14.715 = 44.145 \text{ kN}$$

$$R = \sqrt{P_H^2 + P_V^2} = \sqrt{1948.78 + 4808.03} = 82.20 \text{ kN}$$

$$\tan\alpha = \frac{69.34}{44.145}$$

$$\tan\alpha = 1.57$$

$$\therefore \alpha = 57.5^\circ$$ Ans

The line of action of R must pass through the centre of cylinder.

3.21. A 3 m diameter roller gate retains water on both sides of a spillway crest as shown in Fig 3.37. Determine (a) the magnitude direction and location of the resultant hydrostatic thrust acting on the gate per unit length, and (b) the horizontal water thrust on the spill way per unit length.

Solution

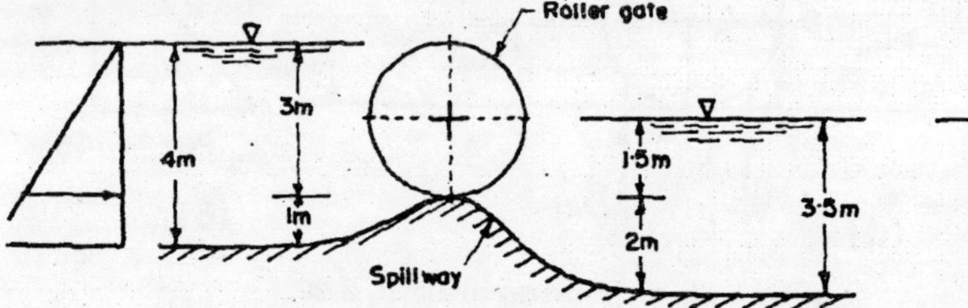

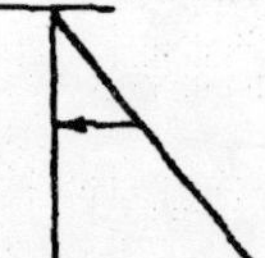

Fig. 3.37. Roller gate on spillway P. 3.21.

Thrust on the gate Horizontal component on left side

$$\therefore P_{H_L} = \frac{1}{2}\rho g 3^2 = \frac{1}{2}\times 9.81\times 9 = 44.14 \text{ kN/m}$$

Horizontal Component on the right side

$$\therefore P_{H_R} = \frac{1}{2}\times 9.81\times(1.5)^2 \quad = 11.03 \text{ kN/m}$$

Net horizontal component on the gate

$$P_H = P_{H_L} - P_{H_R} = 44.14 - 11.03 = 33.11 \text{ kN/m}$$

Vertical component on the left side

$$P_{V_L} = \rho g \times \frac{\pi}{4}(3)^2 \times \frac{1}{2}\times 1 = 34.67 \text{ kN/m}$$

Vertical component on the Right side

$$P_{V_R} = \rho g \times \frac{\pi}{4}(3)^2 \times \frac{1}{4} \times 1 = 17.34 \text{ kN/m}$$

Net vertical Component on the gate

$$\text{Pv} = P_{V_L} + P_{V_R} = 34.67 + 17.34 = 50.01 \text{ kN/m}$$

R = Resultant hydrostatic thrust on the gate

$$= \sqrt{P_H^2 + P_V^2}$$

$$= \sqrt{(33.11)^2 + (50.01)^2} = 60 \text{ kN/m}$$

$$\alpha = \tan^{-}\left(\frac{33.11}{50.01}\right) = \tan^{-1}(0.662) = 33.5^{\circ} \qquad \text{Ans}$$

Resultant hydrostatic thrust acting at an angle of 33.5° to the vertical and passes through the centre of the gate (normal to the surface).

∴ Depth of centre of pressure = r + r cos α

= 1.5 (1 + cos 33.5°)

= 2.75 m below the free surface of left side

Horizontal thrust on the spillway

From pressure diagrams (Fig 3.37),

Thrust from left hand side

$$= \frac{1}{2}(\rho g \times 3 + \rho g \times 4) \times 1 = \frac{1}{2} \times 9.81 \times 7 = 34.335 \text{ kN/m}$$

Thrust from right hand side

$$= \frac{1}{2}(\rho g \times 1.5 + \rho g \times 3.5) \times 2 = \frac{1}{2} \times 9.81 \times 5 = 49.05 \text{ kN/m}$$

Resultant thrust (horizontal) on the spillway

= 49.05 – 34.335 = 14.72 kN/m towards left.

3.22. Determine the horizontal, vertical and total pressures acting on the Tainter gate in Fig 3.38.

Solution

Fig 3.39(b) shows the geometric arrangement and dimensions of this gate. They were calculated as follows.

H = 1.25 sin 45° = 0.884 m

b = 1.25 cos 45° = 0.884 m

a = 1.25 – 0.884 = 0.366 m

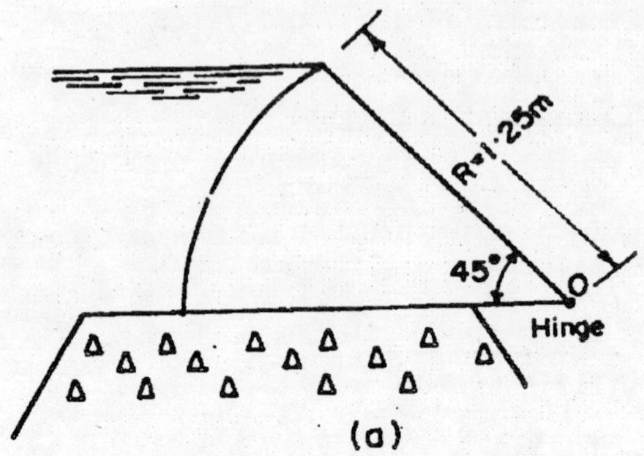

Fig. 3.38a. Tainter gate P. 3.22.

The horizontal pressure P_H on a 1 m section of this gate is shown in the pressure diagram in Fig 3.38 (b) and is calculated as follows

$$P_H = \rho g \frac{H^2}{2} \times 1$$

$$= 9.81 \times \frac{(0.884)^2}{2} = 3.83 \text{ kN}$$

acting at a distance $\bar{x} = \dfrac{0.884}{3} = 0.295$ m from the hinge.

The vertical pressure P_v is due to the weight of the wedge of water ABC in Fig 3.38 (b). The force P_v can be calculated by determining the area of the trapezoid ABCO and subtracting from it the area of the circular sector ACO : that is

$$A_{trap} = \frac{0.366 + 1.25}{2} \times 0.884 = 0.714\ m^2$$

$$A_{sector} = \frac{\pi (2.5)^2}{4} \times \frac{45°}{360°} = 0.614 m^2$$

Fig. 3.38 (b&c). Tainter gate P. 3.22.

Consequently, the area of the water producing P_v can be found as follows.

$$A_{ABC} = A_{trap} - A_{sector}$$

$$= 0.714 - 0.614 = 0.1 \text{ m}^2$$

If the gate is considered to be 1 m in length then

$$P_V = A_{ABC} \times 1 \times \rho g$$

$$= 0.1 \times 1 \times 9.81 = 0.981 \text{ kN}$$

P_H is now completely defined by its value, direction and point of application. P_v is defined only from the point of view of value and direction Its point of application will be through the centre of gravity of the area ABC. This centre of gravity is not readily found.

However, since the resultant force P must act at right angles to the surface of the gate, it must also act through the centre of the circle or point o using this fact leads to the free body diagram of.

From similar triangles of this figure. the moment arm of P_v in relation to o, $\bar{y}$ can be calculated as follows.

$$\frac{\bar{y}}{P_H} = \frac{\bar{x}}{P_V}$$

$$\bar{y} = \frac{\bar{x} P_H}{P_V} = \frac{0.295 \times 3.83}{0.981} = 1.15 m$$

The total force P is from Fig P 3.28(c).

$$P = \sqrt{P_H^2 + P_V^2} = \sqrt{(3.83)^2 + (0.981)^2} = 3.92 \text{ kN}$$

3.23. Fig. 3.39 is a schematic drawing of a tainter gate with the upstream water elevation 0.75 m above the top of the gate and water spilling over as shown. Determine the total force acting on this gate and the horizontal and vertical components of this force. The radius of the gate is 3.5 m.

Solution

From the geometrics of the figure the following dimensions are found

$$b = 3.5 \cos 55° = 2.01 \text{ m}$$

$$a = 3.5 - 2.01 = 1.49 \text{ m}$$

$$h_1 = 3.5 \sin 55° = 2.87 \text{ m}$$

$$h = 2.87 + 0.75 = 3.62 \text{ m}$$

The horizontal pressure diagram will be trapezoidal with the top unit pressure 0.75 ρg and the bottom unit pressure 3.62 ρg. This diagram can be divided into the rectangle and triangle as shown in Fig 3.39(b). If P_1 and P_2

are the resultant horizontal forces for the rectangular and triangular portion respectively, then

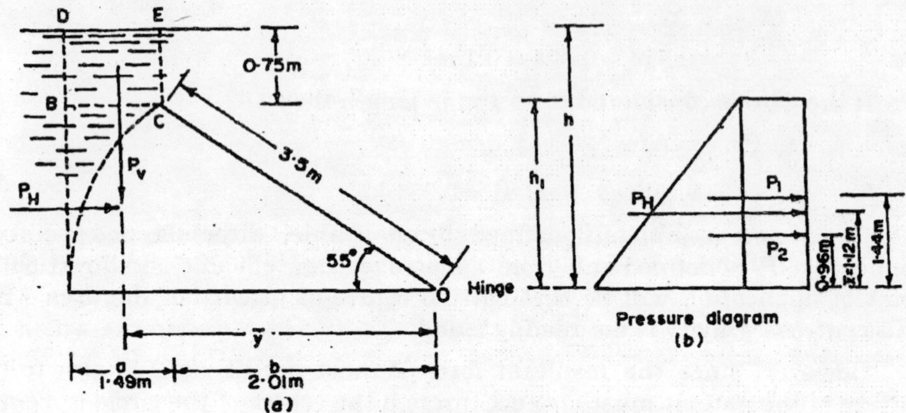

Fig. 3.39. Tainter gate P. 3.23.

$$P_1 = 0.75 \quad \rho g \times 2.87$$

$$= 0.75 \times 9.81 \times 2.81 = 21.11 \text{ kN per meter of gate.}$$

$$P_2 = \frac{2.87}{2} \times \rho g \ x\, 2.87$$

$$= \frac{1}{2} \times (2.87)^2 \times 9.81 \quad = 40.41 \text{ kN per meter of gate}$$

Their respective moment arms in relation to the hinge O are

$$\bar{x}_1 = \frac{2.87}{2} = 1.44 \text{ m}$$

$$\bar{x}_2 = \frac{2.87}{3} = 0.96 \text{ m}$$

Net resultant horizontal force P_H is

$$P_H = P_1 + P_2$$

$$= 21.11 + 40.41 = 61.52 \text{ kN per meter of gate}$$

Its moment arm $\bar{x}$ is found by equating moments about the hinge as follows.

$$61.55\ \bar{x} = 21.11\ \bar{x}_1 \times 1 + 40.41\ \bar{x}_2 \times 2$$

$$= 21.11 \times 1.44 + 40.41 \times 0.96 = 30.39 + 38.79 = 69.18$$

$$\bar{x} = \frac{69.18}{61.52} = 1.12 \text{ m}$$

P_v = Weight of the water with volume equal to the area ADEC by one meter per unit section of the gate

$$= (1.49 \times 0.75)\ \rho g + \left[\frac{1.49 + 3.5}{2} \times 2.87 - \frac{\pi}{4}(7)^2 \times \frac{55}{360}\right] \times \rho g$$

$$= 10.96 + [7.16 - 5.88] \times 9.81 = 10.96 + 12.55 = 23.51 \text{ kN}$$

$$\therefore \frac{\bar{y}}{6.27 \times 9.81} = \frac{1.2}{2.4 \times 9.81} \quad \therefore \bar{y} = 2.926\ m$$

$$P = \sqrt{P_H^2 + P_V^2} = \sqrt{(61.52)^2 + (23.51)^2} = 65.8 \text{ kN/m} \qquad \text{Ans}$$

3.24 A sluice gate consists of a quadrant of a circle of radius 2.5 m pivoted at its centre O (Fig 3.40). Its centre of gravity is at G as shown. When the water is level with the pivot O, calculate the magnitude and direction of the resultant pressure on the gate due to the water and the turning moment required to open the gate. The width of the gate is 3 m and it has a weight of 50 kN.

Solution

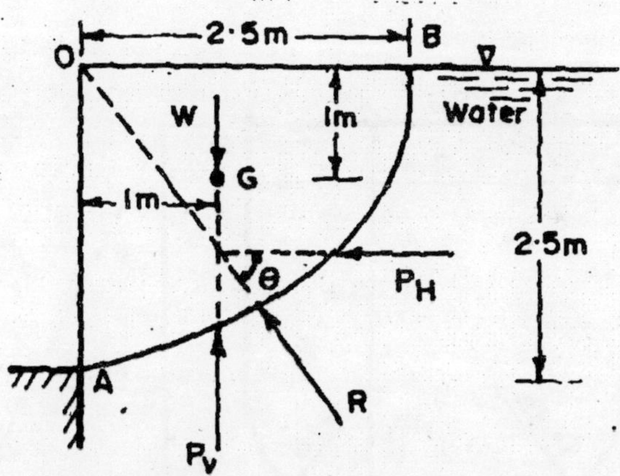

Fig. 3.40. Sluice gate P. 3.24.

Horizontal component P_H

P_H = Resultant pressure on projection OA of curved surface

$$= \rho g \times Area\ OA \times \frac{OA}{2}$$

$$= 9.81 \times (3 \times 2.5) \times \frac{2.5}{2} = 91.96 \text{ kN}$$

Vertical component P_v = weight of fluid which would occupy OAB

= ρg × volume of cylindrical sector OAB

$$= 9.81 \times \frac{\pi}{4}(2.5)^2 \times 3 = 144.46 \text{ kN}$$

Resultant Pressure R = $\sqrt{P_H^2 + P_V^2}$

$$= \sqrt{(91.96)^2 + (144.46)^2} = 171.25 \text{ kN}$$

$$\therefore \tan\theta = \frac{P_V}{P_H} = \frac{144.46}{91.96} = 1.57$$

$$\therefore \theta = 57.52^\circ$$ Ans

Since the surface is part of a cylinder R will act through O. Therefore R has no moment about the pivot O, the only turning moment is that due to the weight of the gate.

Turning moment required to open gate

$$= w \times 1$$

$$= 50 \times 1 = 50 \text{ kN-m}$$ Ans

3.25 An open tank 3m x 1m in cross section (Fig 3.41) holds water to a depth of 3m. Determine the magnitude, direction and line of action of the forces exerted upon the plane surfaces AB & CD and the curved surface BC of the tank.

Solution

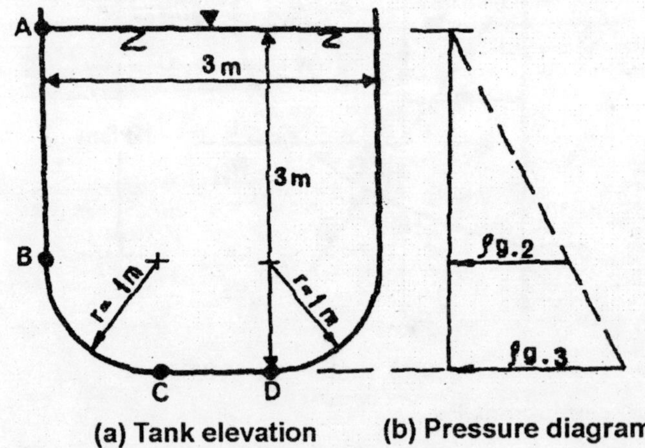

(a) Tank elevation (b) Pressure diagram

Fig. 3.41. Open tank P. 3.25.

Force on face AB per meter length = Area of pressure diagram.

$$= \frac{1}{2} \times \rho g \times 2^2 = \rho g \times 2 = 19.62 \text{ kN/m}$$

The force on face AB acting normal to the face AB at a depth of 2/3 times (2) = 1.33 m from the water surface.

Force on curved surface BC per meter length

$$P_H = \rho g \times 2 \times 1 + \frac{1}{2}\ \rho g \times (1)^2$$

$$= \rho g \left[2 + \frac{1}{2} \right]$$

$$= \rho g \times 2.5 = 24.525 \text{ kN/m}$$

Vertical component P_v = weight of water above the surface

$$\therefore P_V = \rho g \times 2 \times 1 \times 1 + \rho g \times \frac{\pi}{4} (1)^2 \times 1$$

$$= \rho g \left[2 + \frac{\pi}{4} \right]$$

$$= 9.81 \times 2.785 = 27.32 \text{ kN/m}$$

Resultant thrust $= \sqrt{(24.525)^2 + (27.32)^2} = 36.713$ kN/m

$$\theta = \tan^{-1}\left(\frac{P_v}{P_H}\right) = \tan^{-1}\left(\frac{27.32}{24.525}\right) = 48°5$$

Resultant thrust acting at an angle 48°5', to the horizontal and passing through the centre of curvature of the surface BC.

Force on surface CD per meter length.

Uniform pressure intensity on CD = $\rho g \times 3$

Total thrust on CD = Uniform pressure times area

$$= \rho g \times 3 \times 1 \times 1 = 29.43 \text{ kN/m}$$

Total thrust on CD acting vertically downwards (normal to CD) through the mid point of the surface CD.

3.26. The profile of the inner face of a dam is a parabolic with equation $y = 0.3\, x^2$ (Fig 3.42). The dam retains water to a depth of 30 m above the base. Determine the hydrostatic thrust on the dam per unit length, its inclination to the vertical and the point at which the line of action of this thrust intersects the horizontal base of the dam.

Solution

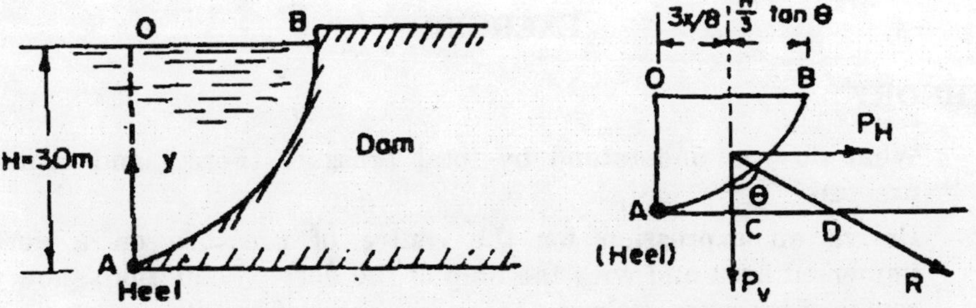

Fig. 3.42. Parabolic profile of the inner face of a dam. (a) and force diagram P. 3.26.

$$P_H = wA\bar{x}$$

$$= w(H \times 1) \times \frac{H}{2} = \frac{wH^2}{2}$$

$$9.81 \times \frac{30^2}{2} = 4414.5 \text{ kN}$$

$$= 4.4145 \text{ MN}$$

P_V = Volume of water in OAB times ρg

Here $y = 0.3x^2$

$$x = \sqrt{y/0.3}$$

when $y = 30$ *therefore* $x = \sqrt{\frac{30}{0.3}} = 10m$

$$\therefore P_V = \frac{2}{3} \times 30 \times 10 \times 1 \times 9.810 = 1962 \text{ kN}$$

$$= 1.962 \text{ MN}$$

$$\therefore R = \sqrt{P_H^2 + P_V^2}$$

$$= \sqrt{(4.4145)^2 + (1.962)^2} = 4.83 \text{ MN}$$

$$\theta = \tan^{-1}\frac{P_V}{P_G} = \tan^{-1}\frac{1.962}{4.415} = \tan^{-1}(0.44) = 23.96^\circ \text{ with horizontal}$$

$$= 66^\circ \text{ with vertical.}$$

The resultant R cuts at D and vertical force cuts at C.

$$\therefore AD = AC + CD$$

$$= \frac{3}{8} \times 10 + \frac{H}{3}\tan\theta$$

$$= 3.75 + 10 \times 2.249$$

$$= 3.75 + 22.49 = 26.24 \text{ m from heel}$$

Ans

EXERCISES

A. THEORY

3.1. What do you understand by total pressure (Force) and centre of pressure?

3.2. Derive an expression for the centre of pressure on a vertical immersed area and with the help of the derived equation, show that centre of pressure is always below its centre of gravity.

3.3. Derive the expression for the centre of pressure for an inclined immersed plane surface. Thus prove that the centre of pressure is always below the centroid of the area.

3.4. Define centre of pressure. Derive on expression for the total pressure and the centre of pressure of a plane lamina immersed in a liquid.

3.5. Prove that the centre of pressure is below the centre of gravity.

3.6. If a plane lamina is submerged in water, prove that the centre of pressure will be below its centre of gravity. What is the condition when the two coincide ?

3.7. Explain the terms (i) total pressure (ii) centre of pressure and (iii) pressure diagrams.

3.8. What do you understand by the term pressure diagrams ? How is it useful to find the total pressure?

3.9. Derive an expression for total pressure and centre of pressure on a plane lamina immersed below and inclined to the free surface.

3.10. Explain the graphical method of finding the hydrostatic force and the position of centre of pressure for a plane surface. What are the limitations in using this method?

3.11. Show that the vertical component of liquid surface on a curved surface equals the weight of liquid above the curved surface.

3.12. How does resultant pressure differ from total pressure?

3.13. Explain a method by which the total pressure and the centre of pressure on any non-planer surface immersed in a fluid can be calculated.

3.14. Develop form first principle a general expression for a position of the centre of pressure of a fluid on one side of a sloping flat plate completely immersed in the fluid.

3.15. Prove that for a lock gate, the reaction between the two gates is equal to the reaction at the hinge, i.e. R = P

Also prove that the reaction between the two gates,

$$P = \frac{F}{2\sin\alpha}$$

Where α = inclination of the gate with normal to the side of the lock.

B. UNSOLVED PROBLEMS

3.1. A cylindrical tank 600 mm in diameter with its axis vertical is filled to a depth of 1.5 m with water. What is the total pressure on the curved surface? What is the resultant pressure on this surface [20.7 kN, 0]

3.2. A circular lamina 1.25 m in diameter is immersed in water so that the distance of its perimeter measured vertically below the water surface varies between 600 mm and 1.5 m. Find the total force due to the water acting on one side of the lamina, and the vertical distance of the centre of pressure below the surface. [12.65 kN, 1.098 m]

3.3. A pipe line which is 4 m in diameter contained a gate valve. The pressure at the centre of the pipe is 2 bar. If the pipe is filled with oil of specific gravity 0.87; find the force exerted by the oil upon the gate and position of the centre of pressure. [2.5 MN, 23 m]

3.4. A circular opening, 3 m diameter, in a vertical side of a tank is closed by a disc of 3 m diameter which can rotate about a horizontal diameter. Calculate (a) the force on the disc, and (b) the torque required to maintain the disc in equilibrium in vertical position when the head of water above the horizontal diameter is 4 m. [282.74 kN, 4.14 m, 39.58 kN m]

3.5. A cubical tank has sides of 1.5 m. It contains water for the lower 0.6 m depth. The upper remaining part is filled with oil of specific gravity 0.9. Calculate for one vertical side of the tank (a) total pressure, and (b) position of centre of pressure. [15.16 kN, 1.005]

3.6. A caisson for closing the entrance to a dry dock is of trapezoidal form 16 m wide at the top and 12 m wide at the bottom and 8 m deep. If the water on the outside is 1 m below the top level of the caisson and the dock is empty, find the total water pressure on it and the depth of centre of pressure. [3.164 MN, 4.56 m, below the water surface]

3.7. A vertical rectangular gate 2 m wide and 2.5 m high is subjected to water pressure on one side, the water surface being at the top of the gate. The gate is hinged at the bottom, and is held by a horizontal chain at the top. Calculate the tension in the chain [61.31 kN, 1.67 m, 20.43 kN]

3.8. A vertical square area 1m times 1 m is submerged in water with upper edge of 0.5 m below the water surface. Locate the horizontal line of the surface of the square such that the force on the upper portion equals the force on the lower portion. (0.618 m)

3.9. An isosceles triangle of base 3 m and altitude 6 m is immersed vertically in water with its axis of symmetry horizontal. If the head of water on its axis is 9 m, locate the centre of pressure both laterally and vertically. [9.04 m]

3.10. A circular plate 3m in diameter is submerged in water, its greatest and least depths below the water surface being 2m and 1 m respectively. Find (a) the total pressure on the front face on the plate and (b) the position of the centre of pressure. [104.013 kN, 1.5417 m]

3.11. A circular plate 2 m in diameter is immersed in water in an inclined direction so that the distance of its perimeter measured vertically below the water surface varies between 1 m and 2.5 m. Find the total force due to water acting on one side of the plate and the vertical distance of the centre of pressure below the water surface. [53.9 kN, 1.8304 m]

3.12. (a) An open steel tank of base 4 m square has its sides sloping outwards such that its top is 7 m square. If the tank is 2 m high and is filled with water, determine the total thrust and its location (i) on the base, and (ii) on one of the sloping sides.

(b) If the four sizes of the tank sloping inwards so that its top is 1 m square, find the thrust and its top is 1 m square, find the thrust and its location on the base when it is filled with water [(a) (i) P = 314 kN, $\bar{h}$ = 0.91 m, (ii) 122.75 kN]

(b) Due to hydrostatic paradox, P = 314 kN, $\bar{h}$ = 1.243m]

3.13. A tank of oil has a right triangular panel near the bottom as shown in Fig 3.47. Calculate the total resultant force on the triangular panel and the location of the centre of pressure on the panel. [2.54 MN, The resultant force acts at 0.444 m down and 0.111 m to the right of the centroid]

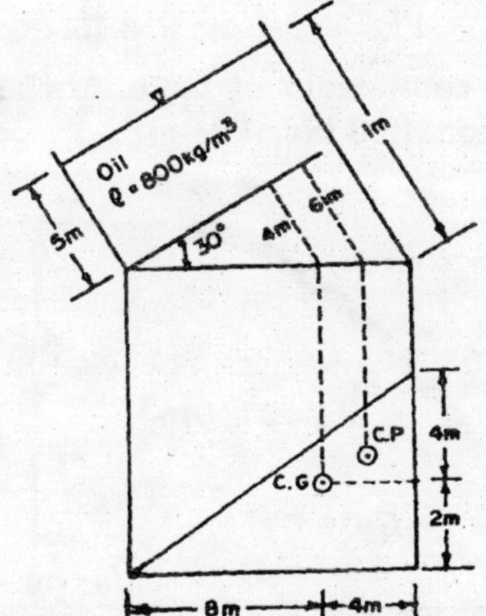

Fig. 3.47. Oil tank P. 3.13.

3.14. Gate AB is Fig 3.48 is 1.2 m long and 0.8 m wide. Calculate force F on the gate and its centre of pressure position x. [F = 38.75 kN, X = 0.615 m from point A.]

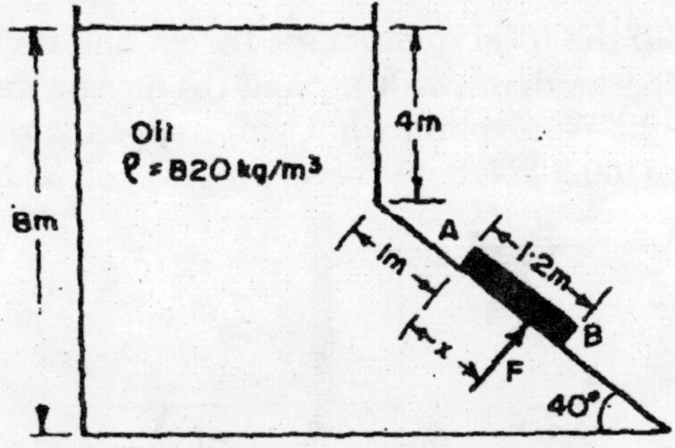

Fig. 3.48. Gate P. 3.14.

3.15. The cross section of a gate is shown in Fig 3.49. Its dimension normal to the plane of the paper is 10 m, and its shape is such that $x = 0.2\ y^2$. The gate is pivoted about O. Find the horizontal and vertical forces and the clockwise moment acting on the gate if the water depth is 2 m. [195.8 kN, 52.2 kN, 143 kNm]

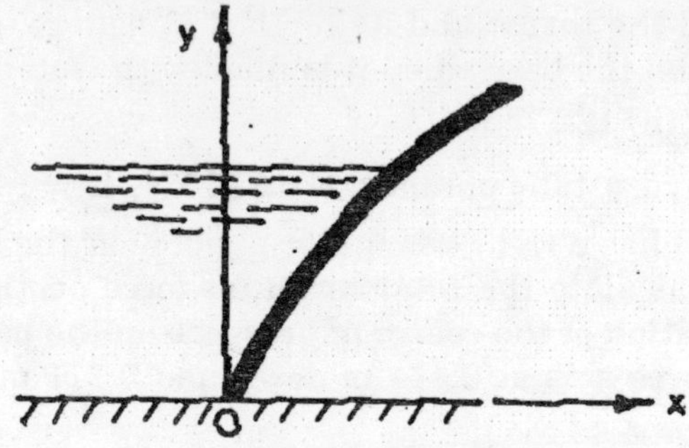

Fig. 3.49. Gate P. 3.15.

3.16. Find the vertical component of force on the curved gate of Fig 3.50 and its line of action. [9.6 kN, 0.24 m]

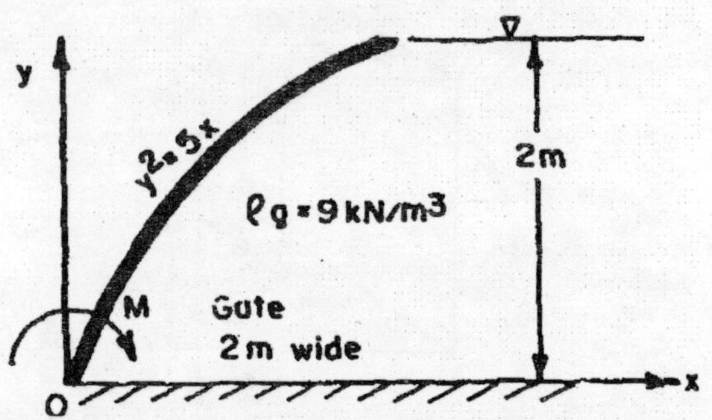

Fig. 3.50. Cylindrical gate P. 3.16.

3.17. A radial gate, 2 m long, hinged about a horizontal axis, closes rectangular sluice of a control dam by the application of a counter weight W (Fig. 3.51)

Determine (a) the total hydrostatic thrust and its location on the gate when the storage depth is 4 m, and (b) for the gate to be stable, the counter weight W. Explain what will happen if the storage increases beyond 4 m. [100.7 kN, 3.49 below water level, 53.3 kN]

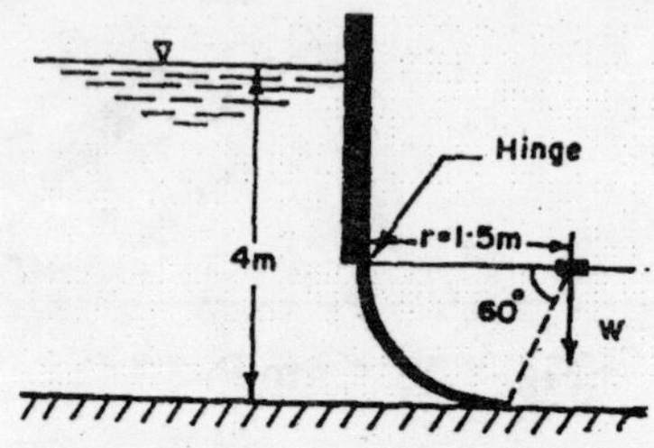

Fig. 3.51. Radial gate P. 3.17.

3.18. A sector gate of radius 3 m and length 4 m retains water as shown in Fig 3.52 Determine the magnitude, direction and location of the resultant hydrostatic thrust on the gate [132.435 kN, 108.45 kN, 171.175 kN, 39°31' with horizontal].

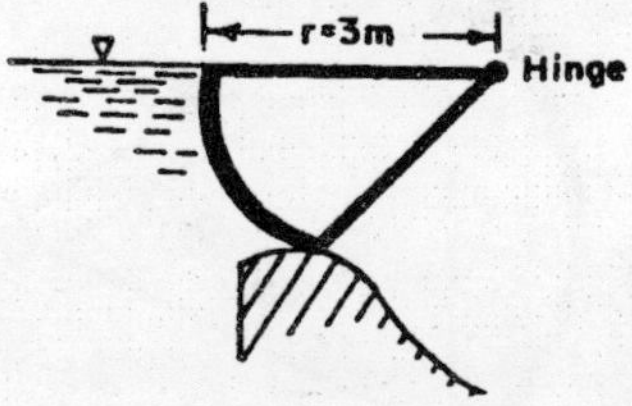

Fig. 3.52. Sector gate P. 3.18.

3.19. A sector shaped sluice gate having a radius of curvature of 5.4 m is as shown in Fig 3.53. The centre of curvature C is 0.9 m vertically below the lower edge A of the gate and 0.6 m vertically above the horizontal axis passing through O about which the gate is constructed to turn. The mass of the gate is 3000 kg per meter run and its centre of gravity is 3.6 m horizontally from the centre O. If the water level is 2.4 m above the lower edge of the gate find per meter run (a) the resultant force acting on the axis at O, (b) the resultant moment about O. [36.2 kN, 89 kNm]

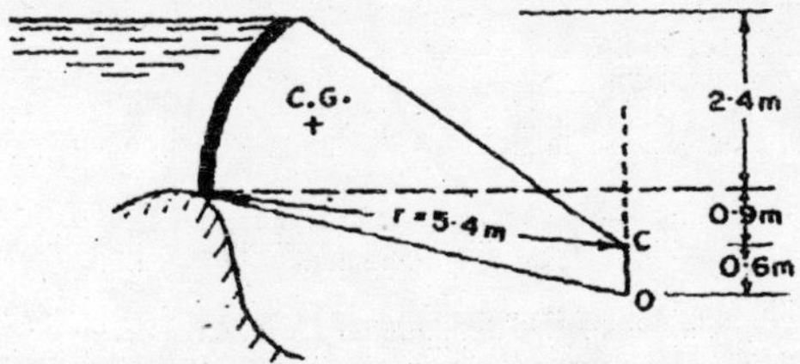

Fig. 3.53. Sector gate P. 3.19.

3.20. A dam has a parabolic shape $y = y_o\left(\frac{x}{x_o}\right)^2$ as shown in the Fig. 3.54 having x_o = 6 m and y_o = 9m. The fluid is water. Compute the horizontal, vertically and resultant thrust exerted by water per meter length of dam as well as their line of action. [393.305 kN, 353.16 kN, 531.576 kN, 41.6335°.]

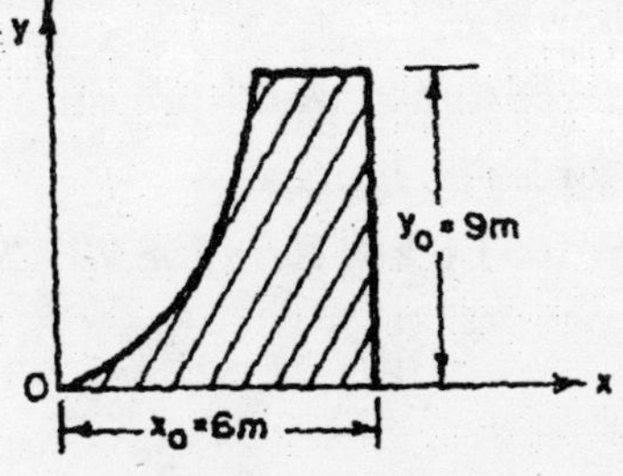

Fig. 3.54. Dam P. 3.20

3.21. A tainter gate of 90° sector is subjected to water pressure as shown in Fig. 3.55. Determine the horizontal and vertical pressure acting on the face of the gate. [245.25 kN, 70 kN, 255 kN, 15°55' 43"].

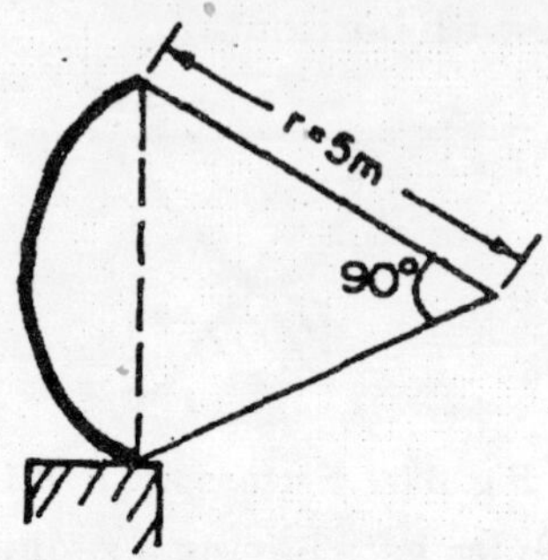

Fig. 3.55. Tainter gate P. 3.21.

3.22. The Fig. 3.56 shows the cross section of a dam with a parabolic face, the vertex of the parabola being at O. The axis of the parabola is vertical and 13 m from the face at the water level.

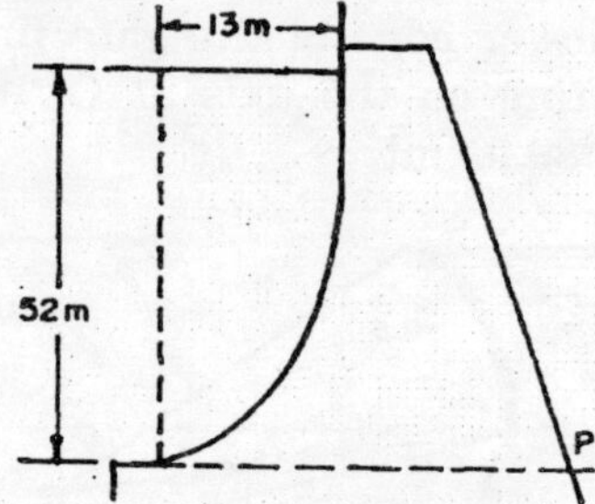

Fig. 3.56. Dam P. 3.22.

Estimate the resultant force per meter run due to the water, its inclination to the vertical and how far from O its line of action cuts the horizontal OP. The centroid of the half parabolic cross section of water is 4.75 m from the vertical through O [4.415 MN/m width 13.26 MN/m width, 56.8 m form O]

3.23. The gates of a lock which is 8 m wide include an angle of 120° in the closed position. Each gate is held on two hinges, one placed 0.8 m and another at 6 m from the bottom of the lock. If the water levels are 9 m and 3 m on the upstream and downstream sides respectively, determine the following:

(a) the resultant force due to water pressure

(b) magnitude of forces on hinges.

[h = 3.25 m, P = 1631.6 kN, R_T = 768.73 kN, R_B = 862.86 kN].

4

Buoyancy and Floatation

4.1. INTRODUCTION

Buoyancy is that phenomenon which determines whether an object floats or sinks. When an object floats either above or below the surface in a static fluid, the resultant buoyant force equals the weight of the fluid displaced and acts vertically in the direction of the free surface through the centroid of the displaced volume. The resultant buoyant force is nearly independent of the relative pressure that the fluid exerts on the exterior surface of the object because of its depth in the fluid, and equals the difference between the force acting above the object and that acting below the object. Thus if the buoyant force that the fluid exerts is greater or equal to the downward weight of the object, it will float at or in equilibrium below the surface, and if it is not, the object will sink toward the bottom.

4.2. BUOYANCY

A body immersed partially or fully in a fluid experiences a vertical upward force. This vertical upward force on a floating or submerged body is known as buoyant force. The tendency of a submerged body to rise in a fluid because of the upward fluid pressure which opposes the downward force of gravity, is known as buoyancy.

Consider the submerged body PQRS in Fig. 4.1. The resultant horizontal pressure on the body is equal to zero because the horizontal pressures on the vertical projection from both sides are equal in magnitude and opposite in direction. On the two ends of a vertical prism of cross section dA are two vertical pressure forces, p_1 dA and p_2 dA as shown. The difference between the upward and downward pressure forces is the buoyant force dF_B on the vertical prism.

$$
\begin{aligned}
dF_B &= p_2\, dA - p_1\, dA \\
&= \rho g (z_2 - z_1) dA \\
&= \rho g h\, dA = \rho\, g\, dV
\end{aligned}
$$

where dV = volume of the vertical prism.

$= hdA$

The buoyant force F_B on the entire submerged body PQRS is obtained by integrating dF_B. Therefore

$$F_B = \int dF_B = \int \rho g dV = \rho g V \tag{4.1}$$

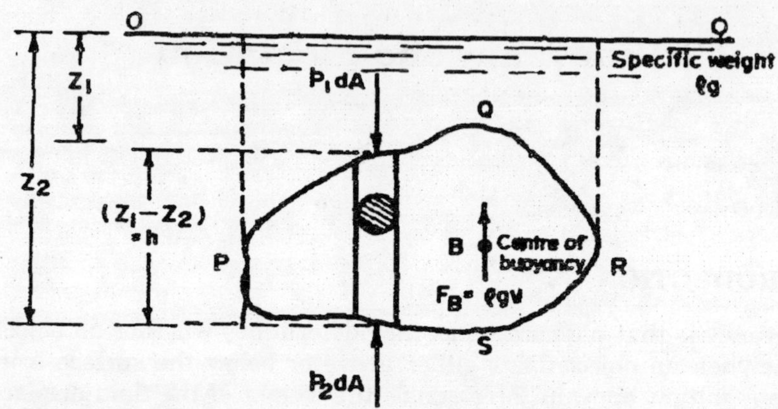

Fig. 4.1. Buoyant force on a submerged body

The buoyant force acts vertically upward through the centre of buoyancy which coincides the centroid of the displaced volume.

4.3. CENTRE OF BUOYANCY

The point of application of the buoyancy force on the body is known as centre of buoyancy.

4.4. EQUILIBRIUM OF FLOATING BODIES

From the analysis in section 4.2 it should be clear that if a body is floating freely and is therefore in equilibrium the following two conditions must be satisfied.

(1) The up thrust (F_B) must be equal to the weight (W) of the body (F_B = W), and

(2) The centre of gravity of the body and the centre of buoyancy must be in the same vertical line, so that their moment about any axis is zero.

The equilibrium of a body may be stable, unstable or neutral, depending upon whether, when given a small displacement, it tends to return to the equilibrium position, move further from it or remain in the displaced position. For a floating body, such as a ship, stability is of major importance.

4.5. METACENTRE

Metacentre is defined as a point at which the line of action of the buoyancy force will meet the normal axis of the body when the body is given a small angular displacement (Fig. 4.2).

4.6. METACENTRIC HEIGHT

The distance between the centre of gravity of the floating body and the metacentre i.e. the distance MG or GM is known as the metacentric height (Fig. 4.2).

4.7. STABILITY OF FLOATING BODIES-METACENTRE AND METACENTRIC HEIGHT

Fig. 4.2(a) shows a body floating in equilibrium. The weight W acts through the centre of gravity G and the buoyant thrust F_B acts through the centre of buoyancy B of the displaced fluid in the same straight line as W.

When the body is slightly tilted through a small heel angle θ, the centre of buoyancy moves from B to B_1, where B_1 is now the centroid of the immersed trapezium in Fig. 4.2(b), the upthrust acts vertically through B_1 and intercepts the line BG produced in M thus defining the position of the metacentre. The distance MG between the centre of gravity G of the body and the metacentre M is called metacentric height and it must always have a positive value if the body is to be stable.

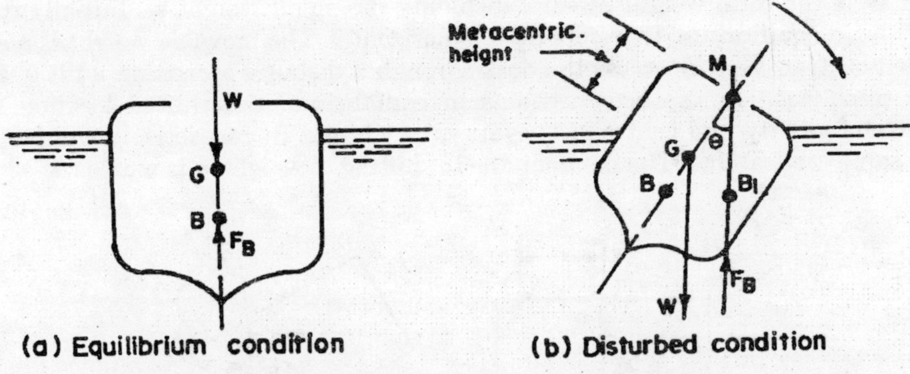

Fig. 4.2. Metacentric and metacentric height

The position of the metacentre relative to the centre of gravity of body governs its stability. Therefore, the conditions governing the stability of floating bodies in which the centre of gravity is above the centre of buoyancy, is as follows.

(a) Stable equilibrium (Fig. 4.3(a)) When the metacentre, M, is above the centre of gravity G of the body is in stable equilibrium. This condition can be achieved when BM > BG.

(b) Unstable equilibrium [Fig. 4.3(b)] : When the metacentre, M, is below the centre of gravity, G, the body is in unstable equilibrium. This condition therefore, exists when BM < BG.

(c) Neutral equilibrium [Fig. 4.3.(c)] : When the metacentre M coincides with the centre of gravity the body is in neutral equilibrium. This condition of neutral equilibrium exists when BM = BG.

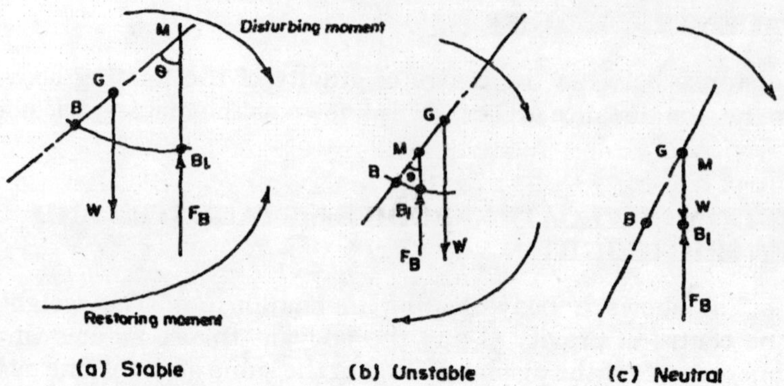

Fig. 4.3. Condition of equilibrium

4.8 EXPERIMENTAL METHOD OF DETERMINATION OF METACENTRIC HEIGHT

Let, w is the movable weight placed at the centre of the deck of the ship and W is the total weight of ship, including the small weight w. Initially the ship is, in equilibrium, therefore deck is horizontal. The movable weight is now rolled off transversely across the deck through a distance x causing a tilt θ. In the tilted position the body remains in equilibrium, the shifted location of centre of gravity and centre of buoyancy, i.e. G_1 and B_1 respectively will be in the same vertical line. The moment due to shifting of weight w is w x cos θ.

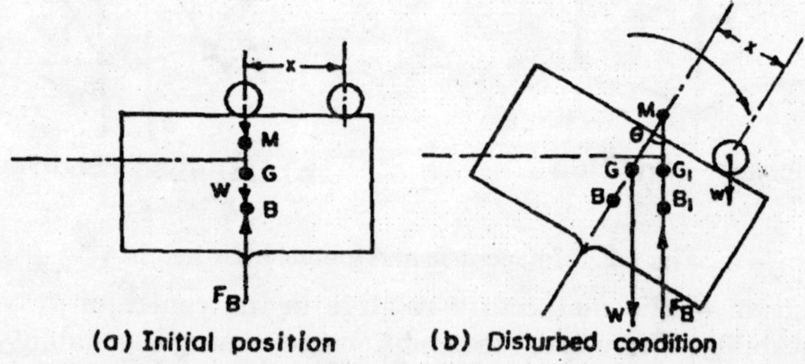

Fig. 4.4. Experimental determination of the metacentric height.

The moment due to shifting of G to G_1 is given by

$$\text{W GG}_1 = \text{W GM} \sin\theta$$

Equating two moments

$$\text{W GM} \sin\theta = \text{w}\, x \cos\theta$$

$$\text{GM} = \frac{\text{w}x}{\text{W} \tan\theta}$$

$$= \frac{\text{w}x}{\text{W}\,\theta} \tag{4.2}$$

4.9 ANALYTICAL METHOD FOR METACENTRIC HEIGHT

In Fig 4.5, P Q is the original water line plane and B the centre of buoyancy in the equilibrium position. When the vessel is displaced by a small angle θ. The centre of gravity relative to the vessel remains in the same position but the centre of buoyancy shifts from B to B_1 as a result of the alteration in the shape of the displaced fluid. RS is the water line plane in the displaced position. For small angles of tilt, $BB_1 = BM\,\theta$.

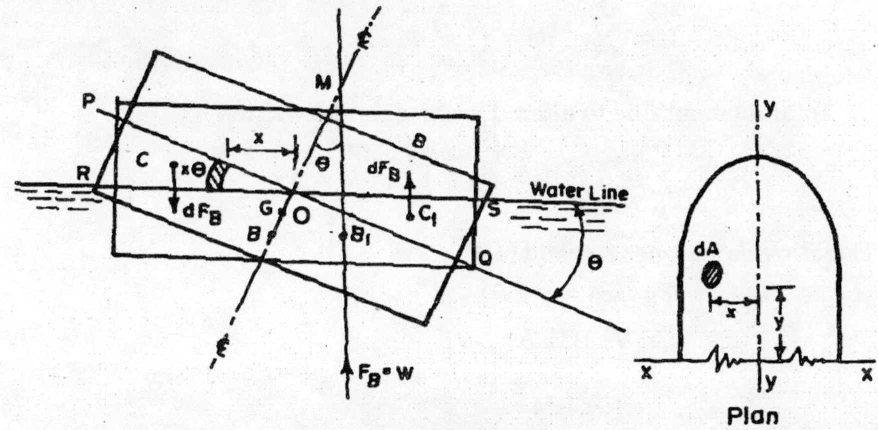

Fig. 4.5. Stability of vessel.

In relations to Fig 4.5 the original stable water line PQ is tilted through a small angle θ. liquid which occupies region OPR in the stable position now fills region OQS. Thus, while the buoyancy on the left of the vessel axis is reduced by an amount dF_B due to reduced volume of liquid displaced by the left side of the vessel. It is increased on the right by an amount dF_B due to increase in volume of liquid displaced by the right side of the vessel. In other words we can say that the loss of buoyancy dF_B associated with emergence of the wedge OPR, is sensibly equal to the gain in buoyancy dF_B due to immersion of the wedge OQS. These set up a restoring moment $dF_B\,CC_1$, where CC_1 is distance between the centroids of the wedges OPR and OQS. Hence the couple produced by the movement of the wedge OPR to OQS must be equal to the couple due to the movement of F_B from B to B_1.

$$F_B\,BB_1 = dF_B\,CC_1 \tag{4.3}$$

If dA is a small area in the water line plane at a distance x from the axis of rotation yy, it will generate a small volume, shown shaded, when the vessel is tilted.

Volume swept out by dA = dA x corresponding depth of wedge

$$= dA\,x\,\theta$$

Summing all such volumes and multiplying by the specific weight ρg of the liquid gives the buoyant force.

Since all elements exert a moment in the same sense, no matter whether they be in the immersed or emergent wedge, the total moment is

$$\sum_{p}^{Q} \rho\, g\theta x^2\, dA = \rho\, g\, \theta \sum x^2\, d\, A$$

Righting Couple due to fluid wedges.

The Symbol I (Second moment of area) depends entirely on the shape of the water plane and may, more precisely, be expressed in terms of the calculus as:

$$\text{I} = \sum_{p}^{Q} x^2\, dA = \int_{p}^{Q} x^2\, dA$$

$\therefore$ Total moment due to altered wedges can be written as

$$\sum_{p}^{Q} \rho g\, \theta\, x^2\, dA = \rho\, g\, \theta \int_{p}^{Q} x^2\, dA = \rho\, g\, \theta I \qquad (4.4)$$

Equating equations (4.3) and (4.4)

$$\text{F}_\text{B}\ \text{BB}_1 = \rho \text{g}\, \theta \text{I}$$

$$(\rho \text{g V})\ \text{BM}\ \theta = \rho\, \theta \text{I}$$

$$\therefore\ \text{BM} = \frac{\text{I}}{\text{V}} \qquad (4.5)$$

The distance BM is known as metacentric radius

Thus the metacentric height

$$\text{GM} = \text{BM} - \text{BG}$$

$$= \frac{\text{I}}{\text{V}} - \text{BG}$$

When G lies above B, and

$$\text{GM} = \frac{\text{I}}{\text{V}} + \text{BG}$$

When G lies below B, in which case the vessel is permanently stable.

When the vessel oscillates about the yy axis the movement is known as rolling (heeling). Pitching occurs when the movement about the xx axis and the corresponding second moment of area $I_{xx} = \int y^2\, dA$ must be used in above equations.

In practice equation (4.5) may be used for all cases where the angle of heel θ measured in radians is approximately equal to tan θ and thus the optimum angle of heel is between 10° and 15°. When the angle of heel θ exceeds this limit then the movement of B_1 parallel to BM is no longer negligible.

Thus, for large angles of heel, the movement of the centre of buoyancy normal and parallel to BM must be considered and is given by

$$\text{BM} = \frac{\text{I}}{\text{V}}\left(1 + \frac{\tan^2 \theta}{2}\right) \qquad (4.6)$$

4.10 THE PERIOD OF ROLL OF A VESSEL

The rolling period of vessel may be determined approximately by making certain simplifying assumptions. The calculation requires a knowledge of the metacentric height and the moment of inertia, I, of the vessel about a force and aft axis through G. The restoring moment, when a vessel is heeled, is assumed to have the same value as that deduced from hydrostatics, the effect of the motion of the surrounding water being neglected.

It is to be noted that the righting moment is opposite in the sense to the angle of heel, θ, and that the moment of inertias may be expressed in terms of the corresponding radius of gyration k, we deduce that

Righting moment = Inertia torque

$$mg\ GM\ \theta = -\ mk^2\alpha$$

$$gGM\ \theta = -\ k^2\frac{d^2\theta}{dt^2} \qquad \left[Beacuse\ \alpha = \frac{d^2\theta}{dt^2}\right]$$

$$\frac{d^2\theta}{dt^2} + \frac{g}{k^2}\ GM\ \theta = \text{O}$$

Since this corresponds to simple harmonic motion

$$\text{Periodic time, T} = 2\pi\sqrt{\frac{Displacement}{Acceleration}}$$

$$T = 2\pi\sqrt{\frac{\theta}{GM\ \theta g / k^2}}$$

$$T = 2\ \pi\sqrt{\frac{k^2}{GM\ g}} \qquad (4.7)$$

Equation (4.8) gives reasonable agreement with the actual value of the periodic time for rolling oscillation, but is less accurate when applied to pitching. This is mainly due to ignoring the effect of the movement of the liquid on the oscillation, which in fact, tends to damp the oscillation, causing it to sloe down and eventually to die out.

From the equation (4.8) also gives that although a larger metacentric height will improve stability, it produces a short periodic time of oscillation, which results in discomfort and excessive stress on the structure of the vessel.

ILLUSTRATIVE EXAMPLES

4.1 A rectangular barge is 20 m long, 6 m wide and 3 m deep. It has a draft of 2 m when fully loaded. The centre of gravity of the barge is on the axis of symmetry at the water surface. Determine the stability condition of the barge and the metacentric height.

Solution

$$I = \frac{1}{12}bd^3 = \frac{1}{12} \times 20 \times 6^3 = 360 m^4$$

$$V = 20 \times 6 \times 2 = 240 m^3$$

$$BM = \frac{I}{V} = \frac{360}{240} = 1.5 m$$

G from the bottom given

$$\therefore BG = OG - OB = 2 - 1 = 1 \text{ m}$$

$$GM = BM - BG = 1.5 - 1 = 0.5 \text{ m}$$

Since BM – BG = 0.5 > O the floating barge is stable. Ans

4.2 A rectangular pontoon floating in sea water is 21 m long, 7 m wide, 2.4 m deep and its weight 1.5 MN. It carries on its upper deck, at the normal vertical, a load of 1 MN. The centre of gravity of load 2.4 m above deck and the centre of gravity of pontoon is 1.2 m below the deck. Find the metacentric height. The density of sea water is 1025 kg/m³.

Solution

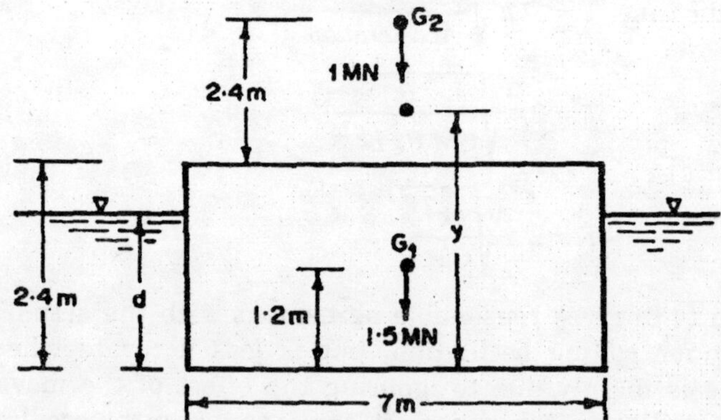

Fig. 4.6. Rectangular pontoon P. 4.2.

Let y be the distance of centre of gravity of both the pontoon and deck load above base of pontoon. Then the value of y can be calculated as follows

$$1.5 \times 1.2 + 1 \times 4.8 = 2.5 y$$

$$y = \frac{1.8 + 4.8}{2.5} = \frac{6.6}{2.5} = 2.64 \text{ m}$$

Let d be the draft, $W = F_B$

$$\therefore (1.5 + 1) \times 10^6 = \rho g \times 7 \times 21 \times d$$

$$2.5 \times 10^6 = 1025 \times 9.81 \times 7 \times 21 \times d$$

$$\therefore d = 1.69\,\text{m}$$

$$\therefore BG = y - \frac{d}{2} = 2.64 - \frac{1.69}{2} = 1.794$$

$$BM = \frac{I}{V} = \frac{\frac{1}{12}\times 21\times 7^3}{21\times 7\times 1.69} = \frac{600.25}{248.43} = 2.416\text{ m}$$

$$GM = BM - BG = 2.416 - 1.794 = 0.622\text{ m}$$ Ans

4.3 A rectangular pontoon, 6 m long and 3 m wide has the displacement of 150 kN. Find the maximum height of centre of gravity of this float in sea water in the stable equilibrium and with its sides vertical. Take density of sea water is 1030 kg/m³.

Solution

$$w = \frac{W}{V}$$

$$1030 \times 9.81 = \frac{150\times 1000}{6\times 3\times d}$$

$$\therefore d = 0.825\text{ m}$$

Hence, centre of buoyancy B is 0.4125 above the base that is $\frac{d}{2}$.

$$I = \frac{1}{12}bd^3 = \frac{1}{12}\times 6\times (3)^3 = 13.5$$

$$V = 6 \times 3\times 0.825 = 14.85$$

$$BM = \frac{I}{V} = \frac{13.5}{14.85} = 0.90 m^3$$

For the floating body to be in stable equilibrium, Metacentre M should be above centre of gravity G or, BM > BG. In the limiting case, BM may be equal to BG. Such a case will give the maximum allowable height of G.

∴ Maximum height of G above base

$$= \frac{d}{2} + BG \text{ or } BM = 0.4125 + 0.909 = 1.32\text{ m}$$ Ans

If the height of G above base is greater than 1.32, the body will not float in stable equilibrium with its side vertical but it will topple over by rolling about the (horizontal) longitudinal axis yy.

4.4 A barge 20 m long by 6.5 m wide has a draft of 2 m when floating in upright position. It's centre of gravity is 2.3 m above the bottom. (a) What is its initial metacentric height? (b) If a 30 kN weight is shifted 3 m across the barge, to what distance does the water line rise on the side? Find the righting moment for this tilt.

Solution

(a) $$I = \frac{1}{12}bd^3 = \frac{1}{12} \times 20 \times (6.5)^3 = 457.7 \text{ m}^4$$

$$V = 20 \times 6.5 \times 2 = 260 \text{ m}^3$$

$$BM = \frac{I}{V} = \frac{457.7}{260} = 1.76 \ m$$

G from the bottom given

$$BG = OG - OB = 2.3 - 1 = 1.3 \text{ m}$$

$$GM = BM - BG = 1.76 - 1.3 = 0.46 \text{ m}$$

(b) $$GM = \frac{wx}{w \tan\theta}$$

$$0.46 = \frac{30 \times 3}{(\rho g \times 260) \times \tan\theta}$$

$$\tan\theta = \frac{90}{173.276} = 0.0767$$

$$\theta = 4.386^\circ$$

Rise of water line on side $= \frac{6.5}{2} \tan\theta = 3.25 \times 0.767 = 0.249$ m

Righting moment for this lift $= W.GM \ \tan\theta$

$= 2550.6 \times 0.46 \times 0.0767 = 89.99 = 90$ kNm **Ans**

4.5 A steel pontoon has the dimensions as shown in Fig 4.7(a). Its weight is 800 kN and it is one of several which support a temporary bridge which crosses a river. The centre of gravity of the pontoon is at the geometrical centre of the plant and 1.5 m above the base. The bridge trusses impose a total maximum load of 2000 kN on the centre line of the pontoon deck. Determine the metacentric height of the pontoon when it is loaded as shown in Fig 4.7(b).

Solution

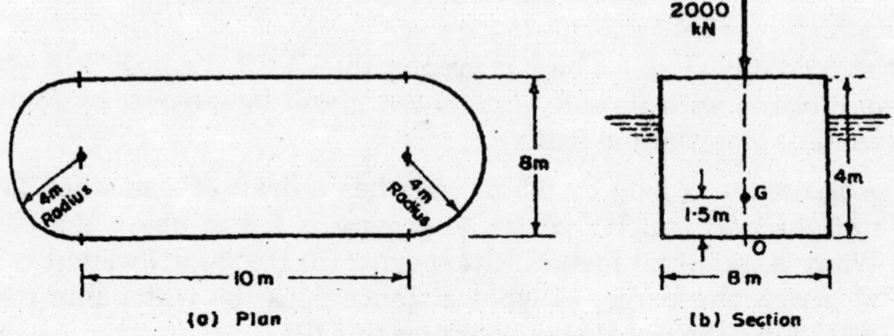

Fig. 4.7. Steel portion. P. 4.5

Weight of pontoon = 800 kN

Weight of Superimposed load = 2000 kN

Total Weight = 2800 kN

$$w = \frac{W}{V}$$

$$9.81 = \frac{2800}{V}$$

$$V = 285.42 \ m^3$$

At the water line, $A = 10 \times 8 + \frac{\pi}{4}(8)^2 = 130.26 \ m^2$

Submerged depth $= \frac{V}{A} = \frac{285.42}{130.26} = 2.19 m$

Height of centre of buoyancy $= \frac{2.19}{2} = 1.095 \ m$

$$BG = OG - OB = 1.5 - 1.095 = +0.404$$

$$I = \frac{1}{12} \times 10 \times (8)^3 + \frac{\pi}{64}(8)^4 = 627.728 m^4$$

$$BM = \frac{I}{V} = \frac{627.728}{285.42} = 2.2 m$$

$$GM = BM - BG = 2.2 - 0.404 = 1.796 \text{ m}$$ Ans

4.6 A barge displaces 10^6 kg has the horizontal cross section at the water line as shown in Fig 4.8. Its centre of buoyancy is 2 m below the water surface, and its centre of gravity is 0.5 m below the water surface. Determine its metacentric height for pitching and for rolling.

Solution

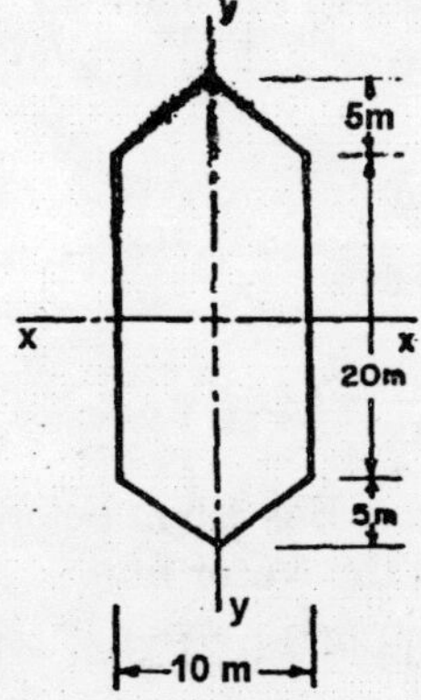

Fig. 4.8. Barge P. 4.6.

$$BG = 2 - 0.5 = 1.5 \text{ m}$$

$$V = \frac{10^5}{1000} = 1000\ m^3$$

$$I_{xx} = \frac{1}{12} \times 10 \times 20^3 + 2 \times \frac{1}{36} \times 10(5)^3 + \left(\frac{1}{2} \times 10 \times 5\right) \times 2\times\left(10 + \frac{1.5}{3}\right)^2$$

$$= 6666.67 + 69.44 + 6805.55 = 13541.65 \text{ m}^4$$

$$I_{yy} = \frac{1}{12} \times 20 \times 10^3 + \left(\frac{1}{12}\right)(5)(5)^3$$

$$= 1666.67 + 208.33 = 1875 \text{ m}^4$$

For Pitching (about xx axis)

$$GM = \frac{I_{xx}}{V} - BG = \frac{13541.65}{1000} - 1.5 = 13.54 - 1.5 = 12.04$$

For Rolling (about yy axis)

$$GM = \frac{I_{xx}}{V} - BG = \frac{1875}{1000} - 1.5$$

$$= 1.875 - 1.5 = 0.375$$

Ans

4.7 A ship having a length of 60 m, a beam of 8 m and a displacement of 15 MN, floats in sea water of density 1025 kg /m³. A weight of 150 kN is moved a distance of 6 m, across the deck causing the ship to heel through 3°. Determine its metacentric height. If the second moment of area at the water surface about the fore and aft axis is 75% of that of the circumscribing rectangle, and the position of the centre of buoyancy is 1.5 m below the water line, determine position of the centre of gravity of the ship.

Solution

$$GM = \frac{wx}{W \tan\theta} = \frac{150 \times 10^3 \times 6}{15 \times 10 \times \tan 3^\circ} = 1.1448 \text{ m}$$

$$I = 0.75\left[\frac{1}{12} lb^3\right] = 0.75 \times \frac{1}{12} \times 60 \times (8)^3 = 1920\ m^3$$

$$V = \frac{W}{\rho g} = \frac{15 \times 10^6}{1025 \times 9.81} = 1491.758$$

$$BM = \frac{I}{V} = \frac{1920}{1491.758} = 1.287$$

Ans

Since the centre of buoyancy B is 1.5 m below the waterline the metacentre is 1.5 – 1.287 = 0.2129 m below the water line.

The centre of gravity is 1.1448 + 0.2129 m = 1.357 m below the water line.

4.8 A hollow wooden cylinder of relative density 0.6 has an outer diameter of 800 mm and inner diameter of 400 mm and has its ends open. It is required to float in oil of specific gravity 0.9. Calculate the maximum height of cylinder so that it shall be stable when floating with its axis vertical. Find also the depth to which it will sink.

Solution

$$W_{oil} = S\rho g \times V$$

$$= 0.6 \times 9810 \times \frac{\pi}{4}\left[(0.8)^2 - (0.4)^2\right] \times h_{max}$$

$$W_{oil} = 2218.969\ h_{max} \qquad (1)$$

$$\text{Weight of oil displaced} = S\rho g \times \frac{\pi}{4}\left[(0.8)^2 - (0.4)^2\right] \times d$$

$$= 0.9 \text{ x } 9810 \text{ x } .37699 \text{ d} = 3328.44 \text{ d} \qquad (2)$$

Equation (1) and (2)

$$2218.969\ h_{max} = 3328.44\ d$$

$$\therefore d = 0.666\ h_{max}$$

$$BG = \frac{h_{max}}{2} - \frac{d}{2} = 0.5\ h_{max} - 0.333\ h_{max} = 0.166\ h_{max}$$

$$I = \frac{\pi}{64}\left[(0.8)^4 - (0.4)^4\right] = 0.0188$$

$$V = \frac{\pi}{4}\left[(0.8)^2 - (0.4)^2\right] \times 0.666\ h_{max} = 0.251\ h_{max}$$

$$BM = \frac{I}{V} = \frac{0.0188}{0.251\ h_{max}} = \frac{0.0748}{h_{max}}$$

For maximum value of h_{max} Bm & BG must be equal, hence

$$\frac{0.0748}{h_{max}} = 0.166\ h_{max}$$

$$h_{max}^2 = \frac{0.0748}{0.166} = 0.45$$

$$h_{max} = 0.67\ m$$

$$\therefore d = 0.666 \times h_{max} = 0.666 \times 0.67 = 0.4472\ m$$

Ans

4.9 A cylindrical buoy 1.8 m in diameter, 1.2 m long and weighing 10 KN floats in a salt water of density 1025 kg/m³. It's centre of gravity is 0.45 m from the bottom. If a load of 2 kN is placed on the top, find the maximum height of its centre of gravity if the buoy is to remain in stable equilibrium.

Solution

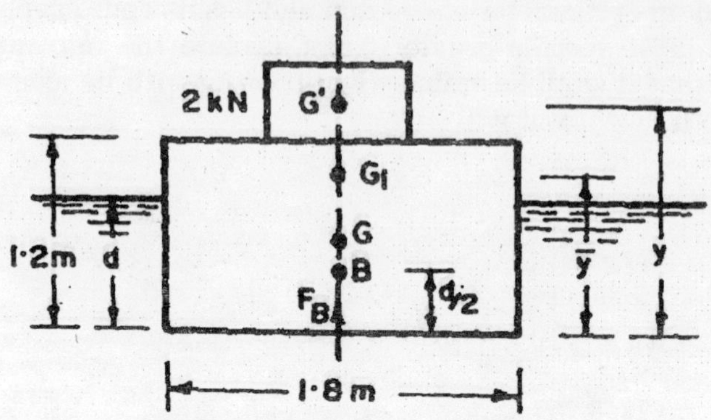

Fig. 4.9. Cylindrical buoy P. 4.9.

F_B = Weight of buoy and load

V = Volume of salt water displaced

$$\therefore F_B = \rho g V$$

$$(10 + 2) = 1025 \times 9.81 \times \frac{\pi}{4}(1.8)^2 d$$

$$d = \frac{12000}{1025 \times 9.81 \times 2.54} = 0.4689 \text{ m}$$

B is at a distance $\frac{d}{2}$ above the base i.e. 0.234 m the buoy and load are to just remain in stable equilibrium, the metacentre must coincide with the centre of gravity of the system G_1.

$$BM = BG_1 = \frac{I}{V} = \frac{\frac{\pi}{64}(1.8)^4}{\frac{\pi}{4}(1.8)^2 \times 0.4689} = 0.4318 \text{ m}$$

$$\therefore \bar{y} = \frac{d}{2} + BG_1 = \frac{0.4689}{2} + 0.4318 = 0.666 \text{ m above the base.}$$

To find the position of G' above the base i.e. y, take moments about the base

$$2000 \times y = 12000 \times 0.666 - 10000 \times 0.45$$

$$y = 3.9978 - 2.25 = 1.7478 \text{ m} \qquad \text{Ans}$$

4.10 A rectangular pontoon is 6 m in plan and it is 1.2 m deep. When loaded on the deck (at normal vertical) with a load weighing 40 kN, it floats in sea water with a draught of 0.6 m. The centre of gravity of pontoon is 0.6 m above its base and the centre of gravity of the load is 1.2 m above the deck. The load is moved transversely along the deck through a distance x meters. Determine the maximum value of x if the angle of heel is not to exceed 2°. Take density of sea water is 1025 kg/m³.

Solution

$$\text{Displacement} = S\rho g \times \text{Volume}$$

$$= 1025 \times 9.81 \times 0.6 \times 6 \times 6 = 217.1934 \text{ kN}$$

$$\text{Weight of pontoon} = 217.1934 - 40 = 177.1934 \text{ kN}$$

If centre of gravity G of weight W and rolling load by y meters above the base of the pontoon, we have

$$217.1934\, y = 40 \times 2.4 + 177.1934 \times 0.6$$

$$y = 0.9315$$

$$BG = 0.9315 - 0.3 = 0.6315$$

$$BM = \frac{I}{V} = \frac{\frac{1}{12} \times 6 \times (6)^3}{6 \times 6 \times 0.6} = 5$$

$$GM = BM - BG = 5 - 0.6315 = 4.368$$

Now,
$$GM = \frac{wx}{W \tan\theta}$$

$$4.368 = \frac{40 \times x}{217.1934 \times \tan 2^\circ}$$

$$x = \frac{4.368 \times 217.1934 \times 0.0349}{40} = 0.828 \text{ m}$$ Ans

4.11 A ship has a total displacement of 100 MN is sea water of density 1025 kg/m³. Its second moment of area at the water surface about the fore and aft axis is 24300 m⁴ and its centre of gravity is 2 m above the centre of buoyancy. If the periodic time for rolling oscillation is 10 Sec, determine the metacentric height and the relevant radius of gyration above the fore and aft axis through the centre of gravity.

Solution

$$W = \rho g V$$

$$100 \times 10^6 = 1025 \times 9.81 \times V$$

$$V = 9945.05 \text{ m}^3$$

$$BM = \frac{I}{V} = \frac{24300}{9945.05} = 2.44\, m$$

Metacentric height GM = BM – BG = 2.443 – 2 = 0.443 m Ans

$$T = 2\pi \sqrt{\frac{k^2}{GM \times g}}$$

$$10 = 2\pi \sqrt{\frac{k^2}{0.433 \times 9.81}}$$

$$k^2 = 11.0166 \; k = 3.319$$ Ans

4.12 (a) On moving a mass of 4000 kg a distance of 9 m across the deck of a vessel of 2000 $\times 10^3$ kg displacement the angle of list is changed from + 0.015 to – 0.015 radians. Find the metacentric height.

(b) Find the periodic time of rolling of the ship in (a) part assuming the radius of gyration of the ship about a longitudinal axis through its centre of gravity to be 3 m

Solution

$$GM = \frac{mx}{M \tan\theta} = \frac{400 \times 9}{2000 \times 10^3 \times 0.03} = 0.6 \text{ m}$$

$$T = 2\pi \sqrt{\frac{k^2}{GM\, g}}$$

$$= 2\pi \sqrt{\frac{9}{0.6 \times 9.81}} = 2\pi \sqrt{1.529}$$

$$= 2\,\pi \times 1.236 = 7.769 \text{ s.}$$ Ans

4.13 A rectangular sectioned pontoon weighs 2 MN and has a length of 15 m. Its required to have a free board of 600 mm when the pontoon is vertical, and its centre of gravity is 300 mm, above the centre of the cross section. The metacentric height is to be 1.2 m when the angle of heel is 10°. Determine the breadth and height of the pontoon when it is floating in fresh water.

Solution: Considering 10° as a large angle of heel, the movement of the centre of buoyancy normal and parallel to BM.

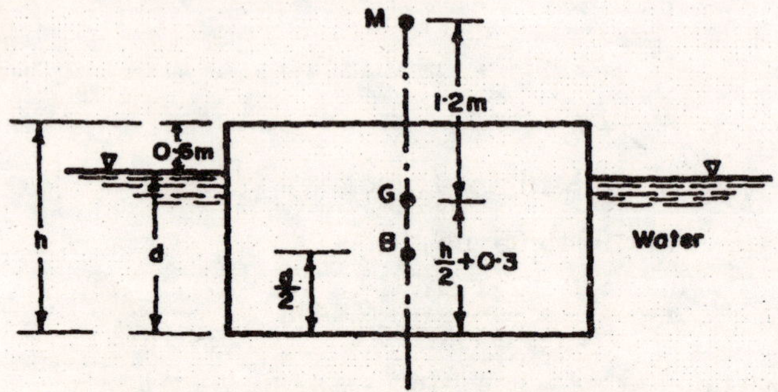

Fig. 4.10. Rectangular pontoon P. 4.13.

Here GM = 1.2 m

h = d + 0.6

$$BM = \frac{h}{2} + 0.3 + 1.2 - \frac{d}{2}$$

$$= \frac{h}{2} + 0.3 + 1.2 - \left(\frac{h - 0.6}{2}\right)$$

$$1.2 + 0.3 + 0.3 = 1.8 \text{ m} \quad (1)$$

$$BM = \frac{I}{V}\left[1 + \frac{\tan^2\theta}{2}\right]$$

$$= \frac{15\, b^3}{12 \times 15\, bd}\left[1 + \frac{\tan^2 10^\circ}{2}\right]$$

$$= \frac{b^2}{12d} \times 1.0156 = 0.0846\, \frac{b^2}{d} \quad (2)$$

From (1) and (2)

$$0.0846\, \frac{b^2}{d} = 1.8, \qquad b^2 = 21.268\, d \quad (3)$$

Since the pontoon is floating freely

$$W = F_B$$

$$2 \times 10^6 = 10^3 \times 9.81 \times bd \times 15$$

$$bd = \frac{2 \times 10^3}{9.81 \times 15}$$

$$\frac{b^2}{21.268} = 13.59$$

$$b^3 = 289.0655$$

$$b = 6.612 \text{ m}$$

$$\therefore d = 2.055 \text{ m.}$$

∴ Breadth of pontoon = 6.612 m

Height of pontoon = 2.055 + 0.6 = 2.65 m.

It should be noted that if the effect of the movement of the centre of buoyancy parallel to BM had been ignored.

$$BM = \frac{I}{V} = \frac{b^2}{12d}$$

$$\therefore \frac{b^2}{12\, d} = 1.8$$

$$b^2 = 21.6\ d$$

Since the pontoon is floating freely

$$W = F_B$$

$$2 \times 10^6 = 10^3 \times 9.81 \times bd \times 15$$

$$\therefore b^3 = 293.577$$

$$b = 6.64\ m$$

$$\therefore d = \frac{b^2}{21.6} = 2.045\ m$$

$\therefore$ Height of the pontoon = 2.045 + 0.6 = 2.645 m

Breath of the pontoon = 6.64 m

EXERCISES

A. THEORY

4.1 Define the terms buoyancy, centre of buoyancy and metacentre.

4.2 State the conditions for stable equilibrium of a body floating partially immersed in liquid.

4.3 Show that, if B is the centre of buoyancy and M is the metacentre for rolling of a partially immersed floating body, $BM = \frac{I}{V}$. Where I is second moment of inertia of the surface of floatation about the longitudinal axis, and V is the immersed volume.

4.4 What is the difference between the centre of gravity, centre of buoyancy, centroid and centre of pressure ?

4.5 Define metacentre, metacentric height and how they are measured.

4.6 Show from consideration of the force exerted on an immersed surface in a fluid that the resultant force on a body immersed in a fluid is vertical and equal to the weight of fluid displaced.

4.7 What are the three conditions in which a solid body can be in equilibrium.

4.8 Define the term "metacentre" and show how the stability of a floating body depends upon the position of the metacentre and the centre of gravity.

4.9 A floating body oscillates about its equilibrium position. Show that the periodic time of rolling is given by

$$T = 2\pi\sqrt{\frac{k^2}{GM\ g}}$$

4.10 A ship is inclined at a small angle from the vertical. show, by means of a sketch, the forces acting on the ship, and explain what is meant by the "righting moment".

4.11 Describe briefly the experimental method of determination of the metacentric height of a floating object.

4.12 Explain the three states of equilibrium of a floating body in terms of the magnitude and direction of righting couple acting on the body as a result of slight angular displacement.

B. UNSOLVED PROBLEMS

4.1 The barge is 2.75 m deep, 10 m wide and 30 m long and draws 2 m of water. If the centre of gravity of the barge is 1.5 m above the bottom at G, determine the righting moment when it has an angle of heel of 5 degrees. [1888.18 kNm]

4.2 A floating rectangular pontoon weighing 490.5 kN with centre of gravity at the water line rolls 10° when an overturning torque of 59.8 KNm is applied. How high may a 49050 N load be raised on a light scaffolding. [GM 0.7 m, 7.7 m above water line]

4.3 A barge has a uniform rectangular cross section of width 2L and vertical draft of height H. Determine (a) the metacentric height for a small tilt angle and (b) the range of ratio $\frac{L}{H}$ for which the barge is statically stable. Assume that the centre of mass G of the barge is exactly at the waterline, as shown in Fig. 4.11.

$$[GM = .\ \frac{L^2}{3H} - \frac{H}{2},\ L\ 1.225\ H]$$

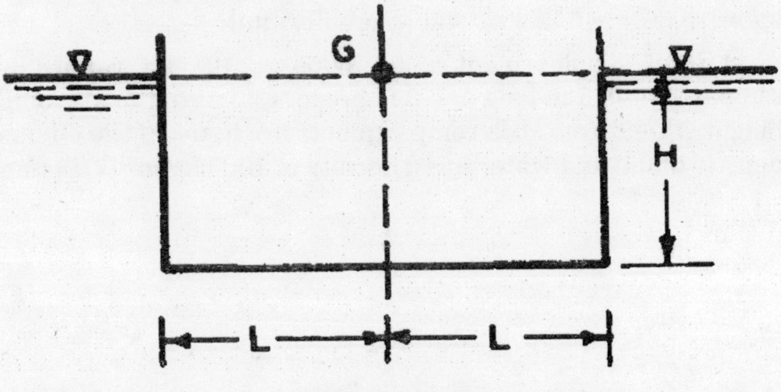

Fig. 4.11. P. 4.3.

4.4 A ship which has been loaded in an up river fresh water harbour proceeds downstream into salt water, consuming 40000 kg of fuel and stores on the way. Length of the ship 80 m. Displacement 7 x 10^6 kg as originally loaded. Distance of centre of gravity from bow 49 m as originally loaded. Longitudinal metacentric height 1000 m. Water plane area 1000 m^2 which may be assumed constant. If 0.97 m^3 of salt water and 1 m^3 of fresh water each have a mass of 1000 kg. (Density of salt water is 1030 kg/m^3 and water is 1000 kg/m^3) (a) Deduce the change of draught from the condition as originally loaded. (b) How much cargo can now be taken on board to bring the ship to its original draught ? (c) If the centre of gravity of the 40 x 1000 kg of fuel and stores consumed is 40 m from the bow of the ship, and if the additional cargo has to be placed at 60 m from the bow, what will be the change of trim (defined as the algebraic difference of changes of draught at bow and stern of ship)? [242 mm, 249981 kg, 0.2828 mm down by the stern]

4.5 A rectangular pontoon weighing 900 KN floats in sea water. It is 15 m long, 9 m wide and 4 m deep. Find the metacentric height. The centre of gravity of pontoon may be taken as its geometrical centre. Take density of sea water as 1030 kg/m^3 [8.56 m]

4.6 A floating dock may be considered essential of two sides pontoons 110 m long, 3.5 m wide and 12 m deep, attached to a bottom pontoon 150 m long, 25m wide and 4 m deep. The bottom of the side pontoon is 1 m above that of the bottom pontoon.

Discuss its stability, when a ship is being lifted, and state at what stage the righting moment per unit angle of heel has its minimum value.

Determine the height of the metacentric above the centre of buoyancy of the dock for this condition, excluding the free surface effects of the flooded compartments.

4.7 A barge is 6 m long, 4 m wide and 2 m high its weight 150 KN along with its cargo. The centre of barge with cargo is at 1.5 m above the bottom of the barge. If the barge carrying cargo is to be floated in sea water, determine the metacentric height of the floating barge. Does the barge float in stable equilibrium? The density of sea water is 1030 kg/m^3. [GM = + 0.964 m, stable equilibrium].

4.8. A ship of displacement mass 5000 x 10^3 kg carries oil fuel in compartments 6 m long and 3 m broad. Estimate the loss of metacentric height when three such compartments are in use if the oil has a relative density of 0.9 and water has a density of 1000 kg/m^3 [7.29 mm].

5

Kinematics of Fluid

5.1 INTRODUCTION

Kinematics is that part of mechanics which deals with quantities derivable from units of space and time. Although forces are the causes responsible for accelerated or decelerated motion, Kinematics is solely concerned with the effects of the motion on the displacement, time, velocity, acceleration or any other quantity derivable from displacement and time. Before undertaking the study of forces responsible for the motion of fluids, it is desirable to acquire some understanding of the types of motion from the kinematics point of view.

The subject has two main aspects

(a) The development of methods and techniques for describing and specifying the motions of fluids.

(b) The determination of the condition for the kinematics possibility of fluid motions i.e. the exploration of the consequences of continuity in the motion.

5.2 DESCRIPTION OF FLUID MOTION

There are two important and commonly used methods of specifying fluid motion. The first, known technically as the Lagrangean method, may be described as historical since we follow the history of each individual fluid particle i.e., at each instant of time the location, density state of stress, etc are specified for a certain fluid particle of fixed identity. This method is usually used in the dynamics of discrete particles.

The second method, know technically as the Eulerian, focuses attention on a fixed point of space, and specifies at each instant of time the density., pressure etc. of the fluid particle which happens to occupy that point at that instant. This is also described as the cinematographic method.

In fluid mechanics it is usually easier to use the Eulerian method to describe a flow-either experimental or analytical investigations. There are, however, certain instances in which the Lagrangean method is more convenient. In fluid machinery (such as turbines and pumps) Lagrangean method is useful in which the fluid particles gain or lose energy as they move along their flow paths.

5.3 FLUID FLOW CLASSIFICATIONS

There are many criteria according to which fluid flow may be classified.

(a) Steady flow and unsteady flow

(b) Uniform flow and Non uniform flow

(c) One-two-there dimensional flow

(d) Laminar flow and turbulent flow

(e) Rotational flow and irrotational flow

5.3.1 Steady flow and unsteady flow

Many engineering problems refer to a steady condition of operation.

In general the parameters such as velocity, pressure and density, which describe the behaviour and state of a fluid, are not constant in a particular set of circumstances. They may vary from one point to another, or from one instant of time to another, or they may vary with both position and time.

Steady flow is defined as that in which the various parameters at any point do not change or vary with time. Mathematically, this can be stated as

$$\frac{\partial}{\partial t}(\;) = 0 \tag{5.1}$$

Using the density for illustration, we may then write

$$\frac{\partial \rho}{\partial t} = 0$$

Flow in which change with time do occur is termed unsteady or nonsteady flow. Mathematically, this can be stated as

$$\frac{\partial}{\partial t}(\;) \neq 0 \tag{5.2}$$

Using the density for illustration, we may then write

$$\frac{\partial \rho}{\partial t} \neq 0.$$

Not surprisingly, steady flow is simple to analyse than unsteady flow and we shall confine our attention mainly to steady conditions.

5.3.2 Uniform flow and non uniform flow

Flow is described as uniform if the velocity at a given instant is the same in magnitude and direction at every point in the fluid. If, at the given instant, the velocity changes from point to point, the flow is described as non uniform. Mathematically,

$$\frac{\partial V}{\partial S} = \frac{\partial(\text{Velocity vector})}{\partial(\text{Space coordinates})} = 0 \tag{5.3}$$

$$\frac{\partial u}{\partial x} = \frac{\partial v}{\partial y} = \frac{\partial w}{\partial z} = 0$$

and
$$\frac{\partial V}{\partial S} = \frac{\partial(\text{Velocity vector})}{\partial(\text{Space coordinates})} \neq 0 \tag{5.4}$$

$$\frac{\partial u}{\partial x} = \frac{\partial v}{\partial y} = \frac{\partial w}{\partial z} \neq 0.$$

5.3.3 One, two and three dimensional flow

A one dimensional flow has spatial variations in one direction only. Such a flow is also said to be uniform at every cross section normal to the main direction of flow. Only one independent space variable is needed to describe the variation.

We usually designate x to be that variable. Thus V = f(x) is one dimensional. The flow of real fluids cannot be completely one dimensional since the velocity at a boundary must be zero with respect to the boundary. They may, however, be assumed to be one dimensional in many instances. Whenever the flow is one dimensional actual or assumed-a one dimensional analysis may be applied to the system, and the average value of the velocity is used to describe the flow at a section.

Similarly, a two dimensional flow is one in which spatial variations exist in two directions, or variations exist along some Planar surface. Two Cartesian independent space variables are needed to describe the variation.

Thus V = f(x,y) is two dimensional.

Three dimensional flow is of flow in which the fluid or flow parameters vary in the x, y and z direction in a Cartesian Coordinates system. Thus, gradient of the fluid or flow parameters exist in all three directions. Thus v = f (x,y,z) is three dimensional.

Most real flow are three dimensional. In practice it is nearly always possible to consider the flow to be one - or two dimensional. This greatly simplifies the equation of flow. Minor adjustments to these simplified equation can often be incorporated to allow for a two – or three dimensional flow.

Mathematically,

Steady	Unsteady	Types of flow
v = f (x)	v = f (x, t)	One dimensional
v = f (x, y)	v = f (x, y, t)	Two dimensional
v = f (x, y, z)	v = f (x, y, z, t)	Three dimensional

5.3.4 Laminar flow and turbulent flow

When a given flow is at a relatively low velocity or with a highly viscous fluid, any disturbances that might be introduced into the flow field due to vibrations, irregularities at a boundary surface, etc, are relatively quickly damped out by the fluid viscosity and the flow is referred to as laminar. Fluid may be regarded as moving in layers, or lamina, with the only exchange across adjacent layers due to molecular motion. As the flow velocity is increased or the viscosity decreased, the disturbance would continue to be observed for greater down-stream distances and eventually a transition state will be reached above which at least some of the disturbances will amplify. This transition state depends upon the fluid viscosity, the velocity and other details of the flow geometry such as the local pressure gradient.

Turbulent flow is therefore characterized by local irregularities in the fields of mechanical properties such as velocity, pressure, or temperature. The fluid particles move in all three directions most often in an irregular fashion.

Turbulent flow is most prevalent in nature. Laminar flow may often only be developed under laboratory conditions and usually occurs when velocities are extremely small.

The type of flow is identified by Reynold number. Reynolds number represents the ratio of inertia forces to the viscous forces that exist in the flow field and is dimensionless.

The flow through a pipe is always laminar if the corresponding Reynolds number $\left(\text{Re} = \dfrac{\rho v D}{\mu}\right)$ is less than 2000 and for all practical purposes the flow may be assumed to pass through a transition to full turbulent flow in the range of Reynold number from 2000 to 4000. The flow can be assumed to be turbulent if the Reynold number exceeds 4000.

5.3.5 Rotational flow and Irrotational flow

A flow is said to be irrotational if every fluid element does not rotate about its own axis which is perpendicular to the plane of rotation. The condition of irrotationality is satisfied if the average of the angular velocities of two mutually perpendicular linear element of every fluid element in that plane is zero (Fig 5.1).

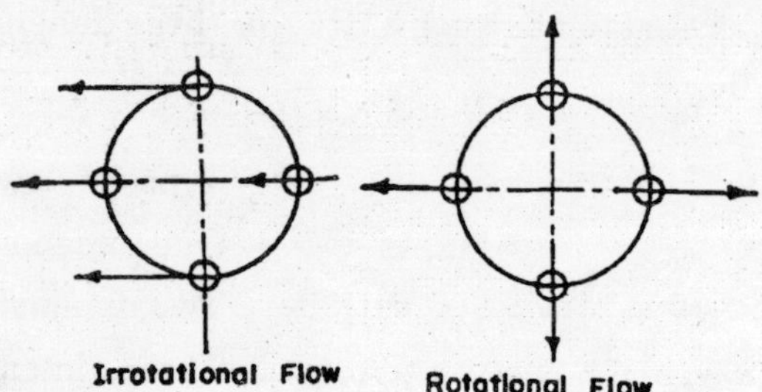

Fig. 5.1. Irrotational and Rotational flow.

A fluid is said to be rotational if every fluid element rotates about its own axis as they move throughout the space (Fig 5.1).

A well known example of irrotational motion is that of the carriages of the Big (Ferris) Wheel in a fair ground: although each carriage follows a circular path as the wheel revolves, it does not rotate with respect to the earth - the passenger fortunately remain upright and continue to face in the same direction.

5.4 FLOW LINES

Fluid mechanics is a highly visual subject. The patterns of flow can be visualized in a different ways and we can view these sketches or photographs and learn a great deal qualitatively and often quantitatively about the flow.

Four basic types of line patterns are used to visualize flows. There are streamlines, path lines, streaklines and timelines.

5.4.1 Streamline

A streamline is an imaginary line in the fluid to which the velocity vector, at every point through which it passes, is a tangent, at the instant considered. Hence there can be no flow across a streamline since there will be no component of velocity normal to the line.

The concept of the streamline is very useful, specially in ideal flow, because it enables the fluid flow to be conceived as occurring in patterns of streamline. These patterns may be described mathematically so that the whole system of analysis may be based on it. It requires, however, a mathematical definition of a streamline (Fig 5.2).

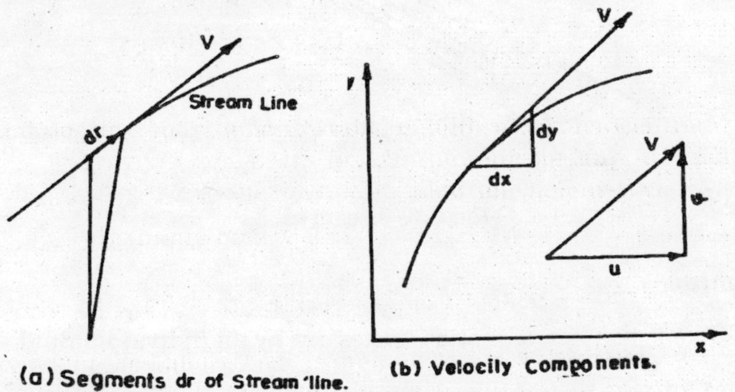

Fig. 5.2. Streamline and velocity components.

In vector notation, since the velocity must be tangent or parallel to the line, and since the cross product of parallel vectors is zero, the equation for the streamline is

$$V \times dr = 0 \tag{5.5}$$

Expansion of this cross product shows that

$$\frac{u}{dx} = \frac{v}{dy} = \frac{w}{dz} \tag{5.6a}$$

or, $$\frac{dx}{u} = \frac{dy}{v} = \frac{dz}{w} \tag{5.6 b}$$

is the equation of streamline, along with v x dr = 0

From Fig 5.6(b) in the x-y plane,

$$\frac{dy}{dx} = \frac{v}{u}$$

and thus $$\frac{dz}{dy} = \frac{w}{v} \tag{5.6c}$$

and $$\frac{dx}{dz} = \frac{u}{w}$$

5.4.2 Stream tube

The collection of streamline may form an imaginary flow passage known as a stream tube (Fig 5.3) Since streamlines always have the direction of velocity, stream tubes will be considered as solid tubes through which the fluid passes.

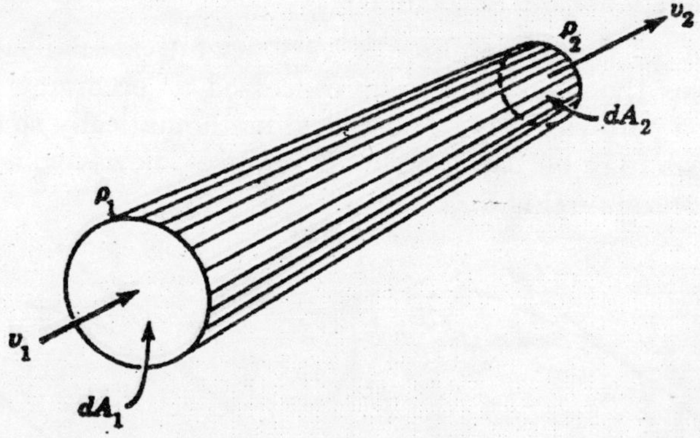

Fig. 5.3. Stream tube

5.4.3 Path line

A pathline is the line of motion traces out by an individual fluid particle in a finite time.

The Pathline is defined by integration of the relation between velocity and displacement in equation

dx = udt, dy = vdt, dz = wdt.

In other form

$$\frac{dx}{dt} = u\,(x, y, z, t)$$

$$\frac{dy}{dt} = v\left(x,\ y,\ z,\ t\right) \tag{5.7}$$

$$\frac{dz}{dt} = w\left(x,\ y,\ z,\ t\right)$$

with initial conditions x (t_o) = x_o, y (t_o) = y_o and z (t_o) = z_o. Integrating with respect to time using the condition (x_o, y_o, z_o, t_o). Then eliminate time to give the pathline function f(x,y,z,t). The integration may be laborius.

For steady flow, the particle velocity at any point in the flow field will not depend upon time, and all particles that pass through this point will follow the same trajectories. Thus pathline are identical to streamlines for steady flow.

5.4.4 Streak line

A streakline is the locus of particles which have earlier passed through a prescribed point. In other words a streakline is an instantaneous locus of all fluid particles that have passed through a given point.

A streakline is a line joining the instantaneous positions of succession of particles which have issued from one source or passed through one point.

In experimental work, streaklines are obtained by injecting dye, smoke or the subsequent flow pattern.

To compute the streakline, take the integrated result of equation (5.3) and retain time as a parameter. Find the integration constant which cause the path line to pass through (x_o, y_o, z_o) for a sequence of times $\zeta < t$. Then eliminate ζ from the result to obtain the streaklines.

A streamline is convenient to calculate mathematically, while the other two are easier to generate experimentally. It is to be noted that a streamline is instant aneous lines, while the path line and streakline are generated by the passage of time.

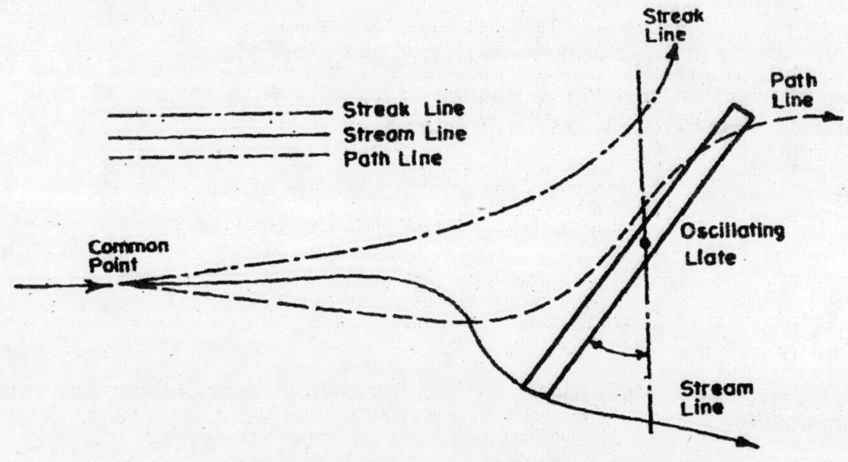

Fig. 5.4. Unsteady past flow an oscillating plate.

A path line can be found by a time exposure of a single marked particle moving through the flow. Streamline are difficult to generate experimentally in unsteady flow unless one marks a great many particles and notes their direction of motion during a very short time interval. In steady flow the situation simplifies greatly:

Streamline, path lines, and streakline are identical in steady flow.

The different flow lines for unsteady flow are explained in Fig. 5.3. The flow in Fig. 5.4 is unsteady and periodic due to the oscillating of plate against the oncoming stream. We see that the dash – dot streakline does not coincide with either the streamline or pathline passing through the same release point. This is due to characteristic of unsteady flow.

5.5 VELOCITY AND ACCELERATION

Velocity and accelerating are vector quantities, having both magnitude and direction. However, often the direction is know or assumed and only the magnitude is to be determined. Then, the scalar components in particular directions are used. In this book most equation and analyses are in terms of scalar components, although vectors are used where they make physical principles more understandable.

In general flow, the velocity of a fluid particle is a function of both the position of the point and time, i.e.

$$V = f(x, y, z, t)$$

The components of velocity in the three direction are

$$u = \frac{dx}{dt}$$

$$v = \frac{dy}{dt} \tag{5.8}$$

$$w = \frac{dz}{dt}$$

where u, v and w are its components in x, y, and z directions.

Acceleration is the rate of change of velocity with respect to time. The components of acceleration in tree directions are

$$a_x = \frac{du}{dt}$$

$$a_y = \frac{dv}{dt} \tag{5.9}$$

$$a_z = \frac{dw}{dt}$$

Applying these operations to the preceding expressions for velocity components give

$$a_x = \frac{\partial u}{\partial x}\frac{dx}{dt} + \frac{\partial u}{\partial y}\frac{dy}{dt} + \frac{\partial u}{\partial z}\frac{dz}{dt} + \frac{\partial u}{\partial t}$$

$$= \frac{\partial u}{\partial x} u + \frac{\partial u}{\partial y} v + \frac{\partial u}{\partial z} w + \frac{\partial u}{\partial t}$$

Rearranging and including the other components obtained similarly

$$a_x = u \frac{\partial u}{\partial x} + v \frac{\partial u}{\partial y} + w \frac{\partial u}{\partial z} + \frac{\partial u}{\partial t}$$

$$a_y = u \frac{\partial v}{\partial x} + v \frac{\partial v}{\partial y} + w \frac{\partial v}{\partial z} + \frac{\partial v}{\partial t} \quad (5.10)$$

$$a_z = u \frac{\partial w}{\partial x} + v \frac{\partial w}{\partial y} + w \frac{\partial w}{\partial z} + \frac{\partial w}{\partial t}$$

where a_x, a_y and a_z are the acceleration components in the x, y and z direction respectively.

It is usual to write the differential operator appearing in these equations as

$$\frac{D}{Dt} = u \frac{\partial}{\partial x} + v \frac{\partial}{\partial y} + w \frac{\partial}{\partial z} + \frac{\partial}{\partial t} \quad (5.11)$$

and then we have the set of equations

$$a_x = \frac{Du}{Dt}, a_y = \frac{Dv}{Dt}, a_z = \frac{D_w}{Dt}$$

The equation (5.11) is termed the material derivative or substantial derivative.

The material derivative concept is very useful in analysis involving various fluid parameters, not just the acceleration.

5.5.1 Convective acceleration and Local acceleration

The equation (5.10) results from two things. One the fluid particles may be accelerated by the convective act of moving from one position to another where the velocity is different and, two, the particles may be accelerated locally because the flow is changing with time at each point. The convective acceleration terms are

$$u \frac{\partial u}{\partial x} + v \frac{\partial u}{\partial y} + w \frac{\partial u}{\partial z}$$

$$u \frac{\partial v}{\partial x} + v \frac{\partial v}{\partial y} + w \frac{\partial v}{\partial z} \quad (5.12)$$

$$u \frac{\partial w}{\partial x} + v \frac{\partial w}{\partial y} + w \frac{\partial w}{\partial z}$$

and the local acceleration terms are

$$\frac{\partial u}{\partial t}, \frac{\partial v}{\partial t}, \frac{\partial w}{\partial t}$$

In general we can write the equation (5.10) as

Substantial Acceleration	=	Convective Acceleration	+	Local Acceleration

5.5.2 Tangential acceleration and normal acceleration

In a steady flow the velocity vector of all fluid particles passing any fixed point on a streamline remains invariant at all times. However, the velocity vector of any one fluid particle may change at successive point along a streamline because of either the convergence or divergence of the stream tube. The acceleration of fluid particles in a steady flow is, therefore, expressed in terms of the change of velocity with respect to the displacement.

Fig 5.5 shows a particle of fluid moving a distance ds along a curved path from A to B during an infinitesimal time interval dt in a field of steady flow. The increase in the magnitude of velocity u at A to u + du at B – is due to the convergence of the stream tube. For small dθ, sin dθ is approximately dθ and cosθ is approximately 1. Therefore, ∂_{u_t} is approximately du and du_n is approximately udθ (neglecting the higher order of differentials dudθ). To recall the kinematic relationship, $\frac{ds}{dt} = u$ the tangential component of acceleration becomes,

$$a_t = \frac{\partial u_t}{\partial t} \times \frac{ds}{ds} = u\frac{\partial u}{\partial s} = \frac{\partial}{\partial s}\left(\frac{u^2}{2}\right) \tag{5.13}$$

Which is due to the change in the magnitude of velocity along the streamline.

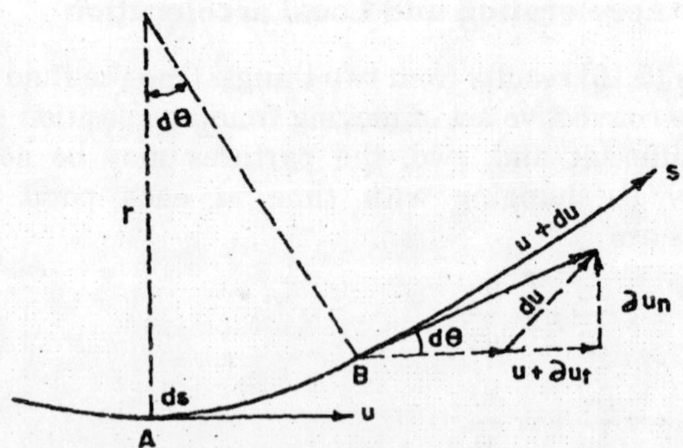

Fig. 5.5. Motion of a fluid particle along a curvilinear paths

The normal component of acceleration a_n is

$$a_n = \frac{\partial u_n}{\partial t} = \frac{u\,d\theta}{\partial t} = \frac{u\,(ds/r)}{\partial t} = \frac{u^2}{r} \tag{5.14}$$

Which is due to the change in the direction of velocity vector.

When the flow is unsteady, in addition to the change of velocity vector of the particle along its path, velocity vector also change from time to time at any fixed point. Thus the two components of acceleration at any point in an unsteady two dimensional flow field can be written as

$$a_t = \frac{\partial}{\partial s}\frac{u^2}{2} + \frac{\partial u}{\partial t} \tag{5.15}$$

and $$a_n = \frac{u^2}{2} + \frac{\partial u_n}{\partial t} \tag{5.16}$$

Where the second term $\frac{\partial u_n}{\partial t}$ represents the change of the velocity vector of the fluid particles with respect to time at any fixed point in the flow field.

Examples of streamline pattern and their corresponding types of acceleration in steady flows (Fig 5.6).

5.6 PRINCIPLE OF CONTINUITY-CONSERVATION OF MASS FLOW

The continuity equation is actually a mathematical statement describing the conservation of flow rates of fluid mass in fixed region. In Fig. 5.7 we bring under observation any fixed region within the fluid and since, in the absence of nuclear reaction, matter is neither created nor destroyed within that region.

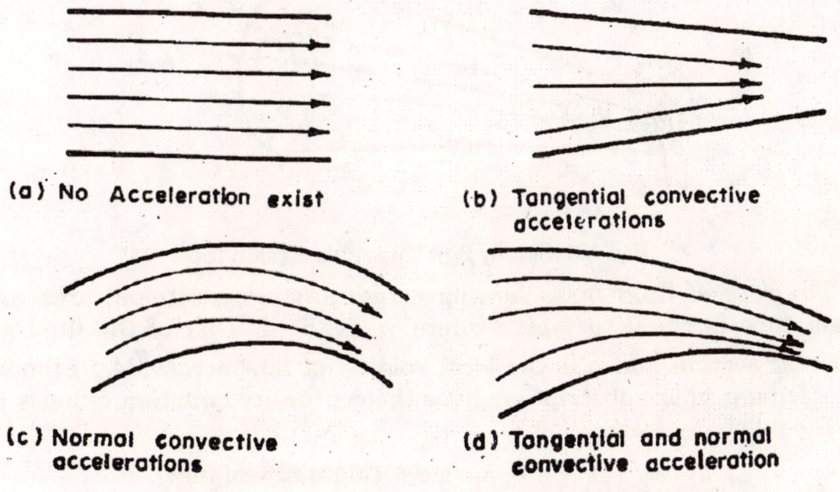

Fig. 5.6. Streamline patterns and types of acceleration

Fig. 5.7. Conservation of Mass flow

Rate at which mass enters the region = Rate at which mass leaves the region + Rate at which fluid mass stored within the region.

When the flow is steady, the rate at which the fluid mass is stored within the region is zero. The continuity equation for steady flow then becomes

Rate at which mass enters the region = Rate at which mass leaves the region.

Which, using mathematical symbols, may be written as

$$\frac{dm_{entering}}{dt} = \frac{dm_{leaving}}{dt} \tag{5.17}$$

A more general statement written as

$$\left(\frac{dm}{dt}\right)_{\text{any section}} = \text{Constant} \tag{5.18}$$

5.6.1 One dimensional continuity equation

In Fig. 5.8 there is shown a stream tube of infinitesimal crosssection. During a differential amount of time dt, fluid mass flowing into an arbitrary chosen region through the left section 1 will either stay within the region or flow out of the right section 2.

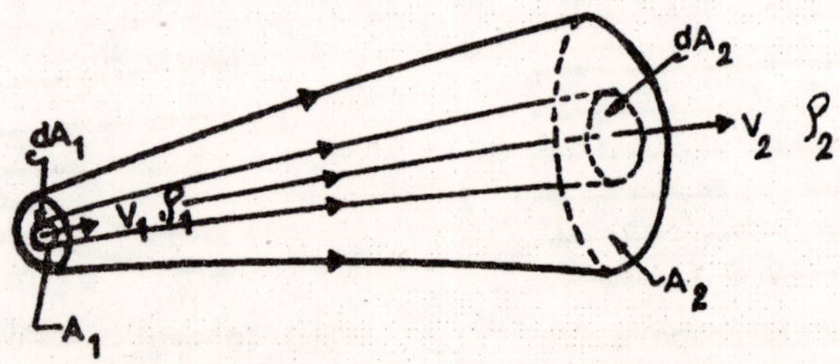

Fig. 5.8. Steady flow through a stream tube.

The rate of fluid mass flowing across any crosssectional area dA in a stream tube is equal to $\rho u dA$, where ρ is the density of the fluid at that particular section and u is the local velocity of flow across dA in the stream tube. Using proper subscription gives the continuity equation of mass rate of flow the following from.

$$\rho\, u_1\, dA_1 = \rho\, u_2\, dA_2 = \rho\, u\, d\, A \text{ (Mass rate of flow)}$$

If the fluid is incompressible, that is the continuity equation may be further simplified to $\rho_1 = \rho_2 = \rho$ = Constant,

$$u_1\, dA_1 = u_2\, dA_2 = u\, dA = Q \text{ (Volume rate of flow)}$$

dQ is the differential volume rate of flow at any section in the stream tube.

The total volume rate of flow Q across any section of a channel or a conduit is determined by integrating the differential volume rate of flow in all stream tubes

$$Q = \int_A u\,dA = u\,A$$

where $u = \dfrac{Q}{A}$, is the average velocity of flow across the section of the channel

When the continuity equation for the volume rate of steady flow of incompressible fluid is written for the entire section of flow channel

$$Q = u_1\,A_1 = u_2\,A_2 = u\,A$$

Which is one of the most important relationships in fluid mechanics

The equation is usually referred to as the one dimensional equation of continuity. In most engineering problems of fluid flow in pipes and open channel, the one dimensional approach is used if the velocity distribution across the flow section is very nearly uniform.

5.6.2 Continuity equation for three dimensional flow using Cartesian Coordinates

The general form of the continuity equation by using the basic concepts can be written in the form as

$$M_x + M_y + M_z = M_{x+dx} + M_{y+dy} + M_{z+dz} + M_{stored} \qquad (5.19)$$

The control volume ABCDEFGH in Fig. 5.9 is taken in the form of a small rectangular prism with sides dx, dy and dz in the x, y, and z directions respectively. The average values of the component velocities in these directions are u, v, and w.

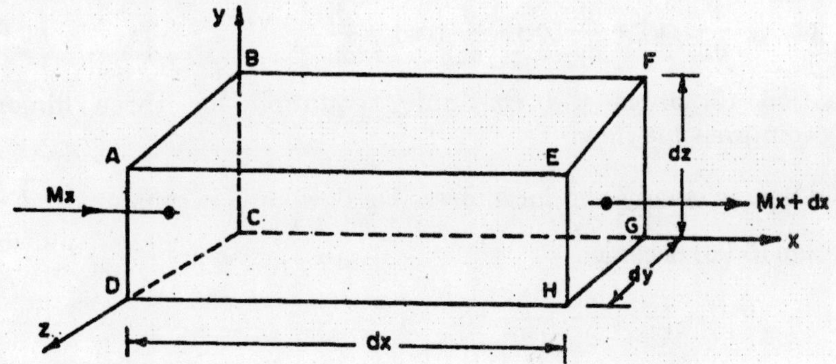

Fig. 5.9. Continuity equation for three dimension.

The mass rate of flow in the x direction at the left face ABCD is given by

$$M_x = \rho\,u\,dydz \qquad (5.20)$$

The mass rate of flow at the right face EFGH is given by

$$M_{x+dx} = \left(\rho + \frac{\partial p}{\partial x}dx\right)\left(u + \frac{\partial u}{\partial x}dx\right) dy\,dz$$

$$M_{x+dx} = \rho u\, dy\, dz + \rho \frac{\partial u}{\partial x} dx\, dy\, dz + \frac{\partial \rho}{\partial x}\frac{\partial u}{\partial x}(dx)^2\, dy\, dz$$

$$= \rho u\, dy\, dz + \left(\rho \frac{\partial u}{\partial x} + u\frac{\partial p}{\partial x}\right) dV + \textit{small term neglected}$$

$$= \rho u\, dy dz + \frac{\partial}{\partial x}(\rho u) dV \tag{5.21}$$

The net inflow of mass in the x direction per unit time is the difference between the two equations (5.20) and (5.21), which is

$$M_x - M_{x+dx} = -\frac{\partial}{\partial x}(\rho u)\, dV \tag{5.22}$$

Similarly, for the y and z direction

$$M_y - M_{y+dy} = -\frac{\partial}{\partial y}(\rho v)\, dV \tag{5.23}$$

$$M_z - M_{z+dz} = -\frac{\partial}{\partial z}(\rho w)\, dV \tag{5.24}$$

The rate of increase of mass in the region is given by

$$M_{stored} = \frac{\partial}{\partial t}(\rho dV) \tag{5.25}$$

Putting in equation (5.23)

$$-\frac{\partial}{\partial x}(\rho u) dV - \frac{\partial}{\partial y}(\rho v)\, dV - \frac{\partial}{\partial z}(\rho w)\, dV = \frac{\partial}{\partial t}(\rho)\, dV$$

or, $$\frac{\partial}{\partial x}(\rho u) + \frac{\partial}{\partial y}(\rho v) + \frac{\partial}{\partial z}(\rho w) = \frac{\partial}{\partial t} \tag{5.26}$$

Equation (5.26) is the continuity equation for three dimensional unsteady compressible flow.

If the flow is steady and incompressible flow i.e. ρ is constant and $\frac{\partial p}{\partial t} = 0$, and equation (5.26) becomes

$$\frac{\partial u}{\partial x} + \frac{\partial v}{\partial y} + \frac{\partial w}{\partial z} = 0 \tag{5.27}$$

For two dimensional flow, there will be no components of velocity in z direction so that equation (5.27) becomes

$$\frac{\partial u}{\partial x} + \frac{\partial v}{\partial y} = 0 \tag{5.28}$$

5.7 DEFORMATION OF A FLUID ELEMENT

In general at a fixed instant the velocity vector which describes a fluid motion will vary throughout space. If we examine the motion of the six faces of an elemental cube of fluid having sides dx, dy and dz we will find that four types of deformation or movement are possible as a consequence of spatial variation in velocity:

(a) translation, (b) linear deformation,

(c) angular deformation, (d) rotation.

These may, of course, occur singly or in combination.

Fig. 5.13 shows velocity components at each corner of the three faces of the elemental fluid cube which lies in the x-y, y-z and x-z planes respectively. During a time dt the faces of the cube will move through space, and the cube will change its form. The four types of movement of the faces in the x-y plane are shown in Fig. 5.14.

(a) In Fig. 5.14 (a) the translation in the x and y direction are udt and vdt respectively there would, of course, be a translation in the z direction also.

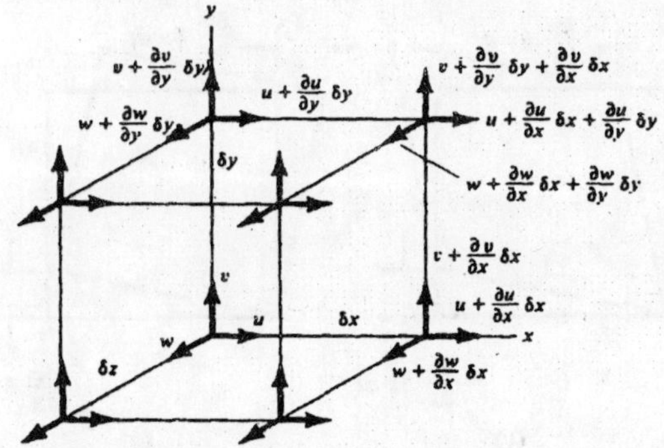

Fig. 5.13. Instantaneous velocity at some corners of an elemental cube of a fluid.

(c) Angular deformation

(b) In Fig. 5.14 (b) the linear deformation in the x and y direction are $\left(\frac{\partial u}{\partial x}\right)$ dx dt and $\left(\frac{\partial v}{d_y}\right)$ $dy\ dt$, respectively, with a corresponding linear deformation of $\left(\frac{\partial w}{\partial z}\right)$ dz dt in the z direction.

(c) In Fig. 5.14(C) the net angular deformation is the difference between the angular deformation of the individual sides dx and dy, considering positive angles as counter clockwise.

Net angular deformation

$$= \left(\frac{\frac{\partial v}{\partial x} dx\ dt}{dx}\right) - \left(\frac{-\frac{\partial u}{\partial y} dy\ dt}{dy}\right)$$

$$= \left(\frac{\partial v}{\partial x} + \frac{\partial u}{\partial y}\right) dt$$

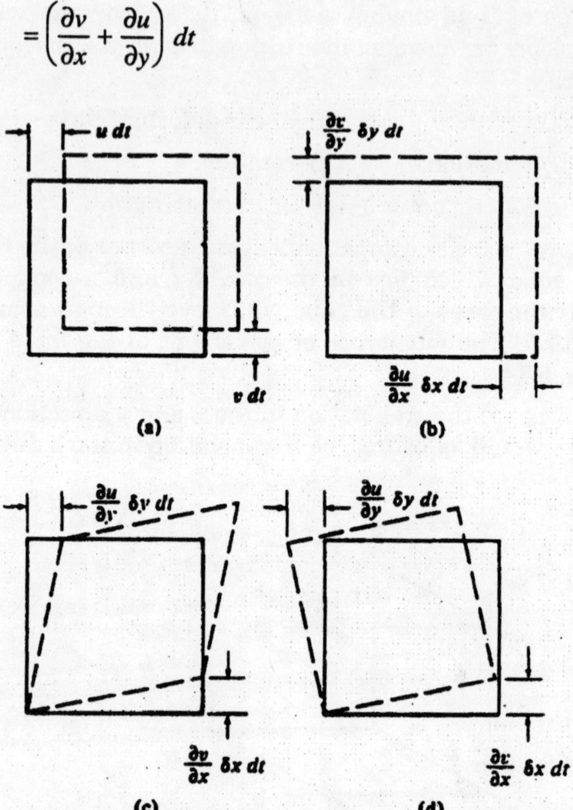

Fig. 5.14. Movement of the face of an elemental cube of sides dx, dy and dz in the x-y plane in a time dt.

The rate of distortion is defined as one half the rate of angular deformation:

$$\epsilon_{xy} = \epsilon_{yx} = \frac{1}{2}\left(\frac{\partial u}{\partial y} + \frac{\partial v}{\partial x}\right) \tag{5.50a}$$

Similarly in y-z and x-z plane

$$\epsilon_{yz} = \epsilon_{zy} = \frac{1}{2}\left(\frac{\partial v}{\partial z} + \frac{\partial w}{\partial y}\right) \tag{5.50 b}$$

and $$\epsilon_{zx} = \epsilon_{xz} = \frac{1}{2}\left(\frac{\partial w}{\partial x} + \frac{\partial u}{\partial z}\right) \tag{5.50 c}$$

Also,

$$\epsilon_{xx} = \frac{\partial u}{\partial x},\ \epsilon_{yy} = \frac{\partial v}{\partial y}\ \ and\ \ \epsilon_{zz} = \frac{\partial w}{dz} \tag{5.50 d}$$

(d) Rotation of the elemental face in Fig. 5.14(d) is the average of the rotation of the sides dx and dy, clockwise again considered negative, and is

$$\frac{1}{2}\left(\frac{\partial v}{\partial x}-\frac{\partial u}{\partial y}\right) dt$$

The rate of rotation is the angular velocity about an axis parallel to the z axis,

$$\text{or,}\quad w_z=\frac{1}{2}\left(\frac{\partial v}{\partial z}-\frac{\partial u}{\partial y}\right) \qquad (5.51a)$$

Similarly

$$w_y=\frac{1}{2}\left(\frac{\partial u}{\partial z}-\frac{\partial w}{\partial x}\right) \text{ in the x-z plane} \qquad (5.51b)$$

and

$$w_x=\frac{1}{2}\left(\frac{\partial w}{\partial y}-\frac{\partial v}{\partial z}\right) \text{ in the y-z plane} \qquad (5.51c)$$

In vector notation the rotation of a fluid at a point is expressed as

$$W=\frac{1}{2}\,Curl\ V \qquad (5.52)$$

$$=\frac{1}{2}\begin{bmatrix} i & J & k \\ \frac{\partial}{\partial x} & \frac{\partial}{\partial y} & \frac{\partial}{\partial z} \\ u & v & w \end{bmatrix}$$

$$=w_x\,i+w_y\,j+w_z\,k$$

The condition for the flow to be irrotational, the angular velocity in equations (5.51) and (5.52) are zero.

The distinction between the rotational and irrotational flow is important because, for example, it will be shown later that the generation of lift by such surface as aerofoils is associated with irrotational flow. Also a useful and practical procedure of determining flow nets can only be applied to irrotational flow. The Bernoullis equation derived for a streamline applies to all streamline in the flow field only if the flow is irrotational.

5.8 CIRCULATION AND VORTICITY

Circulation is defined as the line integral of the tangential velocity component of velocity around a closed curve. It is positive in the anticlockwise direction. Mathematically, the line integral of the velocity is the sum of the products of the element dS of the curve and the corresponding component of velocity tangent to the curve. Referring to the close curve of Fig. 5.15, and using the symbol Γ (Greek capital gamma) to denote circulation, we may write.

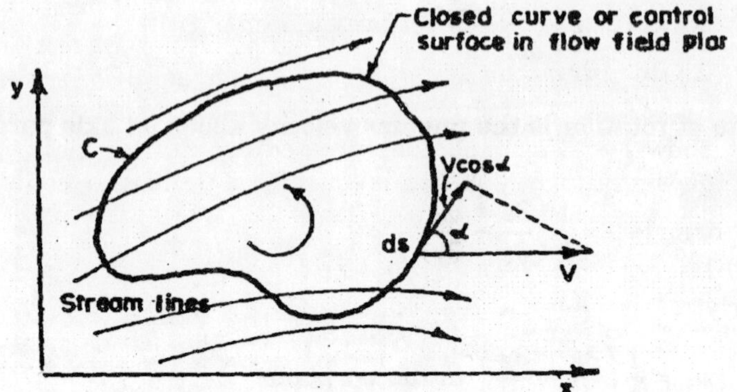

Fig. 5.15. Circulation for the rectangular element

$$\Gamma \equiv \oint_c V \cos \propto ds \tag{5.53}$$

In vector notation this may be written in terms of the scalar product as

$$\Gamma \equiv \oint_c V\, dr \tag{5.54}$$

Where v denotes the vector velocity and r the vector radius from any fixed region. If u, v and w denote respectively the vector components in the x–, y– and z– direction, the cartesian form of above equations (5.53) and (5.54) becomes

$$\Gamma = \oint_c (u\, dx + v\, dy + w\, dz) \tag{5.55}$$

Although the calculation of circulation around an arbitrary curve in a flow field is generally a tedious step by step integration, the principle can be applied easily and fruitfully to specific closed curves such as squares and circles. Calculation of the circulation around the basic differentially sized square element of Fig. 5.16 yields a concept having great general significance because it yields the point value of the circulation in a flow. This value is needed, for examples in calculations of pressure-velocity-head relations (expressed through Bernoullis equation) in two dimensional ideal incompressible flows.

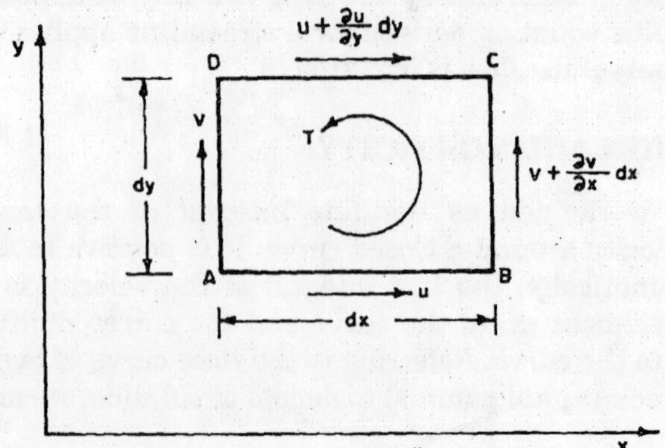

Fig. 5.16. Circulation for the rectangular element.

5.8.1 Circulation for the rectangular element

Let us now consider the circulation Γ around a small square element in the x-y plane, as shown in Fig. 5.16. Employing equation (5.55) with the velocity components shown in Fig. 5.16 and proceeding counter clockwise from the origin, remembering that circulation is considered positive in counter clock wise direction.

$$\text{Circulation} \quad \Gamma = \tau_{AB} + \tau_{BC} + \tau_{CD} + \tau_{DA}$$

$$\Gamma = u\, dx + \left(v + \frac{\partial v}{\partial x} dx\right) dy - \left(u + \frac{\partial u}{\partial y} dy\right) dx - v\, dy$$

$$= u\, dx + v\, dy + \frac{\partial v}{\partial x} dx\, dy - u\, dx - \frac{\partial u}{\partial y} dx dy - v\, dy$$

$$= \frac{\partial v}{\partial x} dx\, dy - \frac{\partial u}{\partial y} dx dy$$

$$= \left(\frac{\partial v}{\partial x} - \frac{\partial u}{\partial y}\right) dx\, dy \tag{5.56}$$

The circulation per unit area in the x-y plane is therefore given by

$$\frac{\Gamma}{Area} = \frac{\partial v}{\partial x} - \frac{\partial u}{\partial y} \tag{5.57}$$

5.8.2 Circulation for the circle

Considering a small circular circuit of radius r (Fig. 5.17).

$$\Gamma = \oint V_s\, ds = \oint w r . r d\,\theta = r^2 \oint w\, d\,\theta$$

$$r^2 = \overline{w}\, 2\,\pi \tag{5.58}$$

where $\overline{w}$ is the average value of the angular velocity w about the center for all particles on the circle.

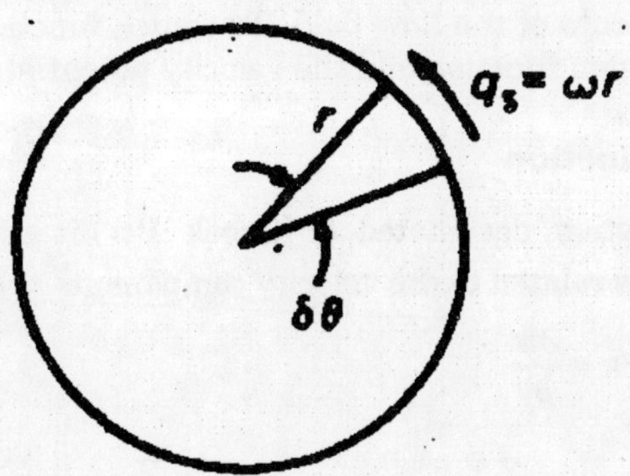

Fig. 5.17. Circulation for circular element.

The circulation per unit area given by

$$\frac{\Gamma}{Area} = \frac{r^2 \overline{w}\, 2\pi}{\pi r^2} = 2\,\overline{w} \tag{5.59}$$

5.8.3 Vorticity

Vorticity at a point is defined as the circulation round an infinitesimal circuit there to the area of that circuit. In other words circulation per unit area.

Vorticity is denoted by a symbol ζ (Greek `zeta').

$$\text{Vorticity} = \frac{\text{Circulation}}{\text{Area}}$$

For rectangular element refer equation (5.59)

$$\zeta = \frac{\partial v}{\partial x} - \frac{\partial u}{\partial_y}$$

For the circle refer equation (5.59)

$$\zeta = 2\,\overline{w}$$

That is the vorticity at a point is twice the average angular velocity of particles at that point. If the vorticity is zero at all points in a region then the flow in that region is said to be irrotational. Flow in regions where the vorticity is other than zero is said to be rotational.

5.9 STREAM FUNCTION AND VELOCITY POTENTIAL

To solve the problems of fluid mechanics in which steady incompressible two dimensional fluid motion and the basic fundamental equation like continuity and Eluier's equation. We must obtain solutions for the three partial differential equations for the three unknowns (the velocity components u and v and the fluid pressure p) subject to certain boundary conditions.

The complexity of solving three simultaneous partial differential equations can be reduced by the introduction of a single function capable of describing the two velocity components of the flow field. Two such functions are available for this purpose: the stream function and the velocity potential function.

5.9.1 The stream function

The stream function, designated ψ (Greek `Psi') is a function of the space coordinates x,y and is related to the velocity components such that

$$u = \frac{\partial \psi}{\partial y} \tag{5.60}$$

$$v = -\frac{\partial \psi}{\partial x}$$

where, the velocity components u is given by the partial derivative with respect to y and the velocity components v is given by the partial derivative with respect to x.

In a steady, incompressible, two dimensional flow, the continuity equation is written as

$$\frac{\partial u}{\partial x} + \frac{\partial \psi}{\partial y} = 0 \tag{5.61}$$

The above equation (5.61) suggests that if we define a function (x,y), which relates the velocities as written in equation (5.60), than the continuity equation is identically satisfied. This conclusion can be verified by simply substituting the expressions for u and v into equation (5.61) so that

$$\frac{\partial}{\partial x}\left(\frac{\partial \psi}{\partial y}\right) + \frac{\partial}{\partial y}\left(-\frac{\partial \psi}{\partial x}\right) = \frac{\partial^2 \psi}{\partial x \partial y} - \frac{\partial^2 \psi}{\partial y \partial x} = 0.$$

Thus, the equation of continuity for an incompressible, two dimensional and steady flow is the necessary and sufficient condition for the existence of a stream function. The flow need be two dimensional only in the sense that it requires only two spatial coordinates to describe the motion. Therefore stream function exist both for plane flow and for axially symmetric flow. A corollary to this is that the existence of a stream function is a necessary condition for a physically possible flow (i.e., one that satisfies the continuity equation). If one wishes to use the stream function approach to solve three dimensional flow, two stream functions are required.

Another particular advantage of using the stream function is related to the fact that ψ is constant along a streamline.

The equation for the stream line for the x-y plane is written as (equation (5.6))

$$\frac{dy}{dx} = \frac{v}{u}$$

or, $$u\ dy - v\ dx = 0.$$

Substituting the expression for u and v into the differential equation of the streamlines, we obtain

$$\frac{\partial \psi}{\partial y} dy + \frac{\partial \psi}{\partial x} dx = 0 \tag{5.62}$$

As defined in calculus, the total differential $d\psi$ of the stream function ψ is given by

$$d\psi = \frac{\partial \psi}{\partial x} dx + \frac{\partial \psi}{\partial y} dy \tag{5.63}$$

Comparing both the equations (5.62) and (5.63) it therefore follows that

$$d\psi = 0$$

or, ψ = Constant (5.64)

Hence the stream function is constant along a streamline. Different value of the constant will define different streamline (5.18).

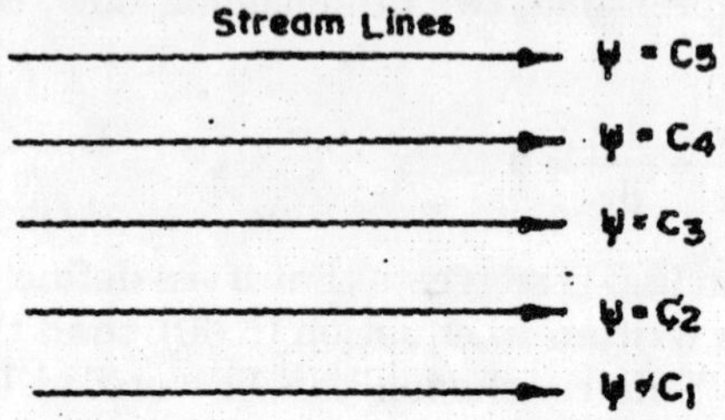

Fig. 5.18. Stream functions is constant along a streamline

The physical significance of ψ follows from consideration of the difference in the values of ψ for two stream lines as shown in Fig. 5.19.

Let us consider two closely spaced streamlines as are shown in Fig. 5.19 (a). The lower streamline is designated ψ and the upper one $\psi + d\psi$. Let dq represents the volume rate of flow (per unit width perpendicular to the x-y plane) passing between the two streamlines. We know that flow never crosses streamlines.

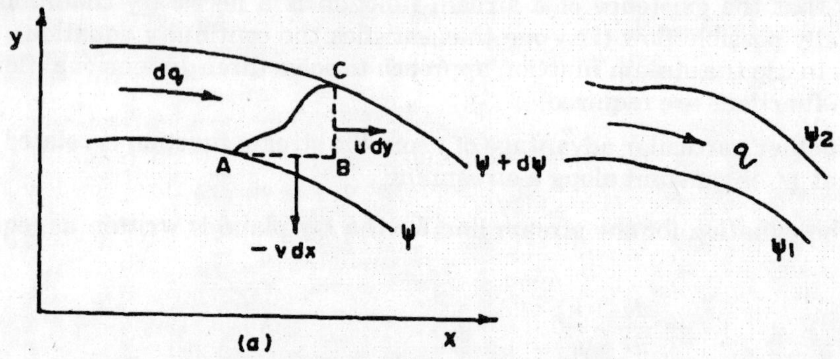

Fig. 5.19. The flow between two streamlines.

From conservation of mass we know that the inflow, dq, crossing the arbitrary surface AC of Fig. 5.19(a) must equal the net outflow through the surfaces AB and BC. Thus

$$dq = udy - vdx$$

or in terms of stream function

$$dq = \frac{\partial \psi}{\partial x}\, dx + \frac{\partial \psi}{\partial y}\, dy \tag{5.65}$$

Comparing the equations (5.63) and (5.65) we get

$$dq = d\psi \tag{5.66}$$

Thus, the volume rate of flow q, between two streamlines such as ψ_1 and ψ_2 of Fig. 5.19 (b) can be determined by integrating equation (5.66) to yield

$$q = \int_1^2 d\psi = \psi_2 - \psi_1$$

If the upper streamline, ψ_2 has a value greater than the lower streamline, ψ_1, q is positive, which indicates that the flow is from left to right. For $\psi_1 > \psi_2$ the flow is from right to left.

The stream function ψ for the incompressible flow in the r, θ plane can be written as

$$v_r = \frac{1}{r}\frac{\partial \psi}{\partial \theta} \tag{5.67}$$

$$v_\theta = -\frac{\partial \psi}{\partial r}$$

The stream function concept can be extended to axis-symmetric flow, such as flow in pipes or flow around bodies of revolution and to two dimensional compressible flows, and two dimensional flow under a barrier that is called sluice gate.

5.9.2 Velocity Potential

A velocity Potential function Φ (Greek `Phi') for steady, irrotational flow in the xy plane is defined as a function of x, y and z such that the partial derivative of Φ with respect to displacement in any chosen direction is equal to the velocity in that direction. Therefore, for the x and y directions,

$$u = \frac{\partial \Phi}{\partial x}$$

$$v = \frac{\partial \Phi}{\partial v} \tag{5.68}$$

$$w = \frac{\partial \Phi}{\partial z}$$

The difference between Φ and ψ function lies in the fact that Φ function exits only for irrotational flows, while ψ function are not restricted to irrotational flow.

It is to be noted. however, that the velocity potential can be defined for a general three dimensional flow, whereas the stream function is restricted to two dimensional flows.

The continuity equation for an incompressible flow is written as

$$\frac{\partial u}{\partial x} + \frac{\partial v}{\partial y} + \frac{\partial w}{\partial z} = 0 \tag{5.69}$$

The equation (5.59) suggests that if we define a function Φ (x,y,z), in equation (5.58), then the continuity equation is identically satisfied. This continuity can be verified by simply substituting the expression for u, v and W into equation (5.59) so that

$$\frac{\partial}{\partial x}\left(\frac{\partial \Phi}{\partial x}\right)+\frac{\partial}{\partial y}\left(\frac{\partial \Phi}{\partial y}\right)+\frac{\partial}{\partial z}\left(\frac{\partial \Phi}{\partial z}\right)=0$$

$$\frac{\partial^2 \Phi}{\partial x^2}+\frac{\partial^2 \Phi}{\partial y^2}+\frac{\partial^2 \Phi}{\partial z^2}=0 \tag{5.70}$$

Some times one defines

$$\nabla^2 = \frac{\partial^2}{\partial x^2}+\frac{\partial^2}{\partial y^2}+\frac{\partial^2}{\partial z^2}$$

and calls it the Laplace operator. For ϕ to be an acceptable function, in the sense that an irrotational and incompressible flow field will ensue, we require that ϕ satisfy

$$\nabla^2 \Phi = 0 \tag{5.71}$$

This equation arises in many different areas of engineering and physics and is called Laplace equation. Thus, inviscid, incompressible, irrotational flow fields are governed by Laplace equation. This type of flow is commonly called a potential flow. Although the concept of the velocity potential is applicable to both unsteady and steady flow, we will confine our attention to steady flow

If the motion is two dimensional i.e. there is no flow velocity in, say the z direction the process of obtaining velocity potentials is greatly simplified.

The vorticity of the rectangular element (refer equation 6.57)

$$\zeta = \frac{\partial v}{\partial x}-\frac{\partial u}{\partial y}$$

For irrotational flow $\zeta = 0$

Therefore, $\dfrac{\partial v}{\partial x}-\dfrac{\partial u}{\partial y}=0$

Using the equation (5.58)

$$\text{or} \quad \frac{\partial}{\partial x}\left(\frac{\partial \Phi}{\partial y}\right)-\frac{\partial}{\partial y}\left(\frac{\partial \Phi}{\partial x}\right)=0$$

$$\frac{\partial^2 \Phi}{\partial x \partial y}-\frac{\partial^2 \Phi}{\partial y \partial x}=0$$

Which demonstrates that a potential flow satisfies the condition of irrotationality. In other words, the existence of a velocity potential implies that flow is irrotational.

An alternative way of making the same statement is as follows:

$$u = \frac{\partial \psi}{\partial y}, \; v = -\frac{\partial \psi}{\partial x}$$

so, $$\frac{\partial u}{\partial y} = \frac{\partial^2 \psi}{\partial y^2} \; and \; \frac{\partial v}{\partial x} = -\frac{\partial^2 \psi}{\partial x^2}$$

Therefore,

$$\frac{\partial u}{\partial y} - \frac{\partial v}{\partial x} = \frac{\partial^2 \psi}{\partial y^2} + \frac{\partial^2 \psi}{\partial x^2} = 0$$

This is the Laplace equation, which must be satisfied if the flow is to be a potential flow.

If the velocity potential ϕ exits the flow must be either irrotational or Laplacian.

When the velocity potential function ϕ is equated to a series of constants, equation for a family of equipotential lines are the result. By the very nature of the definition of the velocity potential function [equation (5.68)], equipotential lines are orthogonal to streamlines, Let us consider a velocity potential function along any equipotential line, ϕ = constant as

$$\phi = \Phi(x, y) = C$$

The total change of ϕ expressed in terms of convective change in

$$d\phi = \frac{\partial \phi}{\partial x} dx + \frac{\partial \phi}{\partial y} dy = 0$$

Substituting from equation (5.68) for $\frac{\partial \phi}{\partial x}$ and $\frac{\partial \phi}{\partial y}$ in this equation yields

$$u \, dx + v \, dy = 0 \tag{5.72}$$

So for the equipotential lines

$$\frac{dy}{dx} = -\frac{u}{v}$$

For any flow pattern the velocity potential function ϕ is related to the stream function ψ by means of the two velocity components, u and v, at any point (x,y) in the Cartesian Coordinates system in the form of the following two equations.

$$u = \frac{\partial \phi}{\partial x} = \frac{\partial \psi}{\partial y} \tag{5.73}$$

and $$v = \frac{\partial \phi}{\partial y} = -\frac{\partial \psi}{\partial x} \tag{5.74}$$

5.10 FLOW NET

We have previously shown that lines of constant ψ are streamlines; that is

$$\frac{dy}{dx} = \frac{u}{v} \tag{5.75}$$

and lines of constant ϕ are equipotential lines, that is,

$$\frac{dy}{dx} = -\frac{u}{v} \tag{5.76}$$

A comparison of equation (5.75) and (5.76) shows that lines of constant ϕ (called equipotential lines) are orthogonal to lines be drawn that consists of a family of streamline of constant ψ (streamlines) at all points where they intersect. (Recall that two lines are orthogonal if the product of their slopes is minus one).

For any potential flow field a ``flow net" can be drawn that consists of family of streamlines and equipotential lines. The flow net is useful in visualizing flow patterns and can be used to obtain graphical solution by sketching in streamlines and equipotential lines and adjusting the lines until the lines are approximately orthogonal at all points where they intersect. An example of a flow net is shown in Fig. 5.20.

This is a network of mutually perpendicular streamlines and equipotential lines. The streamline which show the direction of flow at any point are so spaced that there is an equal rate of flow dq discharging through each tube. The discharge dq is equal to the change in ψ from one streamline to the next. The equipotential lines are then drawn everywhere normal to the streamlines. The spacing of the equipotential line are selected in such a way that the change in velocity potential from one equipotential lines to the next is constant. Furthermore, the changes in both sets of lines are made equal, that is, d ψ = d ϕ. As a result they form approximate squares (Fig. 5.20).

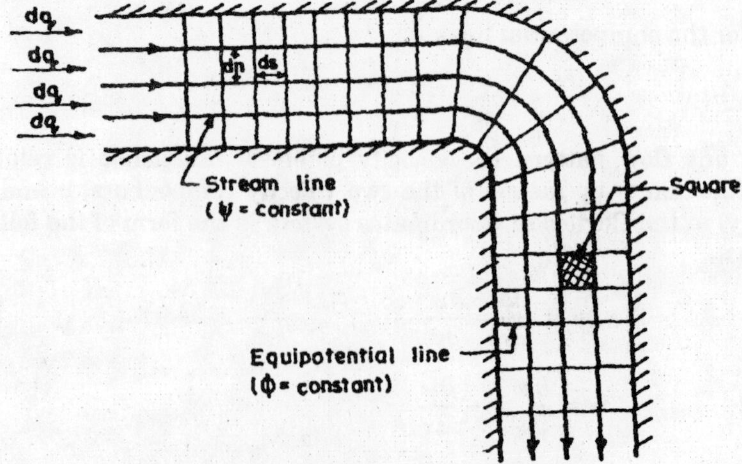

Fig. 5.20. Flow net with in a closed conduct.

From the continuity relationship, the distance between both sets of lines must therefore be inversely proportional to the local velocities. Thus the following relation is a key to the proper construction of any flow net :

$$\frac{v_1}{v_2} = \frac{dn_2}{dn_1} = \frac{dS_2}{dS_1}$$

Where dn and ds are respectively the distance between streamlines and that between equipotential lines.

Some characteristics of conventional flow nets can be enumerated as follows:

(a) Flow nets are based upon the assumption of irrotational flow, It is not necessary that the flow be steady.

(b) The ψ lines intersect ϕ lines, including fixed boundaries, normally.

(c) There is only one possible pattern of flow for a given set of boundary conditions and the flow net, if correctly prepared, will represent this pattern.

(d) In region of uniform flow the squares are of equal size. In diverging flow they increase, and in converging flow they decrease in size in the direction of flow, the spacing of the ϕ – and ψ lines being equal and inversely proportional to the velocity at any point.

(e) The unit figures of the resulting grid approximate to squares, having equal median lines and 90° angles (except at stagnation points and points of theoretically infinite velocity).

(f) The sizes of the squares decrease towards the inside of curves, that is, with decrease of radius of the streamlines in curved flow, since, in irrotational flow around a curved path, the product of velocity and radius is constant.

V_2 = Constant

$$\therefore \quad dn \;\alpha\; \frac{1}{V} \;\alpha\; r.$$

5.10.1 Methods of drawing flow nets

There are three approximate methods for drawing flow nets:

(a) Graphical method.

(b) Numerical method.

(c) Experimental analogy method.

Attention is directed primarily to the graphical and numerical methods, which require no special equipment for their application.

Graphical Method

This method is very simple. This method may be used to determining the flow pattern for any given set of boundary conditions.

The flow net consists of a family of streamlines and a family of velocity potential lines drawn to represent a pattern of two dimensional irrotational flow. At any point

$$V = \frac{dq}{dn} = \frac{d\psi}{dn} = \frac{d\phi}{ds}$$

Where dn and ds are respectively, the streamline and equipotential line spacing. If the constant intervals d ψ and dϕ are made equal, dn equals ds in all parts of the flow net, which then appears as a net work of units which are very nearly squares. In each unit the median lines are equal and the angles are right angles, for the equipotential lines interest the streamlines, including the boundaries at 90°.

These characteristics enable a flow net to be constructed by means of trial and error sketching of squares upon a diagram representing the known boundaries of flow. When the complete flow net or any part of it appears to consist of squares, a check can be made by drawing both diagonals of the squares. These diagonals of the squares should form smooth curves which intersect each other normally. Another check which may also be used to test the accuracy of the flow net is that in each square the sides are tangent in it.

Numerical Analysis

Numerical methods of integration, based upon the calculus of finite differences, can be used to determine patterns of flow in cases which, because of the complexity of their boundary forms, cannot be solved by ordinary analytical methods. The general approach requires the assumption of a network of values of ψ, or ϕ and these values are symmetrically adjusted so as to satisfy the Laplace equation and the boundary conditions. The method yields only particular solutions, but it is simple both in principle and in practice and it is readily adaptable to machine computation. This method is simple, rapid, self-checking and readily programmed for the digital computer.

The methods of numerical analysis have been successfully applied to numerous problems in two dimensional flow, including flow past immersed bodies, free efflux and weir flows, cavitation flow (in conjunction with measured pressures) and percolation. It is also applicable to three dimensional axi-symmetric flows.

Experimental analogies

Generally three methods can be used.

(1) Membrane Analogy; (2) The viscous flow Analogy

(3) Electrical Analogy.

(1) Membrane analogy

A thin elastic membrane, such as a rubber sheet or a soap film, can be stretched over a suitably shaped frame so as to form of three dimensional surface whose contours in plane view, correspond closely to the streamline of a flow patterns.

The equation of the membrane surface can be written f (x, y, z) = o or z =f (x, y). At each point on the surface, the internal elastic or surface tension are in equilibrium. It is assumed that the membrane is so thin that its rigidity is negligible and that gravity forces need not be considered; further, the internal stress, is assumed to vary where and in all directions constant and deflections are assumed to be so small that secondary stresses due to them need not be considered.

(2) The viscous flow analogy

The viscous flow between horizontal parallel plates a small distance apart, the velocity distribution across the gap is parabolic. For small `creeping' velocities, the mean velocity across the gap in any region is very nearly proportional to the pressure gradient so that approximately $u = -k\dfrac{\partial p}{\partial x}$, $v = -k\dfrac{\partial p}{\partial y}$ substitution of these values in the continuity equation

$$\frac{\partial u}{\partial x} + \frac{\partial v}{\partial y} = 0$$

gives
$$\frac{\partial^2 p}{\partial x^2} + \frac{\partial^2 P}{\partial y^2} = 0$$

The pressure being thus analogous to the velocity potential of irrotational flow.

(3) Electrical analogy

The flow of an electric current in a two dimensional conductor is analogous to irrotational flow. The electric potential V is the counterpart of ϕ and the components E_x and E_y of the field strength vector E are the counterparts of u and v respectively. The relevant equations in electric field theory are

$$\frac{\partial^2 v}{\partial x^2} + \frac{\partial^2 v}{\partial y^2} = 0$$

$$E_x = \frac{\partial v}{\partial x},\ E_y = \frac{\partial v}{\partial y} = 0$$

A sheet of conducting material or a shallow bath of electrolyte is shaped so that its boundary is similarly to that of the flow pattern to be investigated. If a voltage drop is established along the conductor between the flow entrance and exit boundaries, a potential pattern geometrically similar to that in the field is set up.

A voltmeter or potentiometer probe is used to locate lines of constant potential. If the voltage drop is established across the flow, between the limits representing solid boundaries, the probe can be used to locate a set of lines orthogonal to the first set, that is, streamlines. (Fig. 5.21) This method has been extended to axially symmetrical three dimensional flow problems.

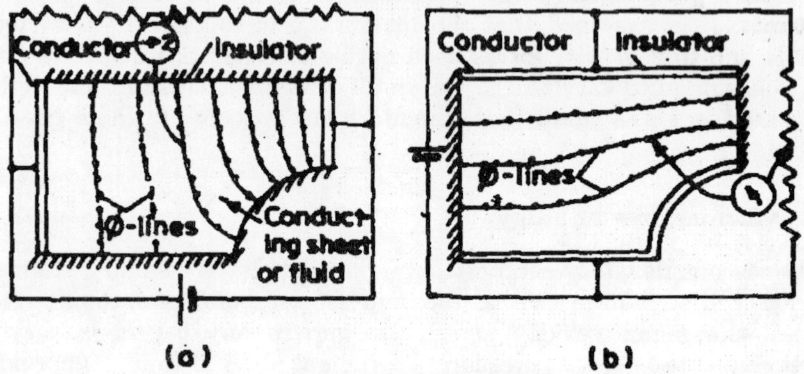

Fig. 5.21. Electric analogy (a) Equipotential lines (b) Streamlines for a flow past a cylindrical in a channel.

5.10.2 Uses of flow net

Flow net may be used for all irrotational flows with geometrical similar conditions, once the flow net is drawn, the spacing between the adjacent streamlines is determined and the application of the continuity equation gives the velocity of flow at any point, if the velocity of flow at any reference point is known. further the flow net analysis assists in the determination of the efficient boundary shapes, for which the flow does not separate from the boundary surface.

The flow net may be used in cases of real fluid with sufficient accuracy. The real fluid behaves like an ideal fluid in most of the cases, except fluid friction or viscosity. Hence in flow cases where viscosity effect of friction is comparatively unimportant, the flow net analysis can be effectively used.

5.10.3 Limitations of flow net

A flow net always indicates some velocity at the boundary, but a real fluid must have zero velocity adjacent to the boundary on account of the fluid friction or viscosity. As such the flow net analysis cannot be applied in the region close to the boundary where the effects of viscosity are predominant. The flow net analysis cannot be used for the flow diverges and separates near the boundaries, because the disturbed flow in the rear of the body (wake formation) is not indicated by a flow net.

ILLUSTRATIVE EXAMPLES

5.1 Petrol flow at 5 m/s through a 200 mm diameter pipe. The pipe then branches into two pipes of diameters 100 mm and 150 mm, respectively. The mass flow rate through the 100 m pipe is found to be 50 kg/s. Determine the velocity of flow in the 150 mm pipe. The relative density of petrol is 0.78.

Solution

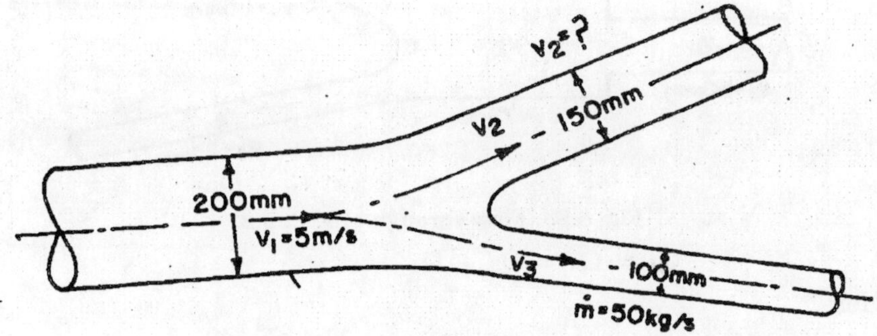

Fig. 5.22. Flow through pipe P. 5.1.

$$m = S\,\rho\,Q_3$$

or, $50 = 0.78 \times 1000\ Q_3$

$\therefore Q_3 = 0.0641\ m^3/s$

$Q_1 = a_1 v_1$

$$Q_1 = \frac{\pi}{4}(0.2)^2 \times 5 = 0.157\ m^3/s$$

Here,

$Q_1 = Q_2 + Q_3$

$0.157 = Q_2 + 0.0641$

$Q_2 = 0.0929\ m^3/s$

Therefore, $Q_2 = a_2\,v_2 = \frac{\pi}{4}(0.15)^2 \times v_2$

$0.029 = 0.0176 \times v_2$

$\therefore v_2 = 5.26$ m/s Ans

5.2 Water flows through a pipe AB of diameter 50mm, which is in series with a pipe BC of diameter 75mm in which the velocity is 2 m/s.

At C the pipe forks and one branch CD is of unknown diameter such that the velocity is 1.5 m/s. The other branch CE is of diameter 25 mm and condition are such that the discharge in the pipe BC divides so that the discharge in the pipe CD is equal to two times of discharge in CE. Calculate, (a) discharge in pipe AB,

(b) velocity in pipe AB (c) discharge in pipe CD, (d) velocity of pipe CE, (e) diameter of pipe CD.

Solution

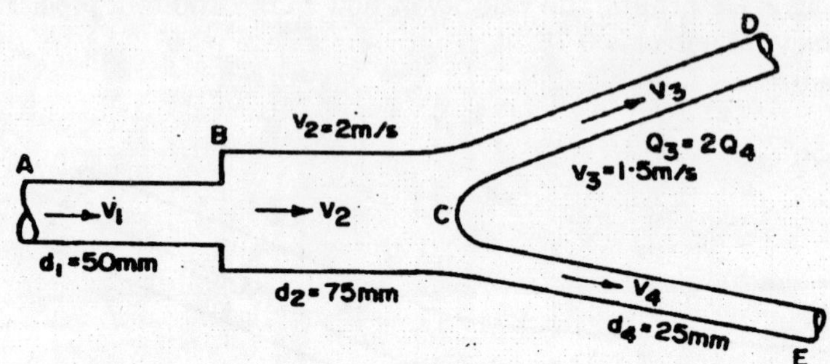

Fig. 5.23. Flow through pipe P. 5.2.

$$Q_1 = Q_2 = Q_3 + Q_4$$

Here,

(a) $$Q_1 = Q_2 = \frac{\pi}{4} d_2^2 \; v_2 = \frac{\pi}{4} (0.075)^2 \times 2$$

$$= 8.8357 \times 10^{-3} \text{ m}^3\text{/s}$$ Ans

(b) $$Q_1 = \frac{\pi}{4} d_1^2 \; v_1$$

$$8.8357 \times 10^{-3} = \frac{\pi}{4} (0.05)^2 \times v_1$$

$$v_1 = 4.5 \; m/s$$ Ans

(c) $$Q_2 = Q_3 + Q_4$$

Given $$Q_3 = 2\,Q_4$$

Therefore, $$Q_2 = Q_3 + \frac{Q_3}{2}$$

$$Q_2 = 1.5\,Q_3$$

$$8.8357 \times 10^{-3} = 1.5\,Q_3$$

$$Q_3 = 5.89 \times 10^{-3} \; m^3/s$$ Ans

and $$Q_4 = \frac{Q_3}{2} = \frac{5.89 \times 10^{-3}}{2} = 2.945 \times 10^{-3} \; m^3/s$$ Ans

(d) $$Q_4 = \frac{\pi}{4} d_4^2 \; v_4$$

$$2.945 \times 10^{-3} = \frac{\pi}{4} (0.025)^2 \; v_4$$

$$\therefore \; v_4 = 6 \; m/s$$ Ans

(e) $Q_3 = \frac{\pi}{4} d_3^2 v_3$

$$5.89 \times 10^{-3} = \frac{\pi}{4} d_3^2 \times 1.5$$

or, $d_3^2 = 4.9995 \times 10^{-3}$

$\therefore \quad d_3 = 0.0707\ m = 70.7\ mm$ **Ans**

5.3 Kerosene flows through a pipe line which contracts from 400mm diameter at A to 275 mm diameter at B and then forks, one branch being 150 mm diameter discharging at C and the other branch 200 mm diameter discharging at D. If the velocity at A is 2.25 m/s and the velocity at D is 4 m/s, what will be the discharge at C and D and velocities at B and C.

Solution

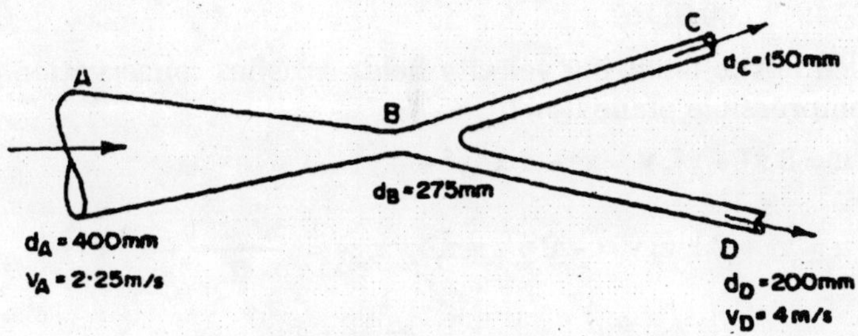

Fig. 5.24. Flow through pipe P. 5.3

Discharge at A, $Q_A = \frac{\pi}{4} d_A^2 v_A$

$$= \frac{\pi}{4} (0.400)^2 \times 2.25 = 0.2827\ m^3/s$$

Discharge at D, $Q_D = \frac{\pi}{4} d_D^2 v_D$

$$= \frac{\pi}{4} (0.2)^2 \times 4 = 0.1256\ m^3/s$$ **Ans**

From the continuity of flow equation

$$Q_A = Q_B = 0.2827\ m^3/s$$

$$\frac{\pi}{4} d_B^2 v_B = 0.2827$$

$$Q_B = \frac{\pi}{4} d_B^2 v_B$$

$$0.2827 = \frac{\pi}{4}(0.275)^2 \times v_B$$

$$v_B = \frac{0.2827}{0.05939} = 4.7596\ m/s$$ Ans

From continuity equation

$$Q_B = Q_C + Q_D$$

$$0.2827 = Q_C + 0.1256$$

$$Q_C = 0.2827 - 0.1256 = 0.1571\ m^3/s$$ Ans

$$Q_C = \frac{\pi}{4} d_C^2\ v_C$$

$$0.1571 = \frac{\pi}{4}(1.5)^2 \times v_C$$

$$\therefore v_C = \frac{0.1571}{0.01767} = 8.89\ m/s$$ Ans

5.4 **Which of the following velocity fields satisfies conservation of mass for incompressible plane flow?**

(a) $u = 3x^2 + y^2$, $v = x^3 - 4x(y^2 - 3y)$

(b) $u = y^2 + 2xz$, $v = -2yz + x^2yz$, $w = \frac{x^2 z^2}{2} + x^3 y^4$

(c) $u = xyzt$, $v = -xyzt^2$, $w = \frac{x^2 z^2}{2} + x^3 h^4$

(d) $u = 2xy - x^2 + y$, $v = 2xy - y^2 + x^2$

(e) $u = 2x^2 - xy + z^2$, $v = 4x^2 - 4xy + y^2$, $w = -2xy - 3yz + y^2$

Solution

	Velocity	Terms	Sum	Possible ?
(a)	$u = 3x^2 + y^2$	$\frac{\partial u}{\partial x} = 6x$	$\neq 0$	No
	$v = x^3 - 4x(y^2 - 3y)$	$\frac{\partial u}{\partial y} = -4x(2y-3)$		
(b)	$u = y^2 + 2xz$	$\frac{\partial u}{\partial x} = 2z$		
	$v = -2yz + x^2yz$	$\frac{\partial u}{\partial y} = -2z + x^2 z$	$= 0$	Yes
	$w = -\frac{x^2 z^2}{2} + x^3 y^4$	$\frac{\partial w}{\partial z} = -\frac{2zx^2}{2}$		

(c) $u = x y z t$ $\qquad \frac{\partial u}{\partial x} = y z t$

$v = - x y z t^2$ $\qquad \frac{\partial v}{\partial x} = - x z t^2$ $\qquad = 0$ Yes

$w = \frac{x^2}{2}\left(xt^2 - yt\right)$ $\qquad \frac{\partial w}{\partial x} = z\left(xt^2 - yt\right)$

(d) $u = 2xy - x^2 + y$ $\qquad \frac{\partial u}{\partial x} = 2y - 2x$ $\qquad = 0$ Yes

$v = 2xy - y^2 + x^2$ $\qquad \frac{\partial v}{\partial y} = 2x - 2y$

(e) $u = 2x^2 - xy + z^2$ $\qquad \frac{\partial u}{\partial x} = - 4x - y$

$V = 4x^2 - 4xy + y^2$ $\qquad \frac{\partial v}{\partial y} = - 4x + 2y$ $\qquad \neq 0$ No

$w = - 2xy - 3yz + y^2$ $\qquad \frac{\partial w}{\partial z} = - 3$

5.5 **Calculate the circulation around the path in the x-y plane boundary by the lines y = o, x =6, y =4, and x = o for the flow field given by**

$$u = y^2 - x^2 \text{ and } v = 2xy.$$

Solution

$u = y^2 - x^2 \qquad v = 2xy$

$\frac{\partial u}{\partial x} = -2x \qquad \frac{\partial v}{\partial y} = +2x$

Therefore, $\frac{\partial u}{\partial x} + \frac{\partial v}{\partial y} = -2x + 2x = 0$

$\frac{\partial u}{\partial y} = 2y \qquad \frac{\partial v}{\partial x} = 2y$

Therefore, $\frac{\partial v}{\partial x} - \frac{\partial u}{\partial y} = 2y - 2y = 0$

Flow is irrotational.

$$\tau = \int_0^6 - x^2\, dx + \int_0^4 12y\, dy + \int_6^0 \left(16 - x^2\right) dx + \int_0^6 o\, dy$$

$$= - 72 + 96 + 72 - 96 = 0$$

***Note*: If the flow is irrotational, the circulation should be zero.**

5.6 **A flow is described by the stream function**

$$\psi = \frac{x^3}{3} - \frac{y^2}{2} - xy$$

What is the circulation around the square enclosed by x = ± 2, $y = \pm 2$?

Solution

$$\psi = \frac{x^3}{3} - \frac{y^2}{2} - xy$$

$$u = \frac{\partial \psi}{\partial y} = -y - x$$

$$v = -\frac{\partial \psi}{\partial y} = -x^2 + y = y - x^2$$

$$\frac{\partial u}{\partial x} = -1$$

$$\frac{\partial v}{\partial y} = +1$$

Therefore, $\frac{\partial u}{\partial x} + \frac{\partial v}{\partial y} = -1 + 1 = 0$

$$\frac{\partial v}{\partial x} = -2x$$

$$\frac{\partial u}{\partial y} = -1$$

Therefore, $\frac{\partial v}{\partial x} + \frac{\partial u}{\partial y} = -2x + 1 \neq 0$

The flow is rotational.

$$\tau = \int u\, dx + v\, dy$$

$$= \int_{-2}^{+2} (-x+2)\, dx + \int_{-2}^{+2} (y-4)\, dy + \int_{2}^{-2} (-x-2)\, dx + \int_{2}^{-2} (y-4)\, dy$$

$$= \left(-\frac{x^2}{2} + 2x\right)_{-2}^{+2} + \left(\frac{y^2}{2} - 4y\right)_{-2}^{+2} + \left(-\frac{x^2}{2} + 2x\right)_{2}^{-2} + \left(\frac{y^2}{2} - 4y\right)_{2}^{-2}$$

$$= 8 - 16 + 8 + 16 = +16 \text{ m}^2$$ **Ans**

5.7 **A two dimensional velocity field is given by**

$$u = x^2 - y^2 + x \quad \textit{and} \quad v = -(2xy + y)$$

At (2,1) compute (a) Acceleration a_x and a_y (b) and maximum acceleration

Solution

(a) $u = x^2 - y^2 + x$ *and* $v = -(2xy + y)$

$$\frac{\partial u}{\partial x} = 2x + 1 \qquad \frac{\partial v}{\partial x} = -2y$$

$$\frac{\partial u}{\partial y} = -2y \qquad \frac{\partial v}{\partial y} = -(2x + 1)$$

$$a_x = \frac{Du}{Dt} = u\frac{\partial v}{\partial x} + v\frac{\partial v}{\partial y}$$

$$= \left(x^2 - y^2 + x\right)(2x + 1) + (-2xy - y)(-2y)$$

$$= (4 - 1 + 2)(4 + 1) + (-4 - 1)(-2)$$

$= 25 + 10 = 35$ **Units** **Ans**

$$a_y = \frac{Du}{Dt} = u\frac{\partial v}{\partial x} + v\frac{\partial v}{\partial y}$$

$$= \left(x^2 - y^2 + x\right)(-2y) + (-2xy - y)(-2x + 1)$$

$$= (4 - 1 + 2)(-2) + (-4 - 1)(-4 - 1)$$

$= -10 + 25 = 15$ **Units** **Ans**

(b) **For maximum acceleration**

$$\tan\alpha = \frac{a_y}{a_x} = \frac{15}{35} = 0.42857$$

$\therefore \alpha = \tan^{-1}(.42857) = 23.3^\circ$ **Ans**

5.8 **An idealized velocity field is given by**

$$u = 3xt, \qquad v = -yt^2 \textit{ and } w = 2zx.$$

At the point (1, - 1,0), compute the total acceleration

Solution

$$u = 3xt, \qquad v = -yt^2 \textit{ and } w = 2zx.$$

$$\frac{\partial u}{\partial x} = 3t \qquad \frac{\partial v}{\partial x} = 0 \qquad \frac{\partial w}{\partial x} = 2z$$

$$\frac{\partial u}{\partial x} = 0 \qquad \frac{\partial v}{\partial y} = t^2 \qquad \frac{\partial w}{\partial y} = 0$$

$$\frac{\partial u}{\partial z} = 0 \qquad \frac{\partial v}{\partial z} = 0 \qquad \frac{\partial w}{\partial z} = 2x$$

$$\frac{\partial u}{\partial t} = 3x \qquad \frac{\partial v}{\partial t} = -2yt \qquad \frac{\partial w}{\partial t} = 0$$

$$a_x = \frac{du}{dt} = u = \frac{\partial u}{\partial x} + v\frac{\partial u}{\partial y} + w\frac{\partial u}{\partial z} + \frac{\partial u}{\partial t}$$

$$= 3xt\,(3t) + \left(-yt^2\right)(0) + 2z\,(0) + 3x$$

$$= 9\,xt^2 + 3x = 9t^2 + 3 \; at \; (1, -1, 0)$$ Ans

$$a_y = \frac{dv}{dt} = u = \frac{\partial v}{\partial x} + v\frac{\partial v}{\partial y} + w\frac{\partial v}{\partial z} + \frac{\partial v}{\partial t}$$

$$= 3xt\,(0) + \left(-yt^2\right)(-t)^2 + 0 + (-2yt)$$

$$= +\,yt^4 - 2yt = 2t - t^4 \; at \; (1, -1, 0)$$ Ans

$$a_z = \frac{dw}{dt} = u\;\frac{\partial w}{\partial x} + v\frac{\partial w}{\partial y} + w\frac{\partial w}{\partial z} + \frac{\partial w}{\partial t}$$

$$= (3xt)\,(2z) + 0 + 2zx\;(2x) + 0 = 6xzt + 4zx^2$$

$$= 0 \text{ at } (1, -1, 0)$$ Ans

5.9 **The velocity components for a two dimensional incompressible flow are**

$$u = 2x + y \qquad and \qquad v = -x - 2y$$

(a) Is continuity satisfied? If so, find the stream function.

(b) Is the flow irrotational? If so, find the potential function.

Solution

$$u = 2x + y \qquad and \qquad v = -x - 2y$$

$$\frac{\partial u}{\partial x} = 2 \qquad \frac{\partial v}{\partial y} = -2$$

Therefore, $\frac{\partial u}{\partial x} + \frac{\partial v}{\partial y} = 2 - 2 = 0$

Continuity equation satisfied.

$$u = \frac{\partial \psi}{\partial y} = 2x + y \tag{1}$$

$$v = -\frac{\partial \psi}{\partial x} = -x - 2y \tag{2}$$

Integrating equation (1)

$$\int \partial \psi = \int (2x + y)\,\partial y$$

$$\psi = 2xy + \frac{y^2}{2} + f(x) \tag{3}$$

Differentiating partially w. r. t. x

$$\frac{\partial \psi}{\partial x} = 2y + f'(x)$$

$$x + 2y = 2y + f'(x) \qquad \text{[From equation (2)]}$$

$$x = f'(x)$$

Integrating

$$f(x) = \frac{x^2}{2} + C$$

Putting in equation (3)

$$\psi = 2xy + \frac{y^2}{2} + \frac{x^2}{2} + C \qquad \text{Ans}$$

(b) $u = 2x + y \qquad v = -x - 2y$

$$\frac{\partial u}{\partial y} = 1 \qquad \frac{\partial v}{\partial x} = -1$$

Therefore, $\dfrac{\partial v}{\partial x} - \dfrac{\partial u}{\partial y} = -1 - 1 = -2 \neq 0.$

The flow is rotational and is not derivable form a potential function.

Note – The stream function is a consequence of the continuity equation and thus is applicable to both rotational and irrotational flows. The potential function, however, is a consequence of the condition for zero vorticity and thus is applicable to only irrotational flows. All flows must satisfy continuity, and thus it is possible for velocity components to satisfy the continuity requirements but not satisfy the condition for being irrotational. In this instance, the velocity components would not be derivable from a potential function ϕ since the flow would be rotational.

5.10 A flow field is characterised by stream function $\psi = y^3\, x$. Determine the velocity potential ϕ for the flow.

Solution

For a given stream function ψ the flow must be possible, but whether it is Laplacian or not Laplacian be checked for before attempting to evaluate velocity potential ϕ.

$$\psi = y^3\, x$$

$$\frac{\partial \psi}{\partial x} = y^2 \qquad \frac{\partial \psi}{\partial y} = 3y^3\, x$$

$$\frac{\partial^2 \psi}{\partial x^2} = 0 \qquad \frac{\partial^2 \psi}{\partial y^2} = 6yx.$$

The Laplace equation

$$\frac{\partial^2 \psi}{\partial x^2} + \frac{\partial^2 \psi}{\partial y^2} = 0 + 6yx. \neq 0.$$

Therefore, ψ is not laplacian, i.e. flow exists but is not irrotational. Hence velocity potential ϕ cannot be defined for this flow.

5.11 **Given the stream function**

$$\psi = \frac{y^3}{3} - x^2 y$$

(a) What is the velocity vector which describes the flow?

(b) Show that the flow is Laplacian.

(c) What is the velocity potential ϕ ?

Solution: (a)

$$\psi = \frac{y^3}{3} - x^2 y$$

$$u = \frac{\partial \psi}{\partial y} = \frac{3y^2}{3} - x^2 = y^2 - x^2$$

$$v = -\frac{\partial \psi}{\partial x} = +2\,xy$$

Therefore, $u = y^2 - x^2$ *and* $v = 2\,xy$ **Ans**

(b) $$\psi = \frac{y^3}{3} - x^2 y$$

$$\frac{\partial \psi}{\partial x} = -2xy \qquad \frac{\partial \psi}{\partial y} = y^2 - x^2$$

$$\frac{\partial^2 \psi}{\partial x^2} = -2y \qquad \frac{\partial^2 \psi}{\partial y^2} = +2y$$

Therefore, $$\frac{\partial^2 \psi}{\partial x^2} + \frac{\partial^2 \psi}{\partial y^2} = -2_y + 2y = 0$$

The flow is Laplacian.

(c) $$u = \frac{\partial \phi}{\partial x} = y^2 - x^2 \qquad (1)$$

$$v = \frac{\partial \phi}{\partial y} = 2xy \qquad (2)$$

Integrating equation (1)

$$\phi = xy^2 - x^3/3 + f(y) \qquad (3)$$

Differentiating partially w. r. t. y

$$\frac{\partial \phi}{\partial y} = 2\,xy + f'(y)$$

$$2\,xy = 2\,xy + f'(y)$$

$$f'(y) = 0$$

$$f(y) = C$$

Putting in equation (3)

$$\phi = xy^2 - \frac{x^3}{3} + C \qquad \text{Ans}$$

5.12 Show that the velocity potential $\phi = 2\left(x^2 - y^2\right)$ represents a steady motion and find a corresponding expression for the stream function ψ. What would be the magnitude and direction of flow at the coordinate (4,2) ?

Solution

$$\phi = 2\left(x^2 - y^2\right)$$

$$\frac{\partial \phi}{\partial x} = 4\,x \qquad \frac{\partial \phi}{\partial y} = -4\,y$$

$$\frac{\partial^2 \phi}{\partial x^2} = 4 \qquad \frac{\partial^2 \phi}{\partial y^2} = -4$$

Therefore, $\dfrac{\partial^2 \phi}{\partial x^2} + \dfrac{\partial^2 \phi}{\partial y^2} = 4 - 4 = 0.$

The flow is Laplacian, hence represent steady motion.

We know that

$$u = \frac{\partial \psi}{\partial y} = \frac{\partial \phi}{\partial x} = 4x \qquad (1)$$

$$v = -\frac{\partial \psi}{\partial x} = \frac{\partial \phi}{\partial y} = -4y \qquad (2)$$

From (2) $-\dfrac{\partial \psi}{\partial x} = -4y$

$$\partial \psi = 4y\;\partial x$$

Integrating $\psi = 4\,xy + f(y) \qquad (3)$

Differentiating partially w. r. t. y

$$\frac{\partial \psi}{\partial y} = 4x + f'(y)$$

$$4x = 4x + f'(y) \qquad \text{[from (1)]}$$

$$f'(y) = 0$$

$$f(y) = C$$

Putting in equation (3)

$$\psi = 4\,xy + C \qquad \textbf{Ans}$$

(b) $u = 4x = 4 \times 4 = 16$

$$v = 4y = -4 \times 2 = -8$$

$$V = \sqrt{u^2 + v^2}$$

$$= \sqrt{(16)^2 + (-8)^2} = \sqrt{256 + 64} = \sqrt{320} = 17.88\ m/s \qquad \textbf{Ans}$$

$$\tan\theta = \frac{-8}{16} = -0.5\ \theta = -26.565 = 26^\circ\, 33'\, 54'' \qquad \textbf{Ans}$$

EXERCISES

A. Theory

5.1 Differentiate between Eulerian and Lagrangian method of representing fluid motion.

5.2 Define the following and give one practical example for each:

(a) Unsteady flow (b) Uniform flow. (c) Turbulent flow (d) Three dimensional flow.

5.3 Explain the terms (a) Streamline; (b) Path line; (c) Streak line; (d) Stream tube; (e) Time line.

5.4 Distinguish between the following:

(a) Steady flow and unsteady flow; (b) Uniform flow and non uniform flow; (c) Comp-ressible flow and incompressible flow; (d) Laminar and turbulent flow; (e) Circulation and vorticity; (f) Rotational and irrotat-ional flow; (g) Streamline, streak line and path line; (h) Convective, local and total acceleration; (i) Tangential acceleration and normal acceleration.

5.5 Define the following terms:

(a) Irrotational flow; (b) Uniform flow; (c) Laminar flow; (d) Steady flow; (e) Stream line; (f) Stream function; (g) Velocity Potential; (h) Circul-ation; (i) Vorticity. Give the mathematical expression wherever necessary.

5.6 What is a streamline and flow how do you distinguish it from pathline and streak line? Write the equation for stream line.

5.7 Define streamline, pathline and streak line and state when the three lines merge into each other.

5.8 Derive the equation of streamline.

5.9 What are the condition for flow to be irrotational ?

5.10 What do you understand by rotational and irrotational flow ? Drive the condition of irrotational from the first principles.

5.11 Derive, form first principles, equation of continuity for unsteady flow of compressible fluid.

5.12 Derive the equation of continuity for one dimensional flow.

5.13 Derive the continuity equation for steady three dimensional flow.

5.14 Derive the equation of continuity for a compressible flow and show that it tends to divergence of velocity equal to zero for incompressible fluids. State clearly the assumptions made.

5.15 Derive, from first principles, the continuity equation of fluid flow in cylindrical coordinates.

5.16 Derive from first principles, the continuity equation of fluid flow in spherical coordinates.

5.17 Develop the general form of the continuity equation in Cartesian coordinates for three dimensional flow, expressing as a differential equation. Hence, deduce the equation for steady incompressible fluid flow.

5.18 Explain what is meant by circulation and vorticity as applied to potential flow system.

5.19 Derive the equation for the vorticity in a two dimensional flow for a circular section.

5.20 Derive the equation for the vorticity in a two dimensional flow for a rectangular section.

5.21 Define circulation and vorticity. What is the relation between circulation and vorticity?

5.22 Explain the meaning and use of the stream function in dealing with problems of two dimensional fluid motion.

5.23 State the three important properties of stream function and show how linear velocity components may be expressed as a function of the partial derivatives of the stream function.

5.24 Explain the meaning of the term velocity potential. Express (a) the equation of conti-nuity and (b) the Cartesian velocity components, in terms of the velocity potential and state how the two properties, stream function ψ and velocity potential ϕ taken together, completely define a fluid flow.

5.25 Define the terms vorticity, velocity potential and stream function as applied to the flow an inviscid incompressible fluid and show that if the flow is irrotational $\nabla^2 \phi = 0$, and is the velocity potential.

5.26 Discuss the role of stream function in fluid mechanics problems. Attach physical significance to it and show that constant stream function lines are nothing but a streamline.

5.27 State the properties of stream function.

5.28 Show that streamlines are constant stream function lines.

5.29 Show that stream function ψ = constant lines are streamlines.

5.30 Describe, from first principles, the condition for irrotational flow. Prove that, for potential flow, both the stream function and velocity potential satisfy the Laplace equation.

5.31 What is a flow net? Describe, in brief, two methods of drawing the flow net.

5.32 Prove that the streamlines and equipotential lines in a flow net are mutually orthogonal.

5.33 What is flow net ? Explain its uses.

5.34 Define velocity potential and stream function show that :

(a) Existence of the velocity potential implies that flow is irrotational.

(b) They represent orthogonal function in a two dimensional flow field.

(c) Each function is a solution of Laplace's equation for two dimensional irrotational flow.

5.35 Explain briefly the conditions for drawing the flow net.

5.36 Write down the characteristics of conventional flow nets.

5.37 Write the three name of experimental analogy for drawing flow net explain any one.

5.38 What are the uses of flow net ?

5.39 What are the limitations of flow net?

5.40 Show that for irrotational flow

$$\nabla^2 \phi = 0, \quad \text{and} \quad \nabla^2 \psi = 0.$$

B. UNSOLVED PROBLEMS

5.1 Water flow through a pipe AB 1.2 m diameter at 3m/s and then passes through a pipe BC 1.5m diameter. At C, the pipe branches. Branch CD is 0.8m in diameter and carries one third of the flow in AB. The flow velocity in branch CE is 2.5 m/s. Find the volume rate of flow in AB, the velocity in BC, the velocity in CD and the diameter of CE. [3.392 m^3/s, 1.919 m/s, 2.249 m/s, 1.07 m].

5.2 An air duct is of rectangular cross section 300 mm × 450 mm. Determine the mean velocity in the duct when the rate of flow is 0.42 m^3/s. If the duct tapers to a cross section 150 mm x 400 mm, what will be the mean velocity in the reduced section assuming that the density remains unchanged [3.11 m/s, 7 m/s].

5.3 In a 600 mm duct carrying air the velocity profile was found to obey the law $u = 0.45 - 5 r^2$ m/s where u is the velocity at radius r. Calculate the volume rate of flow of the air and the mean velocity [0.0628 m^2/s, 0.222 m/s].

5.4 The velocity along the center line of a nozzle of length L is given by

$$u = 2t\left(1 - \frac{x}{2L}\right)^2$$

where u is the velocity in m/s, t is the time in seconds from the commencement of flow, x is the distance from the inlet to the nozzle. Find the convective acceleration and the local acceleration when t = 3 s, $x = \frac{L}{2}$ and L = 0.8 m. [18.99 m/ s^2, 1.125 m/s^2]

5.5 For steady, incompressible flow, are the following values of u and v possible?

(a) $u = 4xy + y^2$ and $v = 6xy + 3x$ (b) $u = 2x^2 + y^2$ and $v = -4xy$ (c) $u = x^2 y + y^2$ and $v = x^2 - y^2 x$.

[Not possible, possible, possible]

5.6 If $u = x^2 + y^2 + z^2$ and $v = -(xy + yz + xz)$,

determine the velocity components w which will satisfy continuity equation for incompressible fluid [$w = -xz + z^2/2 + f(x, y, t)$].

5.7 The two velocity components in a three dimensional flow field for an incompressible fluid are as follows.

$u = x^3 + y^3 + 2z^2$ and $v = -x^2y - yz - xy^3$. Find the third velocity component w.

$$\left[w = -2x^2z + 3xy^2z + \frac{z^2}{2} + f(x, y, t)\right]$$

5.8 If $u = yz + t$, $v = xz - t$ and $w = xy$, determine the acceleration a_x at the point

(2, 1, 3) at t = 0.5 s. [$a_x = 19.5$ m/s^2]

5.9 In a two dimensional fluid flow, the component of velocity in x direction is

$$u = \frac{y^3}{3} - x^2y + 2x$$

Determine the component of velocity in y direction for a possible flow.

[$v = xy^2 - 2y + C$]

5.10 A flow field is given by in the z, y and z direction are as follows

$$u = x^2 y,\ v = y^2 z \text{ and } w = -(2xyz + yz^2).$$

Determine the acceleration at the point

(2, 1, 3) [$a_x = 28$, $a_y = -3$ $a_z = 105$, $a_R = 108.71$ units]

5.11 A flow field is given by $u = xy^2$, $v = -2y^2 z^2$ and $w = \frac{2z^2}{3} - y^2 z$;

Determine the acceleration at the point (1,2,3) [874.50 units]

5.12 Determine whether the specified flows are rotational or irrotational.

(a) $u = y$, and $v = -\frac{3}{2}x$ (b) $u = xy^2$, $v = x^2 y$

(c) $u = \frac{y^3}{3} + 2x - x^2 y$ and $v = xy^2 - 2y - \frac{x^3}{3}$

(d) $u = y^2 - x^2 + y$ and $v = 2xy + x$

(e) $u = x^2 y + y^2$ $v = x^2 - y^2 x$

[rotational, irrotational, irrotational, irrotational, rotational].

5.13 Given the velocity potential $\phi = \frac{y^3}{3} - x^2 y$

(a) What is the velocity vector which describes the flow ?

(b) Show that vorticity is everywhere zero.

(c) What is the stream function.

$$\left[u = -2xy,\ v = y^2 - x^2,\ \psi = \frac{x^3}{3} - xy^2 + c \right]$$

5.14 The x and y components of velocity in an ideal, two dimensional flow are given by

$$u = 3x^2 - 3y^2 \text{ and } v = -6xy.$$

Find (a) Velocity potential and (b) Stream function.

[$\phi = x^3 - 3xy^2 + C$, $\psi = 3x^2 y - y^3 + C$]

5.15 The stream function is given by $\psi = 3x^2 - 3y^2 + 2x$

Find the velocity potential.

[$\phi = -(6xy + 2y) + C$]

6

Dynamics of fluid flow

6.1 INTRODUCTION

As was discussed in the previous chapter, there are many situations involving fluids in which the fluid can be considered as static. In general, however, the use of fluid involves motion of some type. In fact, dictionary definition of the word "fluid" is "free to change in from". In this chapter we investigate the behaviour of fluid in motion by using the principle of conservation of energy and principle of conservation of momentum.

In this chapter we will discuss in some detail the use of Newton's second law as it is applied to fluid particle motion that is ideal in some sense. In 1738, Daniel Bernoulli employed Newton's laws of motion to establish a theory governig fluid flow. Because of the complexity of the problem, he did not include viscosity in the development. Although this equation is one of the oldest in fluid mechanics and the assumptions involved in its derivation are numerous, it can be effectively used to predict and analyse a variety of flow situations. However, if the equation is applied without proper respect for its restrictions, serious errors can arise. Indeed, the Bernolluis equation is appropriately called "the most used and the most abused equation in fluid mechanics".

We will now pay our attention to the law of conservation of momentum. Strictly speaking, momentum is not conserved in most situations. Rather we shall relate changes of linear momentum of given mass of fluid with the forces that act upon it. Some of the forces that act on the fluid are transmitted by walls and other boundaries of the fluid. The momentum equation that will be derived will permit us to calculate the froces that flowing fluid exert on walls and other boundaries of the fluid. This is so because such forces are equal in magnitude, but opposite in sign, to the forces which these boundaries exert on the fluid.

An important category of fluid mechanics problem that are readily solved with the help of the moment of momentum equation involves machines that rotate or tend to rotate around a single axis.

6.2 EULER'S EQUATION OF MOTION

Euler's equation is based on the Newton's second law of motion. The integration of the Euler's equation give the Bernoulli's equation. Students should note carefully that the derivation of Euler`s equation is based on the following assumptions.

(1) Only pressure and gravity forecs are to be considered.

(2) Flow is along the streamline.

(3) Flow is steady.

(4) Flow is with negligible viscosity.

(5) Flow is ideal.

(6) Flow is incompressible.

(7) Velocity is uniform over the section.

In Fig 6.1 a stream tube of fluid of length ds and area dA is isolated as a free body from the moving fluid. The external forces are due to the pressure acting on the tube and the gravitational (weight of the stream tube).

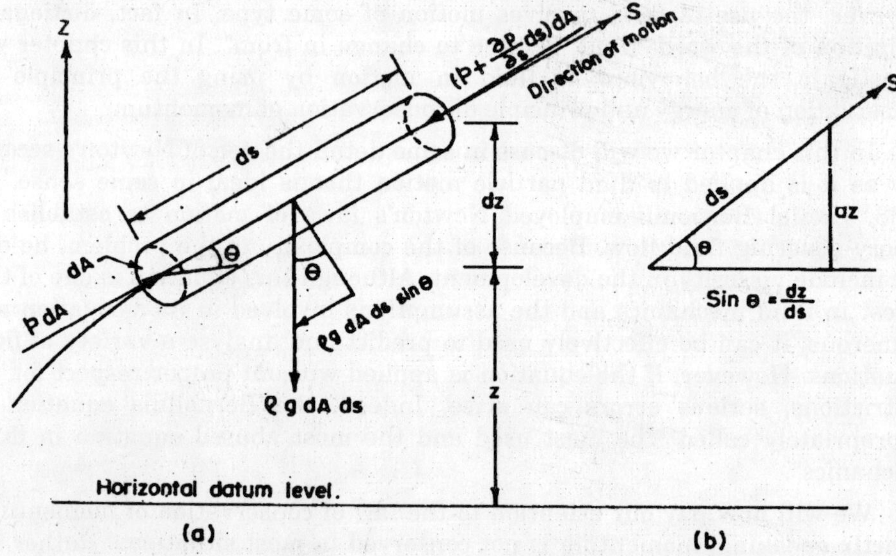

Fig. 6.1. Forces acting on an element of stream tube

The length ds is so small that curvature of the stream lines over this distance may be neglected.

The external forces which act on the chosen free body of fluid element causing the acceleration of flow are as follows.

1. The resultant force due to pressure acting on the two end surfaces in the direction of flow is

$$p\,dA - \left(p + \frac{\partial p}{\partial s}ds\right)dA = -\frac{\partial p}{\partial s}dsdA \tag{6.1}$$

where the direction of motion is considered to be positive.

2. The component of gravity force (ρg dAds) of the fluid element is shown at an angle θ with the direction of flow (Fig. 6.1a)

The Component of the gravity force of the fluid element in the direction of motion is, therefore,

$$-\rho\, g\, dA\, ds\, \sin\theta = -\rho g\, dA\, ds \frac{dz}{ds} \quad \text{[Fig 6.1(b)]} \tag{6.2}$$

Thus, the resultant external force acting on the free body in the direction of flow is the sum of the pressure force and gravitational force is

$$\Sigma F_s = -\frac{\partial p}{\partial s} ds\, dA - \rho g\, dA\, ds \frac{dz}{ds} \tag{6.3}$$

The sum of these forces equal to mass multiplied by acceleration.

$$\text{mass} = m = \rho\, dA\, ds$$

$$a_s = \frac{\partial v}{\partial t} = \frac{\partial s}{\partial t} \times \frac{\partial v}{\partial s} = v \frac{\partial v}{\partial s}$$

Therefore, $$ma_s = \rho dA\, ds\, v \frac{\partial v}{\partial s} \tag{6.4}$$

According to Newton's law of motion. Using equation (6.3) and (6.4)

$$-\frac{\partial p}{\partial s} ds\, dA - \rho\, g\, dA\, ds \frac{dz}{ds} = \rho dA\, ds\, v \frac{\partial v}{\partial s}$$

or, $$-\frac{1}{\rho}\frac{\partial p}{\partial s} - g\frac{dz}{ds} = v\frac{\partial v}{\partial s}$$

or, $$\frac{1}{\rho}\frac{\partial p}{\partial s} + g\frac{dz}{ds} + v\frac{\partial v}{\partial s} = 0 \tag{6.5}$$

The equation (6.5) is known as Euler's equation of motion. The equation (6.5) is written in a differential form. The equation (6.5) can also be written in a common form.

$$\frac{dp}{\rho} + g\, dz + v\, d\, v = 0 \tag{6.6}$$

6.2.1 Bernoull's equation-Integration of Euler's equation along a streamline for steady flow

The equation (6.6) cannot be integrated completely with respect to s unless ρ is either constant or a known function of p. For a fluid of constant density (incompressible), however, the result of integration is

$$\frac{p}{\rho} + \frac{v^2}{2} + g\, z = \text{Constant} \tag{6.7}$$

Which is known as Bernoulli's equation.

The Bernollis equation is a powerful and useful equation because it relates pressure changes to velocity and elevation changes along a streamline. However, it gives correct results only when applied to a flow situation where all the seven assumptions are reasonable.

An alternative form of writing equation (6.7) is

$$\frac{p}{\rho g} + \frac{v^2}{2g} + z = H \tag{6.8}$$

In the equation (6.8), each term has the dimensions of energy per unit weight, such as Nm/N or more simply m, indicating a scaler quantity. It is thus convenient to refer to each term as a head. Therefore, H is called the total energy head, and is the sum of pressure velocity and potential heads. Fig 6.2 is helpful to visualise the relationship among the the types of energy. It is to be noted that because of the assumption that no energy is lost or added, the total head remains height of each head term varies as predicted by Bernoulli's equation.

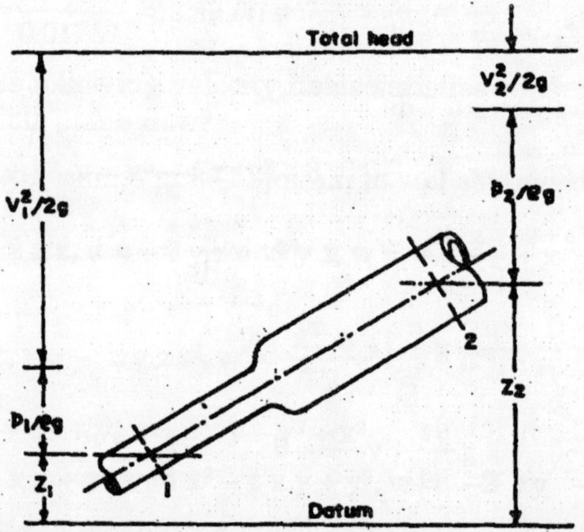

Fig. 6.2. Pressure head, potential head, velocity head and total head

The equation (6.7) can be applied for a single streamline. The sum of the three terms is constant along any streamline, but the value of the constant may be different for different streamlines in a given stream.

If equation (6.7) is intergrated along the streamline between any two points indicated by suffixes 1 and 2,

$$\frac{p_1}{\rho g} + \frac{v_1^2}{2g} + z_1 = \frac{p_2}{\rho g} + \frac{v_2^2}{2g} + z_2 \tag{6.9}$$

6.2.2 Limitations on Bernoulli's equations

Although Bernoulli's equation is applicable to a large number of practical problems, there are several limitations that must be understood in order to apply it properly.

1. There can be no energy lost due to friction.

2. There can be no heat transferred into or out of the fluid.

3. The equation is only applicable to incompressible flow.

4. There can be no mechanical devices between the two sections of interest which would add or remove energy from the system, since the equation states that the total evergy in the fluid is constant.

6.2.3 Modification to Bernoulli's equation

It is evident that in the flow of a real fluid, say in the flow through a pipe system, there will be losses of energy due to friction separation eddying etc: Therefore the total head will not be constant but decrease in the directon of flow as a result of energy dissipation. The modification can be made simply by considering that the total head at section 1 is equal to the total head at section 2 further down stream, plus the losses between the two sections and this is expressed by

$$\frac{p_1}{\rho g} + \frac{v_1^2}{2g} + z_1 = \frac{p_2}{\rho g} + \frac{v_2^2}{2g} + z_2 + \textit{losses} \tag{6.10}$$

in which the losses are also in Nm/N or in m head of he fluid concerned.

A further modification can be made to take into account an addition or subtractions of energy between the two sections. Pumps, which convert mechanical energy into hydraulic energy and turbines, which perform the reverse function, are typical examples of this type of energy.

In the turbulent flow of a real fluid the velocity distribution at any section adjacent to or between solid boundaries is not uniform, and in order to express the velocity head in terms of the average velocity v it is necessary to introduce a dimensionless coefficient α whereby the velocity head is $\alpha\left(v^2/2g\right)$. The value of α is dependent on the velocity distriubution and is obtained by the elemental stream tubes by the kinetic energy head based on the average velocity.

The value of α is always greater than unity. The value may be as high as 2.0 (laminar flow in pipes) but for normal turbulent flow in pipes it is usually only slightly in excess of unity ranging between 1.01 and 1.10. Irregular sections produce higher values.

6.2.4 The physical significance of Bernoulli's equation

The physical significance of Bernoulli's equations can be most easily understood by examining Fig 6.3 on which each term is shown for sections along an irregular tube. In Fig. 6.3(a) the total head or total evergy is a constant for flow without friction and is represented by the total head or total energy line as a horizontal line at a constant distance from the datum plane. With friction this is not true and the total head line in Fig. 6.3(b) is shown as a dashed curved line dropping off to the right. It will be noticed that when the velocity in the tube increases, the sum of the potential and pressure head must decrease and that for a decreased velocity, the sum of the potential and pressure heads increases. For steady flow, $\frac{v_3^2}{2g}$ is a constant, verquiring $\frac{p_3}{\rho g}$ to be constant for the case of no friciton (Fig. 6.3(a)) in the case of friction (Fig. 6.3(b)), $\frac{p_3}{\rho g}$ must decrease since $\frac{v_3^2}{2g}$ is constant.

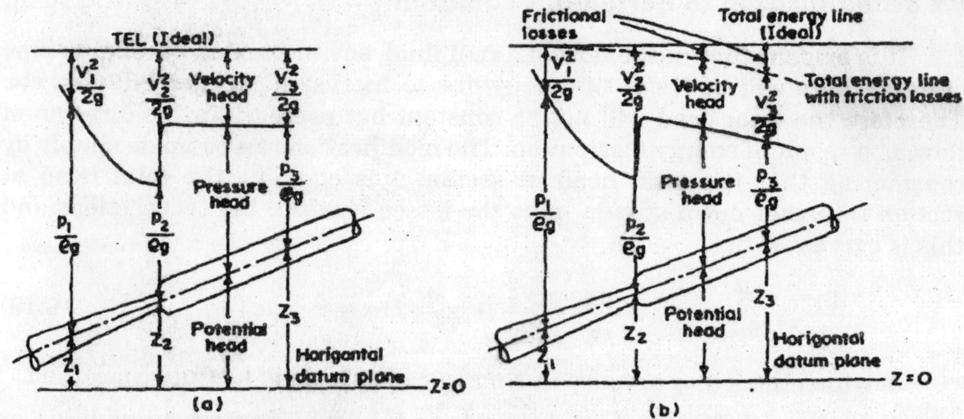

Fig. 6.3 (a, b). Head variation in a tube

6.3 APPLICATIONS OF BERNOULLI'S EQUATION

Bernoulli's equation is a basic one and it has the widest application in fluid mechanics. Many types of devices using principles invlolved in the Bernoulli equation have been develo-ped to measure flow rates and flow velocities. In this article we will consider ideal flow meters – those devoid of viscous, compressibity, and other effects.

An effective way to measure the flow rate through a pipe is to place some type of restri-ction within the pipe and to measure the pressure difference between the high pressure, low velocity upstream section and the high velocity, low pressure downstream section. Three commonly used types of flow meter are illustrated; the venturimeter, the orifice meter, and the nozzle meter. The operation of each is based on the same physical principles an increase in velocity causes decrease in pressure. The differences between them is a matter of cost, accuracy and how closely their actual operation obeys the idealized flow assumptions.

6.3.1 Venturimeter

The venturimeter is a device which has been used over many years for measuring flow rate along a pipe.

A venturimeter, shown in Fig 6.4, Consists of short converging conical tube leading to a cylindrical portion, called the throat, of smaller diameter than that of the pipeline, which is followed by a diverging section (divergent diffuser cone) in which the diameter increases again to that of the main pipeline. Continuity requires a greater velocity at the throat than at the inlet, there is consquently a differnce of pressure between inlet and throat and measurement of this pressure differnce enables the rate of flow through the meter to be calculated.

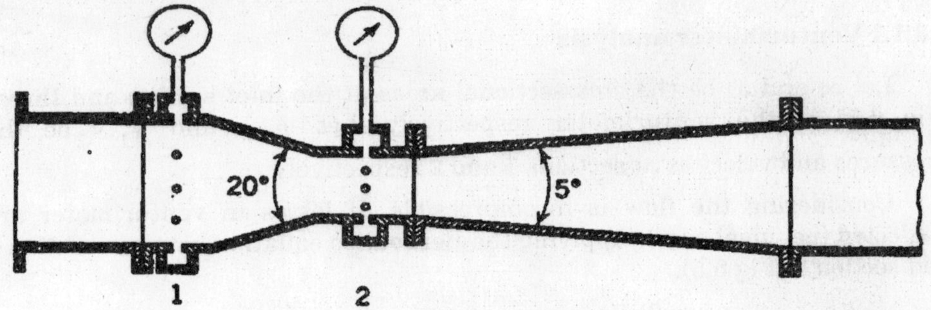

Fig. 6.4. Venturimeter

The angle between the converging walls of the upstream cone is about 20°, that of the downstream cone about 6°. In so far as the metering qualities are concerned an exact value is not important. The diveging cone section often has an included angle between 5° and 15°, so that the expanding flow does not appreciably separate from the walls. This angle is somewhat of a compromise, because a more gradual expanding section would certainly restrict the sepatation losses; it would also result in a longer section in which the flow was moving very rapidly and which would be subject to high wall shear stresses. There is no need to have a long, tapering converging cone inlet section, because separation is not a critical factor here. The contraction section is designed so that there is no separation in the throat. Most of the head loss that occurs in a well designed venturimeter is due to friction losses along the walls rather than losses associated with separated flows and the inefficient mixing motion that accompanies such flow.

The size of the venturimeter is expressed in terms of inlet diameter (d_1) and throat diameter (d_2). It may be noted that d_1 is also the diameter of the pipeline to which the venturimeter is connected. The ratio d_2/d_1 (known as throat ratio) should be such that the pressure at throat does not become negative. More commonly the diameter of throat d_2 is kept half of the inlet i.e. pipe diameter d_1. However, the value of throat diameter may be kept 1/3 to 3/4th of pipe diameter d_1. To obtain the more accuracy in measurement the throat diameter d_2 to be reduced, causes velocity to increse at throat, correspondingly a large pressure difference is established between inlet and throat. But on the other hand pressure becames very low at the threat say vapour pressure. In such situation boiling occures (though the temperature need not be high) vapour bubbles form, and then collapse as the fluid moves into a region of higher pressure (lower velocity). This pressure can produce dynamic effects that cause very large pressure transients in the vicinity of the bubbles. Pressures as large as 690 MPa are believed to occur. At this point liquid gives out dissolved air and begins to vaporize and the continiuty of flow breaks. This is not desirable and it is always avoided by keeping the throat diameter sufficiently large. Hence the throat diameter is decided by keeping in view the above factors.

6.3.1.1 Venturimeter analysis

Let a_1 and a_2 be the crosssectional areas at the inlet section and throat (Fig 6.5) of the venturimeter respectively. Let p_1, p_2 and v_1, v_2 be the pressures and velocites at sections 1 and 2 respectively.

Considering the flow is incompressible. If losses in venturimeter are neglected (i.e. ideal case). Applying the Bernoulli's equation between section 1 and section 2 (Fig 6.5).

$$\frac{p_1}{\rho g} + z_1 + \frac{v_1^2}{2g} = \frac{p_2}{\rho g} + z_2 + \frac{v_2^2}{2g}$$

Since $z_1 = z_2$ for horizontal venturimeter Therefore,

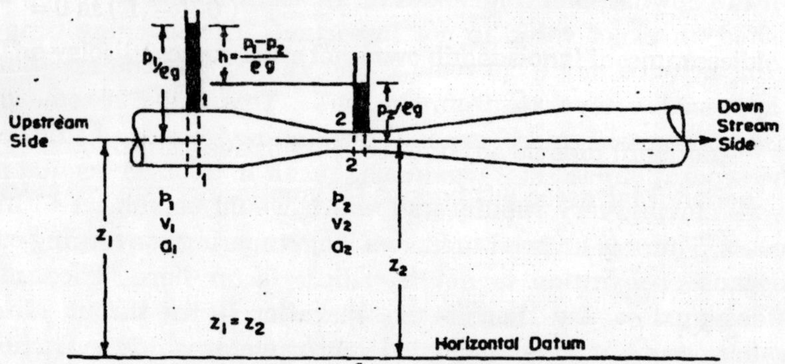

Fig. 6.5. Horizontal venturimeter

$$\frac{p_1}{\rho g} + z_1 + \frac{v_1^2}{2g} = \frac{p_2}{\rho g} + z_2 + \frac{v_2^2}{2g}$$

or, $$\frac{P_1 - p_2}{\rho g} = h = \frac{v_2^2}{2g} - \frac{v_1^2}{2g} \qquad (6.11)$$

where $h = \frac{p_1 - p_2}{\rho g}$ = Difference in piezometic heads at section 1 and 2.

Now applying equation of continuity between section 1 and 2.

$$Q_1 = a_1 v_1 = a_2 v_2$$

or, $$v_1 = \frac{a_2}{a_1} v_2$$

Substititing the value of v_1 in equation (6.11)

$$\frac{v_2^2}{2g} - \frac{a_2^2}{a_1^2}\frac{v_2^2}{2g} = h$$

or, $$\frac{v_2^2}{2g}\left[1 - \frac{a_2^2}{a_1^2}\right] = h$$

or, $$\frac{v_2^2}{2g}\left[\frac{a_1^2 - a_2^2}{a_1^2}\right] = h$$

or, $$v_2^2 = \frac{a_1^2\, 2gh}{a_1^2 - a_2^2}$$

or $$v_2 = \frac{a_1\sqrt{2gh}}{\sqrt{a_1^2 - a_2^2}} \tag{6.12}$$

Since losses have not been considered the velocity v_2 in the equation (6.13) gives theoretical value. Therefore, the theoretical discharge

$$Q_{th} = a_2\ v_2$$

$$Q_{th} = \frac{a_1\, a_2\sqrt{2gh}}{\sqrt{a_1^2 - a_2^2}} \tag{6.13}$$

The equation (6.13) is often written as

$$Q_{th} = K\sqrt{h}$$

where $$K = \frac{a_1\, a_2\ \sqrt{2gh}}{\sqrt{a_1^2 - a_2^2}}$$

K is called the meter constant and is dependent upon the geometry of the venturimeter. In pracitce it is convenient to express the meter constant in terms of the ratio of the throat area to that of the inlet and to introduce

$$\frac{a_2}{a_1} = m \quad \text{(area ratio)}$$

therefore, $$K = a_2\sqrt{2g}\ /\ \sqrt{1-(a_2/a_1)^2} = \frac{a_2\sqrt{2g}}{\sqrt{1-m^2}}$$

When a real fluid flows through a venturimeter it is apparent that losses due to friction will occur between the inlet and the throat (although small), resulting in a pressure differnce between the inlet and the throat which is somewhat greater than that expected from an ideal fluid and used in the above equation. As the use of this value h in equation (6.13) would give too high a value of Q a coefficient of discharge C_d is introduced.

$$\frac{\text{Actual discharge}}{\text{Theoretical discharge}} = \frac{Q_{act}}{Q_{th}}$$

Therefore,

$$Q_{act} = C_c \; Q_{th} = C_d \frac{a_1 \, a_2 \sqrt{2gh}}{\sqrt{a_1^2 - a_2^2}} \tag{6.14}$$

[It is to be noted that h is the difference of piezometric head of the fluid in the meter, not the difference of levels of the manometer fluid].

The C_d also accounts for effects of non-uniformity of velocity over section 1 and 2. Although C_d varies somewhat with flow rate, the viscocity of the fluid and surface roughness, a value of about 0.98 is usual with fluids of low viscocity (Fig 6.6).

Fig 6.7 is a graphical representation of the variation of $\sqrt{h} = \sqrt{\dfrac{p_1 - p_2}{\rho g}}$ with Q act which from equation (6.21), should be a straight line through the origin, if the value of Cd is constant over thre range of measurement.

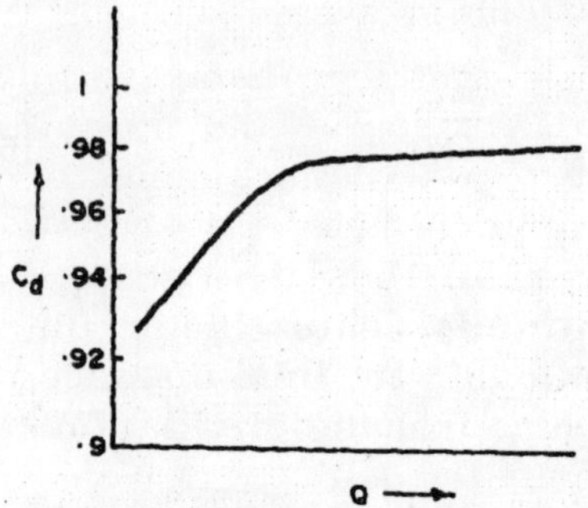

Fig. 6.6. Variation of c_d with Q for venturimeter.

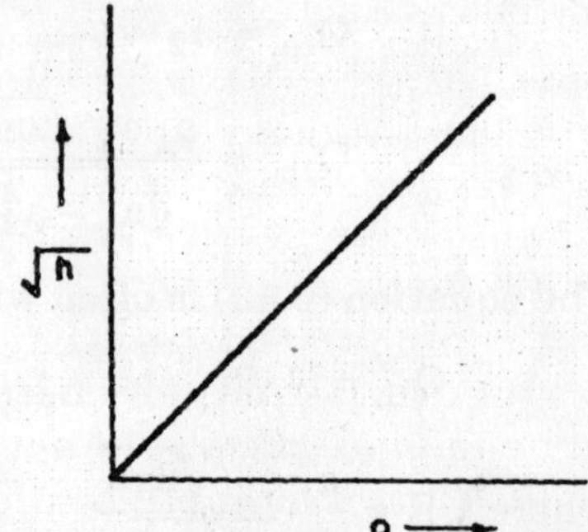

Fig. 6.7. Variation of $\sqrt{h}$ with Q for venturimeter.

Since C_d is merely a convenient means of expressing the head loss $h_{f\,1\text{-}2}$, an exact relation between these variable can be obtained by using the Bernoulli's equation including an allowance for friction loss, that is

$$\frac{P_1}{\rho g} + \frac{v_1^2}{2g} = \frac{p_2}{\rho g} + \frac{v_2^2}{2g} + h_f$$

for horizontal venturimeter $(z_1 = z_2)$

where h_f is the head loss due to friction in the converging cone (i.e. at section 1 and 2).

Then

$$\frac{v_2^2}{2g} - \frac{v_1^2}{2g} = \frac{p_1 - p_2}{\rho g} - h_f$$

$$\frac{v_2^2 - v_1^2}{2g} = h - h_f$$

Using the same procedure outlined above,

$$Q_{act} = \frac{a_1 \, a_2 \sqrt{2g\left(h - h_f\right)}}{\sqrt{a_1^2 - a_2^2}} \tag{6.15}$$

Equating equations (6.14) and (6.15)

$$C_d \, \frac{a_1 \, a_2 \sqrt{2gh}}{\sqrt{a_1^2 - a_2^2}} = \frac{a_1 \, a_2 \sqrt{2g\left(h - h_f\right)}}{\sqrt{a_1^2 - a_2^2}}$$

or, $C_d \, \sqrt{h} = \sqrt{h - h_f}$

$$\text{or,} \quad C_d = \sqrt{\frac{h - h_f}{h}} \tag{6.16}$$

or, $C_d^2 = \dfrac{h - h_f}{h}$

or, $h\left(C_d^2 - 1\right) = -h_f$

$$\text{or,} \quad h_f = h\left(1 - C_d^2\right) \tag{6.17}$$

But from equation (6.14)

$$h = \frac{Q^2}{C_d^2 \, 2g}\left(\frac{a_1^2 - a_2^2}{a_1^2 \, a_2^2}\right) = \frac{1}{C_d^2}\left(\frac{v_2^2 - v_1^2}{2g}\right)$$

Therefore, $h_f = \dfrac{1}{C_d^2}\left(\dfrac{v_2^2 - v_1^2}{2g}\right)\left(1 - C_d^2\right)$

$$= \left(\frac{1}{C_d^2} - 1\right)\left(\frac{v_2^2 - v_1^2}{2g}\right)$$

$$= \left(\frac{1}{C_d^2} - 1\right)\left(\frac{Q^2}{2g}\right)\left(\frac{1}{a_2^2} - \frac{1}{a_1^2}\right)$$

$$\text{or,} \quad h_f = \left[\frac{1}{C_d^2} - 1\right]\left[1 - \left(\frac{d_2}{d_1}\right)^4\right]\left[v_2^2 / 2g\right] \tag{6.18}$$

The equation (6.14) can be writion as,

$$Q = C_d \, \frac{a_1 \, a_2 \sqrt{2gh}}{\sqrt{a_1^2 - a_2^2}}$$

$$= C_d \; K \sqrt{h} \tag{6.19}$$

The distinction between the meter constant (K) and the coefficient of discharge of (C_d) should be made and should not be mistaken one for the another. It should be remembered that the meter constant (K) depends purely on the geometry of the meter, while the coefficient of discharge (C_d) depends on the losses of energy in the converging cone.

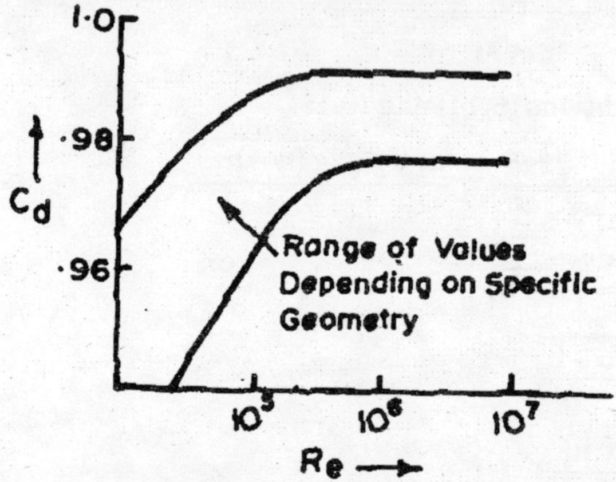

Fig. 6.8. Variation of C_d v/s Re for the venturimeter

The coefficient of discharge C_d is a function of Reynolds number is given in Fig 6.8. The Reynolds number is computed at the condition of the throat. Since the boundry layer forms on the solid walls of the passages, it is conceivable that the larger the diameter of the passage, the less the influence of the viscous layer on the entire flow. It is for this reason that one also suspects that C_d, besides being a function of Reynolds number (Re), is a function of the diameter ratio d_2/d_1.

6.3.1.2 Vertical/Inclined Venturimeter

Venturimeter can also be used for measuring the flow rate through a pipe which is laid either in vertical or in an inclined position. A venturimeter connected in an inclined pipe as shown in Fig. 6.9.

Applying Bernoulli's equation between section 1 and section 2 for no loss energy, we get

$$\frac{p_1}{\rho g} + z_1 + \frac{v_1^2}{2g} = \frac{p_2}{\rho g} + z_2 + \frac{v_2^2}{2g}$$

or, $$\frac{v_2^2}{2g} - \frac{v_1^2}{2g} = \left(\frac{p_1}{\rho g} + z_1\right) - \left(\frac{p_2}{\rho g} + z_2\right)$$

or, $$\frac{v_2^2}{2g} - \frac{v_1^2}{2g} = h \tag{6.20}$$

The same expression which we obtained in the equation (6.11), except the difference of elevation is included in the equation (6.20).

In the Fig. 6.9, we are using U tube manometer, the value of h is calculated with the help of a formula

$$h = \times \left(\frac{S_m}{S} - 1\right) \qquad (6.21)$$

Here $S_m > S$.

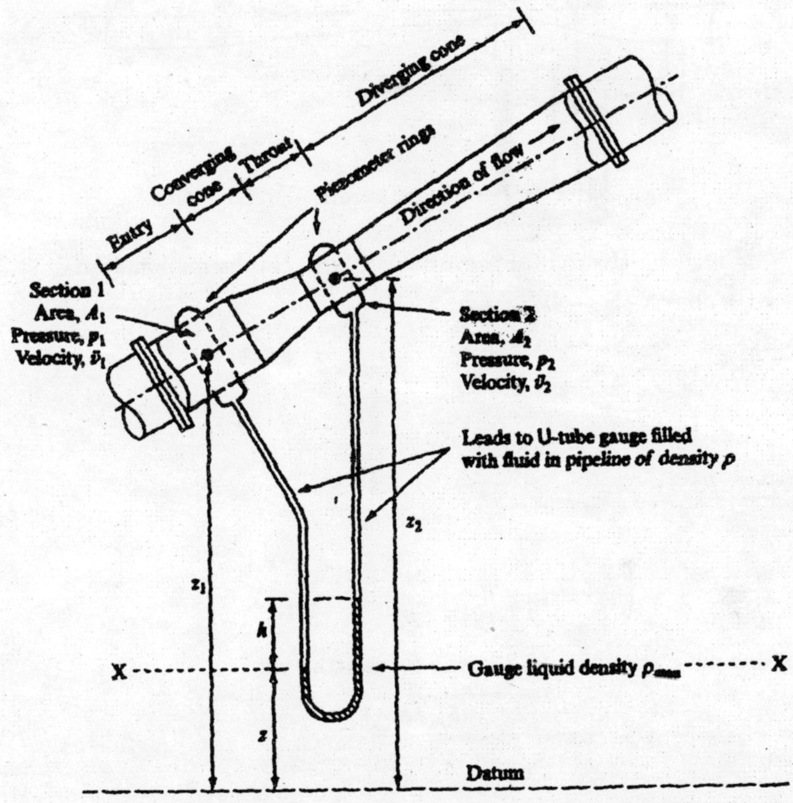

Fig. 6.9. Inclined venturimeter with U tube manometer.

If we use the inverted type of U tube manometer, the value of h is calculacted with the help of a formula

$$h = \times \left(1 - \frac{S_o}{S}\right) \qquad (6.22)$$

Here S > So

If we use U tube manometer for horizontal, vertical and inclined the value of h is given by the formula

$$h = \times \left(\frac{S_m}{S} - 1\right)$$

This relationship holds good for any orientation of venturimeter.

6.3.1.3 Use of differential manometer in venturimeter

(a) Horizontal venturimeter

Applying the principle of manometer for the Fig 6.10

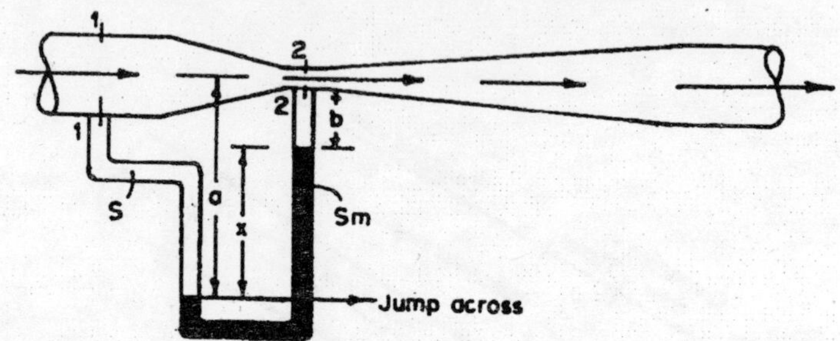

Fig. 6.10. Horizontal venturimeter with U tube manometer.

$$h_1 + aS - x\,S_m - b\,S = h_2$$

or, $h_1 - h_2 = x\,S_m - (a - b)\,S$

$$= x\,S_m - xS$$

$$= x\left(S_m - S\right)$$

or, $\dfrac{p_1 - p_2}{\rho g} = x\left(S_m - S\right)$

or, $$\frac{p_1 - p_2}{S\rho g} = x\left(\frac{S_m}{S} - 1\right) = h \tag{6.23}$$

The flow rate equation (6.13) can be written as

$$Q_{act} = C_d \frac{a_1\, a_2 \sqrt{2g\, x\left(S_m / S - 1\right)}}{\sqrt{a_1^2 - a_2^2}} \tag{6.24}$$

(b) Inclined venturimeter

Using the principle of manometer for the Fig. 6.11

$$h_1 + aS - xS_m - \left(b + z\right)S = h_2$$

or, $h_1 - h_2 = x\,S_m - \left(b + z\right)S - as$

$$= x\,S_m - \left(a - b\right)S + zS$$

$$= x\left(S_m - S\right) + \left(z_2 - z_1\right)S$$

$$\frac{p_1 - p_2}{\rho g} = x\,s\left(\frac{S_m}{S} - 1\right) + \left(z_2 - z_1\right)S$$

or, $$\left(\frac{p_1}{S\rho g} + z_1\right) - \left(\frac{p_2}{S\rho g} + z_2\right) = x\left(\frac{S_m}{S} - 1\right)$$

or, $$h = x\left(\frac{S_m}{S} - 1\right) \tag{6.25}$$

This equation can also be used for inclined venturimeter.

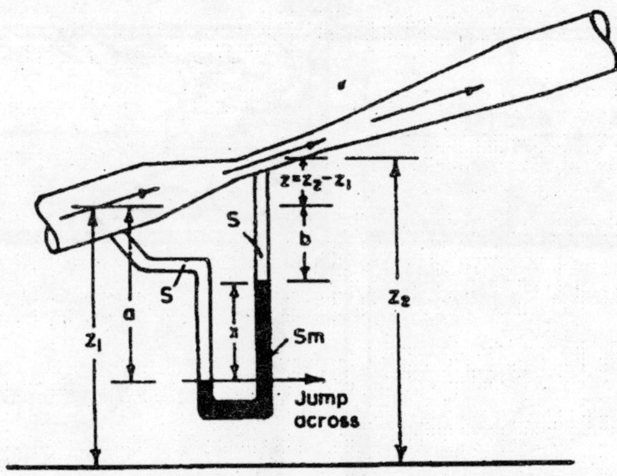

Fig. 6.11. Inclined venturimeter with U tube manometer

6.3.2 Orifice meter

The venturimeter is unquestionably the best instrument for measuring flow in pipes as it causes very little energy loss. Unfortunately, the insturment is expensive and in circumstances in which initial cost matters but head less in the meter do not, the orificemeter can be used.

A simple type of orifice meter consists essentially of a circular plate in which is machined a concentric hole of diameter d_o, as shown in Fig 6.12, installed in a pipe of daimeter d_1. It is to be observed that the section at which the streamlines come closest together is not at the orifice plate but at a section approximately $d_1/2$ downstream form it. This minimum crosssection of the stream tube is known as the vena contracta. Pressure connections are made at sections 1 and 2; section 1 is at a distance of pipe diameter d_1 from the orifice and section 2 is at the vena contracta. The pressue at piont (2) is minimum and velocity is maximum, because the vena contracta area is minimum. Downstream of section 2 at the place where the flow lines breaks down into a highly turbulent area wherein the flow is returned to its original area. If the second pressure tapping is located here, Bernoulli's equation cannot be applied because of the very large energy loss caused by the flow divergence from the narrow vena contracta, to the full pipe bore. The area at vena contracta, a_2, is less than area of the orifice, a_o. Thus the ratio of area at vena contracta to area of orifice is known as coefficient of contraction ($Cc = a_2/a_o$). The coefficient of contraction (Cc) is less than one.

Applying Bernoulli's equation between section 1 and section 2 in Fig 6.12

$$\frac{p_1}{\rho g} + \frac{v_1^2}{2g} = \frac{p_2}{\rho g} + \frac{v_2^2}{2g} \qquad [where,\ z_1 = z_2]$$

or, $$\frac{v_2^2}{2g} - \frac{v_1^2}{2g} = \left(\frac{p_1 - p_2}{\rho g}\right) = h \tag{6.26}$$

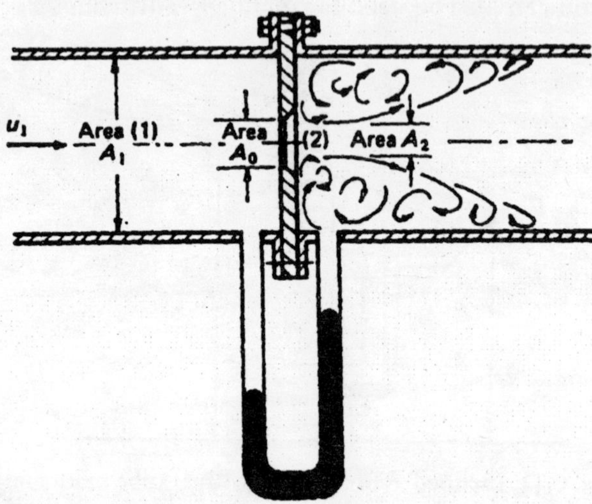

Fig. 6.12. Orifice meter with U tube

Applying continuity equation

$$Q = a_o\ v_o = a_1\ v_1 = a_2\ v_2$$

Taking $a_o\ v_o = a_2\ v_2$

$$v_o = \frac{a_2}{a_o} v_2 = C_c\ v_2 \tag{6.27}$$

and $a_o\ v_o = a_1\ v_1$

$$v_o = \frac{a_1}{a_o} v_1 \tag{6.28}$$

Therefore, from the equation (6.20) and (6.21).

$$\frac{a_1}{a_o} v_1 = C_c\ v_2$$

$$v_2 = \frac{a_1}{a_o\ C_c} v_1$$

Substituting the value of v_2 in equation (6.26)

$$\frac{v_1^2}{2g}\left[\frac{a_1^2}{a_0^2\ C_c^2} - 1\right] = h$$

or, $$v_1^2 = \frac{2gh\ a_0^2\ c_c^2}{a_1^2 - a_0^2\ c_c^2}$$

or, $$v_1 = \frac{a_o C_c \sqrt{2gh}}{\sqrt{a_1^2 - a_0^2 C_c^2}} \tag{6.29}$$

The equation (6.29) respresents the theoretical velocity, as no allowance for losses was made in Bernoulli's equation at the beginning of the derivation. To correct this for the fact that due to losses the actual pressure differnce $(p_1 - p_2)$ will be somewhat greater, the coefficient of velocity c_v is introduced,

Thus

$$C_v = \frac{actual\ velocity}{theoretical\ velocity}$$

and therefore the actual velocity in the pipe is given by

$$v_{1act} = \frac{C_c C_v C_o \sqrt{2gh}}{\sqrt{a_1^2 - a_0^2 C_c^2}}$$

and the actual discharge

$$Q = a_1 v_1$$

$$= \frac{a_1 a_o C_c C_v \sqrt{2gh}}{\sqrt{a_1^2 - a_0^2 C_c^2}}$$

$$= \frac{a_1 a_o C_d \sqrt{2gh}}{\sqrt{a_1^2 - a_0^2 C_c^2}} = \frac{C_d a_o \sqrt{2gh}}{\sqrt{1 - a_0^2 / a_1^2 C_c^2}} \tag{6.30}$$

where C_d is the coefficient of discharge of the orifice. The above expression can be simplified for the discharge through the orificemeter by using a coefficient C_o expressed as

$$C_o = C_d \frac{\sqrt{1 - a_0^2 / a_1^2}}{\sqrt{1 - a_0^2 / a_1^2 C_c^2}} = C_d \frac{\sqrt{a_1^2 / l_0^2 - 1}}{\sqrt{a_1^2 / a_0^2 - C_c^2}}$$

Therefore, $$Q = \frac{C_o a_o \sqrt{2gh}}{\sqrt{1 - a_0^2 / a_1^2}} \tag{6.31}$$

The value of C_o is a function of the Reynolds number and d_o/d_1. Typical value of C_o are given in Fig 6.13. Note also that the value of C_o depends on the specific construciton of the orificemeter (i.e, the placement of pressure taps, whether the orifice plate edge is square or beveled, etc.).

At high values of Re (greater than 10^5) the coefficient of discharge is nearly constant for given values of m. Also C_o decreasing with increasing values of m, this being mainly due to the increase of the coefficient of contraction.

The orificemeter described above are known as sharp-edged orifices and usually when chamfered on one side must be used with the square edge facing the flow. In general for high values of Re and m, the coefficient of discharge (C_o) is approximately 0.6 to 0.61 indicating that losses are fairly high.

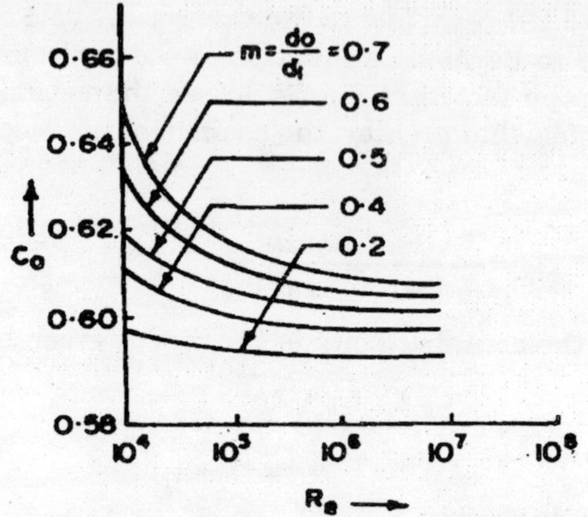

Fig. 6.13. Variation of c_o v/s Re

6.3.3 Flow nozzle or Nozzle meter

The flow nozzle is shown in Fig. 6.14 is essentially a venturimeter with divergent part omitted and the basic equation are the same as those for the venturimeter. The dissipation of energy downstream of the throat is greater than for a venturimeter but this disadvantage is often offset by the lower cost of nozzle.

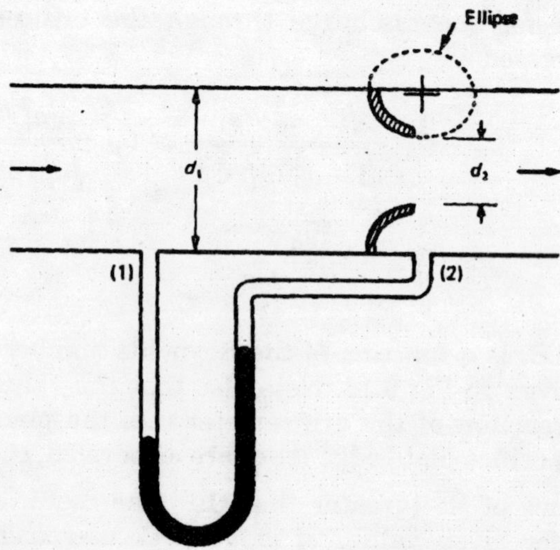

Fig. 6.14. Flow nozzle with U tube manometer.

The flow nozzle is a gradual contraction of the flow stream followed by a short straight cylindrical section is shown in Fig. 6.14. Three variation of which are shown in Fig. 6.15. This is placed between flanges of pipe sections rather than a simple (and less expensive) plate with a hole as in an orificemeter. The resulting flow pattern for the nozzle meter is closer to ideal than the orifice meter flow. There is only a slight vena contracta and the secondary flow separation is less severe, but there still are viscous effect. There are accounted for the use of the nozzle discharge coefficient C_n.

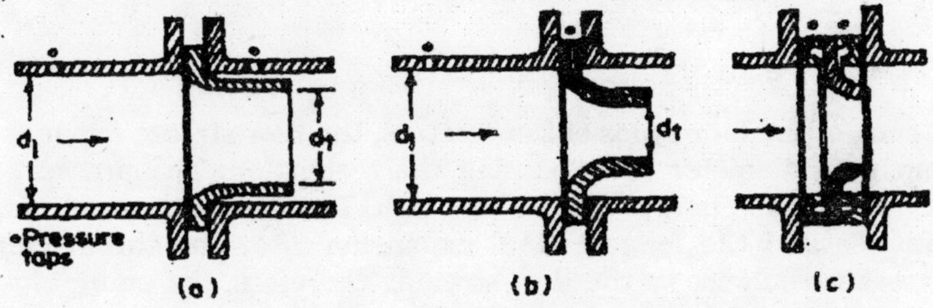

Fig. 6.15. Typical nozzle meter construction.

The value of C_n is very close to one. The value of C_n is a function of the diameter ratio $m = d_t / d_1$ and the Reynolds number. Typical values obtained form experiments are shown in Fig 6.16. Again, precise value of C_n depends on the specific details of the nozzle design. The ASME recommended a equation of C_n :

$$C_n = 0.9975 - 0.00653 \left(\frac{10^6}{R_e}\right)^a \tag{6.32}$$

where a = 0.5 for Reynold number less than 10^6 and a = 0.2 for Reynolds number more than 10^6. The Reynolds number is based on the throat diameter of the nozzle in equation (6.32).

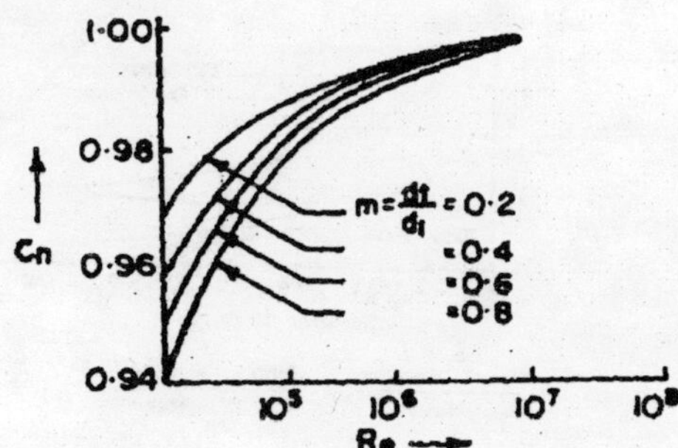

Fig. 6.16. Variation of C_n v/s Re

The flow rate is calculated with the help of a formula

$$Q = \frac{C_n\ a_1 \sqrt{2gh}}{\sqrt{1 - \dfrac{at^2}{a_1^2}}}$$

The nozzle meter is more efficient (less energy dissipated than the orifice meter).

6.3.4. Flow tubes

Several proprietary design for modified flow meters called flow tubes are available. These can be used for application similar to those for which the venture, nozzle and orifice meter are used, but flow tubes have somewhat lower pressure loss (higher pressure recovery).

6.3.5 Pressure recovery

In each of the four types of flow meters, the flow stream expands back to the main pipe diameter after passing the restriction. The pressure drop is proportional to the energy loss. P_1-P_3 = ρgh. The careful streamlining of the venturimeter and the long gradual expansion after the throat cause very little excess turbulence in the flow stream. Therefore, the energy loss is low and the pressure recovery is high. The lack of gradual expansion causes the nozzle to have a lower pressure recovery, while that for an orifice is still lower. The best pressure recovery is obtained form the flow tube. Fig 6.17 shows the comparison among the several types of flow meter with regard to pressure recovery.

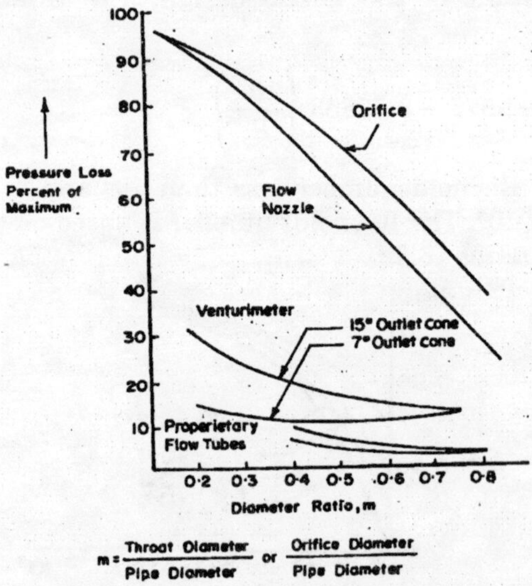

Fig. 6.17. Comparison of pressure loss for flow meters.

6.4 VELOCITY MEASUREMENTS

The velocity at a point or a number of point through out a section in a fluid stream is often needed in order to establish the velocity profile. This velocity profile may be used to obtain the average velocity thorugh out the section from an integration of the velocity profile in order to determine the flow rate in the fundamental studies of boundary layers or wakes.

A point velocity is almost impossible to measure, since any sensing device occupies a finite region. However, if the area of flow occupied by the sensing element is very small compared with the total are of the flow stream, we may consider the measured velocity to be essentially a point velocity.

6.4.1 Static, stangnation and dynamic pressures

A useful concept associated with Bernoulli's equation deals with the stagnation and dynamic pressures. These pressures arise from the conversion of kinetic energy in a flowing fluid into a pressure rise as the fluid is brought to rest. Each term of the Bernoulli equation (6.33) has the dimensions of force per unit area, N/m².

$$p + \frac{1}{2}\rho\, v^2 + \rho\, g\, z \;=\; Constant \tag{6.33}$$

The first term, p is the actual thermodynamic pressure of the fluid as it flows. To measure its value one could move along with the fluid, thus being "static" relative to the moving fluid. Hence, it is normally termed the static pressure. Another way to measure the static pressure would be to drill a hole in a flat surface and fasten a piezometer.

The third term in equation (6.33), ρgz is termed the hydrostatic pressure. It is not actually a pressure, but does represent the change in pressure possible due to potential energy variations of the fluid as a result of elevation changes.

The second term in equation (6.33), $\frac{\rho v^2}{z}$ is termed the dynamic pressure. Its interpretation can be seen in Fig. 6.18 by considerning the pressure at the end of a small tube inserted into the flow and pointing upstream. After the initial transient motion has died out, the liquid will fill the tube to a height h_2 as shown in Fig. 6.19. The fluid in the tube, including that at its tip, (2) will be stationary. That is, $v_2=0$, or point (2) is a stagnation point.

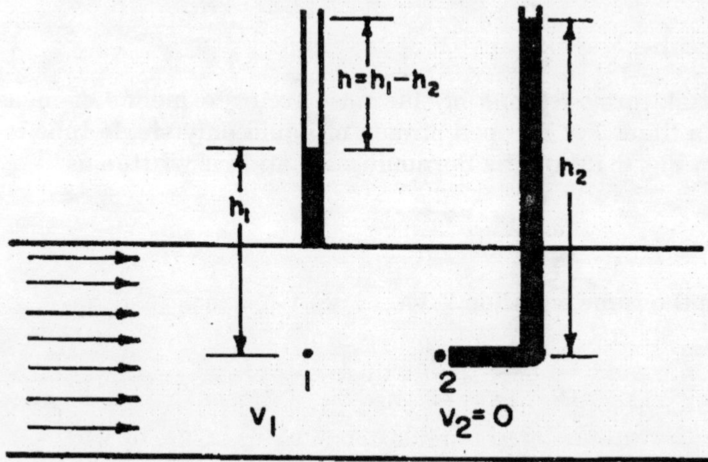

Fig. 6.18. Measurement of static and stagnation pressure

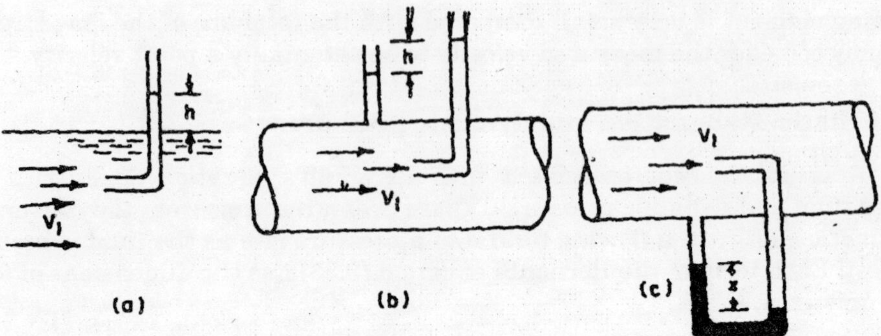

Fig. 6.19. Plot tube in (a) an open liquid stream (b) a pipe, and (c) a pipe connected to a menometer.

Appiying Bernoulli's equation between point (1) and (2), using $v_2 = 0$ and assuming that $z_1 = z_2$, we find that

$$p_2 = p_1 + \frac{\rho v_1^2}{2} \tag{6.34}$$

Hence, the pressure at the stagnation point is greater than the static pressure, P_1, by an amount $\frac{\rho v_1^2}{2}$ the dynamic pressure. The equation (6.34) can be written in words as:

Stagnation pressure = Static pressure + Dynamic pressure (6.35)

If elevation effects are neglected, the stagnation pressure, is the largest pressure obtainable along a streamline. It represernts the conversion of all of the kinetic energy into a pressure rise.

Knowledge of the values of the static and stagnation pressures in a fluid implies that the fluid speed can be calculated. This is the principle on which the Pitot static tube is based.

6.4.2 Pitot tube

Pitot tube provides one of the most accurate means of measuring the velocity of a fluid. For an open stream of liquid only single tube is necessary as shown in Fig. 6.19(a). The Bernoulli's equation is written as

$$\frac{p_2}{\rho g} = \frac{p_1}{\rho g} + \frac{v_1^2}{2g}$$

This is the same equation 7.34.

$$\frac{p_2}{\rho g} - \frac{p_1}{\rho g} = \frac{v_1^2}{2g}$$

or, $$h = \frac{v_1^2}{2g}$$

or, $$v_1 = \sqrt{2gh} \tag{6.36}$$

If the stream velocity at a point in a pipe is to be measured (Fig. 6.19(b)) the same equation is obtained.

In practice, it is difficult to measure the height h above the surface of the moving fluid. The tubes in Fig. 6.19(b) may be connected to a differential U tube manometer as shown in Fig. 6.19 (c) to result in a more convenient system. The h is obtained very easily as we obtained in the case of venturimeter

$$h = x\left(\frac{S_m}{S} - 1\right)$$

Therefore,

$$v = \sqrt{2g\,x\left(\frac{S_m}{S} - 1\right)} \tag{6.37}$$

6.4.3 Pitot- static tube

The tubes recording static pressure and stagnation pressure are frequently combined into one instrusment known as Pitot-static tube (Fig. 6.20). This tube consists of two concentric cylinders bent into an L shape and with various head forms, such that the inner cylindrical is at the stagnation pressure and the annular region between cylinders is at the free stream pressure. The centre tube measures the stagnation pressure at its open tip. The outer tube is made with several small holes at an appropriate distance from the tip so that they measure the static pressure. The actual shape and size of Pitot static tubes vary considerably. Some of the more common types are shown in Fig. 6.20.

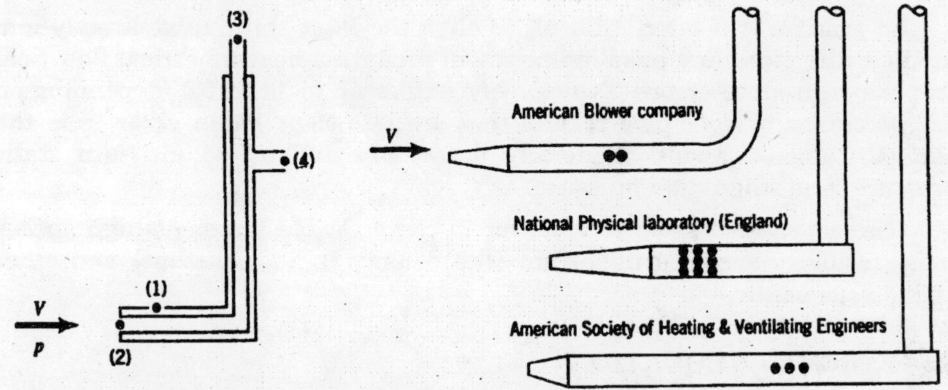

Fig. 6.20. Pilot static tube and typical Pitot-static tube designs.

The Pitot-static tube privides a simple, relatively inexpensive way to measure fluid speed. Its use depends on the ability to measure the static and stagnation pressures. Care is needed to obtain these values accurately. The indeed static pressure is subject to some uncertainty because of the static holes may be exposed to a pressure slightly different from that of the free stream. If the flow is turbulent, the high- frequency velocity and pressure fluctutions have the effect of indicating a velocity higher than the time average velocity and it is common to use a coefficeint c_v, so that

$$v = C_v \sqrt{2gh} \tag{6.38}$$

Unless an instrument is carefully calibrated, c_v is usually assumed to be unity.

The pressure along the surface of an object varies from the stagnation pressure at its stagnation point to value that may be less than the free stream pressure. A typical pressure varation for a Pitot-tybe is indicated in Fig 6.21. Clearly it is important that the pressure taps be properly located to ensure that the pressure measured is actually the static pressure.

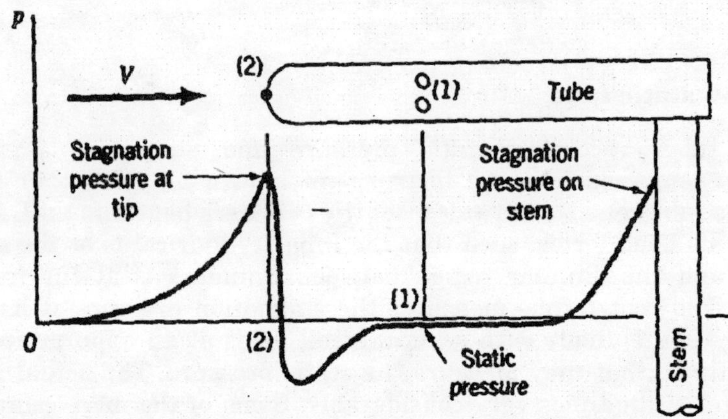

Fig. 6.21. Typical pressure distribution along a Pitot-Static tube.

In practice it is often diffecult to align the Pitot static tube directly into the flow direction. Any misalignment will produce a nosymmetrical flow field that may introduce errors. Typical, yaw angles up to 12 to 20° (depending on the particular design) give results that are less than 1% in error from the perfectly aligned results Generally it is more difficult to measure static pressure than stagnation pressure.

The above discussion is valid for in compressible flows. At high speeds compressibility becomes important (the density is not constant) and other phenomena occur.

6.5 MOMENTUM EQUATION

The momentum equation is based upon Newton's second law of motion applied to the entire force system. It provides a simple means of relating the hydrostatic force and the boundary force to the change of momentum flux-that is, the momentum which passed a point during a unit time intreval. The force and resulting acceleration are analyzed from the overall viewpoint if the change which accur, rather than the viewpoint the details of the flow processes and internal mechonics. There are number of problems which can be solved with the help of momentum equation. Examples of these problems are: forces on a pipe bend; solution of the hydraulic jump, forcec on a propeller blade, the forces of a jet of water impinging on a wall or on the ground, and jet propulsion.

6.5.1 Impulse-Momentum equation

In mechanics, the momentum of a particle is defined as the product of its mass m and it velocity v:

$$\text{Momentum} = mv$$

The particles of a fluid stream will possess momentum, and, wherever the velocity of the stream is changed in magnitude or direction, there will be a corresponding change in the momentum of the fluid particles. In accordance with Newton's second law, a force is required to produce this change, which will be propartional to the rate at which the change of momentum occures.

It may therefore be useful to restate the law:

The rate of change of momentum is proportional to the impressed force, and takes place in the direction of that force.

or, Mathematically,

$$F = \frac{d}{dt}(m\,v) \tag{6.39}$$

$$F\,dt = d\,(mv)$$

$$F\,dt = m\,dv$$

where Fdt = Impulse

mdv = change of momentum.

The equation (6.39) can be written as

$$F = \text{mass} \times \text{acceleration.}$$

$$F = ma \tag{6.40}$$

If the acceleration be equal to the final velocity minus the initial velocity divided by the time interval over which the velocity change took place, the equation takes the form.

$$F = m\left(\frac{v_2 - v_1}{t}\right)$$

$$F = \frac{m}{t}\left(v_2 - v_1\right)$$

$$F = \dot{m}\,(v_2 - v_1) \tag{6.41}$$

The $\dot{m} = \frac{m}{t}$ is known as mass flow rate.

The mass flow rate is also written as

$$\dot{m} = \frac{\text{mass}}{\text{time}} = \text{Density} \times \frac{\text{volume}}{\text{time}}$$

Density × Volume flow rate [i. e. discharge]

Substituting the value of m in equation (6.41)

$$F = \rho\, Q\, (v_2 - v_1)$$

The above equation represents the reaction force on a surface would have on the fluid stream by virtue of a change in velocity, ($v_2 - v_1$). Conversely, The fluid jet would exert a force on the surface of equal magnitude, but opposite in sense, that is, –F. Thus the equation for the force of a jet impinging on a surface is

$$F = \rho\, Q\, (v_1 - v_2) \tag{6.42}$$

6.5.2 Momentum equation for two-and there dimensional flow along a stream line

The student should note carefully that the derivation of momentum equation is based on the fllowing assumptions

(1) The fluid is incompressible.

(2) The velocity change uniformly.

(3) The pipe material is non elastic.

Fig. 6.22 shows a two dimensional problems in which v_1 makes an angle α_1 with the x axis while v_2 makes a corresponding α_2. Since both momentum and force are vector quantites, they can be resolves into y components in the x and y directions and equation (6.41) applied. Thus Fx and Fy are the components of the resultant force on the element of fluid ABCD.

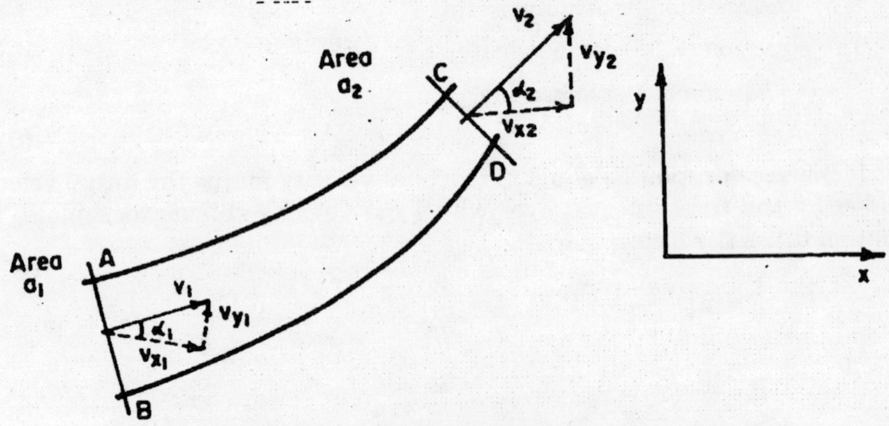

Fig. 6.22. Momentum equation for two dimensional flow.

Fx = Rate of change of momentum of fluid in x direciton

$$= \rho Q \left(v_2 \;\cos\alpha_2 - v_1 \;\cos\alpha_1\right)$$

$$= \rho Q \left(v_{x_2} - v_{x_1}\right) \tag{6.43}$$

Similarly

$$F_y = \rho\, Q \left(v_2 \sin\alpha_2 - v_1 \sin\alpha_1\right)$$

$$= \rho\, Q \left(v_{y_2} - v_{y_1}\right) \tag{6.44}$$

The resultant force is given by

$$F_R = \sqrt{F_x^2 + F_y^2} \tag{6.45}$$

Direction of the resultant force with horizontal is found by

$$\tan\theta = \frac{F_y}{F_x} \tag{6.46}$$

For three dimensional flow, the same method can be used, but the fluid will also have component velocities v_{z1} and v_{z2} in the Z direction and the corresponding rate of change of momentum in this direction will require a force

$$F_z = \rho\, Q \left(v_{z_2} - v_{2_1}\right) \tag{6.47}$$

Therefore, the resultant force for three dimensional is given by

$$F_R = \sqrt{F_x^2 + F_y^2 + F_z^2} \tag{6.48}$$

6.5.3 Momentum correction factor

The momentum equation is based on the assumption that the velocity is constant across any given crosssection. When a real fluid flows past a solid boundary, shear stresses are developed and the velocity is no longer uniform over the crosssection. In a pipe, the velocity will vary from zero at the wall to a maximum at the centre. The momentum per unit time for the whole flow can be found by summing the momentum per unit time through each element if the cross section provided that there are sufficient small for the velocity perpendicular to each element to be taken as uniform. the true rate of momentum flow perpendicular to the cross section is not $\rho\, a\, v^2$ *but* $\beta \rho\, a\, v^2$. Here v is the mean velocity over the crosssection, and β is the momentum correction factor. With constant density, the value of β for the velocity distribution is found by the analysis $u / u_{\max} = (Y / R)^{1/7}$ *is* $49/72 \times \left(\frac{60}{40}\right)^2 = 1.02$ which for most purposes differs negligibly from unity. Disturbances upstream, however, may give a markedly higher value. For fully-developed laminar flow in a circular pipe β = 4/3. In a given situation β is always less than ∝ the kinetic energy correction factor.

6.5.4 Application of the momentum equation

As already stated, the momentum equation is very useful in the bulk solution of many problems in which the net forces causing a change in momentum flux are to be determined. Application of these examples are discussed in detail.

6.5.4.1 Forces on a pipe bend

When the flow is confined within a pipe the static pressure may vary from point to point and forces due to differences of static pressure may be taken into account. Consider the pipe bend illustrated in Fig 6.23 in which not only the direction of flow but also the crosssectional area is changed.

We are assuming here that the bend is in the horizontal plane : Change of elevation are negligible. We also assume that the condition at section 1 and 2 are uniform and that the streamline there are straight and parallel. the pressure and cross sectional are a at section 1 are P_1 and a_1 respectively. The pressure and cross sectional are at section 2 are P_2 and a_2 respectively. The velocities at sections 1 and 2 are v_1 and v_2. Let the pipe bend exert a force on the fluid F_{BF}, with comonents F_{BFX} and F_{BY} in the x and y direction indicated. The forec due to the pressure p_1 and p_2 of the fluid outside the control volume acting on areas a_1 and a_2 at stations 1 and 2 will be $p_1 a_1$ and $p_2 a_2$. The force due to gravity will be zero here.

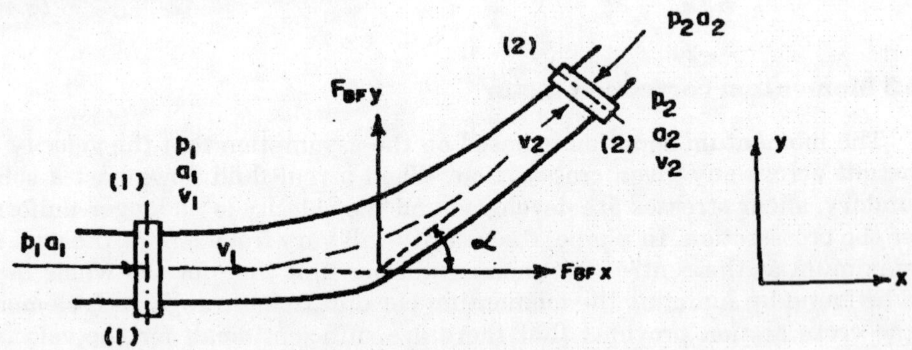

Fig. 6.23. Forces or a pipe bend.

The momentum equation in the x direction on the fluid in the control volume is :

$$\sum F_x = \rho Q \left(v_{x_2} - v_{x_1}\right)$$

$$p_1 a_1 - p_2 a_2 \cos\alpha + F_{BFX} = \rho Q \left(v_2 \cos\alpha - v_1\right) \quad (6.49)$$

The momentum equation in the y direction on the fluid in the control volume is :

$$\sum F_y = \rho Q \left(v_{y_2} - v_{y_1}\right)$$

$$0 - p_2 a_2 \sin\alpha + F_{BFY} = \rho Q \left(v_2 \sin\alpha - 0\right)$$

or, $$-p_2\, a_2 \sin\alpha + F_{BFy} = \rho\, Q\, (v_2 \sin\alpha) \tag{6.50}$$

From the above equations (6.49) and (6.50), F_{BFX} and F_{BFY} may thus be determined.

The force exerted by the fluid on the bend F_{FB} is equal and opposite to the F_{BF}.

It is important to note that sometimes other effects may give rise to additianal forces which will then have to be taken into account. These may include,

(1) the weight of the conduit itself

(2) the weight of the fluid in the conduit, and

(3) inertia forces such as due to value closure or pulsating flow.

The equations (6.49) and (6.50) can be written for force exerted by the fluid on bend as :

$$p_1\, a_1 - p_2\, a_2 \cos\alpha - F_{BFx} = \rho\, Q\, (v_2 \cos\alpha - v_1)$$

or, $$F_{FBx} = p_1\, a_1 - p_2\, a_2 \cos\alpha + \rho\, Q\, (v_1 - v_2 \cos\alpha) \tag{6.51}$$

$$= \text{Static force} + \text{Dynamic force}$$

$$= +v_e\, (\rightarrow) = -v_e\, (\leftarrow)$$

and,

$$0 - P_2\, a_2 \sin\alpha - F_{FBy} = \rho\, Q\, (v_2 \sin\alpha - 0)$$

or, $$F_{FBy} = -P_2\, a_2 \sin\alpha + \rho\, Q\, (-v_2 \sin\alpha) \tag{6.52}$$

$$= +v_e\, (\uparrow) = v_e\, (\downarrow)$$

Now the resultant force on the bend is given by

$$F_R = \sqrt{(F_{FBx})^2 + (F_{FBy})^2} \tag{6.53}$$

and the direciton of resultant force with the horizontal direction is given by

$$\tan\theta = \frac{F_y}{F_x} \tag{6.54}$$

6.5.4.2 Force due to the diflection of a jet by a curved vane

If the stream is deflected by ther curved vane (Fig. 6.24) enrering and leaving tangentially without impact, a force will be exerted between the fluid and the surface of the vane to cause this change of momentum. Since the pressure is atmospheric both where the fluid enters the control volume and where it leaves, therefore, the ``static force" is zero.

It is usually convenient to calculate the components of this force parallel and perpendicular to the direction of the incoming stream by calculating the rate of change of momentum in these two directions. The components can then be conbined to give the magni-tude and direction of the resultant force which the vane exerts on the fluid, and the equal and opposite reaction of the fluid on the vane.

$$F_{Fvx} = Dynamic \quad force = \rho Q \left(v_1 - v_2 \cos \alpha\right) \tag{6.55}$$

$$F_{Fvy} = \rho Q \left(- v_2 \sin \alpha\right) \tag{6.56}$$

$$F_R = \sqrt{\left(F_{Fvx}\right)^2 + \left(F_{Fvy}\right)^2} \tag{6.57}$$

$$\tan \theta = F_{Fvy} / F_{Fvx} \tag{6.58}$$

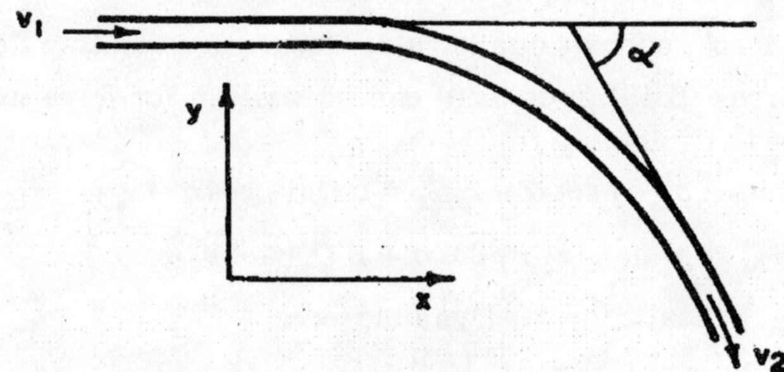

Fig. 6.24. Force exerted on curved vave.

6.5.4.3 Force at a nozzle

As a special case, we may consider the horizontal nozzle illustrated in Fig 6.25. Assum- ing uniform conditions with streamlines straight and parallel at the section 1 and 2 we have

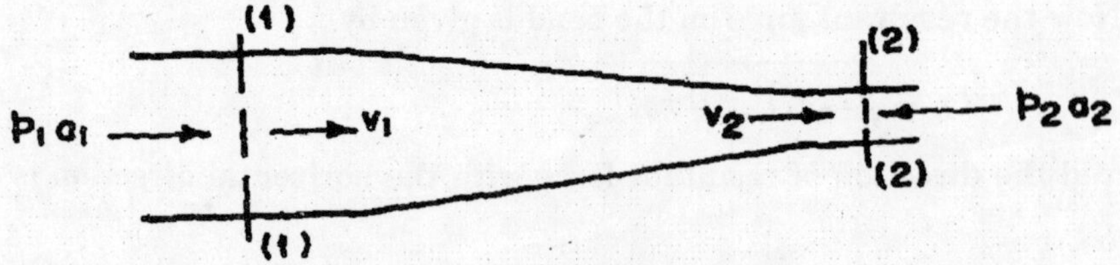

Fig. 6.25. Force at a nozzle.

Force exerted in the direction on the fluid between plates 1 and 2.

$$p_1 \, a_1 - p_2 \, a_2 + F = \rho Q \left(v_2 - v_1\right) \tag{6.59}$$

6.5.4.4 Reaction of a jet

A typical example is the reaction force exerted when a fluid is discharged in the form of a high velocity jet, and which is applied to the propulsion of ships and aircraft through the use of propellers, pure jet engines, and rocket motors. The propulsive force can be determined from the application of the momentum equation to flow through a suitable control volume.

For aircraft or missiles propelled in the atmosphere it is not nesessary to employ a self contained system, the propulsive force being obtained form the reaction of jet of atmospheric air which is taken in and accelerated by means of a propeller, turboprop or jet engine and expelled at the rear of the craft.

6.6 NAVIER-STOKES EQUATIONS

The Navier- Stokes equations are an expression of Newton's laws of motion and a constitutive law that relates the stresses in the fluid with the volume and shape distortions that a material element can undergo in a Newtonian fluid. This relationship has been restricted in this chapter to incompressible fluids, but no appreciable complexity would be introduced if gases had been included in the treatment.

In a fluid motion it is necessary to consider the following two classes of forces : forces acting throughout the mass of the body (gravitational forces) : forces acting on the boundary (pressure and friction). The pressure force is due to the thermodynamic pressure. A viscous fluids in motion has additional surface forces acting on a fluid element-tangential or shearing forces and normal forces which are in addition to the pressure forces-acting upon an inviscid fluid in motion.

Surface forces on a viscous fluid element in motion are shown in Fig 6.26. Normal stresses are designated as σ and shear stresses as τ. The first subscript indicates the plane of the area element by giving the direction of the normal to the area, and the second subscript indicates the direction of the stress component. Normal stresses reduce in magnitude to the fluid pressure as the shear stress is reduced to zero, whereas the viscosity goes to zero.

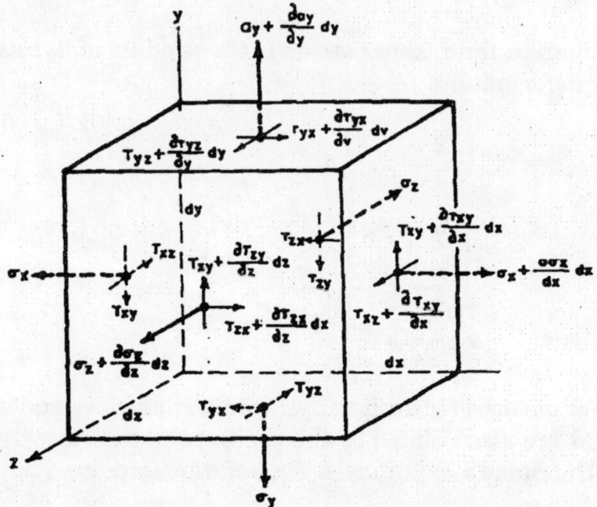

Fig. 6.26. Tangential and normal surface forces on a viscous fluid element in motion.

As inspection of the surface forces in Fig 6.26 indicates that the net surface forces in the x, y and z directions are

$$dF_s(x) = \left(\frac{\partial \sigma_x}{\partial x} + \frac{\partial \tau_{yx}}{\partial y} + \frac{\partial \tau_{zx}}{\partial z}\right) dx\, dy\, dz$$

$$dF_s(y) = \left(\frac{\partial \sigma_y}{\partial y} + \frac{\partial \tau_{zy}}{\partial z} + \frac{\partial \tau_{xy}}{\partial x}\right) dx\, dy\, dz \tag{6.60}$$

$$dF_s(z) = \left(\frac{\partial \sigma_z}{\partial z} + \frac{\partial \tau_{xz}}{\partial x} + \frac{\partial \tau_{yz}}{\partial y}\right) dx\, dy\, dz$$

The body forces per unit mass has components x, y and z as before. Then Newton's second law may be written for a unit volume as

$$\rho\, a_x = \rho\, X + \frac{\partial \sigma_x}{\partial x} + \frac{\partial \tau_{yx}}{\partial y} + \frac{\partial \tau_{zx}}{\partial z}$$

$$\rho\, a_y = \rho\, Y + \frac{\partial \sigma_{xy}}{\partial x} + \frac{\partial \tau_{y}}{\partial y} + \frac{\partial \tau_{zy}}{\partial z} \tag{6.61}$$

$$\rho\, a_z = \rho\, Z + \frac{\partial \sigma_{xz}}{\partial x} + \frac{\partial \tau_{yz}}{\partial y} + \frac{\partial \tau_{z}}{\partial z}$$

It can be shown that by taking moments about the edges for the elemental volume in Fig 6.26 corresponding to the axes,

$$\tau_{xy} = \tau_{yx}$$

$$\tau_{yz} = \tau_{zy}$$

$$\tau_{zx} = \tau_{xz}$$

For a Newtonian fluid, shear stress is the product of dynamic viscosity and rate of angular deformation.

$$\tau_{xy} = \mu \left(\frac{\partial u}{\partial y} + \frac{\partial v}{\partial x}\right)$$

$$\tau_{yz} = \mu \left(\frac{\partial v}{\partial z} + \frac{\partial w}{\partial y}\right) \tag{6.62}$$

and $$\tau_{zx} = \mu \left(\frac{\partial w}{\partial x} + \frac{\partial u}{\partial z}\right)$$

The normal stresses for an isotropic (properties independent of orientation. Newtonian fluid are also related to the pressure p, the viscosity μ, and velocity gradient from the Stoke's hypothesis. The relationship are

$$\sigma_x = -\, p + 2\mu \frac{\partial u}{\partial x} - \frac{2}{3}\, \mu \left(\frac{\partial u}{\partial x} + \frac{\partial v}{\partial y} + \frac{\partial w}{\partial z}\right)$$

$$\sigma_y = -\, p + 2\mu \frac{\partial v}{\partial y} - \frac{2}{3}\, \mu \left(\frac{\partial u}{\partial x} + \frac{\partial v}{\partial y} + \frac{\partial w}{\partial z}\right) \tag{6.63}$$

$$\sigma_z = -p + 2\mu \frac{\partial w}{\partial z} - \frac{2}{3}\mu\left(\frac{\partial u}{\partial x} + \frac{\partial v}{\partial y} + \frac{\partial w}{\partial z}\right)$$

For an inviscid fluid, $\mu = 0$ and it is thus obvious that $\sigma_x = \sigma_y = \sigma_z = -p$. For a viscous fluid in motion, equations (6.62) can be added together so that the pressure may be defined as the arithmatic average of the three normal stresses.

$$-p = \frac{1}{3}\left(\sigma_x + \sigma_y + \sigma_z\right) \tag{6.64}$$

In cartesian coordinates, acceleration can be expressed in terms of the individual components.

$$a_x = u\frac{\partial u}{\partial x} + v\frac{\partial u}{\partial y} + w\frac{\partial u}{\partial z} + \frac{\partial u}{\partial t}$$

$$a_y = u\frac{\partial v}{\partial x} + v\frac{\partial v}{\partial y} + w\frac{\partial v}{\partial z} + \frac{\partial v}{\partial t} \tag{6.65}$$

$$a_z = u\frac{\partial w}{\partial x} + v\frac{\partial w}{\partial y} + w\frac{\partial w}{\partial z} + \frac{\partial w}{\partial t}$$

Substitution of equation (6.65) for the accelerations, equation (6.62) for the shear stresses, and equation (6.63) for the normal stresses in equation (6.61) will give the complete equations of motion for the viscous fluid motion, the Navier Stokes equations. For a constant viscosity fluid and for incompressible flow they become

$$\frac{\partial u}{\partial t} + u\frac{\partial u}{\partial x} + v\frac{\partial u}{\partial y} + w\frac{\partial u}{\partial z} = X - \frac{1}{\rho}\frac{\partial \rho}{\partial x} + v\left(\frac{\partial^2 u}{\partial x^2} + \frac{\partial^2 u}{\partial y^2} + \frac{\partial^2 u}{\partial z^2}\right)$$

$$\frac{\partial v}{\partial t} + u\frac{\partial v}{\partial x} + v\frac{\partial v}{\partial y} + w\frac{\partial v}{\partial z} = Y - \frac{1}{\rho}\frac{\partial \rho}{\partial y} + v\left(\frac{\partial^2 v}{\partial x^2} + \frac{\partial^2 v}{\partial y^2} + \frac{\partial^2 v}{\partial z^2}\right) \tag{6.66}$$

$$\frac{\partial w}{\partial t} + u\frac{\partial w}{\partial x} + v\frac{\partial w}{\partial y} + w\frac{\partial w}{\partial z} = Z - \frac{1}{\rho}\frac{\partial \rho}{\partial z} + v\left(\frac{\partial^2 w}{\partial x^2} + \frac{\partial^2 w}{\partial y^2} + \frac{\partial^2 w}{\partial z^2}\right)$$

Equations (6.66) are commonly called the Navier Stokes equations, named in honour of the French mathematician L.M.H. Navier (1758-1836) and the English mechanician Sir G.G. Stokes (1819-1903), who were responsible for their formulation. These three equations of motion (Equation 6.66), when combined with the conservation of mass equation (continuity equation) $\frac{\partial u}{\partial x} + \frac{\partial v}{\partial y} + \frac{\partial w}{\partial z} = 0$, provide a complete mathematical description of the flow of incompressible, Newtonian fluids. We have four equations and four unknowns (u,v,w and p), and therefore the problem is "well posed" in mathematical terms. Unfortunately, because of the general complexity of the Navier-Stokes equations (they are nonlinear, second order, partial differential equations), they are non amenable to exact mathematical solutions except in a few instances.

However in those in which solutions have been obtained and compared with experimental result; the result have been in close agreement. Thus, the Navier-Stokes equations are considered to be the governing differential equations of motion for incompressible, Newtonian fluids.

6.2.1 Navier-Stokes equation in vector form and meaning of each term

The Navier-Stokes equation (6.66) can be written in vector form as :

$$\frac{DV}{Dt} = g - \frac{1}{\rho}\nabla p + \nu \nabla^2 V \tag{6.67}$$

or, $$\frac{\partial V}{\partial t} + (V.\nabla)V = g - \frac{1}{\rho}\nabla p + \nu \nabla^2 V \tag{6.68}$$

(a) (b) (c) (d) (e)

The dimensions of each term are L/T^2, i.e. each term is an acceleration.

The term (a), $\frac{\partial V}{\partial t}$, represents the local acceleration of the fluid particle at a fixed point in space. For steady flow, this term is zero.

The term (b), $(V.\nabla)V$, is the convective acceleration of the fluid particle, and it predicts how the flow differes from one space location to the next at the same instant of time. Uniform flow has no convective acceleration of course, since uniform means of the same value.

The term (c), (g), represents the acceleration due to gravity.

The term (d), $-\frac{1}{\rho}\nabla p$, is the pressure acceleration due to the pumping action of the flow.

The term (e), $\nu \nabla^2 v$, is the viscous deceleration due to the fluid's frictional resistance to objects moving through it.

The physical significance of the various terms of equation (6.68) can be illustrated to same extent by considering a few examples of fluid flow.

Consider the motion of an inviscid fluid. Such fluids are considered to be ideal, so that the Navier-Stokes equation for inviscid fluid flows would then include terms (a) to (d) to equation (6.68)

$$\frac{\partial V}{\partial t} + (V.\nabla)V = F - \frac{1}{\rho}\nabla p \tag{6.69}$$

and is popularly called Euler's equation, this is a first order nonlinear partial differential equation.

Another example of the Navier-Stokes equation is the case of very slow fluid motion. Such fluid flow problems are mathematically defined as those where the total acceleration $\frac{DV}{Dt}$ [terms (a) and (d) of equation (6.68)] and acceleration due to gravity, term (c), are both zero. The Navier-Stokes equation for very slow motion is then

$$\nabla p = \mu \nabla^2 V$$

and is popularly called Stokes flow, or creep flow. It applies to the analysis of fluid behaviour in lubrication mechanics, capillary flows, and certain molten metals. Parallel flow constitutes that class of motion where all velocity components save one are zero. If that flow were in the x direction, then terms (b) and (c) and equation (6.68) would be zero. The Navier-Stokes equation then becomes.

$$\frac{\partial u}{\partial t} = -\frac{1}{\rho}\frac{\partial p}{\partial x} + \upsilon\left(\frac{\partial^2 u}{\partial y^2} + \frac{\partial^2 u}{\partial z^2}\right)$$

6.2.2 Navier-Stokes in cylindrical polar coordinates

In terms of cylindrical polar coordinates the Navier-Stokes can be written in the r, θ and z coordinates as

In r coordinate

$$\rho\left(\frac{\partial v_r}{\partial t} + v_r\frac{\partial v_r}{\partial r} + \frac{v_\theta}{r}\frac{\partial v_r}{\partial \theta} - \frac{v_\theta^2}{r} + v_z\frac{\partial v_r}{\partial z}\right)$$

$$= F_r - \frac{\partial p}{\partial r}$$

$$+ \mu\left(\frac{\partial^2 v_r}{\partial r^2} + \frac{1}{r}\frac{\partial v_r}{\partial r} - \frac{v_r}{r^2} + \frac{1}{r^2}\frac{\partial^2 v_r}{\partial \theta^2} - \frac{2}{r^2}\frac{\partial v_\theta}{\partial \theta} + \frac{\partial^2 v_r}{\partial z^2}\right) \tag{6.70}$$

In θ coordinate

$$\rho\left(\frac{\partial v_\theta}{\partial t} + v_r\frac{\partial v_\theta}{\partial r} + \frac{v_\theta}{r}\frac{\partial v_\theta}{\partial \theta} - \frac{v_r v_\theta}{r} + v_z\frac{\partial v_\theta}{\partial z}\right)$$

$$= F_\theta - \frac{1}{r}\frac{\partial p}{\partial \theta}$$

$$+ \mu\left(\frac{\partial^2 v_\theta}{\partial r^2} + \frac{1}{r}\frac{\partial v_\theta}{\partial r} - \frac{v_\theta}{r^2} + \frac{1}{r^2}\frac{\partial^2 v_\theta}{\partial \theta^2} - \frac{2}{r^2}\frac{\partial v_r}{\partial \theta} + \frac{\partial^2 v_\theta}{\partial z^2}\right) \tag{6.71}$$

In z co-ordinate

$$\rho\left(\frac{\partial v_z}{\partial t} + v_r\frac{\partial v_z}{\partial r} + \frac{v_\theta}{r}\frac{\partial v_z}{\partial \theta} + v_z\frac{\partial v_z}{\partial z}\right)$$

$$= F_z - \frac{\partial p}{\partial z} + \mu\left(\frac{\partial^2 v_z}{\partial r^2} + \frac{1}{r}\frac{\partial v_z}{\partial r} + \frac{1}{r^2}\frac{\partial^2 v_z}{\partial \theta^2} + \frac{\partial^2 v_z}{\partial z^2}\right) \tag{6.72}$$

Where F_r, F_θ and F_z are the body force components per unit mass in there direction indicated.

In cylindrical coordinates the stresses components for an incompressible fluid are

$$\sigma_r = -p + 2\mu \frac{\partial v_r}{\partial r}$$

$$\sigma_\theta = -p + 2\mu \left(\frac{1}{r}\frac{\partial v_\theta}{\partial \theta} + \frac{v_r}{r} \right)$$

$$\sigma_z = -p + 2\mu \frac{\partial v_z}{\partial z}$$

$$\tau_{r\theta} = \mu \left[r \frac{\partial}{\partial r}\left(\frac{v_\theta}{r} \right) + \frac{1}{r}\frac{\partial v_r}{\partial \theta} \right] \tag{6.73}$$

$$\tau_{\theta z} = \mu \left[\frac{\partial v\theta}{\partial z} + \frac{1}{r}\frac{\partial v_z}{\partial \theta} \right]$$

$$\tau_{rz} = \mu \left[\frac{\partial v_r}{\partial z} + \frac{\partial v_z}{\partial r} \right]$$

To provide a brief introduction to the use of the Navier-Stokes equations, a few of the simplest exact solutions are developed in the next section. These solutions will prove to be relative simple, this is not the case in general.

ILLUSTRATIVE EXAMPLES

6.1 Oil (relative density = 0.85) is flowing in a pipe under the conditions shown in Fig. 6.27. If the total head loss from point 1 to point 2 is 1m, find the pressure at point 2.

Solution

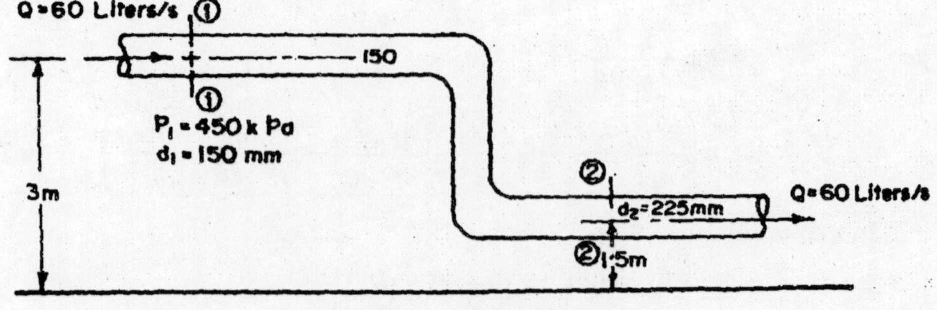

Fig. 6.27. Pipe line P. 6.1.

$$Q = a_1\ v_1 = a_2\ v_2$$

Therefore, $$v_1 = \frac{Q}{a_1} = \frac{60 \times 10^{-3}}{\frac{\pi}{4}(0.15)^2} = 3.395\ m/s$$

$$v_2 = \frac{Q}{a_2} = \frac{60 \times 10^{-3}}{\frac{\pi}{4}(0.15)^2} = 1.509\ m/s$$

Applying Bernoulli's equation between point (1) and (2),

$$\frac{p_1}{s\rho g} + \frac{v_1^2}{2g} + z_1 = \frac{p_2}{s\rho g} + \frac{v_2^2}{2g} + z_2 + h_1$$

$$\frac{450}{0.85 \times 9.81} + \frac{(3.395)^2}{2 \times 9.81} + 3 = \frac{P_2}{s\rho g} + \frac{(1.509)^2}{2 \times 9.81} + 1.5 + 1$$

$$53.96 + 0.5874 + 3 = \frac{p_2}{s\rho g} + 0.116 + 1.5 + 1$$

$$57.55 - 2.616 = \frac{p_2}{s\rho g}$$

$$p_2 = S\rho g\ (54.9379) = 458.1\ kPa$$ Ans

6.2 The centre line of tapered pipe AB slopes down from A to B at an angle of 30° to the horizontal. The distance AB is 5m and the diameter increases uniformly form 100mm at A to 150mm at B. The pipe carries petrol (S = 0.74) and pressure gauges are installed at A and B. Find (a) the flow rate when the reading on the pressure gauges are equal, (b) the pressure difference across AB for the same flow rate when the direction of taper is reversed.

Solution

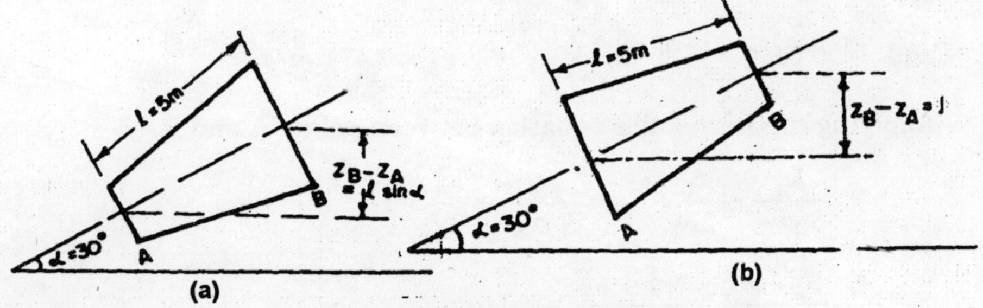

Fig. 6.28. Tapered pipe. P. 6.2.

(a) $Q = v_A\ a_A = v_B\ a_B$ Applying the Bernoulli's equation between A and B.

$$\frac{p_A}{S\rho g} + \frac{v_A^2}{2g} + z_A = \frac{p_B}{S\rho g} + \frac{v_B^2}{2g} + z_B$$

$$\left[Here\ \frac{p_A}{S\,\rho\,g} = \frac{p_B}{S\,\rho\,g} = Given,\right]$$

$$z_B - z_A = \frac{v_A^2 - v_B^2}{2g} = \frac{v_B^2}{2g}\left[\frac{v_A^2}{v_B^2} - 1\right]$$

$$l \sin\alpha = \frac{v_B^2}{2g}\left[\frac{a_A^2}{a_B^2} - 1\right]$$

$$5 \sin 30^\circ = \frac{v_B^2}{2g}\left[\left(\frac{0.15}{0.1}\right)^4 - 1\right]$$

$$2.5 = \frac{v_B^2}{2 \times 9.81}[5.0625 - 1]$$

$$2.5 = \frac{v_B^2 \times 4.0625}{2 \times 9.81}$$

$$v_B^2 = \frac{2.5}{0.207} = 12.0738$$

$$v_B^2 = 3.474\ m/s$$

$$Q = a_B\, v_B = \frac{\pi}{4}(0.15)^2 \times 3.474 = 0.0614\ m^3/s$$

= 61.4 liters/s **Ans**

(b) $Q = a_A\, v_A = a_B\, v_B$

$$0.0614 = \frac{\pi}{4}(0.1)^2 \times v_B \quad \therefore\ v_B = 7.82\ m/s$$

and $0.0614 = \frac{\pi}{4}(0.15)^2 \times v_A \quad \therefore v_A = 3.474\ m/s$

Applying the Bernoulli's equation between points A and B.

$$\frac{p_A}{S\,\rho\,g} + \frac{v_A^2}{2g} + z_A = \frac{p_B}{S\,\rho\,g} + \frac{v_B^2}{2g} + z_B$$

$$\frac{p_B - p_A}{S\,\rho\,g} = \frac{v_A^2 + v_B^2}{2g} + (z_A - z_B)$$

$$= 0.615 - 3.11684 - 2.5 = -2.5 - 2.5 = -5$$

or, $p_A - p_B = 0.74 \times 1000 \times 9.81 \times 5$

$$= 36297\ N/m^2 = 36.297\ kPa$$ **Ans**

6.3 A venturimeter situated in a 100mm pipe has a throat diameter of 25mm. The pipe carries oil of speedfic gravity 0.82, and a mercuty and oil U tube is connected between the inlet and throat of the meter.

Determine from first principle the flow of oil in liters/s when the vertical distance in the height of the mercuty level is 200mm.

Demonstrate that the manometer reading remains unchanged for the same flow irrespective of the inclination of the meter.

Solution

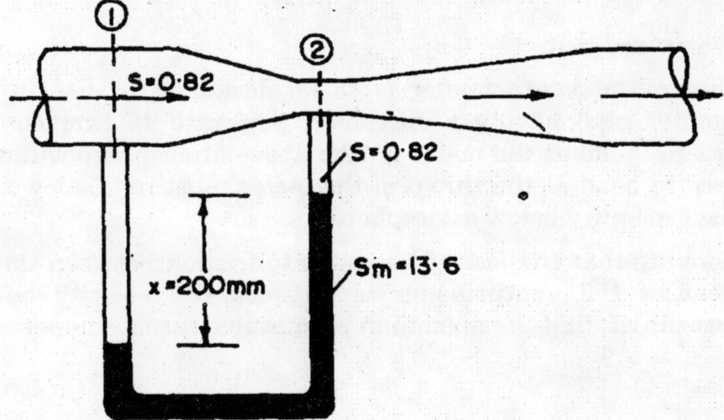

Fig. 6.29. Horizontal Venturimeter. P. 6.3.

Applying the Bernoulli's equation between (1) and (2)

$$\frac{p_1}{S\rho g} + \frac{v_1^2}{2g} + = \frac{p_2}{S\rho g} + \frac{v_2^2}{2g} + z_2 \qquad (z_1 = z_2)$$

or,

$$\frac{v_2^2 - v_1^2}{2g} = \left(\frac{p_1 - p_2}{S\rho g}\right) = h$$

$$= x\left(\frac{S_m}{S} - 1\right) = 0.2\left(\frac{13.6}{0.82} - 1\right) = 3.117 \qquad (1)$$

Using continuty equation

$$Q = a_1 v_1 = a_2 v_2$$

$$\frac{\pi}{4}(0.1)^2 v_1 = \frac{\pi}{4}(0.025)^2 v_2$$

$$v_2 = 4 v_1 \qquad (2)$$

Substituting in equation (1)

$$\frac{\left(v_2^2 - \dfrac{v_2^2}{16}\right)}{2g} = 3.117$$

$$\frac{15}{16} v_2^2 = 61.1569$$

$$v_2^2 = \frac{61.155 \times 16}{15} = 65.232$$

$$v_2 = 8.076 \ m/s$$

$$Q = a_2 \ v_2 = \frac{\pi}{4}(0.025)^2 \times 8.076$$

$$= 3.964 \times 10^{-3} \ m^3/s = 3.96 \ liters/s$$ Ans

For (b) part see text.

6.4 A horizontal venturimeter is to be designed for use in a 150mm diameter pipe having a maximum flow rate 45 liters/s. The static pressure head at the meter is 12m above atmosphere while the static pressure head at the throat of the meter must not be lower than 135 mm of mercury below atmosphere.

Assuming that the loss of head due to friction between the inlet and throat of the venturimeter is 6 times the velocity head at the venturiinlet, find the maximum permissible throat diameter.

Solution

$$\frac{p_1}{\rho g} = 12 \ m \ Gauge$$

$$\frac{p_2}{\rho g} = -(0.135) \times 13.6 = -1.836 \ m \ vacuum$$

$$h = \frac{p_1}{\rho g} - \frac{p_2}{\rho g} = 12 - (-1.836) = 13.836 \ m$$

Using the Bernoulli's equation between points 1 and 2

$$\frac{p_1}{\rho g} + \frac{v_1^2}{2g} = \frac{p_2}{\rho g} + \frac{v_2^2}{2g} + 6\frac{v_1^2}{2g}$$

$$\frac{p_1}{\rho g} - \frac{p_2}{2g} = \frac{v_2^2}{2g} + 6\frac{v_1^2}{2g} - \frac{v_1^2}{2g}$$

$$13.836 = \frac{v_2^2 + 5 \ v_1^2}{2g}$$

$$v_2^2 + 5 \ v_1^2 = 271.46 \quad (1)$$

$$Q = a_1 \ v_1$$

$$45 \times 10^{-3} = \frac{\pi}{4}(0.15)^2 \ v_1$$

$$V_1 = 45 \times 10^{-3} / .01767 = 2.546\ m/s$$

Substituting in equation (1)

$$v_2^2 + 5\,(2.546)^2 = 271.46$$

$$v_2^2 = 239.049 \qquad v_2 = 15.457\ m/s$$

$$Q = a_2\ v_2$$

$$45 \times 10^{-3} = \frac{\pi}{4} d_2^2 \times 15.457$$

$$d_2^2 = 3.70678 \times 10^{-3}$$

$$d_2 = 0.06089 = 60.89\,mm$$ Ans

6.5 **A venturimeter is to be fitted to a pipe of 250 mm diameter where the pressure head is 6 m of water and maximum flow 9 m³ / min. Find the smallest diameter of the throat to ensure that the pressure head does not become negative.**

Solution

$$\frac{p_1}{\rho g} = 6\,m, \frac{p_2}{\rho g} = 0 \qquad \therefore\ H = 6\,m$$

$$a_1 = \frac{\pi}{4}(0.25)^2 = 0.04908\ m^2$$

$$Q = \frac{a_1 \times \sqrt{2gh}}{\sqrt{\left(a_1^2 / a_2^2 - 1\right)}}$$

$$0.15 = \frac{0.04908 \times \sqrt{2 \times 9.81 \times 6}}{\sqrt{\left(a_1^2 / a_2^2 - 1\right)}}$$

$$\sqrt{\left(a_1^2 / a_2^2 - 1\right)} = 3.55$$

$$\frac{a_1^2}{a_2^2} - 1 = 12.606 \qquad \therefore\ \frac{a_1^2}{a_2^2} = 12.606 + 1$$

$$\frac{a_1}{a_2} = 3.688$$

$$a_2 = \frac{a_1}{3.688} = 0\frac{.04908}{3.688}$$

$$\frac{\pi}{4} d_2^2 = \frac{0.04908}{3.688} \qquad \therefore d_2^2 = 0.0169 \qquad \therefore\ d_2 = 0.130\ m$$

$$d_2 = 130\ mm$$ Ans

6.6 **A venturimeter is to be fitted in a 150mm diameter pipeline horizontally at a section where the pressure is 100 kPa. If the maximum flow of water in the pipe is 150 liters/s, find the smallest diameter of the throat so that the pressure at the throat does not fall below 75 kPa vacuum. Assume that 3% of the differential head is lost between the inlet and the throat.**

Solution

$$h = \frac{p_1 - p_2}{\rho g} = \frac{100-(-75)}{9.81} = 17.8389$$

$$h_l = 3\% \;\; of \;\; h = 0.03 \times 17.8389 = 0.5351$$

Therefore, $h - h_l = 17.8389 - 0.5351 = 17.303$

Using the equation

$$Q = \frac{a_1 a_2 \sqrt{2g\,(h - h_l)}}{\sqrt{a_1^2 - a_2^2}}$$

or,
$$Q = \frac{a_1 \sqrt{2g\,(h - h_l)}}{\sqrt{\left(a_1^2 / a_2^2 - 1\right)}}$$

$$150 \times 10^{-3} = \frac{\frac{\pi}{4}(.15)^2 \sqrt{2 \times 9.81 \times 17.303}}{\sqrt{a_1^2 / a_2^2 - 1}}$$

$$\sqrt{a_1^2 / a_2^2 - 1} = \frac{0.3256}{150 \times 10^{-3}} = 2.17$$

$$\frac{a_1^2}{a_2^2} - 1 = 4.71 \qquad \frac{a_1^2}{a_2} = 5.7109$$

$$\frac{(0.1767)^2}{a_2^2} = 5.7109$$

$$a_2^2 = \frac{3.122 \times 10^{-4}}{5.7109} = 5.467 \times 10^{-5}$$

$$a_2 = \left(\frac{\pi}{4} d_2^2\right) = 7.394 \times 10^{-3}$$

$$d_2^2 = 9.414 \times 10^{-3}$$

$$d_2 = 0.097027 = 97 \; mm$$

Ans

Aliter

$$\frac{p_1}{\rho g}+\frac{v_1^2}{2g}=\frac{p_2}{\rho g}+\frac{v_2^2}{2g}+h_l$$

or, $$\frac{v_2^2-v_1^2}{2g}=h-h_l=17.303 \quad (1)$$

Using continuity equation

$$v_1=Q/a_1=\frac{0.15}{\frac{\pi}{4}(0.15)^2}=8.488\ m$$

Substituting in equation (1)

$$v_2^2-(8.488)^2=2\times 9.81\times 17.303$$

$$v_2^2-72.046=339.4848$$

$$v_2^2=339.4848+72.046=411.546$$

$$v_2=20.287$$

$$Q=\frac{\pi}{4}d_2^2\ v_2$$

$$150\times 10^{-3}=\frac{\pi}{4}d_2^2\times 20.287$$

$$\therefore d_2^2=9.41\times 10^{-3}$$

$$d_2=0.097\ m=97\ mm$$ **Ans**

6.7 The inlet and throat diameters of a horizontal venturimeter are 300 mm and 100 respectively. The liquid flowing through the meter is water. The pressure at inlet section is 1.4 bar. The discharge is 150 liters/s. Find the pressure at throat. Assume 4% of differential head is lost between the inlet and the throat. Find also the value of C_d for the venturimeter.

Solution

$$C_d=\sqrt{\frac{h-h_f}{h}}=\sqrt{\frac{h-0.4\ h}{h}}=\sqrt{0.96}=0.9798=0.98$$ **Ans**

Here, $$a_1=\frac{\pi}{4}(0.3)^2 \ \ and\ a_2=\frac{\pi}{4}(0.1)^2=0.07068\ m^2$$

Therefore, $$\frac{a_1}{a_2}=9$$

$$Q=\frac{a_1\sqrt{2g\,(h-0.04h)}}{\sqrt{\left(a_1^2/a_2^2\right)^2-1}}=\frac{0.9798\times .07068\sqrt{2gh}}{\sqrt{(9)^2-1}}$$

$$150 \times 10^{-3} = 0.03429 \sqrt{h}$$

$$\sqrt{h} = 4.373$$

$$h = 19.126 \text{ m}$$

$$h_1 = \frac{p_1}{\rho g} = \frac{1.4 \times 10^5}{10^3 \times 9.81} = 14.27 \ m$$

$$h = h_1 - h_2$$

$$19.126 = 14.27 - h_2$$

$$h_2 = -4.854 \ m \ of \ water$$

$$= \frac{4.854}{13.6} = 0.3569 \ m \ of \ mercury$$

$= 357$ mm of mercury (Vacuum) **Ans**

6.8 A venturimeter is situated in a vertical pipe carrying oil of relative density 0.9. The inlet and throat diameters are 100mm and 50mm respectively, and the throat is 150mm above the inlet.

Determine for a velocity of 3.6 m/s in the 100 mm pipe.

(a) the difference of pressure in N/m² as measured on two independent gauges, one at inlet and the other at throat.

(b) the difference of level of mercury in a vertical differential gauge connected between inlet and throat.

Work from first principles or prove any formula used.

Solution

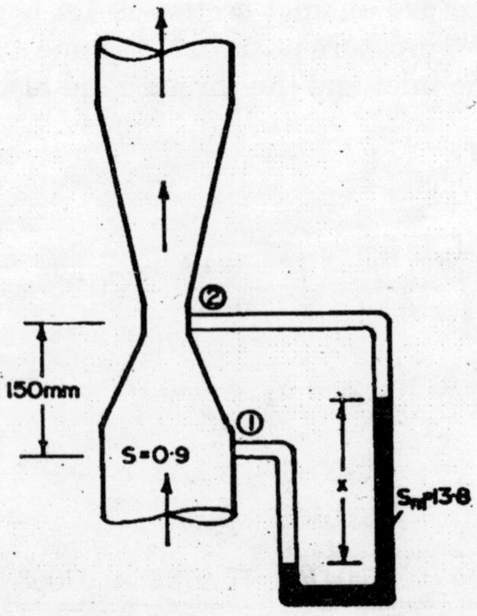

Fig. 6.30. Vertical venturi meter P. 6.8.

Applying Bernoulli's equation between point 1 and 2

$$\frac{p_1}{S\rho g} + \frac{v_1^2}{2g} + z_1 = \frac{p_2}{S\rho g} + \frac{v_2^2}{2g} + z_2$$

$$\frac{p_1 - p_2}{S\rho g} = \frac{v_2^2 - v_1^2}{2g} + (z_2 - z_1) \qquad (1)$$

$$z_1 = 0, \; z_2 = 0.15\ m$$

$$Q = a_1 v_1 = a_2 v_2$$

$$\frac{\pi}{4}(0.1)^2 \times 3.6 = \frac{\pi}{4}(0.05)^2 \times v_2$$

$$v_2 = 14.4\ m/s$$

Therefore, $$\frac{v_2^2 - v_1^2}{2g} = \frac{(14.4)^2 - (3.6)^2}{2 \times 9.81} = 9.908$$

Substituting in equation (1)

$$\frac{p_1 - p_2}{S\rho g} = 9.908 + 0.15 = 10.058$$

$$p_1 - p_2 = 0.9 \times 1000 \times 9.81 \times 10.058 = 88802.082\ N/m^2$$

(b) $$\frac{p_1 - p_2}{S\rho g} + (z_1 - z_2) = h = x\left(\frac{S_m}{S} - 1\right)$$

or, $$\frac{v_2^2 - v_1^2}{2g} = h = x\left(\frac{S_m}{S} - 1\right)$$

$$9.908 = x\left[\frac{13.6}{0.9} - 1\right]$$

x = 0.702 m of mercury = 702 mm of mercury Ans

6.9 A venturimeter has its axis inclined at 60° to the horizontal converges over a length of 500 mm from a diameter of 300 mm to an unknow throat diameter. Petrol of specific gravity 0.78 flows through the meter at the rate of 300 kg/s. The difference of pressure between inlet and throat is 0.4 bar. The loss of head between the main and the throat is 0.2 times the kinetic head of the liquid in the pipe. Calculate the throat diameter.

Solution

Applying Bernoulli's equation between point (1) and (2)

$$\frac{p_1}{S\rho g} + \frac{v_1^2}{2g} + z_1 = \frac{p_2}{S\rho g} + \frac{v_2^2}{2g} + z_2 + h_l$$

$$\frac{p_1 - p_2}{S\,\rho g} + (z_1 - z_2) = \frac{v_2^2 - v_1^2}{2g} + 0.2\,\frac{v_1^2}{2g}$$

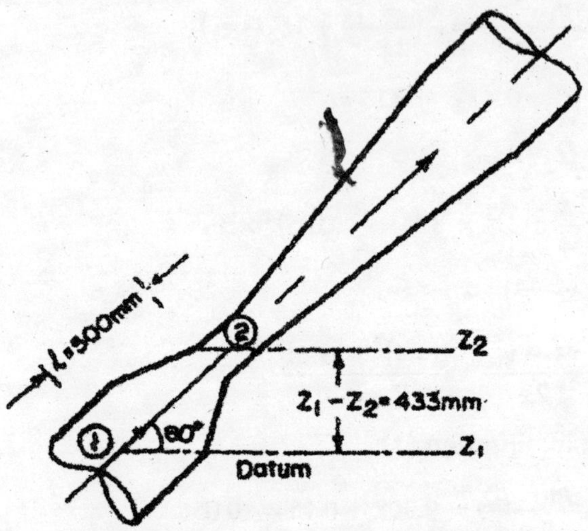

Fig. 6.31. Inclined venturimeter P. 6.9.

$$\frac{0.4\times 10^5}{0.78\times 9810} - 0.433 = \frac{1}{2g}\left[v_2^2 - 0.8\,v_1^2\right]$$

$$2g\,(5.2275 - 0.433) = v_2^2 - 0.8\,v_1^2$$

$$2\times 9.81\times 4.7945 = v_2^2 - 0.8\,v_1^2$$

$$v_2^2 - 0.8\,V_1^2 = 94.068 \qquad (1)$$

$$m = S\,\rho\,Q$$

$$300 = 0.78\times 1000\,Q$$

$$Q = 0.3846\ m^3/s$$

Therefore, $Q = a_1\,v_1$

$$0.3846 = \frac{\pi}{4}\,(0.3)^2\,v_1$$

$$v_1 = 5.37\ m/s$$

Substituting in equation (1)

$$v_2^2 - 0.8(5.37)^2 = 94.068$$

$$v_2^2 - 23.06952 = 94.068$$

$$v_2^2 = 117.1375$$

$$v_2^2 = 10.823 m/s$$

Therefore, $Q = a_2 v_2$

$$0.3846 = \frac{\pi}{4} d_2^2 \times 10.823$$

$$8.5 d_2^2 = 0.3846$$

$$d_2^2 = 0.045245$$

$$d_2 = 0.2127 m$$

$$d_2 = 212.7 mm$$ Ans

6.10 A 500 mm diameter vertical venturimeter discharges water through a throat of 250 mm diameter. The pressure difference measured by a inverted U tube manometer. The vertical distance between inlet and throat section is 500 mm. Determine,

(a) the difference in pressure between these two sections when the dischrge through the meter is 600 liters/s and (b) the manometer deffection x as if the inverted U contains air. Take C_d as unity. Take density of air as 1.2 kg/m³.

Solution

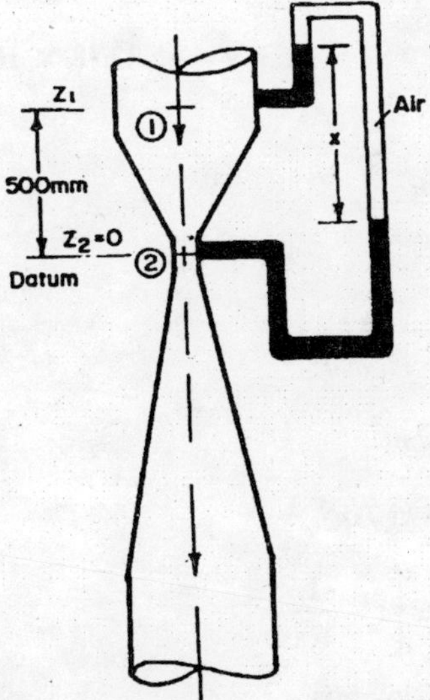

Fig. 6.32. Vertical venturimeter with a inverted U tube manometer. P. 6.10.

$$a_1 = \frac{\pi}{4}(0.5)^2 = 0.19635 m^2$$

$$a_2 = \frac{\pi}{4}(0.25)^2 = 0.04908 m^2$$

$$\frac{a_1}{a_2} = m = 4$$

$$Q = \frac{a_1\, a_2 \sqrt{2gh}}{\sqrt{a_1^2 - a_2^2}} = \frac{a_1 \sqrt{2gh}}{\sqrt{\left(\frac{a_1}{a_2}\right)^2 - 1}}$$

$$600 \times 10^{-3} = \frac{0.19635 \sqrt{2 \times 9.81 \times h}}{\sqrt{16-1}}$$

$$600 \times 10^{-3} = 0.22456 \sqrt{h}$$

$$\sqrt{h} = 2.67$$

$$h = 7.1389 \text{ m}$$

$$h = \frac{p_1}{\rho g} - \frac{p_2}{\rho g} + \left(z_1 - z_2\right)$$

$$7.1389 = \frac{p_1 - p_2}{\rho g} + 0.5$$

$$\frac{p_1 - p_2}{\rho g} = 7.1389 - 0.5 = 6.6389$$

$$\therefore \left(p_1 - p_2\right) = \rho g \times 6.6389$$

$$= 1000 \times 9.81 \times 6.6389 = 65.1278 \times 10^3\ N/m^2$$

65.1278 kPa **Ans**

(b) $$h = x\left[1 - \frac{S_{air}}{S_{water}}\right]$$

$$h = x\left[1 - \frac{\rho_{air}}{\rho_{water}}\right]$$

Taking $\rho_{air} = 1.2\ kg/m^3$

$\rho_{water} = 1000\ kg/m^3$

Therefore, $$7.1389 = x\left[1 - \frac{1.2}{1000}\right]$$

$$7.1389 = x\,[0.999]$$

$$x = 7.146\ m$$ **Ans**

6.11 **An orifice plate with a 150 mm diameter orifice is installed in a pipe line 300 mm diameter carrying oil of relative density 0.85. A mercury U tube manometer connected to the orifice meter shows a deflection of 200 mm. Assuming a coefficient of discharge 0.602 for the orifice meter, determine the discharge.**

Solution

$$a_1 = \frac{\pi}{4}(0.3)^2 = 0.07068\ m^2$$

$$m_o = \frac{a_1}{a_o} = 4$$

$$h = x\left[\frac{S_m}{S} - 1\right] = 0.2\left[\frac{13.6}{0.85} - 1\right] = 3$$

$$Q = \frac{C_d\ a_1\sqrt{2gh}}{\sqrt{a_1^2 / a_0^2 - 1}}$$

$$= \frac{C_d\ a_1\sqrt{2gh}}{\sqrt{m_0^2 - 1}} = \frac{0.602 \times\ 0.07068\sqrt{2 \times 9.81 \times 3}}{\sqrt{16 - 1}}$$

$$= 0.0842\ m^3 / s = 84.2\ liters / s$$ Ans

6.12 An orifice meter is introduced in a 300mm diameter horizontal pipeline carrying water under a pressure of 150kPa. The throat diameter of the meter is 100mm and the pressure at the throat is 400mm of mercury below atmosphere. If 64% of the differential pressure is lost between inlet and vena contracta, determine the flow rate in pipeline and coeffiecient of orifice meter.

Solution

$$C = \sqrt{\frac{h - h_f}{h}} = \sqrt{\frac{h - 0.64h}{h}} = \sqrt{0.36} = 0.6$$ Ans

$$h_1 = \frac{p_1}{\rho g} = \frac{150}{9.81} = 15.29\ m\,of\ water$$

$$h_2 = -0.4 \times 13.6 = -5.44\ m\ of\ water$$

$$h = h_1 - (-h_2) = h_1 + h_2 = 15.29 + 5.44 = 20.73\ m$$

$$Q = C\ \frac{a_1\sqrt{2gh}}{\sqrt{a_1^2 / a_0^2 - 1}} = \frac{a_1\sqrt{2g\ (h - h_1)}}{\sqrt{m_0^2 - 1}}$$

$$a_1 = \frac{\pi}{4}(0.3)^2 = 0.07068\ m^2$$

$$\frac{a_1}{a_o} = m_o = 9$$

Substituting in the discharge formula

$$Q = 0.6 \times \frac{0.07068\ \sqrt{2 \times 9.81 \times 20.73}}{\sqrt{81-1}}$$

$$= 0.09562\ m^3/s \quad = 95.62\ liters/s$$ Ans

6.13 An orifice plate is to be used to measure the rate of air flow through a 2m diameter duct. The mean velocity in the duct will not exceed 15m/s and a water tube manometer, having a maximum difference water levels of 150 mm is to be used. Assuming the coefficient of discharge to be 0.64, determine a suitable orifice diameter to make the full use of manometer range. Take the density of air as 1.3 kg/m³.

Solution

$$h = x\left[\frac{\rho_{water}}{\rho_{air}} - 1\right] = 0.15\ \left[\frac{1000}{1.2} - 1\right] = 124.85 \text{ m of air}$$

$$a_1 = \frac{\pi}{4}(2)^2 = 3.14\ m^2$$

$$Q = a_1\ v_1 = 3.14 \times 15 = 47.12389\ m^3/s$$

$$Q = C_d \times \frac{a_1\ \sqrt{2gh}}{\sqrt{m_0^2 - 1}}$$

$$47.12389 = \frac{0.64 \times 3.14\ \sqrt{2 \times 9.81 \times 124.85}}{\sqrt{m_0^2 - 1}}$$

$$\sqrt{m_0^2 - 1} = 2.11$$

$$m_0^2 - 1 = 4.459$$

$$m_0^2 = 5.459$$

$$m_o = \frac{a_1}{a_2} = 2.3365$$

$$\left(\frac{2}{d_o}\right)^2 = 2.3365$$

$$\frac{2}{d_o} = 1.5285$$

$$d_o = \frac{2}{1.5285} = 1.308\ m$$ Ans

6.14 A 100 mm flow nozzle is installed in a 150 mm pipe to measure flow of oil of specifc gravity 0.85. A mercury U tube manometer connected to the system shows a diflection of 200 mm. The 13% of the differential pressure is lost between inlet and throat of the nozzle, determine the flow rate in the pipe line and coefficient of discharge of the flow nozzle.

Solution

$$h = x\left[\frac{S_m}{S} - 1\right] = 0.2 \left[\frac{13.6}{0.85} - 1\right] = 3$$

$$C = \sqrt{\frac{h - h_f}{h}} = \sqrt{\frac{h - 0.13\,h}{h}} = \sqrt{0.87} = 0.9327$$

$$a_1 = \frac{\pi}{4}(0.15)^2 = 0.01767\ m^2$$

$$\frac{a_1}{a_2} = m_n = 2.25$$

$$Q = \frac{a_1\sqrt{2g\,(h - h_f)}}{\sqrt{m_n^2 - 1}} = C\,\frac{a_1\sqrt{2gh}}{\sqrt{m_n^2 - 1}}$$

$$= 0.9327 \times \frac{0.01767\sqrt{2 \times 9.81 \times 3}}{\sqrt{5.0625 - 1}}$$

$$= 0.06273\ m^3/s = 62.73\ liters/s$$ Ans

6.15 Air flows through a 200 mm diameter pipe and the maximum expected rate of flow is 3m³/min. If the available water manometer has a maximum reading of 200 mm, what would be a suitable diameter for an orifice plate to measure this flow? Assume the density of air to be 1.2 kg/m³, Cv = 0.97 and Cc = 0.65.

Solution

$$h = x\left[\frac{\rho_w}{\rho_a} - 1\right] = 0.2 \left[\frac{1000}{1.2} - 1\right] = 166.466\ m\ of\ air$$

$$a_1 = \frac{\pi}{4}(0.2)^2 = 0.0314\ m^2$$

$$m_o = \frac{a_1}{a_o} = \left(\frac{d_1}{d_o}\right)^2$$

$$Q = \frac{C_c\ C_v\ a_1\sqrt{2gh}}{\sqrt{m_n^2 - 1\ (C_c)^2}}$$

$$\frac{3}{60} = \frac{0.65 \times 0.97 \times 0.0314\sqrt{2 \times 9.81 \times 166.466}}{\sqrt{m_0^2 - 1 \times (0.65)^2}}$$

$$\sqrt{m_o^2 - (0.65)^2} = 22.628$$

$$m_o^2 = 512.02638 + (0.65)^2 \qquad m_o^2 = 512.4488$$

$$m_o = 22.6373 \qquad \frac{d_1^2}{d_o^2} = 22.6373$$

$$\frac{d_1}{d_o} = 4.7578$$

$$d_o = \frac{d_1}{4.7578} = \frac{0.2}{4.7578} = 0.04203 = 42 \; mm$$ Ans

6.16 Water flows at the rate of 10.5 liters/s through a 150 mm diameter pipe in which an orifice meter with a 100 mm diameter orifice is fitted. If the pressure drop across the meter is recorded as a 18mm difference in levels of mercury in a U tube manometer, what whould be the coefficient of discharge Cd ?

If the orifice were 125 mm diameter what whould be the head loss in m of water, for the above values of Q, Cd and Cc ? Also calculate the pressure drop across the meter recorded in U tube manometer.

Solution

$$h = x\left[\frac{S_m}{S} - 1\right] = \frac{18}{1000}\left[\frac{13.6}{1} - 1\right] = 0.2268 \text{ m of water.}$$

$$Q = C_d \frac{a_1 \, a_o \sqrt{2gh}}{\sqrt{a_1^2 - (C_c C_o)^2}}$$

In this problem, the C_c is unknown but this can be expressed in terms of C_d as $C_d = C_c \times C_v$.

Assuming $C_v = 0.95$

then $$C_c = \frac{C_d}{0.95}$$

$$Q = C_d \frac{a_1 \sqrt{2gh}}{\sqrt{\left(\frac{a_1}{a_o}\right)^2 - (C_d / 0.95)^2}}$$

$$0.0105 = C_d \frac{\frac{\pi}{4}(0.15)^2 \sqrt{2 \times 9.81 \times 0.2268}}{\sqrt{(2.25)^2 - \left(\frac{C_d}{0.95}\right)^2}}$$

$$\sqrt{5.0625 - \left(\frac{C_d}{0.95}\right)^2} = C_d \times 3.5502$$

$$5.0625 - \left(\frac{C_d}{0.95}\right)^2 = C_d^2\ 12.60399$$

$$5.0625 = C_d^2 \left[12.603996 + \frac{1}{(0.95)^2}\right]$$

$$5.0625 = C_d^2\ [12.603 + 1.108]$$

$$5.0625 = C_d^2 \times 13.711$$

$$C_d^2 = 0.3692$$

$$\therefore C_d = 0.6076$$

and $$C_c = \frac{C_d}{0.95} = \frac{0.6075}{0.95} = 0.6396$$

Using these values of C_d and C_c with 125mm diameter orifice, the equation of discharge gives

$$Q = \frac{C_d\ a_1 \sqrt{2gh}}{\sqrt{(a_1 / a_o)^2 - C_c^2}}$$

$$= \frac{0.6076 \times \frac{\pi}{4}(0.15)^2 \times \sqrt{2 \times 9.81 \times h}}{\sqrt{(150/125)^2 - (0.6396)^2}} \quad 0.0105 = \frac{0.04755\sqrt{h}}{1.29}$$

$$0.0105 = 0.03688\sqrt{h}$$

$$\sqrt{h} = 0.2848$$

h = 0.081 m = 81 mm of water

$$h = x\ [12.6]$$

$$x = \frac{0.081}{12.6} = 6.437 \times 10^{-3}\ m = 6.437\ mm$$ Ans

6.17 The mean velocity of water a 150mm diameter horizontal pipe is 5m/s. Working from first principles calculate the pressure difference which would be shown by an inverted U tube manometer across the 100mm diameter orifice plate in the pipe. Take the c for orifice meter 0.61.

Solution

Applying Bernoulli's equation between two points 1 and 2.

$$\frac{p_1}{\rho g} + \frac{v_1^2}{2g} = \frac{p_2}{\rho g} + \frac{v_2^2}{2g} + h_l$$

where $h_l = h\left(1 - C_0^2\right)$

$$\left(\frac{p_1}{\rho g} - \frac{p_2}{\rho g}\right) = h = \frac{v_2^2 - v_1^2}{2g} + + h\left(1 - C_o^2\right) \frac{v_2^2 - v_1^2}{2g} = h\,C_o^2 \qquad (1)$$

$$a_1 = \frac{\pi}{4}(0.15)^2 = 0.1767 \text{ m}^2$$

$$Q = a_1 v_1 = 0.1767 \times 5 = 0.08835$$

$$Q = a_2 v_2 \qquad \therefore v_2 \frac{0.08835}{\frac{\pi}{4}(0.1)^2} = 11.25 \text{ m/s}$$

Putting in (1)

$$h\,(0.61)^2 = \frac{(11.25)^2 - (5)^2}{2 \times 9.81}$$

$$h = \frac{5.1764}{(0.61)^2} = 13.91 \text{ m} \qquad \text{Ans}$$

6.18 A 75mm diameter orifice plate is situated in an inclined plane 150 mm diameter at an angle of 45°. The throat pressure tapping is 200 mm above the inlet tapping and $C_d = 0.6$. Petrol of density 780 kg/m³ flows through the meter at rate of 40 liters/s. Determine the pressure difference in k Pa between inlet and throat and (b) the difference of level which would be registered by a vertical mercury manometer, the tubes above the manometer being full of petrol.

Solution

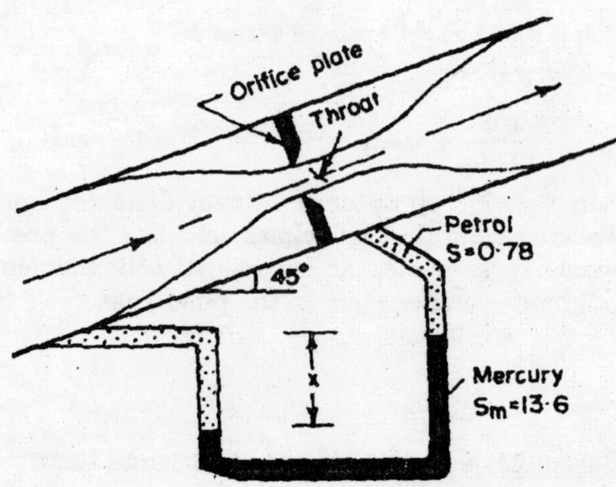

Fig. 6.33. Orifice meter with U tube manometer. P. 6.18.

$$a_1 = \frac{\pi}{4}(0.15)^2 = 0.01767 \quad m^2$$

$$m_o = \frac{a_1}{a_0} = 4$$

$$Q = C\,\frac{a_1\sqrt{2gh}}{m_o^2 - 1} = \frac{0.6\times 0.01767\times\sqrt{2\times 9.81\,h}}{\sqrt{16-1}}$$

$$\sqrt{h} = 3.298$$

$$h = 10.88 \text{ m}$$

$$h = x\left[\frac{S_m}{S-1}\right]$$

$$10.88 = x\left[\frac{13.6}{0.78} - 1\right]$$

$x = 0.662$ m = 662 mm of mercury.

$$\left(\frac{p_1}{S\rho g + z_1}\right)\left(\frac{p_2}{S\rho g + z_2}\right) = h$$

$$\frac{p_1 - p_2}{S\rho g} = h - (z_1 - z_2) \qquad = h + (z_2 - z_1)$$

$$= 10.88 + 0.2 = 11.08$$

$$P_1 - P_2 = s\,\rho\,g\,(11.08)$$

$$= 0.78\times 1000\times 9.81\times 11.08$$

$$= 84781.94 \text{ Pa} = 84.78 \text{ k Pa}$$

Ans

6.19 A pitot tube was placed at the centre of a 200 mm waterpipe when the tube was facing the direction of flow the reading was 420 mm and when turned opposite to the direction of flow the reading was 320 mm. Assuming the mean velocity as two thirds of the center line velocity, find the discharge. Take C = 0.98

Solution

$$h = p_2/\rho g - p_1/\rho g$$

h = Stagnation pressure head – Static pressure head.

$$= 420 - 320 = 100 \text{ mm}$$

$$v_c = v_{max} = C\sqrt{2gh}$$

$$= 0.98\sqrt{2\times 9.81\times 0.1} = 1.3727 \text{ m/s}$$

$$v_{mean} = \frac{2}{3}\times v_{max} = \frac{2}{3}\times 1.3727 = 0.915 \text{ m/s}$$

$$Q = a\ v$$

$$= \frac{\pi}{4}(0.2)^2 \times 0.915 \quad = 0.02875\ m^3/s = 28.75\ liters/s \qquad Ans$$

6.20 A pitot tube is inserted in a pipe of 250 mm diameter. The stagnation pressure at the center of the pipe, recorded by the pitot tube is 20 kPa. The static pressure in pipe is 150 mm of mercury vacuum. Calculater the rate of flow of water through the pipe, if the mean velocity of flow is two third of the center velocity. Take C = 0.98.

Solution

$$\text{Stagnation pressure head} = \frac{20}{9.81} = 2.038\ m$$

Static prssure = 0.15m of mercury vacuum

$$= -0.15 \times 13.6 = -2.04$$

Therefore, h = Stagnation pressure head – static pressure head

$$= 2.038 - (-2.04) = 4.078$$

$$v_c = C\sqrt{2gh} = 0.98\sqrt{2 \times 9.81 \times 4.078} = 8.766\ m/s$$

$$Q = a \times v = \frac{\pi}{4}(0.25)^2 \times 5.84$$

$$= 0.2868\ m^3/s = 286.8\ liters/s \qquad Ans$$

6.21 A pitot tube was used to measure the quantity of water flowing in a pipe of 300 mm diameter. The stagnation pressure at the center line of the pipe is 250 mm of water, mean velocity is 0.78 times the center line velocity and the coefficient of the pitot tube is 0.98, find the rate of flow in liters/s.

Solution

$$h = \frac{p_2\ p_1}{S\rho g} = 250\ \text{mm of water} = 0.25\ \text{m of water}$$

Velocity at center i.e maximum velocity

$$v_{max} = C\sqrt{2gh} = 0.98\sqrt{2 \times 9.81 \times 0.25} = 2.17\ m/s$$

Mean velocity = 0.78 × Maximum velocity

$$= 0.78\ v_{max} = 0.78 \times 2.17 = 1.6929\ m/s$$

$$Q = a\ v$$

$$= \frac{\pi}{4}(0.3)^2 \times 1.6929 \quad = 0.1196\ m^3/s = 119.6\ liters/s \qquad Ans$$

6.22 A submarine submerged in sea water, travels at 16 km/hour. Calculate the pressure at the front stagnation point situated 15m below the sea surface. Take density of sea water 1026 kg/m³

Solution

Static pressure = 15m

$$v = 16 \text{ km/hour } = \frac{16 \times 1000}{3600} = 4.44 \text{ m/s}$$

$$v = \sqrt{2gh}$$

$$4.44 = \sqrt{2 \times 9.81 \times h}$$

$$h = \frac{(4.44)^2}{2 \times 9.81} = 1$$

Stagnation pressure – Static pressure = 1

Stagnation pressure = 1 + Static pressure = 1 + 15 = 16 m

Therefore, $\frac{p}{\rho g} = 16$

$$p = 16 \times 1026 \times 9.81 = 161040.96 \text{ Pa} = 1.61 \text{ bar}$$ Ans

6.23 A pitot tube is mounted at the nose of a fighter aeroplane to indicate the relative speed of the aeroplane. What differential intensity in kPa will the instrument register when the plane is travelling at a speed of 840 km/hr in a wind blowing at 60 km/hr against the direction of the motion of the plane? Take density of air as 1.25 kg/m³.

Solution

$$v = 840 - (-60) = 900 \text{ km/hr } = \frac{900}{3600} \times 1000 = 250 \text{ m/s}$$

$$v = \sqrt{2gh}$$

$$250 = \sqrt{2 \times 9.81 \times h}$$

$$\sqrt{h} = 56.44$$

$$h = 3185.525$$

$$\frac{p_2 - p_1}{\rho g} = 3185.525$$

$$p_2 - p_1 = 3185.525 \times 1.25 \times 9.81$$

$$= 39062.5 \text{ N/m}^2 = 0.390625 \text{ bar} = 39.0625 \text{ kPa}$$ Ans

6.24 A 60° reducing pipe bend tapers from 600 mm diameter inlet to 300 mm diameter at outlet. The gauge pressure at inlet is 140 kPa and the rate of flow of water through the bend is 0.425 m³/s. Neglect friction.

Calculate the net resultant horizontal force exerted by the water on the bend. The bend is in horizontal plane.

Solution

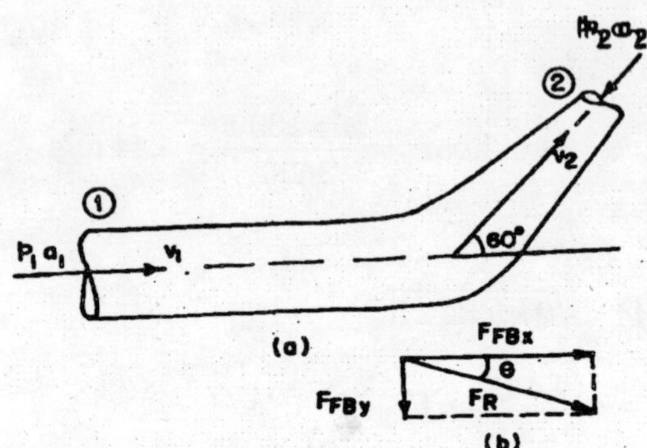

Fig. 6.34. Reducing pipe bend P. 6.24.

$$a_1 = \frac{\pi}{4}(0.6)^2 = 0.2827 \text{ m}^2$$

$$a_2 = \frac{\pi}{4}(0.3)^2 = 0.07068 \text{ m}^2$$

$$v_1 = \frac{Q}{a_1} = \frac{0.425}{0.2827} = 1.503 \text{ m/s}$$

$$v_2 = \frac{Q}{a_2} = \frac{0.425}{0.07068} = 6.01 \text{ m/s}$$

Applying Bernoulli's equation between point 1 and 2

$$\frac{p_1}{\rho g} + \frac{v_1^2}{2g} = \frac{p_2}{\rho g} + \frac{v_2^2}{2g}$$

$$\frac{140}{9.81} + \frac{(1.503)^2}{2\times 9.81} = \frac{p_2}{\rho g} + \frac{(6.01)^2}{2\times 9.81}$$

$$14.27 + 0.115 = P_2/\rho g + 1.84$$

$$\frac{p_2}{\rho g} = 14.385 - 1.84 = 12.54$$

$$p_2 = \frac{12.54 \times 1000 \times 9.81}{1000} = 123.05 \text{ kPa}$$

Force in the x direction is,

$$F_{FBx} = p_1 a_1 \cos\alpha_1 - p_2 a_2 \cos\alpha_2 + \rho Q (v_1 \cos\alpha_1 - v_2 \cos\alpha_2)$$

$$= 140 \times 1000 \times 0.2827 - 123.05 \times 1000 \times 0.07068 \cos 60^\circ$$

$$+ 1000 \times 0.425 \sim (1.503 - 6.01 \cos 60^\circ)$$

$= 39578 - 4348.8273 - 638.35 = 39578 - 4987.17$

$= 34590.82 \text{ N } (\rightarrow)$

Force in the y direction is,

$$F_{FBy} = 0 - p_2\, a_2 \sin\, \alpha_2 + \rho\, Q\,(0 - v_2 \sin\, \alpha_2)$$

$$= -123.05 \times 1000 \times 0.07068 \;\sin 60^\circ \;- 1000 \times 0.425 \times 6.01 \sin 60^\circ$$

$$= -7531.97 - 2212.0454 \;= -9744.019 \;(\downarrow)$$

The resultant force

$$F_R = \sqrt{(F_{FBx})^2 + (F_{FBy})^2}$$

$$= \sqrt{(34590.82)^2 + (-9744.019)^2}$$

$$= \sqrt{1.29 \times 10^9} \;= 35936.73 \text{ N}$$ Ans

The direction of the resultant force with x axis is given as

$$\tan\theta = \frac{F_{FBy}}{F_{FBx}} = -\frac{9744.019}{34590.82} = -0.28169$$

$$\theta = \tan^{-1}(-0.28169) = -15.732^\circ = 15^\circ 43' \, 55''$$ Ans

6.25 A pipe bend tapers from a diameter of 500 mm at inlet to a diameter of 250 mm at outlet and turns the flow through 45°. The pressure at inlet and outlet are 0.4 and 0.23 bar respectively. If the pipe is conveying oil of specific gravity 0.85, calculate the magnitude and direction of the resultant force on the bend when the oil flow rate is 0.45 m³/s. The bend is in horizontal plane.

Solution

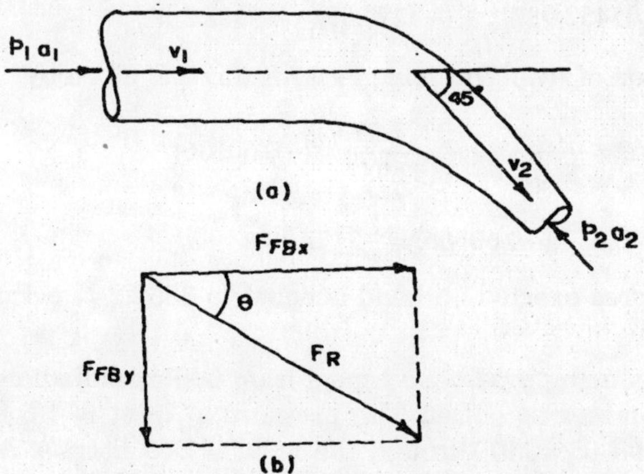

Fig. 6.35. Reducing pipe bend P. 6.25.

$$a_1 = \frac{\pi}{4}(0.5)^2 = 0.19635 \text{ m}^2$$

$$a_2 = \frac{\pi}{4}(0.25)^2 = 0.04909 \text{ m}^2$$

$$v_1 = \frac{Q}{a_1} = \frac{0.45}{0.19635} = 2.292 \text{ m/s}$$

$$v_2 = \frac{Q}{a_2} = \frac{0.45}{0.04909} = 9.167 \text{ m/s}$$

The force in the x direction is,

$$F_{FBx} = p_1\, a_1 \cos\alpha_1 - p_2\, a_2 \cos\alpha_2$$

$$+ S\, \rho\, Q\left(v_1 \cos\alpha_1 - v_2 \cos\alpha_2\right)$$

$$= 0.4 \times 10^5 \times 0.19635 - 0.23 \times 0.04909 \cos 45^\circ$$

$$+ 0.85 \times 1000 \times 0.45\,(2.292 - 9.167 \cos 45^\circ)$$

$$= 7854 - 798.32 - 1602.693 = 7854 - 2401.013$$

$$= 5452.987 \text{ N} \quad (\rightarrow)$$ Ans

The force in the y direction is,

$$F_{FBy} = 0 - p_2\, a_2 \sin\alpha_2 + S\, \rho\, Q\left(0 - v_2 \sin\alpha_2\right)$$

$$= -0.23 \times 10^5 \times 0.04908 \sin 45^\circ + 0.85$$

$$\times 1000 \times 0.45 \times (-9.167 \sin 45^\circ)$$

$$= -798.21 - 2479.38 = -3277.59 \text{ N} \,(\downarrow)$$ Ans

The resultant force

$$F_R = \sqrt{F_{FBx}^2 + F_{FBy}^2}$$

$$= \sqrt{(5452.987)^2 + (-3277.59)^2} = 6362.2\ N$$ Ans

The direction of resultant force F_R with x axis is given by

$$\tan\theta = \frac{F_{FBy}}{F_{FBx}} = -\frac{3277.59}{5452.987} = (-0.60106)$$

$$\theta = \tan^{-1}(-0.60106) = -31^\circ 0' 30'$$ Ans

Total net force exerted on bend is equal to 6362.2 N acting at 31° to the x direction.

6.26 A 30° reducing pipe bend tapers from 600 mm diameter at inlet to 300 mm diameter at outlet. The pressure at inlet is 1.5 bar and the flow rate of oil (S =0.9) through the bend is 500 liters/s. Assuming friction losses to be 20% of the kinetic energy heat at inlet, compute the

magnitude and direction of the resultant force exerted by the fluid on the bend.

Solution

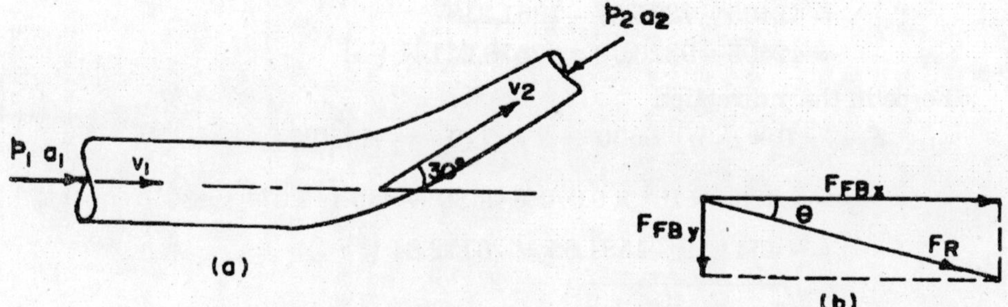

Fig. 6.36. Reducing pipe bend. P. 6.26.

$$a_1 = \frac{\pi}{4}(0.6)^2 = 0.2827\ m^2$$

$$a_2 = \frac{\pi}{4}(0.3)^2 = 0.07068\ m^2$$

$$v_1 = \frac{Q}{a_1} = \frac{0.5}{0.2827} = 1.768\ m/s$$

$$v_2 = \frac{Q}{a_2} = \frac{0.5}{0.07068} = 7.074\ m/s$$

$$h_f = 0.2\frac{v_1^2}{2g} = 0.2 \times \frac{(1.768)^2}{2 \times 9.81} = 0.03188$$

Applying Bernoulli's equation between point 1 and 2

$$\frac{p_1}{S\rho g} + \frac{v_1^2}{2g} = \frac{p_2}{S\rho g} + \frac{v_2^2}{2g} + h_f$$

$$\frac{1.5 \times 10^5}{0.9 \times 10^3 \times 9.81} + \frac{(1.768)^2}{2 \times 9.81} = \frac{p_2}{S\rho g} + \frac{(7.082)^2}{2 \times 9.81} + 0.03188$$

$$16.989 + 0.1593 = \frac{p_2}{S\rho g} + 2.5563 + 0.03188$$

$$\frac{p_2}{S\rho g} = 17.148 - 2.588 = 14.56$$

$$p_2 = S\ \rho g\ (14.56) = 1.285 \times 10^5\ Pa = 1.285 \text{ bar}$$

Force in the x direction

$$F_{FBx} = p_1\ a_1\ \cos\alpha_1 - p_2\ a_2\ \cos\alpha_2 + S\ \rho\ Q\ (v_1 \cos\alpha_1 - v_2 \cos\alpha_2)$$

$$= 1.5 \times 10^5 \times 0.2827 - 1.285 \times 10^5 \times 0.07068 \cos 30^\circ$$

$$+ 0.9 \times 1000 \times 0.5 \left(1.768 - 7.074 \cos 30^\circ\right)$$

$$= 42405 - 7875.57 + 450\,(-4.358)$$

$$= 42405 - 7875.57 - 1961.2187$$

$$= 42405 - 9826.78 = 32578.211 \text{ N } (\rightarrow)$$

Force in the y direction

$$F_{FBy} = 0 - p_2\, a_2 \sin 30^\circ + S\, \rho\, Q\,(0 - v_2 \sin 30^\circ)$$

$$= -1.285 \times 10^5 \times 0.07068 \sin 30^\circ + 450 \left(-7.074 \sin 30^\circ\right)$$

$$= -4541.19 - 1591.65 = -6132.84 \;(\downarrow)$$

$$F_R = \sqrt{\left(F_{FBx}\right)^2 + \left(F_{FBy}\right)^2}$$

$$= \sqrt{(32578.211)^2 + (-6132.82)^2}$$

$$= \sqrt{1.09895 \times 10^9} = 33150.438\ N$$

The direction of the resultant force along the x axis is given by

$$\tan\theta = \frac{F_{FBy}}{F_{FBx}} = -\frac{6132.84}{32578.211} = -0.1882497$$

$$\theta = \tan^{-1}\left(-0.1882497\right) = 10.661^\circ = 10^\circ 39'\ 40''$$ Ans

6.27 Linseed oil with a specific gravity of 0.93 enters the reducing bend (150 mm to 75 mm) show in Fig. 6.37 with a velocity of 3.2 m/s and a pressure of 300 kPa. The outlet end is 2m above the inlet end. Calculate the magnitude and direction of the force on the bend. Neglect energy losses in the bend.

Solution

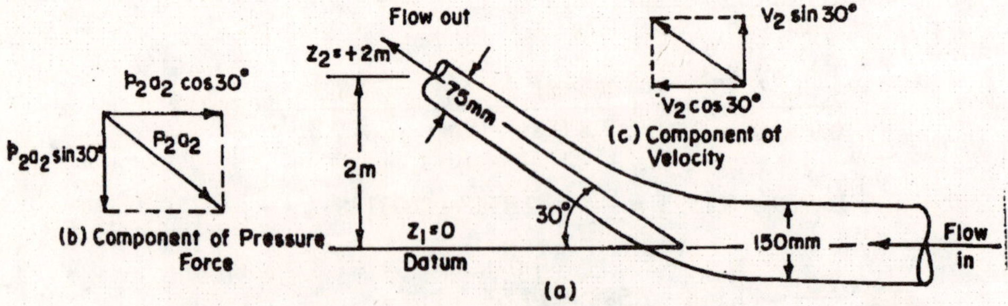

Fig. 6.37. (a) Reducing bend (b) Component of pressure force (c) Component of velocity. P. 6.27.

$$v_1 = 3.2\ m/s,\ a_1 = \frac{\pi}{4}(0.15)^2 = 0.01767\ m^2$$

$$a_2 = \frac{\pi}{4}(0.75)^2 = 4.4178 \times 10^{-3}\ m^2$$

$$v_2 = v_1 \frac{a_1}{a_2} = 3.2 \times 4 = 12.8\ m/s$$

$$Q = a_1\ v_1 = 0.01767 \times 3.2 = 0.0565\ m^3/s$$

Applying Bernolli's equation between point 1 and 2

$$\frac{p_1}{S\ \rho\ q} + \frac{v_1^2}{2g} + z_1 = \frac{p_2}{S\rho q} + \frac{v_2^2}{2g} + z_2$$

$$\frac{300}{0.93 \times 9.81} + \frac{(3.2)^2}{2 \times 9.81} + 0 = \frac{p_2}{S\rho q} + \frac{(12.8)^2}{2 \times 9.81} + 2$$

$$32.88 + 0.5219 = \frac{p_2}{S\ \rho q} + 8.35 + 2$$

$$\frac{p_2}{S\ \rho q} = 33.4019 - 10.35 = 23.0512$$

$$p_2 = 0.93 \times 9.81 \times 23.0512 = 210.3\ \text{kPa}$$

The force in the x direction

$$F_{FBx} = -p_1\ a_1 \cos\alpha_1 - p_2\ a_2 \cos\alpha_2 + S\ \rho\ Q\ (v_1 \cos\alpha_1 - v_2 \cos\alpha_2)$$

$$F_{FBx} = -300 \times 1000 \times 0.01767 + 210.3 \times 1000 \times 4.4178 \times 10^{-3} \cos 30^\circ$$

$$+\ 0.93 \times 1000 \times 0.0565\ (-3.2 + 12.8 \cos 30^\circ)$$

$$= -5301 + 804.592 + 52.545\ (-3.2 + 11.085)$$

$$= -5301 + 804.592 + 414.3239$$

$$= -5301 + 1218.9159 = -4082.08\ N\ (\leftarrow)$$

The force in the y direction

$$F_{FBy} = 0 - p_2\ a_2 \sin\alpha_2 + S\ \rho\ q\ (-v_2 \sin\alpha_2)$$

$$= -210.3 \times 1000 \times 4.4178 \times 10^{-3} \sin 30^\circ$$

$$+ 0.93 \times 1000 \times .0565\ (-12.8 \sin 30^\circ)$$

$$= -464.53 + 52.545\ (-6.4) = -464.53 + 336.288$$

$$= -800.818\ \text{N}$$

$$F_R = \sqrt{(F_{FBx})^2 + (F_{FBy})^2}$$

$$= \sqrt{(-0.4082.08) + (-800.818)^2} = 4159.89\ \text{kN}$$ **Ans**

The direction of the resultant force with the horizontal i,e. x axis.

$$\tan\theta = \frac{800.818}{4082.08} = 0.196$$

$$\theta = \tan^{-1}(0.196) = 11.099° = 11°5'\ 57.3''$$ Ans

6.28 A 600 mm pipe carrying 0.9 m³/s of oil (relative density 0.85) is connected to a 300 mm pipe by a standerd reducer' fitting with a pressure of 2.75 bar, what force is exerted by the oil on the reducer, neglecting any lost head?

Solution

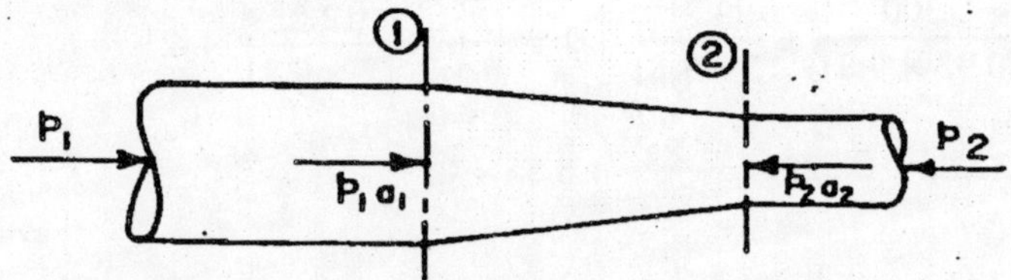

Fig. 6.38. Standard reducer fitting P. 6.28.

$$a_1 = \frac{\pi}{4}(0.6)^2 = 0.2827$$

$$a_2 = \frac{\pi}{4}(0.3)^2 = 0.07068\ m^2$$

$$v_1 = Q/a_1 = \frac{0.9}{0.2827} = 3.18\ m/s$$

$$v_2 = \frac{a_1}{a_2}\ v_1 = 4 \times 3.18 = 12.72\ m/s$$

Applying Bernoulli's theorem between point 1 and 2

$$\frac{p_1}{S\rho q} + \frac{v_1^2}{2g} = \frac{p_2}{S\rho q} + \frac{v_2^2}{2g}$$

$$\frac{2.75 \times 100}{0.85 \times 9.81} + \frac{(3.18)^2}{2 \times 9.81} = \frac{p_2}{S\rho q} + \frac{(12.72)^2}{2 \times 9.81}$$

$$32.97 + 0.5154 = \frac{p_2}{S\rho q} + 8.246$$

$$\frac{p_2}{S\rho q} = 25.2394$$

$$p_2 = 210.458\ kPa$$

$$F_{FNx} = p_1\, a_1 - p_2\, a_2 + S\,\rho Q\,(v_1 - v_2)$$

$$= 2.75 \times 10^5 \times .2827 - 2.104 \times 10^5 \times 0.07068$$

$$+ 085 \times 1000 \times 0.9\,(3.18 - 12.72)$$

$= 77742.5 - 14874.6 - 7298.1 = + 55569.8 \ N \ (\rightarrow)$

$F_{F\,N\,y} = 0$

Hence the force exerted by the oil on the reducer is 55.5698 kN to the right.

6.29 A 600 mm pipe carrying 0.9 m³/s of oil (S = 0.85) has a 90° bend in horizontal plane. The loss of head in the bend is 1m of oil and the pressure at entrance is 3 bar. Determine the resultant force exerted by the oil on the bend.

Solution

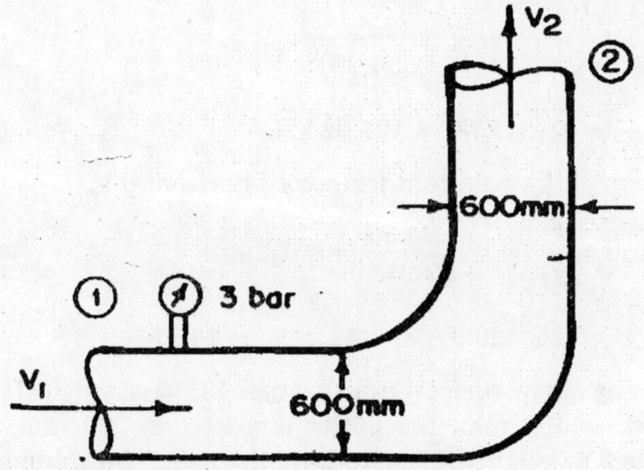

Fig. 6.39. Right angled pipe bend P. 6.29.

$$a = a_1 = a_2 = \frac{\pi}{4}(0.6)^2 = 0.2827 \ m^2$$

$$v_1 = v_2 = \frac{Q}{a} = \frac{0.9}{0.2827} = 3.18 \ m/s$$

Applying Bernoulli's equation between point 1 and 2

$$\frac{p_1}{S\rho g} + \frac{v_1^2}{2g} = \frac{p_2}{S\rho g} + \frac{v_2^2}{2g} + h_f$$

$$\frac{3 \times 10^5}{0.85 \times 10^3 \times 9.81} = \frac{p_2}{S\rho g} + 1$$

$$\frac{p_2}{S\rho g} = 35.97 - 1 = 34.97$$

$p_2 = 34.97 \times 0.85 \times 1000 \times 9.81 = 291.66$ k Pa

$F_{F\,B\,x} = p_1 a_1 - p_2 a_2 \cos 90^\circ + S\rho Q (v_1 - v_2 \cos 90^\circ)$

$= 3 \times 10^5 \times 0.2827 - 0 + 0.85 \times 1000 \times 0.9 \ (3.18)$

$$F_{F\,B\,x} = 84810 + 2432.7 = 87242.7 \text{ N } (\rightarrow)$$

$$F_{F\,By} = p_1\, a_1 \sin 0^\circ - p_2\, a_2 \sin 90^\circ + S\rho\, Q\left(v_1 \sin 0^\circ - v_2 \sin 90^\circ\right)$$

$$= 0 - p_2\, a_2 + 0.85 \times 1000 \times 0.9\ (-v_2)$$

$$= -291.66 \times 0.2827 + 0.85 \times 1000 \times 0.9\ (-3.18) \times 1000$$

$$= -82452.2 - 2432.7 = -82452.2 - 2432.7$$

$$= -84884.98 \text{ N } (\downarrow)$$

The resultant force

$$F_R = \sqrt{(F_{FBx})^2 + (F_{FBy})^2}$$

$$F_R = \sqrt{(87.242)^2 + (84.88)^2}$$

$$= \sqrt{14815.78} = 121.72 \text{ kN}$$ Ans

The direction of the resultant force with horizontal is

$$\tan\theta = \frac{F_{FBy}}{F_{FBx}} = \frac{-84.88}{87.242} = 0.9729$$

$$\theta = \tan^{-1}(0.9729) = 44.213^\circ = 44^\circ 12' 49''$$ Ans

6.30 A reducing elbow turns water through 135° in a vertical plane. The inlet diameter is 400 mm, the outlet diameter is 200 mm, and the outlet center is 2 m below the horizontal inlet axis. The volume of the elbow is 0.2 m³. What is the net force on the horizontal elbow when water flows at 0.4 m³/s and the inlet pressure is 150 kPa gauge? Neglect friction losses.

Solution

$$Q = 0.4\ m^3/s,\ a_1 = \frac{\pi}{4}(0.4)^2 = 0.1256\ m^2$$

$$a_2 = \frac{\pi}{4}(0.4)^2 = 0.0314\ m^2$$

$$v_1 = \frac{Q}{a_1} = \frac{0.4}{0.1256} = 3.18\ m/s.$$

$$v_2 = 4\, a_2 = 4 \times 3.18 = 12.73\ m/s$$

$$Weight\ W = (\rho\, a\, l)g = 1000 \times 0.2 \times 9.81 = 1962\ N$$

$$= 1.962 \text{ kN}$$

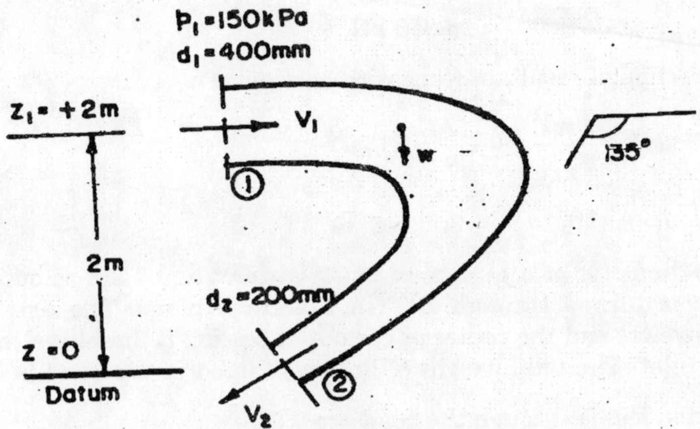

Fig. 6.40. Reducing elbow P. 6.30.

Applying Bernoulli's equation between point 1 & 2

$$\frac{p_1}{\rho g} + \frac{v_1^2}{2g} + z_1 = \frac{p_2}{\rho g} + \frac{v_2^2}{2g} + z_2$$

$$\frac{150}{9.81} + \frac{(3.18)^2}{2 \times 9.81} + 2 = \frac{p_2}{\rho g} + 8.259$$

$$15.29 + 0.515 + 2 = \frac{p_2}{\rho g} + 8.259$$

$$\frac{p_2}{\rho g} = 17.85 - 8.259 = 9.546$$

$$p_2 = 93646.26\ N/m^2 = 93.64 \text{ kPa}$$

The force in the x direction

$$F_{FEx} = p_1\, a_1 - p_2\, a_2 \cos 135° + \rho Q(v_1 - v_2 \cos 135°)$$

$$= 150 \times 1000 \times .1256 - 9364.26 \times 0.0314 \cos 135°$$

$$+ 1000 \times 0.4\,(3.18 - 12.73 \cos 135°)$$

$$= 18840 + 2079.24 + 4872.58 = 25791.82 \text{ N} = 25.79 \text{ kN } (\rightarrow)$$

The force in the y direction

$$F_{FEy} = 0 + p_2\, a_2 \sin 45° - W + \rho\, Q\,(0 + v_2 \sin 45°)$$

$$= + 93646.26 \times 0.0314 \sin 45° - 1962 + 3600$$

$$= + 2079.24 - 1962 + 3600 = + 3717.24\ N = + 3.717\ kN\ (\uparrow)$$

The resultant force

$$F_R = \sqrt{(F_{FEx})^2 + (F_{FEy})^2} = \sqrt{(25.79)^2 + (3.717)^2}$$

$$= \sqrt{678.94} = 26.056 \text{ kN}$$

The direction of resultant force with horizontal is

$$\tan\theta = \frac{F_{FEy}}{F_{FEx}} = \frac{3.717}{25.79} = 0.144$$

$$\theta = \tan^{-1}(0.144) = 8.2^\circ = 8^\circ\ 12'\ 4''$$ Ans

6.31 The diameter of a pipe bend is 2m at inlet and 1.3m at outlet and the flow is turned through 120° in a vertical plane. The axis at inlet is horizontal and the center of the outlet section is 3m above the center of the inlet. The total weight of fluid contained in the bend is 80 kN. The friction losses through the bend are $0.5\ \frac{V_2}{2g}$

Determine the magnitude and direction of the net force exerted on the bend by water flowing through it at 8.5 m³/s when the inlet pressure is 280 kPa gauge.

Solution

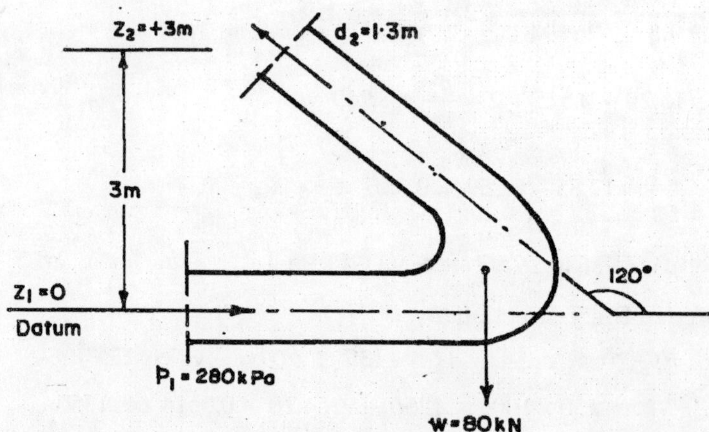

Fig. 6.41. Reducer pipe bend in a vertical plane P. 6.31.

$$a_1 = \frac{\pi}{4}(2)^2 = \pi m^2, \quad = a_2 = \frac{\pi}{4}(1.3)^2 = 1.327\ m^2$$

$$v_1 = \frac{Q}{a_1} = \frac{8.5}{\pi} = 2.7\ m/s$$

$$v_2 = \frac{8.5}{1.327} = 6.4\ m/s$$

Applying Bernoulli's equation between point 1 and 2

$$\frac{p_1}{\rho g} + \frac{v_1^2}{2g} + z_1 = \frac{p_2}{\rho g} + \frac{v_2^2}{2g} + z_2 + h_l$$

$$\frac{280 \times 1000}{1000 \times 9.81} + \frac{(2.7)^2}{2 \times 9.81} + 0 = \frac{p_2}{\rho g} + \frac{(6.4)^2}{2 \times 9.81} + 3 + 0.5 \frac{(6.4)^2}{2 \times 9.81}$$

$$28.54 + 0.3715 = \frac{p_2}{\rho g} + 2.087 + 3 + 1.0435$$

$$\frac{p_2}{\rho g} = 28.8715 - 6.1305 = 22.741$$

$$p_2 = \rho g \times 22.741 = 223.089 \, kPa$$

The force in the x direction

$$F_{FBx} = p_1 \, a_1 - p_2 \, a_2 \cos\alpha_2 + \rho \, Q \, (v_1 - v_2 \cos\alpha_2)$$

$$= 280 \times 1000 \times 3.14 - 223.089 \times 1000 \times 1.327 \cos 120^\circ$$

$$+ 1000 \times 8.5 \, (2.7 - 6.4 \cos 120^\circ)$$

$$= 879645.94 + 148019.55 + 50150 = 1077815.5 \text{ N}$$

$$= 1077.8155 \text{ kN } (\rightarrow)$$

The force in the y direction

$$F_{FBy} = - W - p_2 \, a_2 \sin \alpha_2 - \rho \, Q \, v_2 \sin\alpha_2$$

$$= - 80000 - 223.089 \times 1000 \sin 60^\circ$$

$$+ 1000 \times 8.5 \times 6.4 \sin 60^\circ$$

$$= - 80000 - 256377.38 - 47111.782$$

$$= - 383489.16 \text{ N} = - 383.48916 \text{ kN } (\downarrow)$$

The resultant force is

$$F_R = \sqrt{(F_{FBx})^2 + (F_{FBy})^2} = \sqrt{(1077.8155)^2 + (-383.48916)^2}$$

$$= \sqrt{1161686.3 + 147063.94} = \sqrt{1308750.2} = 1144.00 \text{ kN}$$

The direction of resultant force with reference to horizontal is

$$\tan\theta = \frac{F_{FBy}}{F_{FBx}} = \frac{-383.48916}{1077.8155} = 0.3558$$

$$\theta = \tan^{-1} (0.3558) = 19.58^\circ = 19^\circ \, 35' \, 9''$$ Ans

EXERCISES

A. THEORY

6.1 What are different energies of a fluid? Explain each of them.

6.2 Derive the Bernoulli's equation. State the assumeptions made.

6.3 The Eluer's equation is given by the expression.

$$\frac{1}{\rho}\frac{dp}{ds}+v\frac{dv}{ds}+g\frac{dz}{ds}=0$$

Derive the above equation and from this obtain the Berrnoulli's equation for a incompressible flow.

6.4 Derive from first principles Euler's equation of motion. Mention clearly the assumptions made.

6.5 Under what conditions may Euler's equations of motion be intergrated? Obtain Bernoulli's equation and mention clearly the assumptions made.

6.6 Derive Bernoulli's theorem for incompressible fluid. Also state the significance of Bernoulli's theorem.

6.7 Write Euler's equation in Cartesian Coordinates for three dimensional fluid flow. Therefrom derive Bernoulli's equation for one dimensional flow. State clearly the assumptions made.

6.8 Derive an expression for the free vortex flow and show that Bernoulli's theorem is applicable in this case.

6.9 What is the impulse-moemtum theroem? Derive the same.

6.10 Obtain an expression for Euler's turbine (Moment of momentum or Angular momentum) equation.

6.11 Derive Euler`s equation of motion along a streamline for an ideal fluid stating clearly the assumptions. Explain how this is integrated to get Bernoulli`s equation along a streamline.

6.12 Using moment of momentum therom derive an expression for the Euler's turbine equation.

6.13 What is angular momentum ? How is it related to torque ?

6.14 What is the difference between linear momentum equation and angular momentum equation.

6.15 What are the practical applications of moment of momentum equation ?

6.16 What is kinetic energy correction factor ? Why is it not taken into account in practice ?

6.17 What is momentum correction factor ?

6.18 Derive an expression for the momentum correction factor.

6.19 Derive an expression for the discharge through a Venturimeter.

6.20 Show that the coefficient C_d for the converging cone of a venturimeter is always less than unity and that for the diverging cone, it is always greater than unity.

6.21 Show that the head loss in the inlet section of a venturimeter is $h_l = (1 - C_d^2)\,h$, where C_d is the coefficient of discahrge of the meter and h is the piezometic head difference between the inlet and the throat.

6.22 Show that the head loss between inlet and throat of a venturimeter can be expressed in terms of the velocity head as

$$h_l = \left[\frac{1}{C_d^2 - 1}\right]\left[1\left(\frac{D_2}{D_1}\right)^4\right]\frac{v_2^2}{2g}$$

6.23 Differentiate between the constant and coefficient of a Venturimeter.

6.24 What is an orifice meter ? Where is it used ?

6.25 Derive an expression for the discharge through an orifice meter.

6.26 What is a flow nozzle ? What is difference between a flow nozzle and a Venturimeter.

6.27 What is a pitot tube ? What is the use of pitot tube ?

6.28 Differentiate between a pitot tube and pitot static tube. Draw a typical pressure distribution along a pitot static tube.

6.29 Explain the terms dynamic head, total head, stagnation head, and Piezometer head.

6.30 Explain the procedure of determing discharge rate through a circular pipe with the help of a Prandtl tube.

6.31 Differentiate between stagnation, static and dynamic pressures.

B. UNSOLVED PROBLEMS

6.1 Gasoline (relative density = 0.82) flows in upward direction through an inclined ventutimeter at the rate of 225 liters/s. The meter is inclined at 30° to the horizontal and has the inlet and throat diameter 300 mm and 150 mm respectively. If the throat is 300mm above the inlet and the pressure throat is 300 mm above the inlet and the pressure at the inlet and the throat are 1.5 bar and 0.8 bar abs. Calculate the coefficient of the Venturimeter. [0.96]

6.2 A Venturimeter is used for measurement of discharge of water in horizontal pipeline. If the ratio of upstream pipe diameter to the throat is 2:1, upstream diameter is 300 mm, the difference of pressure between the throat and upstream is equal to 3m head of water and the loss of head through the meter is one eighth of the throat velocity head, calculate the discharge in the pipe. [0.1315 m^3/s]

6.3 The inlet and throat diameters of a horizontal Venturimeter are 300 mm and 100 mm respectively. The flowing liquid through the meter is water. The pressure intensity at inlet is 137.34 kPa while the vacuum pressure head at throat is 370 mm of mercury. Find the rate of flow. Assume that 4 percent of the differential head is lost between the inlet and the throat. Find also the value of C_d for the Venturimeter. [0.1497m^3/s0.98]

6.4 A 600 mm Venturimeter has a throat diameter 200 mm. During a 15 minutes test, the discharge was noted to be 8 x 10^5 liters of water with a mean pressure head of 40m at the inlet section and a vacuum pressure corresponding to 260 mm of mercury at the throat. Compute the coefficient of meter [0.9623].

6.5 If the inlet area if a Venturimeter is n times the throat are and v and p are the velocity and pressure at the throat and the inlet pressure is mp. Show that

$$(m-1)\,\frac{p}{w} = \left(\frac{1-1}{n^2}\right)\frac{v^2}{2g}$$

6.6 If the constant of a Venturimeter is 3 percent less than calculated theoretically, estimate the flow rate in m^3/s through it if the U tube recorded a difference in the mercury level of 200mm. The meter has a throat diameter of 25mm and is installed in 75mm diameter main. [3.367 liters/s]

6.7 A horizontal pipeline contains a venturimeter whose inlet and throat diameter are respectivly 240mm and 140mm. The discharge from the pipeline is weighted in a collecting tank, and it is found that in a 1.5 minutes 10 tonnes of water are discharged. Assuming that the coefficint of the meter is 0.97, determine the head of mercury registered on the differential U-tube manometer which is connected across the pressure taping on the Venturimeter [198mm of mercury].

6.8 A horizontal Venturimeter fitted in a 100 mm diameter oil pipe has a 50 mm diameter throat. If the pressure difference between the inlet and the throat 9.81 kPa and if the relative density of oil is 0.9. Calculate the weight of oil flowing per sec. Take $C_d = 0.99$ [82.698 N/s]

6.9 A Venturimeter measures the flow of oil of relative density 0.85 in a 80mm diameter pipe. The difference of head between inlet and throat in U tube mercury manometer was 250mm when the discharge through pipe is 10 liters/s. Taking $C_d = 0.97$, find the diameter at throat of meter [38.6mm].

6.10 A Venturimeter is to be fitted to a pipe of 250 mm diameter where the pressure head is 7m and the maximum flow 225 liters/s. Find the smallest diameter of the throat to ensure that the pressure head does not become negative. [152 mm].

6.11 Water flows through a vertical Venturimeter in the upward direction. The meter constant is 0.96. The inlet and throat diameter are 150 mm and 75 mm respectively, and the throat is 300 mm and above the inlet. If the presssure difference between the throat and the inlet is 15.2 kPa. Find the flow through the meter in liters/s [21.69 liters/s]

6.12 Find the discharge of water flowing through a pipe 300 mm diameter placed in an inclined position where a Venturimeter is inserted; having a throat diameter 150 mm. The difference of pressure between the main and the throat is measured by a liquid of relative density 0.6 in an inverted U tube which gives a reading of 300 mm. The loss of head between the main and the throat is 0.2 times the kinetic head of the liquid in the pipe [27.08 liters/s]

6.13 A pipe carrying water has 300 mm × 150 mm Venturimeter which is positioned inclined at 30° to the horizontal. The flow is upwards. The converging cone is 450 mm in length and the cd of meter is 0.98. A differential U tube manometer with mercury as indicating fluid is connected to the inlet and to the inlet and to the throat and shows a differential column height of 300 mm (a) Calculate the discharge in the pipe. (b) If the pressure in the inlet section is 50 kPa determine the pressure at the throat. (c) Find the head loss in the converging section of the Venturimeter. [154 liters/s, 10.79 kPa, 0.15 m]

6.14 A Venturimeter of throat diameter 50 mm is fitted into a 125 mm diameter water pipeline. The coefficient of discharge is 0.96. Calculate the flow in the pipeline where the reading on a mercury water differential U tube manometer connected to the upstream and throat sections shows a reading of 200mm. If the energy loss in the downstream divergent cone of the meter is 10 times the velocity head in the pipe. Calculate the total head loss of the meter. [13.43 liters/s h_{lc} = 0.1976m of water, held =0.61m of water, h_l = 0.1976 + 0.61 = 0.8076 m of water].

6.15 Determine the maximum discharge of water that can be carried without cavitation by a horizontal 100 mm × 50 mm venturimeter, which has a coeffiecirnt of discharge of 0.95. The inlet pressure is 10 k Pa (gauge), the vapour pressure of water is 4 π pa (abs) and the local atmosphere pressure is 96kPa (abs.) (27.5 liters/s).

6.16 A Venturimeter inclined at 60° to the horizontal fitted to a 150 mm diameter pipe has a throat diameter of 50mm, the throat being higher than the inlet. The difference of pressure between the throat and inlet which are one meter apart is equivalent to 75mm height of mercury. Calculate the discharge. Assuming C_d = 0.98. [3.36 liters/s]

6.17 A pitot static tube is used to measure the velocity of water in a pipe. The stagnation pressure head is 6 m and the static pressure head is 5 m. Calculate the velocity of flow assuming the coefficient of tube equal to 0.98 [4.34 m/s]

6.18 A pitot static tube placed in the center of a 300 mm pipeline has one orifice pointing upstream and other perpendicular to it. The mean velocity in the pipe is 0.8 times the maximum velocity. Find the discharge through the pipe if the pressure difference between the two orifices is 60mm of water. Take C for pitot tube is 0.98. [0.06m^3/s]

6.19 Find the velocity of flow of an oil through a pipe, when the difference of mercury level in a differential U tube manometer connected to the two tappings of the pitot tube is 100 mm. Take c for pitot tube is 0.98 and relative density of oil is 0.8

6.20 Calculate the velocity of flow of water in a pipe of diameter 300mm at a point, where the stagnation head is 5m and static pressure head is 4m. Given that the coefficient of pitot tube is 0.97 [4.3 m/s].

6.21 A submarine moves horizontaly in sea and has its axis 15m below the surface of water. A pitot tube properly placed just in front of the submarine and along its axis, is connected to the two limbs of a U tube containing mercury. The difference of mercury level is found to be 170mm. Find the speed of submarine knowing that the relative density of mercury is 13.6 and that of sea water is 1.026. [23 km/h.]

6.22 A pitot tube is inserted in a pipe of 300mm diameter. The static pressure of the tube is 100mm of mercury, vacuum. The stagnation pressure at the center of the pipe recorded by the pitot tube is 10 kPa. Calculate the rate of flow of water through the pipe. if the mean velocity of flow is 0.85 times the central velocity. Assuming coefficient of the pitot tube to be 0.98.

6.23 A pitot tube is inserted into an air stream at 95 k Pa (abs.). If a mercury air manometer indicate a positive gauge pressure with a reading of 20mm, calculate the velocity of air. Take density of air 1.22 kg/m^3 and atmospheric pressure 100 kPa (abs) Assuming the pitot tube coefficient is 0.98 [109.8m/s]

6.24 Kerosene (s = 0.81) is flowing in a pipe and the static pressure at a station is 3 kPa. If the stagnation pressure of a pitot tube (c = 0.99) insered at the centreline at that section indicates 4 kPa, what is the centerline velocity at that section? [1.56 m/s]

6.25 Water is flowing on the fixed curved vane or blade at the the rate of 0.014m^3/s. If the cuvred blade is bent by 135° and the valocity is 30 m/s find the magnitude and direction of the resultant force on the bend. [776.N, 22°30']

6.26 A 30° reducing pipe bend reduces from 150 mm to 100 mm diameter is in a horizontal plane. Oil with a specific gravity of 0.86 enters the reducing bend at 150 mm diameter with a velocity of 3.2m/s and a pressure of 150 kPa. Determine the force required to hold the bend in place. Neglect any energy loss in the bend. [1748 N, 23.4°]

6.27 A 115° reducing pipe bend reduces from 200 mm to 80mm diameter is in a horizontal plane. Oil with a specific gravity 0.88 enters the bend with a velocity of 3.5m/s and a pressure of 280kPa. Neglect any energy losses in the bend, find the force required to hold the bend in plane. [10.43 kN, 12.5°].

6.28 A 45° reducing pipe bend reduces from 500mm to 200mm diameter. If the dischrge is 500 liters/s and the upstream pressure is 10 bar, find the magnitude and direction of the force on the bend [174.4kN, 8.26°]

7

Flow through Orifices and Mouth Pieces

7.1 INTRODUCTION

A simple device for measuring flow rate is an orifice. An orifice is a hole, generally rather small, through which a fluid flows. Many orifices are circular in cross section, although square, rectangular, triangular, and other shapes are used. The upstream edge of, the orifice opening may be either sharp or rounded. If this edge is a sharp knife edge, the orifice is called a "standard orifice" for standard orifice, fluid is in contact with only a line (i.e. the sharp edge of the orifice).

A mouthpiece is an attachment in the form of a small tube or pipe fixed to an orifice either externally or internally. The length of the mouthpiece is not more than two to there times of the diameter of an orifice.

In this chapter basic properties of orifices and mouthpieces which find considerable application in flow measurement are dealt in detail. The other practical application in problem related to water intakes for pumping station or treatment plants, flow through nozzles and the flow through control gates on dames.

The orifice flow theory is basically relation to the theory of conversation of energy, or Bernoulli's equation as shown in next article.

7.2 SHARP EDGED ORIFICE DISCHARGING FREE

Sharp edged orifices have some special characteristics which make them useful as flow measuring and control devices. For a true sharp edged orifice is viscosity insensitive: that is its performance is not affected by changes in the temperature of the fluid. Hence flow characteristics of an orifice can be described in terms of dimensionless coefficients.

In practice, it is difficult to obtain a true sharp edge orifice. When orifices are used as flow metering devices, each one must be calibrated to determine its' actual performance under operating conditions.

A very idealized representation of the flow through an orifice is shown in Fig. 7.1.

Applying Bernoulli's theorem between points 1 and 2

$$\frac{p_1}{\rho g} + \frac{v_1^2}{2g} + z_1 = \frac{p_2}{\rho g} + \frac{v_2^2}{2g} + z_2 \tag{7.1}$$

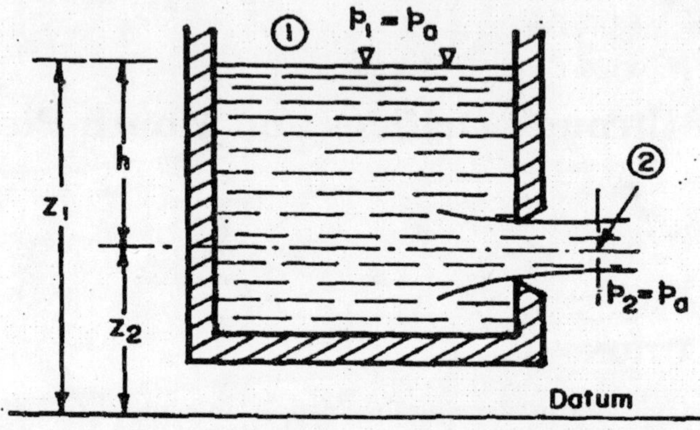

Fig. 7.1. Orifice discharging free.

The reservoir at point 1 is sufficiently large, then the velocity at point *l* is negligible, and hence the kinetic energy or velocity head at this point, can be considered to be negligible.

But $p_1 = p_2 = p_a$ = atmospheric pressure = zero gauge pressure

and $z_1 = z_2 + h$.

Substituting in equation (7.1)

$$0 + 0 + z_1 = 0 + \frac{v_2^2}{2g} + z_2$$

or, $$z_2 + h = \frac{v_2^2}{2g} + z_2$$

or, $$\frac{v_2^2}{2g} = h$$

or, $$v_2 = \sqrt{2gh} \tag{7.2}$$

If equation (7.2) is expressed in general terms for any point in the discharge through the orifice, it yields

$$v = \sqrt{2gh} \tag{7.3}$$

Equation (7.3), known as the Torricelli equation, provides a theoretical expression for the conditions of flow through an orifice. This equation, is applicable only to ideal fluids, however, and adjustment is required to provide for viscosity of the liquid, surface tension and the resistance of the air experienced by real fluids when flowing through an orifice.

Fig. 7.2 shows a typical sharp edged orifice under flow conditions. the curved lines in the illustration represent streamline of fluid flowing through an orifice. Note that the fluid flowing adjacent to or near the surface of the plate must undergo a much greater change of direction than that flowing toward the centre of the orifice. Because of its momentum, the flowing fluid cannot make the sharp turns required for it to exit from the orifice at right angles to the plate. Thus the jet issuing from the orifice contracts. At some point downstream from the point of the orifice opening, the jet achieves a constant cross section and does not contract any more. At this point the area of the jet is minimum, and is called the vena contracta and in Fig. 7.2 it takes place at section. C-C It has been demonstrated empirically that the vena contraction occurs at a distance do/2 downstream from the plane of the opening (do is the diameter of the orifice opening).

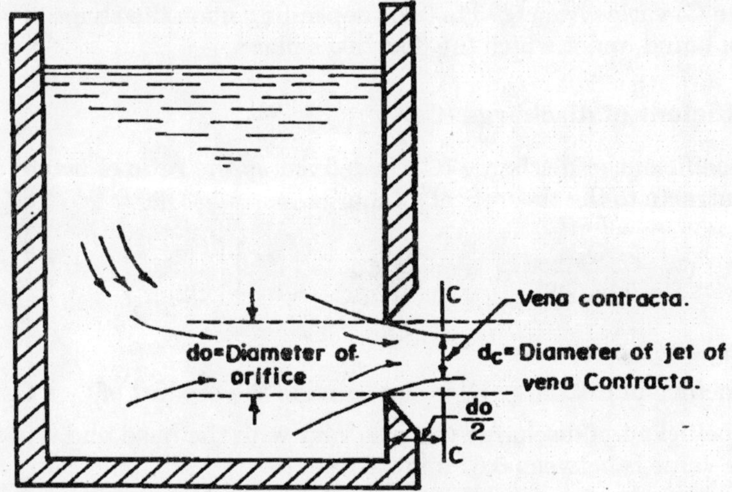

Fig. 7.2. Vena contracta of orifice.

7.3 HYDRAULIC COEFFICIENTS

The terms coefficient of contraction, coefficient of velocity, coefficient of discharge coefficient of resistance are known as hydraulic coefficients.

7.3.1 Coefficient of contraction (C_c)

The coefficient of contraction (C_c) is defined as the ratio of cross sectional area of the jet at the vena contracta to the cross sectional area of the orifice (i.e. actual orifice opening).

$$C_c = \frac{a_c}{a_o} \tag{7.4}$$

The theoretical value of $C_c = \frac{\pi}{\pi + 2} = 0.611$, unfortunately, the area of the vena contracta is not easily calculated by a theoretical method and an element of empiricism must be introduced. In actual practice the value of C_c varies form 0.613 to 0.69 depending upon the size and shape of the orifice and head of the liquid under which the flow takes place.

7.3.2 Coefficient of velocity (C_v)

The coefficient of velocity (C_v) is defined as the ratio of the actual velocity of the jet at vena contracta to the theoretical velocity of flow through the orifice.

$$C_v = \frac{V_c}{\sqrt{2gh}} \tag{7.5}$$

The actual velocity of flow is less than the theoretical one because of the loss of energy taking place due to friction and the curvature of streamlines. The value of the C_v varies from 0.95 to 0.99, depending upon the shape of orifice and the head of liquid under which the flow takes place.

7.3.3 Coefficient of discharge (C_d)

The coefficient of discharge (C_d) is defined as the ratio of actual discharge at vena contracta to the theoretical discharge

$$C_d = \frac{Q_{actual}}{Q_{theoretical}} = \frac{a_c v_c}{a_o \sqrt{2gh}} = \left(\frac{a_c}{a_o}\right)\left(\frac{V_c}{\sqrt{2gh}}\right) \tag{7.6}$$

$$C_d = C_c \times C_v \tag{7.7}$$

The coefficient discharge C_d is also equal to the product of $C_c \times C_v$.

The coefficient of discharge will also vary with the head and type of orifice. Usually its value is between 0.61 and 0.64.

Approximately values empirically determined for the sharp edged coefficients are

$$C_c \approx 0.63 \quad C_v \approx 0.98 \quad \text{and} \quad C_d \approx 0.61$$

The relationship between the three coefficients and head is shown in Fig. 7.3.

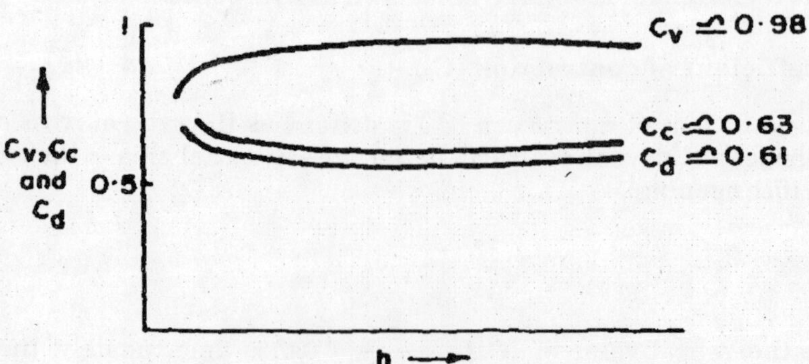

Fig. 7.3. Graph between hydraulic coefficients and head.

7.3.4 Coefficient of resistance (C_r)

The coefficient of resistance (C_r) is defined as the ratio of the loss of kinetic energy as the liquid flows through an orifice and the actual kinetic energy possessed by the flowing fluid.

Theoretical kinetic energy per unit weight of liquid

$$= \frac{v_{th}^2}{2g} = \frac{\left(\sqrt{2gh}\right)^2}{2g} = h$$

Actual kinetic energy per unit weight of liquid

$$= \frac{v^2}{2g} = \frac{C_v^2 \left(\sqrt{2gh}\right)^2}{2g} = h\ C_v^2$$

loss of kinetic energy per unit weight of liquid

$$= h - h\ C_v^2 = h\left(1 - C_v^2\right)$$

Therefore,

$$C_r = \frac{h\left(1 - C_v^2\right)}{h\ C_v^2} = \left(\frac{1}{C_v^2} - 1\right) \qquad (7.8)$$

The C_r thus accounts for the loss of energy that occurs as any liquid flows through an orifice and its value may be determined by knowing the value of the C_v of the orifice.

7.4 EXPERIMENTAL DETERMINATION OF HYDRAULIC COEFFICIENTS FOR AN ORIFICE

7.4.1 Experimental determination of coefficient of contraction

The coefficient of contraction may be determined either directly or indirectly.

In the direct determine of Cc the actual area of the jet at vena contracta by an instrument called the micrometer contraction gauge (Fig. 7.4). This instrument consists of a ring with four fine pitched screws. The screws are equally spaced along the periphery.

The ring is held at the vena contracta and the four screws are adjusted simultaneously so that they just touch the jet. The ring is then removed and the distance between the diametrically opposite screws is measured accurately, which give the diameter of the jet. The mean of two diameters is taken as the diameter of the jet. The area is then calculated; knowing the diameter of the orifice, c_c can be easily determined.

The above method is not reliable. The disadvantages of this method are :

(1) The cross section of the jet is not perfectly circular- hence the diameter measured is not accurate.

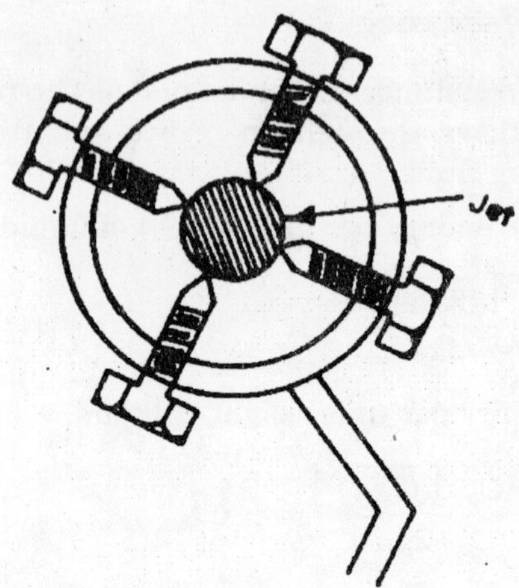

Fig. 7.4. Micrometer contraction gauge

(2) The jet which is coming out form the orifice is constantly to fluctuating and hence it is very difficult to adjust all the four screws simultaneously touching the jet. The indirect determination of the C_c is by using the relation $C_d = C_c \times C_v$

The value of C_d and C_v can be calculated very easily with the help of experiments. From these two experimental values, the C_c can be easily determine. The determination of the two coefficient C_d and C_v is fairly reliable and accurate and hence the indirect method of determining the value C_c is preferred over the direct method.

7.4.2 Determination of coefficient of velocity C_v

There are many methods by which the value of C_v may be determined, some of which are commonly adopted are as follows.

1. Jet distance measurement method (trajectory method)
2. Pitot tube method.
3. Momentum method.

7.4.2.1. Jet distance measurement method (Trajectory method)

Coefficient of velocity can be found by taking observation on the jet liquid issuing from a sharp edged orifice Let us consider the flow through an sharp edged orifice in the side of a vessel (Fig. 7.5).

The head h under which the discharge takes place can be kept constant. If the over flow pipe drips at a more or less uniform rate, the head may be assumed as constant. The position of the vena contracta can be found with the help of a ring with four screws as shown in Fig. 7.4. the section c-c where the area of the jet is the least is the vena contracta. Let us mark the axis of jet, beyond vena contracta, further on the downstream side. In t seconds, let the particle of liquid at point 2 move and occupy the position at point 3 shown on the axis of jet.

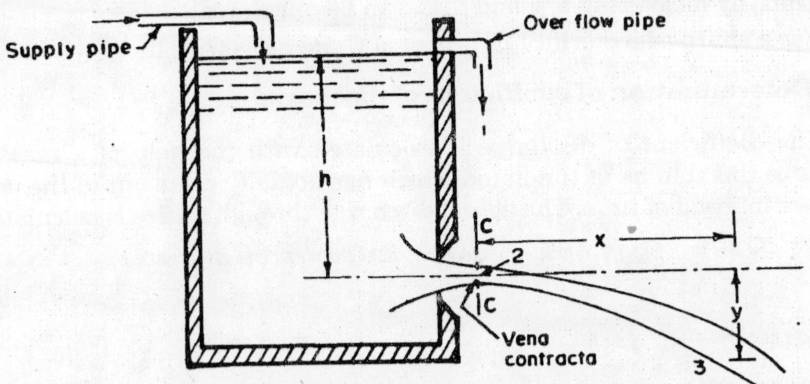

Fig. 7.5. Trajectory from a circular sharp edged orifice

Let, x be the horizontal distance, measured from vena contracta, covered by the particle of liquid in t seconds and y be the vertical distance, measured form vena contracta, covered by the particle of liquid in *t* seconds.

Since the path traced by the particle is a trajectory of the jet from vena contracta of the orifice will be that of a projective under the action of gravity and is given by.

$$x = v\,t \tag{7.9}$$

$$y = \frac{1}{2}\,gt^2 \tag{7.10}$$

From (7.9) we have

$$t = \frac{x}{v}$$

Squaring

$$t^2 = \frac{x^2}{v^2}$$

Substituting the value of t^2 in equation (7.10)

$$y = \frac{1}{2}\,g\,\frac{x^2}{v^2}$$

or, $$v^2 = \frac{gx^2}{2y}$$

or, $$v = \sqrt{\frac{gx^2}{2y}}$$

Now, by the definition of coefficient of velocity

$$C_v = \frac{v}{\sqrt{2gh}} = \frac{\sqrt{gx^2/2y}}{\sqrt{2gh}} = \frac{x}{\sqrt{4\,yh}} \tag{7.11}$$

Thus, by measuring x, y and h, C_v can be calculated,

For a sharp edged orifice the value of C_v increses with h.

7.4.3 Determination of coefficient of discharge

The coefficient of discharge is calculated with the help of a tank, which measures the volume of the liquid which has actually come out of the orifice in a known interval of time. The actual discharge through orifice is calculated as.

$$Q_{act} = C_d \times Area\ of\ orifice \times Theoretical\ discharge$$

$$C_d = \frac{Q_{act}}{a_o\sqrt{2gh}} \qquad (7.12)$$

The coefficient of discharge may also be found by using the experimental obtained value of C_v and C_c. Then using the relation $C_d = C_c \times C_v$.

For small orifices, lower heads or other liquids, the following formula which takes into account all these variables may be used for sharp edged orifices :

$$C_d = 0.592 + \frac{4.5}{\sqrt{R_e}} \qquad (7.13)$$

where, $$R_e = \frac{v\ d_o}{\nu} = \frac{\sqrt{2gh} \times Diameter\ of\ the\ orifices}{Kinematic\ viscosity\ at\ given\ temperature}$$

7.5 SUBMERGED ORIFICE

A submerged or drowned orifice is an orifice which does not discharge into the atmosphere but discharge into same liquid, the whole of the outside of the orifice being under the liquid.

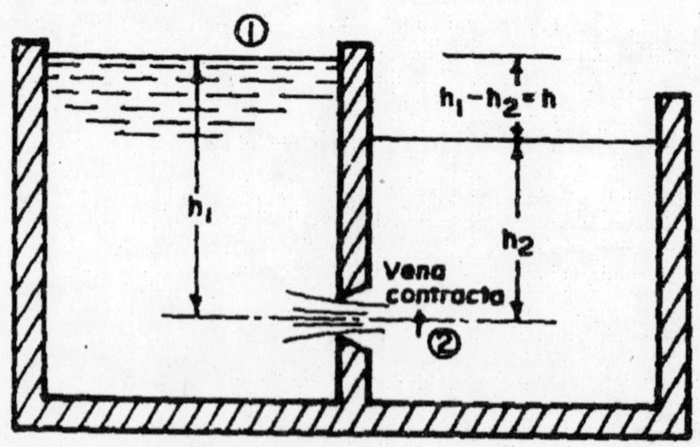

Fig. 7.6. Submerged orifice.

Fig. 7.6 shows a submerged orifice let the heads measured above the centre of the orifice be h_1 and h_2 in the larger and the smaller tank respectively. Let the suffix 1 refer to the water surface in the large tank and the suffix 2 to the vena contracta or the centre of the orifice. Applying Bernoulli's equation between the points 1 and 2 and neglecting the energy losses we have.

$$\frac{p_1}{\rho g}+\frac{v_1^2}{2g}+z_1=\frac{p_2}{\rho g}+\frac{v_2^2}{2g}+z_2 \quad (7.14)$$

Here, p_1 = atmospheric pressure i.e. zero gauge pressure, $\frac{p_2}{\rho g}=h_2$.

Assuming the centre of the orifice as the datum, we have $z_1 = h_1$ *and* $z_2 = 0$.

Substituting all the values in the equation (7.14)

$$0+0+h_1=h_2+\frac{v_2^2}{2g}+0$$

$$\frac{v_2^2}{2g}=h_1-h_2=0$$

Therefore, $v_2=\sqrt{2gh}$

The discharge is calculated as

$$Q=C_d\ a_o\ v_2=C_d\ a_o\ \sqrt{2gh} \quad (7.15)$$

In this case the issuing jet is interfered with the liquid present on the outlet side of the orifice. This reduces the coefficient of discharge. Therefore, the C_d obtained from this method is slightly less than that for free discharge. The average value of C_d may be taken as 0.61.

Application of submerged orifices can be found in lock gates, tide gates, waste water sluices etc.

For submerged mouth pieces the similar formula is used.

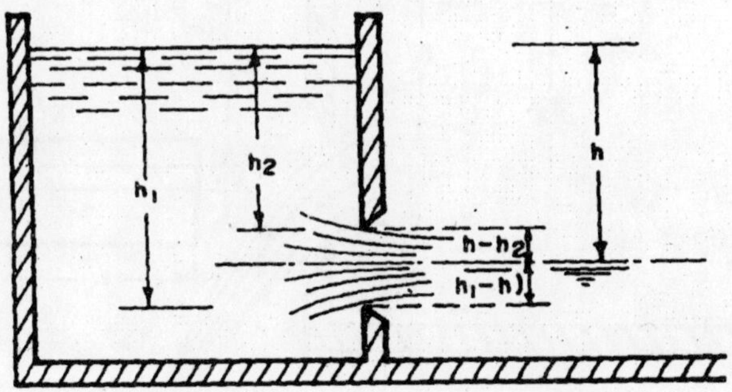

Fig. 7.7. Partially submerged orifice.

7.6 PARTIALLY SUBMERGED ORIFICE

If the liquid level on downstream side (Fig. 7.7) is above the bottom and below the top of orifice, the orifice is known as partially submerged orifice. As such in a partially submerged orifice its upper portion behaves as an orifice discharging free, while the lower portion behaves as a submerged orifice. The discharge is calculated separately for upper and lower portion and added together.

Let, h_1 and h_2 be the heads upto the bottom and top of orifice and let h be the constant difference of liquid levels on the two sides of orifice.

Now discharge a through partially submerged orifice is equal to the discharge through the free portion L (h – h_2) square meters of orifice plus that through the submerged portion L (h_1 – h_2) square meter of orifice.

i.e. $Q = Q_{\text{Free portion}} + Q_{\text{submerged portion}}$

$$= \frac{2}{3} C_{d_1} \sqrt{2g}\, L \left[h^{3/2} - h_2^{3/2}\right] + C_{d_2}\, L\, (h_1 - h) \sqrt{2gh} \qquad (7.16)$$

Here C_d and C_{d2} are coefficients of discharge for free and submerged portion of orifice respectively. C_{d1} may be taken equal to C_{d2} if separate values of C_{d1} and C_{d2} are not given, in such a case, the value of each may be taken as 0.61.

7.7 SHARP EDGED LARGE VERTICAL ORIFICE WITH RECTANGULAR SHAPE

In case of a large orifice, head h cannot be assumed to be uniform over the entire depth of orifice. The velocity of the liquid may no longer be regarded as constant, as the variation in head at different heights of the orifice will be Considerable.

The Large vertical rectangular orifice is shown in Fig. 7.8. Let h_1 and h_2 be the heights from the top and bottom edged of the orifice. Let L be the breadth of orifice.

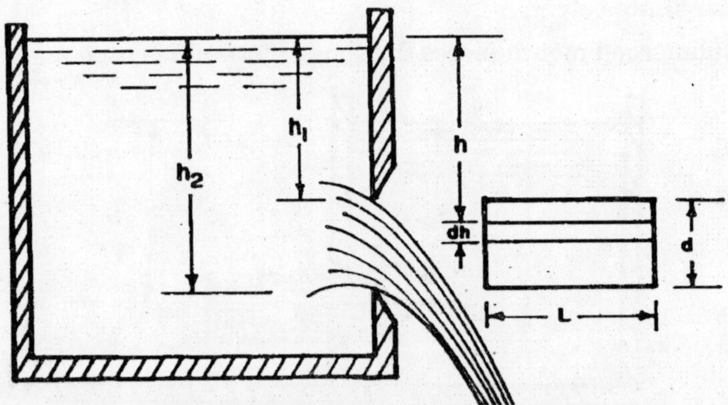

Fig. 7.8. Sharp edged larger vertical rectangular orifice.

Considering the horizontal strip of the orifice of depth h and thickness dh and assume velocity at strip to be proportional to $\sqrt{2gh}$.

Area of Strip = L dh

Velocity of liquid through strip = $\sqrt{2gh}$

Discharge through strip = C_d × area × velocity $= C_d\, L\, dh \sqrt{2gh}$

$$\text{Total discharge } \int_{h_1}^{h_2} C_d\, L\, dh\, \sqrt{2gh} = C_d\, L \sqrt{2g} \int_{h_1}^{h_2} \sqrt{h}\, dh$$

$$= C_d\, L\, \sqrt{2g}\left[h^{3/2}\right]_{h_1}^{h_2} = \frac{2}{3} C_d\, L\, \sqrt{2g}\left[h_2^{3/2} - h_1^{3/2}\right] \qquad (7.17)$$

The average value of Cd may be taken as 0.61. The formula obtained in the equation (7.17) is a accurate formula holds good for small rectangular orifice also; in case of small orifice however we use the approximate formula $Q = C_d\, a_o \sqrt{2gh}$ just to save time in calculation. For small orifice, the difference between discharge given by the accurate formula and the approximate formula is not appreciable. Approximate formula gives slightly larger value of discharge. Still if the small orifice formula is used for a large orifice, then for h = d, the error will be of the order of 1 percent and h=2d, the error will be of the order of only 0.3%. Hence the orifice will be called as small if $h \geq 2d$ and large $h < 2d$ this is the criterion to distinguish between small and large orifices.

7.8 MOUTHPIECES OR TUBES

The discharge obtained with the help of orifice is less due to low value of coefficient of discharge. The coefficient of discharge can be increased with the help of a short pipe which is connected to the orifice. This short piece of pipe of length equal to about 2 to 3 times the diameter of orifice is called the mouthpiece. The standard short tube as shown in Fig. 7.9. It is circular conduit has a square cornered entrance edge.

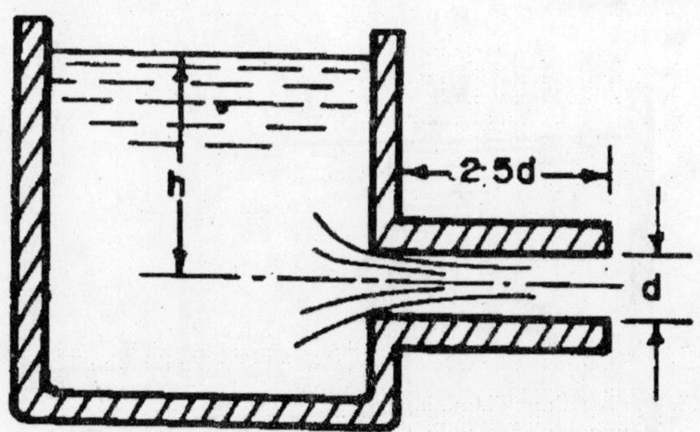

Fig. 7.9. Standard mouthpiece.

The head causing flow from mouthpiece as well as orifice remain same. Diameter at outlet or exit end of the mouthpiece is called the diameter of mouthpiece. The various types of mouth piece are follows.

(a) Mouthpiece of uniform diameter or cylindrical mouth piece.

(b) mouthpiece of uniformly varying diameter.

Mouthpiece is said to be cylindrical when its diameter form inlet to outlet is uniform. Such a mouthpiece may be

1. External cylindrical mouthpiece
2. Reentrant mouthpiece or internal cylindrical mouthpiece

According to nature of discharge

1. Mouthpiece running full and.
2. Mouthpiece running free.

The increase in the discharge would, however, result only if the mouthpiece is running full so that the jet of liquid emerging from the mouthpiece is of the same diameter as that of the mouthpiece. Since the external cylindrical mouthpiece generally run full, only this condition of discharge has been considered of or these mouthpiece.

7.8.1 External cylindrical mouthpiece running full

Consider what happens when flow is started suddenly. The fluid may spring clear of the wall of the mouthpiece in which case the mouthpiece will act like a sharp edged orifice. This is shown in Fig. 7.9. on the other hand, if flow starts more gradually, the jet will undergo a contraction similar to that observed in an orifice. The short distance downstream the jet will re-expand to the full diameter of the tube (Fig. 7.10). Thus the coefficient of contraction for a mouthpiece running full is $C_c = 1.00$.

Let a = area of pipe

a_c = area of flow at vena contracta

Assuming coefficient of contraction C_c to be 0.62 than

$$C_c = \frac{a_c}{a} = 0.62$$

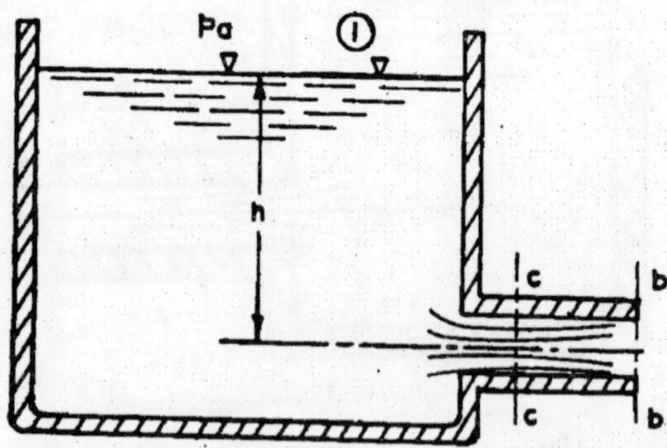

Fig. 7.10. External cylindrical mouthpiece.

Applying the continuity equation between vena contracta and outlet of the mouthpiece.

$$a_c\ v_c = a\ v$$

Therefore, $$v_c = \frac{a}{a_c} v = \frac{v}{0.62} \quad (7.18)$$

The head loss between section cc and bb is just like a flow pattern similar to that after a sudden enlargement and hence h_l between these section is that due to sudden enlargement. Therefore,

$$h_1 = \frac{(v_c - v)^2}{2g} = \frac{\left(\dfrac{v}{0.62} - v\right)^2}{2g}$$

$$= 0.375\,\frac{v^2}{2g} \qquad \text{[From equation 7.18]} \qquad (7.19)$$

Applying Bernoulli's equation between the free surface of the liquid in tank and a section bb just at the outlet of the mouthpiece.

$$\frac{p_1}{\rho g} + \frac{v_1^2}{2g} + z_1 = \frac{p_b}{\rho g} + \frac{v_b^2}{2g} + z_2 + h_1$$

Here

$\frac{p_1}{\rho g} = \frac{p_b}{\rho g}$ = Atmospheric pressure means zero gauge pressure.

$z_1 = h$ and $z_2 = 0$

v_1 is very small hence neglected.

v_b is equal to v at outlet at b-b.

Therefore, $h = \frac{v^2}{2g} + h_1$

or, $$h = \frac{v^2}{2g} + 0.375\,\frac{v^2}{2g} \qquad \text{[From equation 7.19]}$$

or, $$h = 1.375\,\frac{v^2}{2g} \qquad (7.20)$$

or, $$v = \sqrt{\frac{2gh}{1.375}} = 0.8528\,\sqrt{2gh}$$

Therefore, the coefficient of velocity for mouthpiece is

$$C_v = \frac{v}{v_{th}} = \frac{0.8528\,\sqrt{2gh}}{\sqrt{2gh}} = 0.8528$$

Therefore, $C_d = C_c \times C_v = 1 \times 0.8528 = 0.8528$

However, in actual practice the frictional resistance reduces the value of the C_v for the mouthpiece and the actual value of C_v for an external cylindrical mouthpiece is approximately 0.82.

Therefore, $C_d = 1 \times 0.82 = 0.82$ (7.21)

The discharge, $Q = C_d\, a\sqrt{2gh} = 0.82\, a\sqrt{2gh}$

Thus the flow discharged by a standard mouthpiece is about one third greater than that discharge by a sharp edged orifice of the same diameter ($C_d = 0.6$).

In order to find the pressure at the vena contracta apply Bernoulli's equation between the free surface of the liquid in the tank and the section C-C.

$$\frac{p_1}{\rho g}+\frac{v_1^2}{2g}+z_1=\frac{p_c}{\rho g}+\frac{v_c^2}{2g}+z_c$$

Here, v_1 is small hence neglected.

$$z_1 = h \text{ and } z_c = 0$$

Therefore, $h_a + h = h_c + \dfrac{v_c^2}{2g}$

or, $$h_c = h_a + h - \frac{v_c^2}{2g} = h_a + 1.375\frac{v^2}{2g} - \frac{1}{2g}\left(\frac{v}{0.62}\right)^2$$

[Using equations (7.18) and (7.20)]

$$h_c = h_a + 1.375\frac{v^2}{2g} - 2.6014\frac{v^2}{2g} = h_2 - 1.226\frac{v^2}{2g} \quad (7.22)$$

or, $$h_c = h_a - \frac{1.226}{1.375}h = h_a - 0.89\ h \quad (7.23)$$

The equation (7.23) is a theoretical equation. In actual practice the value of C_v is 0.82

Therefore, $v = C_v\sqrt{2gh} = 0.82\sqrt{2gh}$

Substituting in equation (7.22)

$$h_c = h_a - 1.226\,(0.82)^2\,h = h_a - 0.82\,h \quad (7.24)$$

The minimum possible value upto which the pressure at vena contract may be reduced is the absolute zero pressure. Then from equation (7.24), the maximum or the limiting value of the head corresponding to which this minimum pressure may be obtained as.

$$h_c = 0 = h_a - 0.82\ h$$

Taking the standard value of the atmospheric pressure.

$$h = \frac{h_a}{0.82} = \frac{10.336}{0.82} = 12.6 \text{ m}$$

Thus, if the head h is equal to 12.6 m of water then the pressure at vena contra is reduced to absolute zero. However in actual practice the pressure at vena contracta cannot be reduced below the vapour pressure of the flowing liquid. As such in actual practice when the head h becomes equal to only about 12.2 m of water the pressure at vena contracta becomes equal to the vapour pressure of the flowing water.

7.8.2 Flow through convergent divergent mouthpiece

Mouthpiece is said to be of non uniform diameter when its diameter varies gradually from its inlet to its outlet. When diameter decreases gradually from inlet to outlet, the mouthpiece is called the convergent mouthpiece as shown in Fig. 7.11. For a certain convergent proportion (say, angle of convergence θ = 11°.15' average value of $C_d = 0.92$.

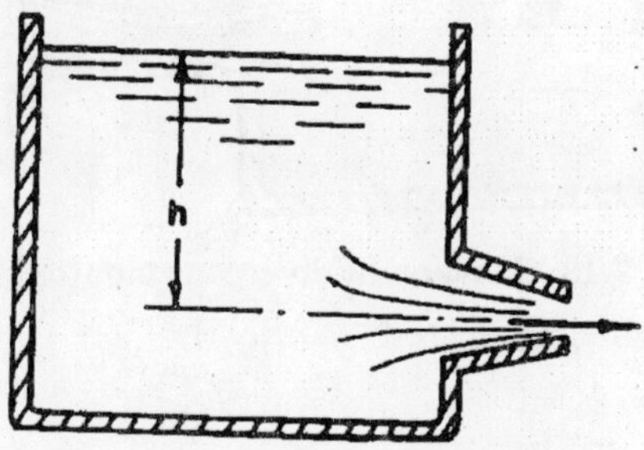

Fig. 7.11. Convergent mouthpiece

When diameter increases gradually from its inlet to outlet average gradually form its inlet to outlet, the mouthpiece is called a divergent mouthpiece as shown in Fig. 7.12. For certain divergence proportions (say $l = 9d$ and angle of divergence $\theta = 5°$ average value of $C_d = 0.97$.

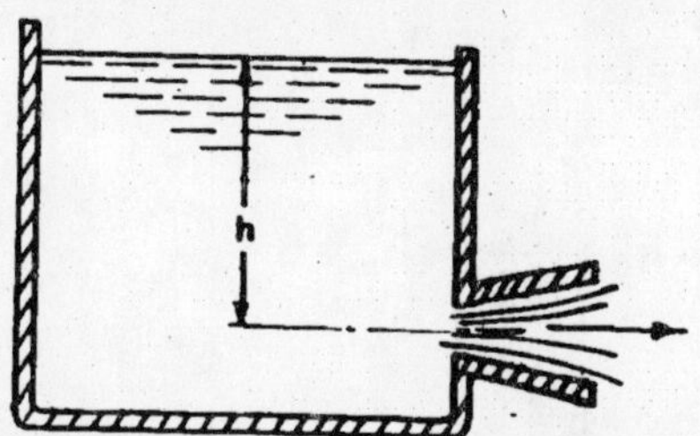

Fig. 7.12. Divergent mouthpiece

If a mouthpiece converges upto vena contracta and then diverged as shown in Fig. 7.13, this type of mouthpiece is known as convergent divergent mouthpiece. As the divergence increases, the velocity at vena contracta i.e. et c-c increases this will cause an increases in the vacuum pressure at the vena contracta and as this cannot be greater than 10.336 m theoretically, or if the liquid be water, 7.8m actually, there is a limit to the amount of divergence if a steadly flow is to be maintained.

Thus the value of the maximum divergence which may be provide for a convergent divergent mouthpiece may be determined.

Applying Bernoulli's equation between the free liquid surface in the reservoir tank and the section c-c and b-b as shown in Fig. 7.13. Neglecting the losses.

$$\frac{p_1}{\rho g} + \frac{v_1^2}{2g} + z_1 = \frac{p_c}{\rho g} + \frac{v_c^2}{2g} + z_c = \frac{p_b}{\rho g} + \frac{v_b^2}{2g} + z_b$$

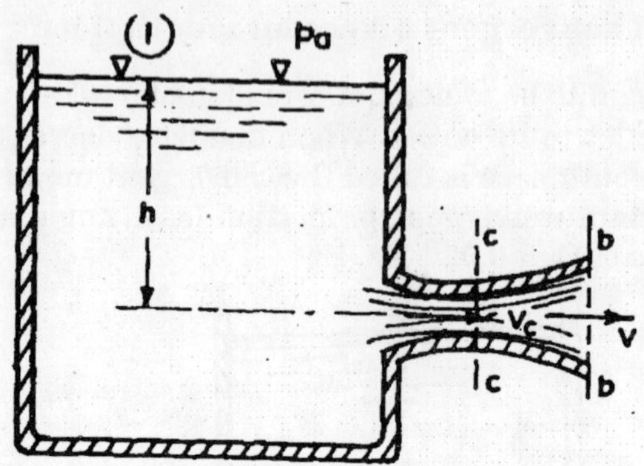

Fig. 7.13. Convergent divergent mouthpiece

Here, v_1 is small hence neglected

$$z_1 = h\ ,\ z_c = z_b = 0$$

$$\frac{p_b}{\rho g} = \frac{p_1}{\rho g} = h_a = \text{atmospheric pressure}.$$

Therefore, $h_a + h = h_c + \frac{v_c^2}{2g} = h_a + \frac{v^2}{2g}$

$$\left[\frac{v^2}{2g} = h\right]$$

or, $h_a + h = h_c + \frac{v_c^2}{2g} = h_a + h$

Therefore, $\frac{v_c^2}{2g} = h_a + h - h_c$

or, $v_c = \sqrt{2g(h_a + h - h_c)}$

but $a_c\, v_c = a\, V$

Therefore, $\frac{a}{a_c} = \frac{v_c}{v} = \sqrt{\frac{2g(h_a + h - h_c)}{2gh}}$

$$= \sqrt{1 + \left(\frac{h_a - h_c}{h}\right)} \quad (7.25)$$

To avoid cavitation at vena contracta, the pressure head should not fall below the 7.8 m of water; hence the maximum ratio of

$$\frac{a}{a_c} = \sqrt{1 + \frac{7.8}{h}} \quad (7.26)$$

The value of C_d for convergent divergent mouthpiece, $c_d = 1$.

7.8.3 Borda's or Re entrant mouthpiece

A Bordas mouthpiece is a special case of a reentrant tube in which a very thin walled tube projects back into a reservoir a length equal to the diameter of the tube (Fig. 7.14).

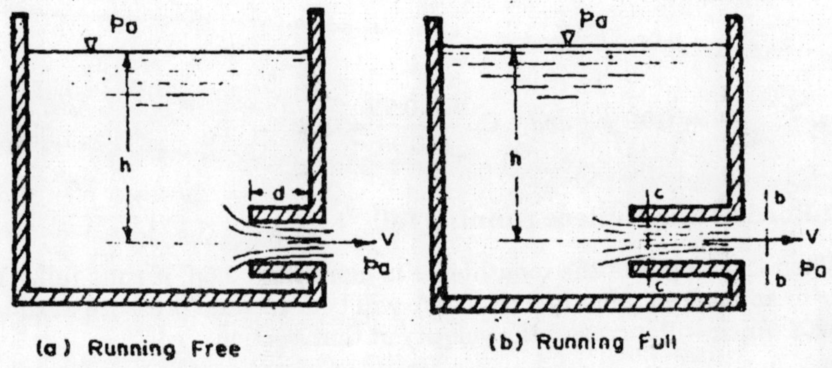

Fig. 7.14. Borda's mouthpiece

7.8.3.1 Borda's mouthpiece running free

In this case the jet after contraction, does not touch the sides of the mouthpiece, it is said to be running free as shown in Fig. 7.14(a).

Pressure of the liquid on the mouthpiece.

$= \rho g h$

Force acting on the mouth piece = pressure × area = $\rho gh\, a$

According to Newton's second law

Net Force = Rate of change of momentum

ρgah = mass × velocity

$$\rho gah = \rho\, a_c\, v_c\, (v_c - o)$$

$$\rho gah = \rho\, a_c\, v_c^2$$

$$ah = a_c \frac{v_c^2}{g}$$

$$a\frac{v^2}{2g} = a_c \frac{v_c^2}{g} \qquad \text{[Hence } v = v_c\text{]}$$

$$a = 2\, a_c$$

$$\text{or,} \quad C_c = \frac{a_c}{a} = 0.5 \qquad (7.27)$$

If the coefficient of velocity therefore,

$$C_d = C_c C_v = 0.5 \times 1 = 0.5$$

Therefore, $$Q = C_d a\sqrt{2gh} = 0.5 a\sqrt{2gh} \quad (7.28)$$

It has been found from experiments that the practical value of coefficient of velocity is 0.98

$$v_c = 0.98 \sqrt{2gh}$$

or, $$\frac{v_c^2}{2g} = 0.96 h \quad and \quad C_c = \frac{0.5}{0.96} = 0.52$$

7.8.3.2 Borda's mouthpiece running full

When the length of the mouthpiece is more than 2.5d, it runs full. This is similar to an external mouthpiece. There will be a vacuum pressure at the vena contracta which will increase, the velocity at that, section.

The flow condition at entrance is same as that for the running free condition.

$$C_c = 0.5 \text{ and } a_c = 0.5 a$$

Using the continuity equation

$$a_c v_c = a v$$

$$v_c = \frac{a}{a_c} v = \frac{v}{0.5} = 2 v \quad (7.29)$$

Applying the Bernoulli's equation between the free surface in the tank (1) and the outlet section of the mouthpiece i.e. at b-b (Fig. 7.14 b).

$$\frac{p_1}{\rho g} + \frac{v_1^2}{2g} + z_1 = \frac{p_b}{w} + \frac{v_b^2}{2g} + z_b + h_l$$

$$\frac{p_1}{\rho g} = \frac{p_b}{\rho g} = \text{Atmospheric pressure, ha.}$$

$$z_1 = h, z_b = 0$$

v_1 is very small hence neglected.

v_b is equal to the outlet velocity v.

Therefore, $$h_a + h = h_a + \frac{v^2}{2g} + h_l$$

$$h = \frac{v^2}{2g} + h_l \quad (7.30)$$

$$h_l = \frac{(v_c - v)^2}{2g} = \frac{(2v - v)^2}{2g} = \frac{v^2}{2g} \quad \text{[Using equation 7.29]}$$

Substituting in equation (7.30)

$$h = \frac{v^2}{2g} + \frac{v^2}{2g}$$

$$\frac{v^2}{g} = h$$

$$v = \sqrt{gh}$$

Therefore, $$C_v = \frac{v}{v_{th}} = \frac{\sqrt{gh}}{\sqrt{2gh}} = \frac{1}{\sqrt{2}} = 0.707$$

Assuming that the value of c_c at the outlet of the mouthpiece is unity

Therefore, $$C_d = \frac{v}{v_{th}} = \frac{\sqrt{gh}}{\sqrt{2gh}} = \frac{1}{\sqrt{2}} = 0.707$$

However in practice the C_d for a Borda's mouthpiece is running full is found to be slightly more than this valve and it is about 0.75.

The pressure at the vena contracta may be found by applying Bernoulli's equation to section at free surface i.e at to vena contracta at section c-c.

$$\frac{p_1}{\rho g} + \frac{v_1^2}{2g} + z_1 = \frac{p_c}{\rho g} + \frac{v_c^2}{2g} + z_c$$

Here $$\frac{p_1}{\rho a} = \text{Atmospheric pressure, ha}$$

v_1 is very small hence neglected

$z_1 = h$ and $z_c = 0$.

Therefore, $$h_a + h = \frac{v_c^2}{2g} + h_c$$

$$h_c = h_a + h - \frac{v_c^2}{2g}$$

but $v_c = 2v$, and $v^2 = gh$

Therefore, $$h_c = h_a + h\ \frac{4\,v^2}{2g} = h_a + h - \frac{2\,gh}{2g} = h_a + h - 2h = h_a - h \qquad (7.31)$$

Thus the pressure at vena contracta is less than atmospheric pressure by an amount equal to the head of liquid in the vessel.

Assuming the liquid to be water and that separation takes place at a absolute pressure of 2.5 m of water, the maximum value of h for steady flow is when.

$$h_c = h_a - h$$

$$2.5 = 10.336 - h$$

$$h = 10.336 - 2.5 \quad = 7.836 \text{ m of water}$$

7.9 FLOW THROUGH AN ORIFICE OR A MOUTHPIECE UNDER VARIABLE HEADS

In our previous discussions of flow we have assumed that a constant head provides the energy required to cause fluid flow. It is obvious that many flow systems do exhibit this condition. However, in practice, we frequently begin with a full reservoir and let the fluid run out through an orifice or mouthpiece or similar flow device. Thus the level of fluid in the reservoir falls and the head goes with it. We must find a method for coping with this situation.

There are several phenomenon which we might wish to evaluate. we might want to calculate the time it would take for the liquid level to fall from the initial height to same second level.

7.9.1 General procedure for calculating time of emptying a tank through an orifice /mouthpiece at its bottom

A typical case is illustrated in Fig. 7.15. We have a reservoir containing an incompressible fluid and an orifice in the bottom of the reservoir through which the fluid is flowing.

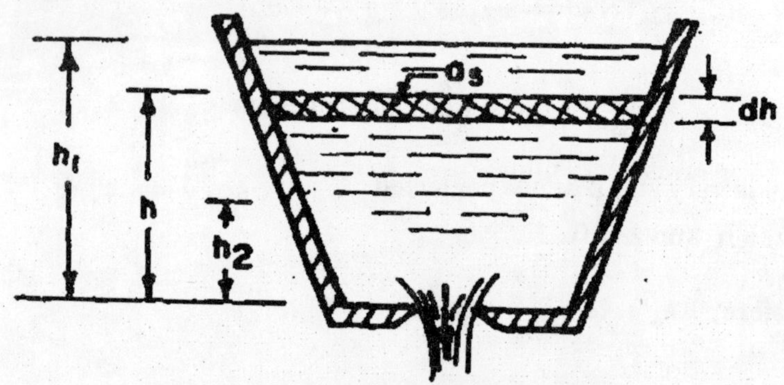

Fig. 7.15. Concept of falling head.

The method of attacking the problem consists of writing equation for flow rate and volume discharge during a given period of time, and solving them for that time.

Step 1. At any intermediate head h between h_1 and h_2, the equation for flow through the orifice would be.

$$Q = C_d \, a_o \sqrt{2gh}$$

Step 2. In the time increment dt, the volume discharge at the above flow rate would be.

$$dv = Q \, dt = C_d \, a_o \sqrt{2gh} \, dt \tag{7.32}$$

Step 3. In the same time increment dt, the level (or head) will have dropped by an increment dh.

Step 4. We can write a second equation for the discharged volume, using the drop in head, dh, and the area of the surface of the liquid at that time, as. Thus

$$dv = a_s \, dh \tag{7.33}$$

Step 5. Equating these two expressions equations (7.32) and (7.34) volumetric increment, We have.

$$dv = a_s \, dh = C_d \, a_o \sqrt{2gh} \, dt$$

Step 6. Solving for time, we obtain.

$$dt = \frac{a_s \, dh}{C_d \, a_0 \sqrt{2gh}}$$

Step 7. In order to evaluate this expression the surface area must be expressed as a function of the head h, $a_s = f(h)$, and this must be integrated between the limit of h_1 and h_2.

$$t = \int_{h_2}^{h_1} \frac{f(h) \, dh}{C_d \, a_o \sqrt{2gh}} \tag{7.34}$$

Equation (7.34) gives the basic form of the solution to the problem of flow under a falling head. As you can see, the problem now becomes one of expressing the surface area in terms of head and evaluating the integral.

7.9.2 Time of emptying cylindrical tank

let us consider a cylinder tank, one of the most common types of reservoirs encountered in engineering works (Fig. 7.16) fluid is discharge through an orifice in the bottom of the reservoir.

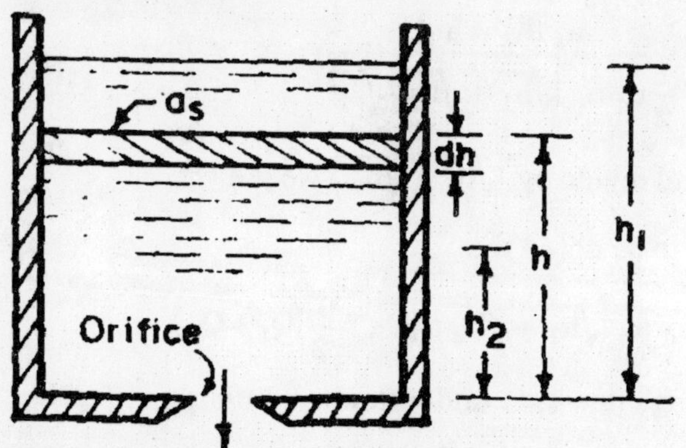

Fig. 7.16. Cylindrical orifice.

Step 1 The flow rate through the orifice is

$$Q = C_d \, a_o \sqrt{2gh}$$

Step 2 The incremental volume is

$$dv = a_s \, dh$$

$$Q = C_d \, a_o \; \sqrt{2gh \, dt}$$

Step 3 Equation and solving for t, we get

$$t = \int_{h_2}^{h_1} \frac{a_s \, dh}{C_d \, a_o \sqrt{2gh}}$$

$$= \frac{a_s}{C_d \, a_o \sqrt{2g}} \int_{h_2}^{h_1} h^{-1/2} \; dh \quad [a_s \text{ is constant}] \tag{7.35}$$

Step 4

$$t = \frac{2 \, a_s}{C_d \, a_o \sqrt{2g}} \times \left(\sqrt{h_1} - \sqrt{h_2}\right) \tag{7.36}$$

7.9.3 Determine the constant head under a head falling

Let us consider a tank as shown in Fig. 7.16. The orifice discharge the same volume of liquid as would be discharged under a head falling from h_1 to h_2, to determine the constant head h_c.

Step 1 Volume discharged under falling head = Volume discharged under constant head.

Step 2 Using the equation (7.35)

$$t = \frac{a_s}{C_d \, a_o \sqrt{2g}} \int_{h_2}^{h_1} \frac{dh}{\sqrt{h}}$$

Step 3 Using the equation (7.36)

$$t = \frac{a_s \left(h_1 - h_2\right)}{\frac{1}{2} c_d \, a_o \left(\sqrt{h_1} + \sqrt{h_2}\right)\sqrt{2g}}$$

Multiply and divide by $\left(\sqrt{h_1} + \sqrt{h_2}\right)$ we get

Step 4 It is to be noted that

$$\frac{1}{2} c_d \, a_o \left(\sqrt{h_1} + \sqrt{h_2}\right) \sqrt{2g} = \frac{1}{2}\left(Q_1 + Q_2\right)$$

or the average flow rate under flow conditions.

Step 5 Then

$$\frac{1}{2} C_d \, a_o \left(\sqrt{h_1} + \sqrt{h_2}\right) \; t = C_d \, a_o \; \sqrt{2gh_c} \; t$$

Step 6 Further more,

$$\frac{1}{2}\left(\sqrt{h_1}+\sqrt{h_2}\right) = \sqrt{h_c} \tag{7.37}$$

With the help of equation (7.37) the value of h_c can be calculated very easily.

7.10 TIME OF EMPTYING (OR FILLING) A TANK WITH INFLOW

In some application encountered in industry, there may be an inflow of liquid at the same time that out flow occurs through the orifice as shown in Fig. 7.17.

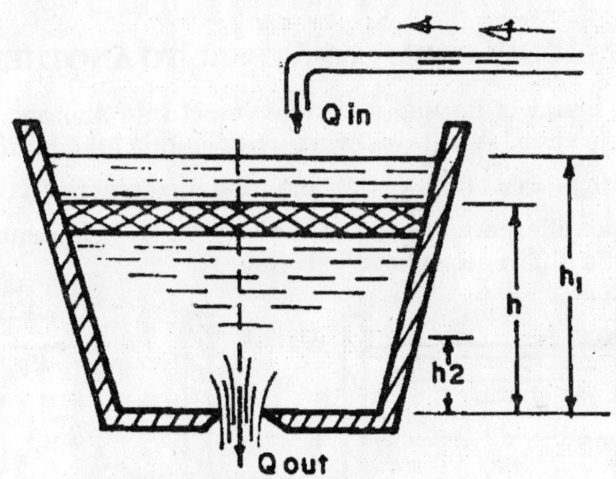

Fig. 7.17. Flow in and out from a reservoir

There are two possible cases.

Case 1 The inflow is greater than the outflow.

$Q_{in} > Q_{out}$ (greater).

Case 2 The outflow is greater than the inflow.

$Q_{out} > Q_{in}$

The same approach is followed as that describe in the article 7.9. that is, we write two equations for the discharge volume, equate them, and solve for the time.

Case 1

$dv = a_s \, dh$

(1) here the incremental volume is expressed in terms of area and head.

$dV = \left(Q_{in} - Q_{out}\right) dt$

(2) In this case, dv is expressed in terms of flow rate and time

$$t = \int_{h_1}^{h_2} \frac{a_s}{Q_{in} - Q_{out}} \, dh = \int_{h_1}^{h_2} \frac{a_s \, dh}{Q_{in} - C_d \, a_o \sqrt{2gh}} \tag{7.38}$$

It can be seen that Q_{in} must be known to evaluate the expression. It is to be noted that in this case, the head would be rising since more liquid is flowing in per unit time than is flowing out.

Case 2

$$dv = a_s\, dh$$

(1) here we have a minus sign, since the head is falling in this case

$$-dV = (Q_{out} - Q_{in})\, dt;$$

(2) in this case, $Q_{out} > Q_{in}$ so that Q_{in} is subtracted from it to yield a positive volume of fluid. It is to be noted that the head would be falling in this case

The equation are solved in the same manner as in case 1.

7.11 FLOW OF LIQUID FROM ONE VESSEL TO ANOTHER

Suppose a liquid is flowing from one vessel into another, so that as the liquid falls on one vessel it will rise by a corresponding amount in the other.

Let the surface area of the vessel be A_1 and A_2 respectively.

let h_1 be the difference of head at the beginning. It is required to find the time taken for the difference of head to become h_2.

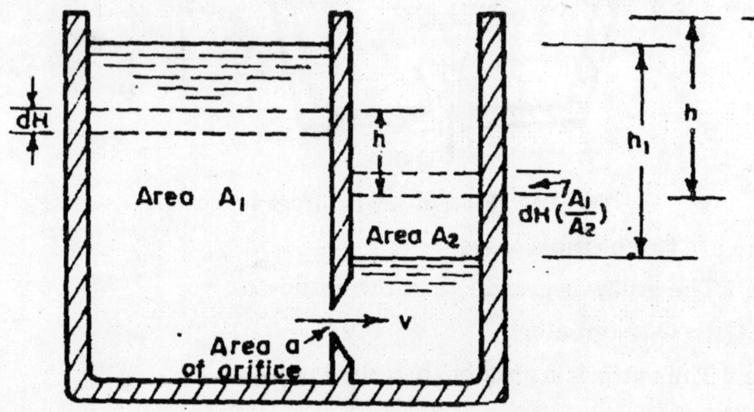

Fig. 7.18. Flow from one vessel to another.

At any particular instant let the head be h, and let the drop of level in the vessel of surface area A_1 during the interval of time dt the liquid level in A_2 will rise, therefore, by the amount dH (A_1/A_2) because the quantity flowing from one tank equals the quantity entering the other.

$$\text{New head } = h - dH - dH\ \frac{A_1}{A_2} = h - dH\left(1 + \frac{A_1}{A_2}\right)$$

Therefore, change of head causing flow

$$dh = dH\left(1 + \frac{A_1}{A_2}\right)$$

$$dH = \frac{dh}{\left(1 + \frac{A_1}{A_2}\right)} \tag{7.39}$$

As quantity flowing from vessel area A_1 equal quantity flowing through orifice, and as dH is negative [head is falling]

$$d\,Q = -A_1\ dH = C_d\ a\ v\ dt$$

$$-A_1\ dh = C_d\ a\ \sqrt{2gh}\ dt$$

Therefore, $dt = -\dfrac{A_1\,dh}{C_d\ a\sqrt{2gh}}$

Substituting the value of dH from (7.39).

Therefore, $dt = -\dfrac{A_1\,dh}{C_d\ a\left(1+\frac{A_1}{A_2}\right)\sqrt{2gh}} = -\dfrac{A_1\,h^{-1/2}\,dh}{C_d\ a\left(1+\frac{A_1}{A_2}\right)\sqrt{2g}}$

Total time taken $\int_0^t dt = -\dfrac{A_1}{C_d\ a\left(1+\frac{A_1}{A_2}\right)\sqrt{2g}}\int_{h_1}^{h_2} h^{-1/2}\,dh$

$$t = -\frac{2A_1}{C_d\ a\left(1+\frac{A_1}{A_2}\right)\sqrt{2g}}\left[\sqrt{h}\right]_{h_1}^{h_2}$$

$$= \frac{2A_1\left[\sqrt{h_1} - \sqrt{h_2}\right]}{C_d\ a\left(1+\frac{A_1}{A_2}\right)\sqrt{2g}} \tag{7.40}$$

If both the vessel have the same surface area $A_1 = A_2$

Then $$t = \frac{A_1\left(\sqrt{h_1} - \sqrt{h_2}\right)}{C_d\ a\sqrt{2g}} \tag{7.41}$$

It is to be noted that the time taken to reduce the difference of liquid level between two vessel of different areas is the same whether the liquid flows from larger reservoir to smaller reservoir or vice versa, provided that the reduction in liquid level is the same in each case.

ILLUSTRATIVE EXAMPLES

7.1 A large tank has a circular sharp edged orifice 28 mm diameter, in a vertical side, at a depth of 1 meter below constant water level. The jet diameter at vena contracta is found to be 22 mm. The water of jet is collected in a vessel having the inside dimensions of 2 m × 0.5 m × 0.5 m The time required to fill this vessel is 5 minutes calculate, C_c, C_v and C_d.

Solution

$$a = \frac{\pi}{4}(0.028)^2 = 6.15 \times 10^{-4}\ m^2$$

$$a_c = \frac{\pi}{4}(0.022)^2 = 3.8 \times 10^{-4}\ m^2$$

$$C_c = \frac{a_c}{a} = \frac{3.8 \times 10^{-4}}{6.15 \times 10^{-4}} = 0.6178$$ Ans

$$v = \sqrt{2gh} = \sqrt{2 \times 9.81 \times 1} = 4.43\ m/s$$

$$c_d = \sqrt{2gh} = \frac{(2 \times 0.5 \times 0.5)/5 \times 60}{6.15 \times 10^{-4}\sqrt{2 \times 9.81 \times 1}} = \frac{1.66 \times 10^{-3}}{2.724 \times 10^{-3}} = 0.6118$$ Ans

$$C_v = \frac{C_d}{C_c} = \frac{0.6118}{0.6178} = 0.99$$ Ans

7.2 A circular orifice, 650 mm² in area made in the vertical side of a tank. If the jet falls vertically 0.5m while moving horizontally through 1.5 m and the discharge is 100 liters/ min of water, determine the horizontal reaction of the jet on the tank.

Solution

$$x = 1.5\ m,\ y = 0.5\ m$$

Therefore, $v = \sqrt{\frac{gx^2}{2y}} = \sqrt{\frac{9.81 \times (1.5)^2}{2 \times 0.5}}$

$$= \sqrt{22.0725} = 4.698\ m/s = 4.7\ m/s$$

$$\dot{m} = \rho Q = \frac{1000 \times 100 \times 10^{-3}}{60} = 1.666 = 1.67\ kg/s$$

Reaction of jet $= m\,v = 1.67 \times 4.7 = 7.849\ N$ Ans

7.3 A tank 1.8 m high, standing on the ground, is kept full of water. There is an orifice in its vertical side at a depth of h meter below the surface. Find the value of h in order that the jet may strike the ground at a maximum discharge from the tank.

Solution

$$y = 1.8 - h$$

and $v = C_c\sqrt{2gh}$

$$x = \sqrt{\frac{2v^2 y}{g}} = \sqrt{\frac{2 \times C_v^2\,(2gh) \times (1.8 - h)}{g}} = 2\,C_v\sqrt{h(1.8 - h)}$$

Thus x will be a maximum when h (1.8 – h) is a maximum or

$$\frac{d}{dh}[h\,(1.8-h)]=0$$

or, $1.8 - 2h = 0$

$h = 0.9$ m Ans

7.4 The water in a tank is 1.8 m deep and over the surface is air at pressure of 75 kPa gauge. Find the rate of flow from an orifice of 50 mm diameter in the bottom of the tank, given that $C_d = 0.6$.

Solution

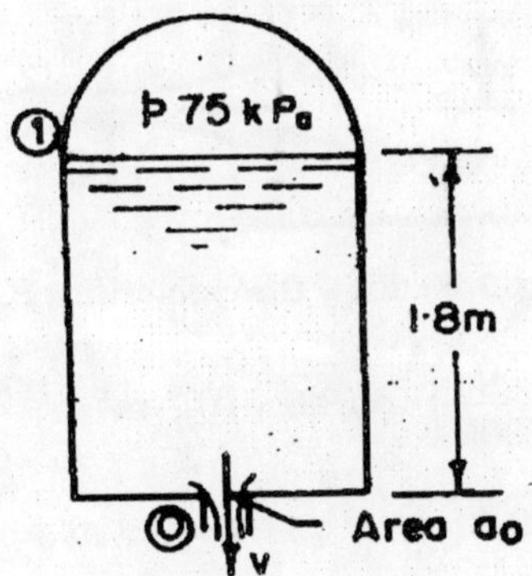

Fig. 7.19. Tank P. 7.4

Applying Bernoulli's equation to the water surface and the outlet at orifice.

$$\frac{p_1}{\rho g}+\frac{v_1^2}{2g}+z_1=\frac{p_o}{\rho g}+\frac{v_o^2}{2g}+z_o$$

Here $\frac{p_o}{\rho g}$ = atmospheric pressure

$z_1 = 1.8$ m, $z_o = 0$

v_1 is very small hence neglected

Therefore, $\frac{75\times10^3}{10^3\times9.81}+1.8=0+\frac{v_o^2}{2g}+0$

$$v_o^2 = 2\times(9.81)\times[7.645+1.8]$$

$$= 2\times9.81\times9.445 = 185.316$$

$v_o = 13.16$

$$Q=C_d\ a_o\ v_o=(0.6)\ \frac{\pi}{4}\left(\frac{50}{1000}\right)^2\times 13.61$$

$$= 0.6\times1.96\times10^{-3}\times13.16=0.016\ \text{m}^3/\text{s} = 16 \text{ liters/s}$$ Ans

7.5 A sharp edged orifice of 50 mm diameter discharges water under a head of 5 m. Find the coefficient of discharge of the measured rate of flow is 12 liter/s. If there is an average pressure within the jet in the plane of the orifice of 30 kPa above atmospheric pressure, calculate the coefficient of contraction. Neglect the friction losses.

Solution

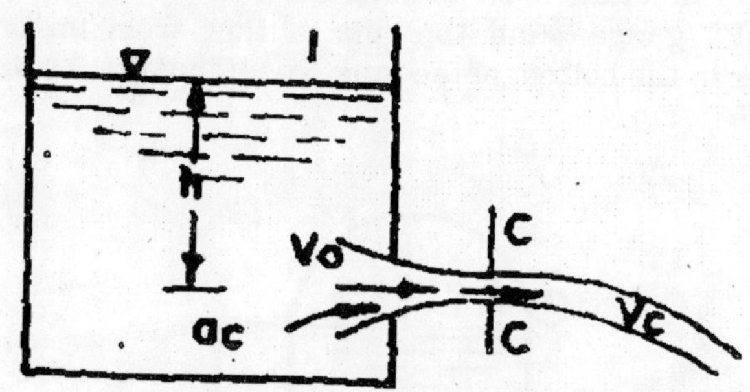

Fig. 7.20. Flow through orifice P. 7.5

$$a_o = \frac{\pi}{4} d^2 = \frac{\pi}{4}\left(\frac{50}{1000}\right)^2 = 19.6 \times 10^{-4} \text{ m}^2$$

$$v = \sqrt{2gh} = \sqrt{2 \times 9.81 \times 5} = \sqrt{98.1} = 9.9 \text{ m/s}$$

$$Q = C_d \ a_o v$$

$$12 \times 10^{-3} = C_d \times 19.6 \times 10^{-4} \times 9.9$$

Assuming $C_v = 1$

$$\therefore C_d = \frac{12 \times 10^{-3}}{194.129 \times 10^{-4}} = 0.618$$ Ans

Applying Bernoulli's theorem between centre of orifice and vena contracta.

$$\frac{p_o}{\rho g} + \frac{v_o^2}{2g} = \frac{p_c}{\rho g} + \frac{v_c^2}{2g}$$

$$\frac{v_0^2}{2g} = \frac{p_c}{\rho g} - \frac{p_o}{\rho g} + \frac{v_c^2}{2g}$$

$$v_0^2 = -\left(\frac{p_o - p_c}{\rho g}\right) 2g + v_c^2 = v_c^2 - 2g\left(\frac{p_o - p_c}{\rho g}\right) = 98.1 - 2\left(\frac{p_o - p_c}{\rho}\right)$$

$$v_0^2 = v_c^2 - 2g\left(\frac{p_o - p_c}{\rho g}\right)$$

$$= 98.1 - 2 \times 30 \qquad = 98.1 - 60 = 38.1$$

$$\therefore v_o = 6.1725$$

The coefficient of contraction is calculated as

$$a_c\ v_c = a_o\ v_o$$

$$\frac{a_c}{a_o} = \frac{v_o}{v_c}$$

$$C_c = \frac{6.1725}{9.9} = 0.623 = 0.623$$ Ans

7.6 A 40 mm diameter orifice in the vertical side of a tank is at a constant level of 2 m above the center of the orifice. if the head loss in the orifice is 200 mm and the coefficient of contraction can be assumed to be 0.625, calculate

(a) The values of the coefficient of velocity and the coefficient of discharge; (b) The discharge through the orifice and (c) The location of the point of impact of the jet on a horizontal plane located 500mm below the centre of the orifice.

Solution

$$h_l = h - \frac{v_a^2}{2g}$$

$$0.2 = 2 - v_a^2/2g$$

or, $v_a = \sqrt{2 \times 9.81 \times 1.8} = 5.94$ m/s

$$v = \sqrt{2gh} = \sqrt{2 \times 9.81 \times 2} = 6.264 \text{ m/s}$$

(a) $C_v = \frac{v_a}{v} = \frac{5.94}{6.264} = 0.948$ Ans

$C_d = C_c\ C_v = 0.625 \times 0.948 = 0.5926$ Ans

(b) $Q = C_d\ a\ \sqrt{2gh}$

$= 0.5926 \times \frac{\pi}{4}(0.04)^2\ \sqrt{2 \times 9.81 \times 2} = 4.66 \times 10^{-3}\ m^s/s = 4.66$ liters/s

(c) $C_c = \frac{x}{\sqrt{4yh}}$

or, $x = C_c\ \sqrt{4yh} = 0.948 \times \sqrt{4 \times 0.5 \times 2} = 1.896$ m Ans

7.7 A 25 mm diameter sharp edged orifice issuing at jet horizontal under a constant head of 1.5 fell 1m vertically and stuck the ground at 2.4 m horizontal from vena contracta. The time required to discharge 100 liters of water was found to be 58 seconds. Find the coefficient of velocity, coefficient of contraction and coefficient of discharge and also the lost of energy at the orifice.

Solution

$$Q = \frac{100}{58} = 1.724 \text{ liters/s}$$

$$a = \frac{\pi}{4}(0.025)^2 = 4.908 \times 10^{-4} \text{ m}^2$$

$$C_v = \frac{x}{2\sqrt{yh}} = \frac{2.4}{2\sqrt{1 \times 1.5}} = 0.9797$$ Ans

$$C_d = \frac{Q_{act}}{Q_{th}} = \frac{1.724 \times 10^{-3}}{a\sqrt{2gh}} = \frac{1.724 \times 10^{-3}}{4.908 \times 10^{-3} \times \sqrt{2 \times 9.81 \times 1.5}}$$

$$= \frac{1.724 \times 10^{-3}}{2.6629 \times 10^{-3}} = 0.647$$ Ans

$$C_c = \frac{C_d}{C_v} = \frac{0.647}{0.9797} = 0.647$$ Ans

$$h_l = h\left(1 - C_v^2\right) = 1.5\left(1 - (0.9797)^2\right) = 0.06 \text{ m of water}$$ Ans

7.8 A submerged rectangular orifice discharges 875 liter/s when the constant difference of water levels on its two sides is 0.25 m. Taking C_d = 0.61, find the dimensions of orifice.

Solution

$$Q = c_d \; a \; \sqrt{2gh}$$

$$875 \times 10^{-3} = 0.61 \times a \times \sqrt{2 \times 9.81 \times 0.25}$$

$$a = \frac{875 \times 10^{-3}}{1.35098} = 0.6476 \text{ m}^2$$

$$a = LD, \text{ therefore, } D = \frac{a}{L} = \frac{0.6476}{1} = 0.6476 \text{ m}$$

Assuming 1 meter width = 0.6476

Therefore depth = 0.6476 m Ans

7.9 A partially submerged rectangular orifice is 1.2 m wide and 0.6 m deep. Its coefficient of discharge is 0.61. The distance form the centre of orifice is 3.7m and the distance from the top edge of the orifice is 3.4m. Calculate the discharge.

Solution

$$Q = Q_1 + Q_2$$

$$Q_1 = \frac{2}{3} C_d \sqrt{2g} \, L\left(h^{3/2} - h_1^{3/2}\right)$$

$$= \frac{2}{3} \times 0.61 \sqrt{2 \times 9.81} \times 1.2 \left[(3.7)^{1/2} - (3.4)^{3/2}\right]$$

$$= 1.8328 \text{ m}^3/\text{s}$$

$$\text{and } Q_2 = C_d \left[L\left(h_2 - h\right)\right] \sqrt{2gh}$$

$$= 0.61\left[1.2\left(4 - 3.7\right)\right] \sqrt{2 \times 9.81 \times 3.7} = 1.871 \text{ m}^3/\text{s}$$

Therefore, $Q = Q_1 + Q_2 = 1.8328 + 1.871 = 3.7038 \text{ m}^3/\text{s}$ Ans

7.10 An external mouthpiece of diameter 75mm diameter fitted to a side of a large vessel. If the head over the mouthpiece is 3m, find the discharge. Take C_d =0.82

Solution

$$Q = C_d \, a \sqrt{2gh} = 0.82 \times \frac{\pi}{4} (0.075)^2 \times \sqrt{2 \times 9.81 \times 3}$$

$$= 0.02779 \text{ m}^3/\text{s} = 27.79 \text{ liters/s}$$ Ans

7.11 Water under a constant head of 6m, is discharging through an external mouthpiece of 100 mm diameter. Determine the pressure of water at the vena contracta. Take atmospheric pressure as 760 mm of mercury.

Solution: Using the equation

$$h_c = h_a - 0.82\,h = 0.76 \times 13.6 - 0.82 \times 6 = 10.336 - 4.92$$

= 5.416 m of water absolute

= 4.92 m of water vacuum gauge

The vacuum pressure at vena contracta should not fall below the vapour pressure.

7.12 An external mouthpiece converges from the inlet upto the vena contracta to the shape of the jet and then it diverges gradually. The diameter at the vena contracta is 40 mm and the head over the center of the mouthpiece is 2.5 m. The head loss in the contraction may be taken as 1 per cent and that in the divergent portion 5 per cent of the total energy head before the inlet. What is the maximum discharge that can be drawn through outlet and what should be the corresponding diameter at the outlet?

Solution

Assuming that the pressure in the system may be permitted to fall upto 7.5m below atmosphere, the liquid conveyed being water.

$$a_c = \frac{\pi}{4} (0.04)^2 = 1.256 \times 10^{-3} \text{ m}^2$$

$h_a - h_c = 7.5$ m

$h = 2m$

Total loss = 1 + 5 = 6% = 0.06

$C_d = 1 - 0.06 = 0.94$

$$Q = C_d \ a_c \sqrt{2gh} = 0.94 \times 1.256 \times 10^{-3} \sqrt{2 \times 9.81 \times 2.5}$$

$= 8.278 \times 10^{-3}$ m³/s = 8.27 liters/s **Ans**

$$\frac{a}{a_c} = \sqrt{1 + \frac{h_a - h_c}{h}} = \sqrt{1 + \frac{7.5}{2.5}} = 2$$

$a = 2 \times a_c = 2 \times 1.256 \times 10^{-3} = 2.5132 \times 10^{-3}$

$$\frac{\pi}{4} d^2 = 2.5132 \times 10^{-3}$$

$$\therefore d^2 = 3.2 \times 10^{-3}$$

$$\therefore d = 0.05656 \text{ m}$$

$= 56.56$ mm **Ans**

7.13 How many liters of water per minute will be discharge freely from a vessel under a constant head 4m through

(a) a sharp edged orifice 50 mm diameter. Take $C_d = 0.61$

(b) an external cylindrical mouthpiece, 50 mm diameter and 150 mm long. Take $C_d = 0.82$

(c) re entrant mouthpiece of 50 mm diameter, running free. Take $C_d =$ 0.52

(d) re entrant mouthpiece of 50 mm diameter, running full. Take $C_d =$ 75

(e) a convergent mouthpiece of inlet diameter 50 mm and outlet diameter 37.5 mm. Take $C_d = 0.92$

(f) a divergent mouthpiece of inlet diameter 50 mm and outlet diameter 62.5 mm. Take $C_d = 0.97$

Solution

(a) $$Q = C_d \ a \sqrt{2gh} = 0.61 \times \frac{\pi}{4} (0.05)^2 \times \sqrt{2 \times 9.81 \times 4}$$

$= 1.9635 \times 10^{-3} \times 5.4039 = 0.0106$ m³/s

$= 0.63663$ m³/min = 636.63 liters/min **Ans**

(b) $$Q = C_d \ a \sqrt{2gh} = 0.82 \times 1.9635 \times 10^{-3} \times 8.8588$$

$= 0.82 \times 0.01739 = 0.01426$ m³/s

$= 0.855795$ m³/min = 855.795 liters/min **Ans**

(c) $Q = C_d\, a\sqrt{2gh} = 0.52 \times [0.01739]$

$= 9.0428 \times 10^{-3}\ m^3/s$

$= 0.542568\ m^3/min = 542.568$ liters/min **Ans**

(d) $Q = C_d\, a\sqrt{2gh} = 0.75 \times (0.01739) = 0.0130425\ m^3/s$

$= 0.78255\ m^3/min = 782.55$ liters/min **Ans**

(e) $Q = C_d\, a\sqrt{2gh} = 0.97 \times \frac{\pi}{4}(0.0375)^2 \times \sqrt{2 \times 9.81 \times 4}$

$= 0.97 \times 1.1044 \times 10^{-3} \times 8.8589 = 9.49 \times 10^{-3}\ m^3/s$

$= 0.569415\ m^3/min = 569.415$ liters/min **Ans**

(f) $Q = C_d\, a\sqrt{2gh} = 0.92 \times \frac{\pi}{4}(0.625)^2 \times \sqrt{2 \times 9.81 \times 4}$

$= 0.92 \times 3.06796 \times 10^{-3} \times 8.8589 = 0.0250094\ m^3/s$

$= 1.5\,002678\ m^3/min = 1500.2678$ liters/min **Ans**

7.14 A pipe enters a vessel horizontal forming a re-entrant sharp edged orifice of 40 mm diameter. If the head above the orifice is 1.5 m and the coefficient of velocity is 0.95 find (a) the coefficient of contraction and discharge when the running free, and (b) the discharge when the orifice is running full, assuming that a vena contracta is formed in the pipe, the coefficient of contraction being the same as in (a), and allowing for the energy loss between the vena contracta and the outlet.

Solution : (a) As per Fig. 7.14

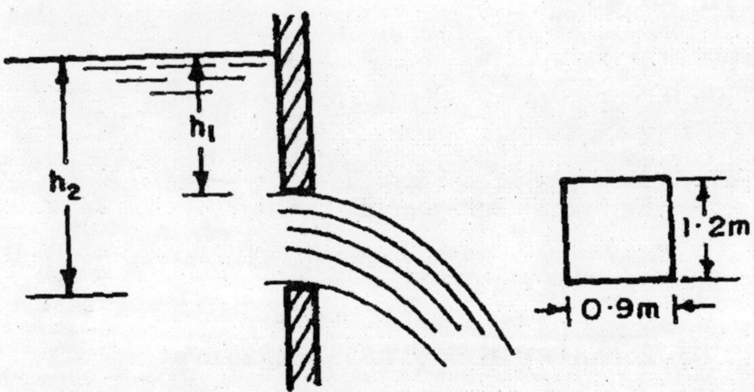

Fig. 7.21. Horizontal vessel P. 7.14.

Rate of change of momentum of jet = Pressure force on orifice

$\rho\, Q\, v = \rho\, g\, h\, a_o$

$\rho\, (a\, v)\, v = \rho\, g\, h\, a_o$

$\rho\, a\, C_v^2\, (2gh) = \rho\, gh\, a_o$

or, $$2a\,C_v^2 = a_o$$

$$\frac{a}{a_o} = \frac{1}{2C_v^2}$$

$$C_c = \frac{1}{2C_v^2} = \frac{1}{2(0.95)^2} = \frac{1}{1.805} = 0.554$$ Ans

$$Q = a\,v = (0.554)\ a_o\ C_v^2\ \sqrt{2gh}$$

$$= 0.554 \times \frac{\pi}{4}\left(\frac{40}{1000}\right)^2 \times C_v^2\ \sqrt{2 \times 9.81 \times 1.5}$$

$$a = 6.96 \times 10^{-4} \times 0.95 \times \sqrt{2 \times 9.81 \times 1.5}$$

$$= 6.96 \times 10^{-4} \times 5.4249$$

$$= 3.5879 \times 10^{-3}\ m^3/s = 3.58\ liters/s$$ Ans

(b) Loss of energy = $h_l = \dfrac{(v_c - v)^2}{2g}$

$$h_l = \frac{v^2}{2g}\left[\frac{1}{C_c} - 1\right]^2$$

$$h_l = \frac{v^2}{2g}\left[\frac{1}{0.554} - 1\right]^2$$

$$h_1 = 0.648\frac{v^2}{2g}$$

Therefore, $h = \dfrac{v^2}{2g} + h_l$

$$1.5 = \frac{v^2}{2g} + 0.648\frac{v^2}{2g} = 1.648\frac{v^2}{2g}$$

$$0.91 = \frac{v^2}{2g}$$

$$v = \sqrt{2 \times 9.81 \times 0.91} = \sqrt{17.858} = 4.225\ m/s$$

$$Q = a\,v = \frac{\pi}{4}(0.04)^2 \times 4.225 \quad 5.31 \times 10^{-3}\ m^3/s = 5.31\ liters/s$$ Ans

7.15 **A convergent divergent mouthpiece having throat diameter of 40 mm is discharging under a constant head of 2.5 meter. Determine the maximum diameter of divergence to avoid separation of flow. The atmospheric pressure is 10 m of water and separation pressure is 2.5 m of water absolute. Find also the maximum discharge.**

Solution: Using the equation (7.26)

$$\frac{a}{a_c} = \sqrt{1+\frac{h_a - h_c}{h}} = \sqrt{1+\frac{10-2.5}{2.5}} = \sqrt{1+3} = \sqrt{4} = 2$$

Here, $a_c = \frac{\pi}{4}(0.04)^2 = 1.256 \times 10^{-3}\ m^2$

$$a = a_c \times 2 = 1.256 \times 10^{-3} \times 2 = 2.513 \times 10^{-3}\ m^2$$

$$\frac{\pi}{4} d^2 = 2.513 \times 10^{-3}$$

$$d^2 = 3.2 \times 10^{-3}$$

$\therefore$ d = 0.0565 = 56.56 mm **Ans**

$$Q_{max} = a_c \sqrt{2gh} = 1.256 \times 10^{-3} \times \sqrt{2 \times 9.81 \times 2.5}$$

$$= 8.796 \times 10^{-3}\ m^3/s\ = 8.796\ liters/s$$ **Ans**

7.16 A convergent divergent mouthpiece is fitted to the side of a tank. If the diameters at the outlet and vena contracta to be 60 mm and 40 mm respectively. Find the maximum head of water for steady flow. Take atmospheric pressure of 10.3 m of water and separation pressure of 25 kpa.

Solution

$$\frac{a}{a_c} = \sqrt{1+\frac{h_a - h_c}{h}}$$

$$\frac{60^2}{40^2} = \sqrt{1+\frac{10.3-2.5472}{h}}$$

$$2.25 = \sqrt{1+\frac{7.7525}{h}}$$

$$5.0625 = 1+\frac{7.7525}{h}$$

$$\frac{7.7525}{h} = 4.0625$$

$$\therefore \quad h = \frac{7.7525}{4.0625} = 1.908\ m$$ **Ans**

7.17 A convergent divergent mouthpiece is fitted to the side of a tank. The discharge through mouthpiece under a constant head 2.5 m is 8.5 liter/s. The head loss in the divergent portion is of the kinetic heat at outlet. Find the throat and exit diameters, if the pressure in the system may be permitted to fall upto 7.5 m below atmosphere. The atmospheric pressure is 1.01325 bar.

Solution

$$Q = a_c\, v_c = a_c \times \sqrt{2 \times 9.81 \times 2.5}$$

$$8.5 \times 10^{-3} = a_c \times 7$$

$$a_c = 1.2136 \times 10^{-3}$$

$$\frac{\pi}{4}\, d_c^2 = 1.2136 \times 10^{-3}$$

$$d_c^2 = 1.545 \times 10^{-3}$$

$$d_c = 0.0393 \text{ m} = 39.3 \text{ mm}$$ Ans

Applying the Bernoulli's equation between free water surface in tank and outlet of mouthpiece (Fig. 7.13).

$$h_a + h = h_a + \frac{v^2}{2g} + 0.5\frac{v^2}{2g}$$

$$h = 1.5\frac{v^2}{2g}$$

$$2.5 = 1.5\frac{v^2}{2g}$$

$$v^2 = \frac{2.5 \times 2 \times 9.81}{1.5} = 32.7$$

$$\therefore v = 5.718$$

$$Q = a\,v$$

$$8.5 \times 10^{-3} = a \times 5.718$$

$$a = 1.4864 \times 10^{-3}, \quad \frac{\pi}{4}\, d^2 = 1.4864 \times 10^{-3}$$

$$d^2 = 1.892 \times 10^{-3}$$

$$d = 0.0435 = 43.5 \text{ mm}$$ Ans

EXERCISES

A. THEORY

7.1 Define an orifice.

7.2 How are the orifice classified according to size, shape, shape of upstream edge and condition of discharge?

7.3 What is the difference between a small and a large orifice.

7.4 What are hydraulic coefficient? Define them.

7.5 Define the $C_d \times C_c$ and C_v.

7.6 Prove the formula $v = \sqrt{2gh}$

7.7 Prove the formula $C_d = C_c \times C_v$

7.8 Show the position of vena contracta with a suitable sketch.

7.9 Derive from Bernoulli's theorem expressions for the theoretical velocity and discharge through a small orifice. How does the actual discharge compare with this theoretical value? Explain the reasons for the difference.

7.10 Define the coefficients used in connection with flow through orifices, explaining why these coefficient are necessary.

7.11 A jet of liquid issues horizontally form an orifice in the vertical side of the tank. Derive an expression for the actual velocity of the jet V. If the jet falls a distance y vertically in a horizontal distance x measured from the vena contracta. If the head of liquid above the orifice is h, what will be the coefficient of velocity?

7.12 What is the difference between orifice and mouthpiece?

7.13 Sketch and explain a Borda's mouthpiece.

7.14 What are different mouthpieces and their uses?

7.15 What is a submerged (drowned) orifice?

7.16 What is a partially submerged orifice?

7.17 Distinguish between (a) External mouthpiece and internal mouthpiece (b) Mouthpiece running free and mouthpiece running full.

7.18 Obtain an expression for absolute pressure head at vena contracta for an external mouthpiece.

7.19 What are convergent, divergent and convergent divergent mouthpieces and where they can be used?

7.20 Explain why the coefficient of discharge of an orifice as low as 0.6 to 0.62? How can its value be increased to 0.8 or more.

7.21 Prove the formula for external cylindrical mouthpiece running full the coefficient of discharge. Why in actual practice the value of coefficient of discharge is taken as 0.82.

7.22 Prove that the flow discharge by a standard mouthpiece is about one third greater than that discharge by sharp edged orifice of the same diameter.

7.23 Derive an expression for the time of emptying a tank of constant cross section, through an sharp edged orifice.

7.24 Derive a formula for the time of emptying a vertical cylindrical tank through an orifice in the bottom.

7.25 Derive an expression for the time of emptying a hemispherical tank through a sharp edged orifice.

7.26 Derive a formula for the time of emptying a tank of variable cross section.

7.27 Derive a formula for the time of emptying a pyramid shaped tank.

7.28 Derive a formula for the time of emptying a conical tank.

7.29 Derive a formula for the time of emptying a swimming tank.

7.30 Derive a formula for the flow of liquid from one vessel to another.

7.31 Derive a formula for the time of emptying (or filling) a tank with inflow.

7.32 Write a general procedure for calculating time of emptying a tank through an orifice/ mouthpiece.

B. UNSOLVED PROBLEMS

7.1 In an experiment to determine the hydraulic coefficients of a 20 mm diameter orifice working under a head of 1 m. The discharge was found to be 0.835 liters/s. It was further observed that coming out of the orifice traveled a distance of 570 mm from the plane of the orifice during a course of fall of 90 mm in the vertical direction from the centre of orifice. Determine the hydraulic coefficients of the profile. [C_C = 0.643, C_v = 0.933, C_d = 0.6]

7.2 Oil flows through a 25 mm diameter orifice under a head of 5.5 m at a rate 3 liter/s. The jet strikes a wall 1.5 m away and 120 mm vertically below the centreline of the contracta jet. Calculate the coefficient of velocity, contraction and discharge. [C_v = 0.923, C_d = 0.588, C_c = 0.637]

7.3 A closed cylindrical tank is 3.5m high and contains an oil of relative density 0.85 to a height of 3m above the bottom. The space above the oil surface contains air under a pressure of 50 kPa. If an 80 mm diameter a orifice is provided on the side of the tank with its centre 250 mm above the bottom, calculate the weight of fluid discharged in one minute. Take C_d =0.6. [19.74 kN/min.]

7.4 A tank has two identical orifices in one of its vertical sides, The upper orifice is 2 m below the water surface and the lower one is 4m below the water surface. Finds the point at which the two jets will intersect, if the coefficient of velocity is 0.9 for both the orifices.

7.5 A large tank has a circular sharp edged orifice 930 mm² in area at a depth of 2.7 m below constant water level. The jet issues horizontally and in a horizontal distance of 2.34m it falls 0.54m. The measured distance is 4.2 dm³/s. Calculate the coefficients of velocity, contraction and discharge. [C_v = 0.97, C_c = 0.624 and C_d = 0.643]

7.6 A Borda mouthpiece of 150 mm diameter discharge water under a head of 3m calculate (a) the discharge and (b) the diameter of the jet at the vena contracta when running free [67.8dm³/s,106mm]

7.7 A vertical triangular orifice in the wall of a reservoir has a base 900 mm long, 600 mm below its vertical and 1.2 m below the water surface. Determine the theoretical discharge. [1.19 m^3/s]

7.8 A rectangular orifice 1.5 m wide and 1 m deep level in the tank is 3 m above the top edge of the orifice. Find the discharge through the orifice Take Cd=0.6 [7.45 m^3/s]

7.9 Water under a constant head of 3.5 m discharges through an external cylindrical mouth piece of 50 mm diameter. Taking $C_c = 0.7$ calculate (a) the discharge, (b) the coefficient of discharge (c) the absolute pressure at vena contracta in meter of water [0.01355 m^3/s, 0.833,676 m of water]

7.10 A submerged orifice 1.5 hide and 0.5 meter deep is provided in one side of a tank. find the discharge in liter/s through the orifice the difference of water level on both the sides of the orifice be 4m. Take $C_d = 0.64$ [4050 liter/s]

7.11 An orifice, in one side of a large tank is rectangular in shape 2 meters broad and 1 meter deep. The water level on one side, of the orifice is 4 meters above its top edge. The water level on the other side, of the orifice, is 0.5 m below its top edge. [$Q_1 = 5.707$ m^3/s, $Q_2 = 5.873$ m^3/s, $Q = 11.58$ m^3/s]

7.12 A cylindrical mouth piece 25 mm in diameter and 80 mm long is attached to tank externally. The head over the mouthpiece is 1 m. Caleulate the discharge. If the same mouthpiece is fitted internally, what will be the new discharge? Further, if the length of the mouthpiece is reduced to 20 mm now will the discharged be affected ? [1.85 lps, 1.54 lps, 1.35 lps, 1.09 lps]

7.13 In order to increases the discharge to atmosphere from an orifice in the side of a tank, a divergent mouthpiece is fitted. The original orifice is if 100 mm in diameter and of such a type that the coefficient of contraction is one and the coefficient if velocity is 0.98. The head over the orifice is 2.5 m calculate the diameter of the outlet of the mouthpiece, if the discharge is to be doubled. The new coefficient may be taken as 0.85. What is the now the pressure in the plane of the original orifice? What is the maximum theoretical discharge of the orifice? [151.9 mm, 2.53 m (abs) 108 lps]

7.14 The throat and the exit diameters of a convergent divergent mouthpiece fitted to a tank contain in water are 75 mm and 125 mm respectively. If separation occurs at a pressure of 25 kPa, and atmospheric Pressure is 100 kPa determine the maximum head over the mouthpiece for a steady uniform flow. How much will be discharge? [h = 1.1384 m, Q = 58 lps]

8

Flow over Notches and Weirs

8.1 INTRODUCTION

A weir is a structure in an open channel over which liquid flows. It may be used for purpose of determining the liquids flow rate. Weirs are the most popular and standard devices for measuring stream flow in open channels. Basically they are obstruction across a flow to cause a unique head discharge relationship.

There is no hard and fast line of distinction between a weir and a notch, but generally speaking, small weirs in thin plates are called notches.

8.2 RECTANGULAR WEIRS

The upper surface or edge of the weir over which the water flows is known as crest or sill of the weir. The over flowing stream of the water over the weir is called nappe or vein. The rate of flow is determined by measuring the height H (head), relative to the crest, at a distance upstream from the crest at least four times the maximum head that is to be employed. The amount of drawdown at the crest is about 0.15 H.

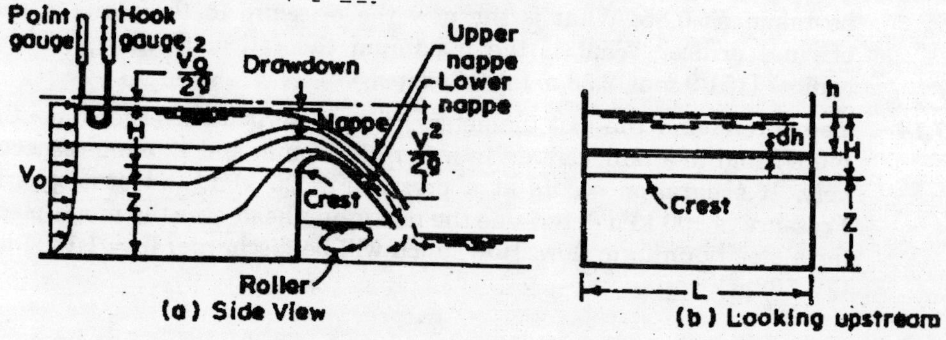

Fig. 8.1. Flow over sharp crested were.

The upstream face of the weir plate should be smooth, and the plate should be strictly vertical. The crest should have a sharp, square upstream edge, and a bevel on the downstream side, so that the nappe springs clear, making a line contact for all but the very lowest heads. If it does not spring clear, the flow cannot be considered as true weir flow and the experimentally determined coefficients do not apply. The velocity at any point in the nappe is related to the energy line as shown in Fig. 8.1. The approach channel should be long enough so that normal velocity distribution exists and the surface should be as free of waves as possible. If the nappe discharge into air the weir is said to have a free discharge and if the discharge takes place above crest or sill, the weir is known as drowned or submerged weir.

There are two types of rectangular weir (Fig. 8.2).

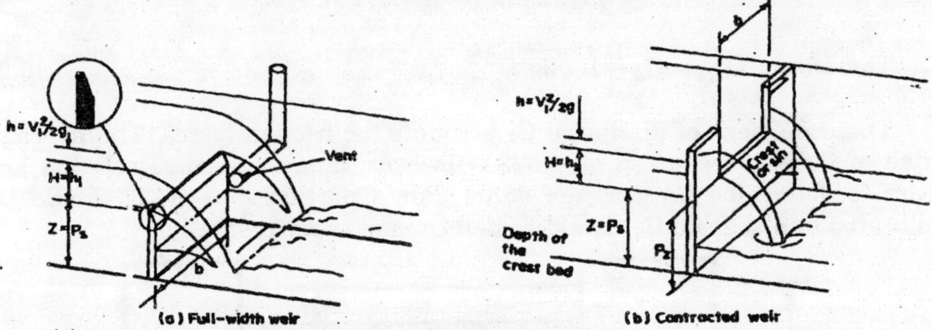

Fig. 8.2. Rectangular Weirs.

(a) 'Un contracted' or full width weirs comprise a plate with a horizontal crest extending from one side of the channel to the other the crest section is as illustrated in the Fig. 8.2(a).

(b) a contracted weir, by contrast, has a crest width which is less than the channel width as illustrated in the Fig. 8.2(b)

Since the operation of a weir is based on the use of a gauged depth to estimate the discharge, we must know how these quantities are related. The actual flow over a weir is quite complex, involving a three dimensional velocity pattern as well as viscous effects. The simplest method of developing a numerical model which represents a weir is to use the Bernoulli's equation as a starting point. An ideal relationship between depth and discharge is obtained. This relationship can then be modified to take account of the differences between ideal and real flows.

8.2.1 Flow over rectangular weir

Let us consider a horizontal ship of width L at a depth h below the free surface and height dh (Fig. 8.3).

$$\text{Area of strip} = Ldh$$

$$\text{Velocity through strip} = \sqrt{2gh}$$

$$\text{Discharge through strip, } dQ = \text{Area} \times \text{velocity} = Ldh\ \sqrt{2gh}$$

Integrating from h = 0 at the free surface to h = H at the bottom of the notch.

$$\text{Total theoretical discharge, } Q_{th} = L\sqrt{2g}\int_0^H \sqrt{h}\, dh$$

$$Q_{th} = L\sqrt{2g}\left[\frac{h^{3/2}}{3/2}\right]_0^H = \frac{2}{3} L\sqrt{2g}\, H^{3/2} \quad (8.1)$$

$$C_d = \frac{Q_{act}}{Q_{th}} = \frac{Q_{act}}{\frac{2}{3} L\sqrt{2g}\, H^{3/2}}$$

$$\text{Therefore, } Q_{act} = C_d \frac{2}{3} L\sqrt{2g}\, H^{3/2} \quad (8.2)$$

The coefficient of discharge C_d accounts for friction losses. The average value of C_d is to be taken as 0.623. However, many empirical relation are given to determine its accurate value. For a particular weir, it should be calibrated in place and C_d should be determined experimentally.

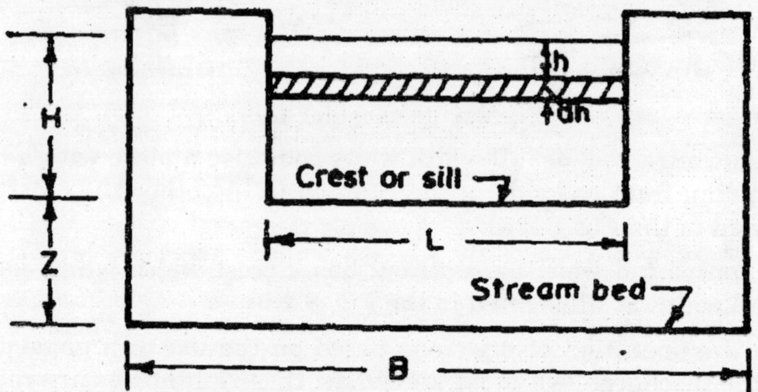

Fig. 8.3. Flow over a rectangular weir.

8.2.2 Flow over rectangular weir with velocity of approach

The liquid approaching the sill or crest has a velocity which is called the velocity of approach v_a. The approach velocity is not considered while deriving an equation (8.2).

The velocity of approach v_a is calculated by the formula.

$$v_a = \frac{\text{Discharge in channel}}{\text{Area in channel}} \quad (8.3)$$

The velocity of approach is used for finding the accurate discharge. the velocity of approach cannot be found until quantity of water is known. A series of approximation must therefore, be used. The procedure is as follows :

Step 1 Discharge is calculated without velocity of approach.

Step 2 The discharge obtained in the step 1 is divided by the area of the channel (Z+H) B on the upstream side.

Step 3 The velocity of approach is obtained by using the equation (8.3)

$$v_a = \frac{Q}{B(Z+H)}$$

Step 4 The process is repeated until the final discharge obtained is 1% of the preceding calculated discharge.

The equation (8.2) is derived neglecting the velocity of approach with integration from 0 to H. However, the effective head which is responsible for the discharge over the weir is the sum of measured head H and the head due to the velocity of approach $\frac{v_a^2}{2g} = ha$. Thus the limit of integration are modified to ha and (H+ha). The discharge is calculated as

$$Q = C_d \sqrt{2g}\, L \int_{ha}^{H+ha} \sqrt{h}\, dh$$

$$= \frac{2}{3} C_d \sqrt{2g}\, L \left[(H+ha)^{3/2} - ha^{3/2}\right] \quad (8.4)$$

where (H+ha) is known as still water head or equivalent static head.

It is to be noted that the value of v_a is considered when its value is given or it can be calculated from the data given; otherwise it is neglected while working out examples, v_a is usually negligible when the waterway in approach channel is very large as compared to the water way of a notch. For suppressed weir, when it is more 1/5th of the height Z of weir, v_a must be considered.

8.2.3 Empirical formulae for discharge over rectangular weir

The cross sectional area of the jet just downstream of the weir will be significantly less than the cross section of the flow at the weir in self due to the converging pattern of stream lines approaching the weir. Therefore, the actual discharge will be less than the theoretical discharge. There are two additional discrepancies in the theory.

(a) the flow approaching the weir will be subjected to viscous forces: this produces two effects – (i) a nonuniform velocity distribution in the channel and (ii) the loss of energy.

(b) the pressure distribution in the flow passing over the weir is not atmospheric.

The value of c_d is not constant. From the dimensional analysis $c_d = f\left[R_e, We, \frac{Z}{H}\right]$. The problems calibration i.e evaluating c_d experimentally become far greater if the nappe clings to the downstream face of the weir. This condition arises due to small discharge. Due to these conditions, the effects of viscosity and surface tension combine to bring about unstable, fluctuating flow conditions. Since the value of H is small at low flows, an accurate determination of C_d becomes impracticable.

A number of empirical formulae have been developed which incorporate C_d.

(a) Francis formula

Experiments by Francis indicated that under the condition depicted in Fig. 8.4 the effect of each side contraction is to reduce the effective width of the nappe 0.1. Hence for such a situation the flow rate may be computed by substituting (L - 0.1n H) for L, where n is the number of end contractions normally 2 but sometimes 1.

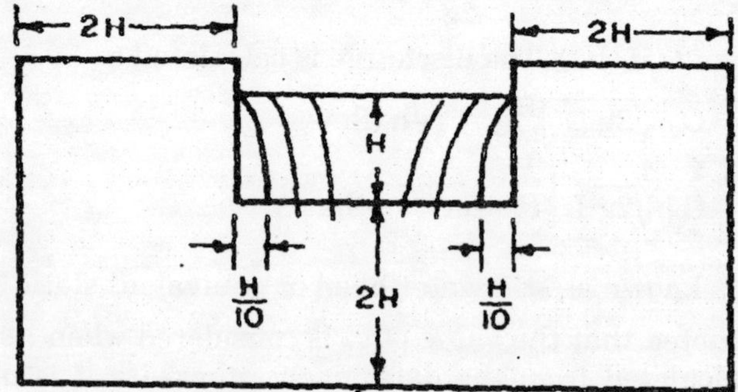

Fig. 8.4. Limiting proportions of standard contracted weir.

$$Q = C_w\,[L - 0.1\,nH]H^{3/2} \tag{8.5}$$

Where C_w = weir coefficient or weir factor $= C_d\,\dfrac{2}{3}\sqrt{2g}$ (8.6)

Using a value of $C_d = 0.623$ in the equation (8.6)

$$C_w = 1.84$$

Substituting in equation (8.5)

$$Q = 1.84\ [(L - 0.1\,n\,H)]\ H^{3/2} \tag{8.7}$$

The discharge formula with velocity of approach is written as

$$Q = 1.84\,[L - 0.1\,n\ (H + h_a)]\ \left|(H + h_a)^{3/2} - h_a^{3/2}\right| \tag{8.8}$$

For suppressed weir, n = 0 the equation (8.7) reduces to

$$Q = 1.84\ L\ H^{3/2} \tag{8.9}$$

(b) Rehbock formula

Rehbock of the Karlsruhe Hydraulic Laboratory in Germany performed small scale but precise experiments covering wide range of conditions found the value of C_d in equation (8.2)

$$C_d = 0.605 + \frac{1}{1000\,H} + 0.08\frac{H}{Z} \tag{8.10}$$

This equation was obtained by fitting a curve to the plotted values of C_d for a great many experiments and is purely empirical. Capillarity is accounted for by the second term while velocity of approach (assumed to be uniform) is responsible for the last term.

If the capillarity effect is neglected then the value of C_d reduces to

$$C_d = 0.605 + 0.08\frac{H}{Z} \tag{8.11}$$

The Rehbock's formula has been found to be accurate within 0.5 per cent for values of Z form 0.1 to 1m and for values of H from 0.024 to 0.6 m with the ratio H/z not greater than 1. It is even valid for great ratios than 1 if the bottom of the discharge channel is lower than that of the approach channel so that backwater does not affect the head. The equation (8.2) is written as.

$$Q = \frac{2}{3} C_d \sqrt{2g}\, L\, H^{3/2}$$

$$= \frac{2}{3}\sqrt{2g}\left[0.605 + 0.08\frac{H}{Z}\right] L\ H^{3/2}$$

$$= \left(1.786 + 0.236\frac{H}{Z}\right) L\, H^{3/2} \tag{8.12}$$

Ackers *et al.* (1978) suggest that the accuracy of the formulae depends, to a discernible extent, on the string of the gauging station for measuring the upstream head H. They recommend that this should be at least 2.67 Z upstream of the weir to avoid undue drawdown effects. The British standards (BS 3680) recommendation is that the station should be between 4H and 5H upstream of the weir.

A number of empirical equations have been developed to enable estimates of C_d to be made.

8.3 FLOW OVER A TRIANGULAR WEIR (V-NOTCH)

The rectangular weirs suffer from a loss of accuracy at low flows.

The V weir largely overcome this problem. The variation of L with height, together with the narrow nappe width in the jet, means that for a given increase in Q, the increase in H for a V weir will be much greater than for a rectangular weir. Conversely, the greater sensitivity limits the range of discharges for which a V weir can be economically applied. The assumptions are the same as for the rectangular weir excepting, of course, the fact that L is not a constant but a function of head.

In Fig 8.5(a) is triangular weir with a vertex angle θ. the ideal rate of discharge though an elementary strip is calculated as. Let us consider a horizontal elementary strip of thickness dh at a depth h below the water surface. If 2x is the width of the strip then

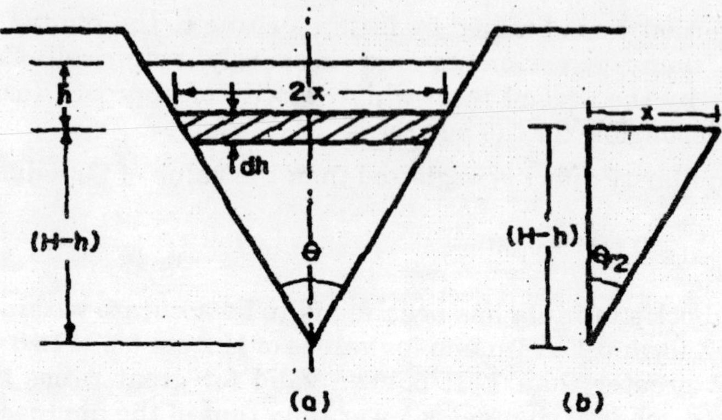

Fig. 8.5. Flow over a triangular weir (V-notch)

From Fig 8.5 (b)

$$\tan\frac{\theta}{2} = \frac{x}{H-h}$$

$$x = (H-h)\tan\frac{\theta}{2}$$

Area of strip = $2x$ dh

$$A = 2\ (H-h)\ \tan\frac{\theta}{2}\ dh.$$

velocity of flow through strip $v = \sqrt{2gh}$

Discharge through strip, dQ = (Area × Velocity) C_d

$$dQ = 2\ (H-h)\ \tan\frac{\theta}{2}\sqrt{2gh}\ \ C_d$$

Total discharge $Q = \int_o^H c_d\ \tan\frac{\theta}{2}\sqrt{2gh}\ 2\left(H-h\right)dh$

$$= \int_o^H 2C_d\ \tan\frac{\theta}{2}\sqrt{2g}\left(H\ \sqrt{h} - h^{3/2}\right)dh$$

$$= 2\ C_d\ \tan\frac{\theta}{2}\sqrt{2g}\left[\frac{Hh^{3/2}}{3/2} - \frac{h^{5/2}}{5/2}\right]_0^H$$

$$= 2\ C_d\ \tan\frac{\theta}{2}\sqrt{2g}\left[\frac{2}{3}H^{5/2} - \frac{2}{5}H^{5/2}\right]$$

$$= 2\ C_d\ \tan\frac{\theta}{2}\sqrt{2g}\left[\frac{4}{15}H^{5/2}\right]$$

$$= \frac{8}{15} C_d \tan \frac{\theta}{2} \sqrt{2g}\, H^{5/2} \tag{8.13}$$

For a given angle θ and assuming C_d is constant, this may be reduced to

$$Q = kH^{5/2} \tag{8.14}$$

In Fig 8.6, experimental values for C_d for water flowing over v-notch with central angles varying from 10 to 90° are plotted.

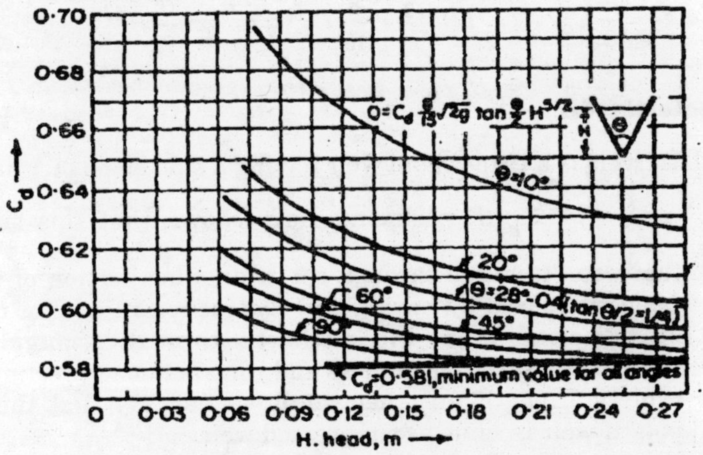

Fig. 8.6. Graph between C_d v/s head

The rise in C_d at less than 0.15 m is due to incomplete contraction. At lower heads the frictional effects reduce the coefficient. At very low head, when the nappe clings to the weir plate, the phenomenon can no longer be classed as weir flow and equations (8.13) and (8.14) are inapplicable.

8.4 FLOW OVER A TRAPEZOIDAL WEIR OR NOTCH

A trapezoidal weir is a combination of rectangular weir and triangular weir. Fig 8.7 shows a trapezoidal weir of still length the L and sides slope at an angle of θ/2 with the vertical.

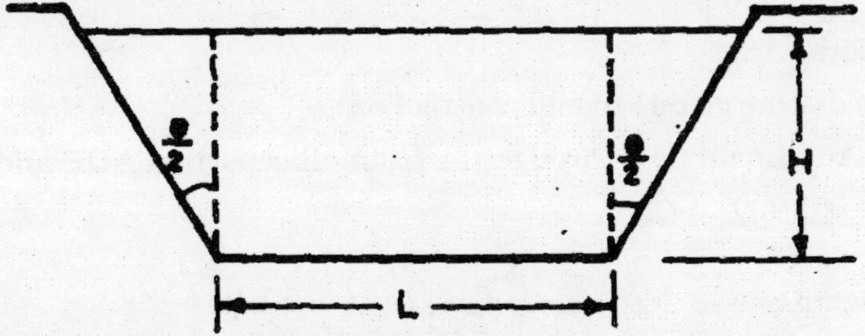

Fig. 8.7. Trapezoidal weir.

The total discharge Q is a calculated as

$$Q = Q_1 + Q_2$$

Where,

Q_1 = Discharge through rectangular portion

Q_2 = Discharge through two triangular portions at the two ends of weir.

If the two triangular portions at the two ends are brought together so that their vertical sides coincide, we get a triangular weir of angle θ and head H. Hence

$$Q = \frac{2}{3} C_{d_1} \sqrt{2g}\ LH^{3/2} + \frac{8}{15} C_{d_2} \sqrt{2g}\ \tan\frac{\theta}{2}\ H^{5/2} \tag{8.15}$$

8.4.1 Cippoletti weir

Cippoletti weir is a trapezoidal weir having a side slope of 1 horizontal to 4 vertical, $\left(i.e. \frac{\theta}{2} = 14° \ 2'\right)$. Experiments have shown that this side slope is required to increase discharge through the triangular portion of weir which would have been decreased by having end contraction in case of rectangular weirs. In the cippoletti weirs the discharge through triangular portion compensates the loss of discharge due to end contractions of rectangular weir having the same sill length. In other words we can say that the discharge from the cippoletti weir is same as suppressed weir.

The discharge through suppressed weir is

$$Q_a = \frac{2}{3} C_d \sqrt{2g}\ LH^{3/2}$$

The discharge over rectangular weir with end contraction suggested by Francis.

$$Q_b = \frac{2}{3} C_d \sqrt{2g}\ (L - 0.2\ H)\ H^{3/2}$$

The increase in discharge through triangular portion ADE and BCF

$$Q_t = \frac{8}{15} C_d \sqrt{2g} \tan\frac{\theta}{2} H^{5/2}$$

Therefore,

Loss of discharge due to end contraction

= Additional flow through the triangular portion ADE and BEF.

$$Q_a - Q_b = Q_t$$

$$\frac{2}{3} C_d \sqrt{2g}\ L\ H^{3/2} - \frac{2}{3} C_d \sqrt{2g}\ (L - 0.2\ H)\ h^{3/2}$$

$$= \frac{8}{15} C_d \sqrt{2g}\ \tan\frac{\theta}{2}\ H^{5/2}$$

$$\frac{2}{3}C_d\sqrt{2g}\,L\,H^{3/2}\left(L-L+\frac{H}{5}\right)=\frac{8}{15}C_d\sqrt{2g}\tan\frac{\theta}{2}H^{5/2}$$

$$\frac{2}{15}C_d\sqrt{2g}\,L\,H^{5/2}=\frac{8}{15}\,C_d\,\sqrt{2g}\tan\frac{\theta}{2}H^{5/2}$$

$$\therefore\ \tan\frac{\theta}{2}=\frac{1}{4}$$

or, $\dfrac{\theta}{2}=\tan^{-1}(0.25)=14°2'$ (8.16)

The discharge for Cippoletti weir is calculated as

$$Q = Q_1 + Q_2$$

$$Q_1=\frac{2}{3}C_d\sqrt{2g}\left(L-\frac{H}{5}\right)H^{3/2}$$

and $Q_2=\dfrac{8}{15}C_d\sqrt{2g}\tan\dfrac{\theta}{2}H^{5/2}$

For $\tan\dfrac{\theta}{2}=\dfrac{1}{4}$

Therefore,

$$Q_2=\frac{8}{15}C_d\sqrt{2g}\times\frac{1}{4}\times H^{5/2}\quad=\frac{2}{15}C_d\sqrt{2g}\;H^{5/2}$$

Therefore,

$$Q=\frac{2}{3}C_d\sqrt{2g}\left(L-\frac{H}{5}\right)H^{3/2}+\frac{2}{15}C_d\sqrt{2g}\;H^{5/2}$$

$$\frac{2}{3}C_d\sqrt{2g}\;H^{3/2}\left[L-\frac{H}{5}+\frac{H}{5}\right]=\frac{2}{3}C_d\sqrt{2g}\;H^{3/2} \quad (8.17)$$

Which is the same as suppressed weir. Cippoletti proposed the formula for discharge

$$Q = 1.86\ L\ H^{3/2} \quad (8.18)$$

It shows that the flow is independent of end contraction.

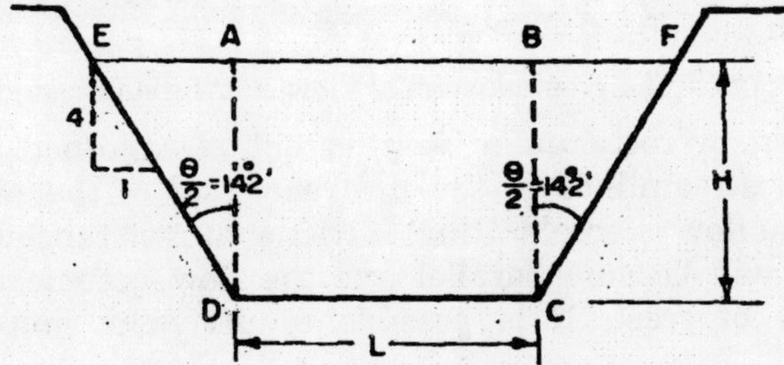

Fig. 8.8. Cippoletti weir.

8.5 LONG BASED WEIRS

In contrast to plate weirs, long based weirs are largely and generally heavily constructed (e.g. from concrete). They are usually designed for use in the field and consequently may have to handle large discharges. An ideal long based weir has the flowing requirements.

(a) Easily and cheaply fabricated (perhaps off site)

(b) Easily installed.

(c) Requires a minimum maintenance.

(d) Produces a minimum afflux (i.e. increase in upstream depth due to the installation of the weir).

(e) Posses a wide modular range.

There are number of different designs.

(1) The rectangular (broad crested) weir.

(2) Round nosed weirs.

(3) Crump weirs.

8.5.1 Broad crested weir

Rectangular weirs are solid weirs of rectangular cross section, which span the full width of a channel (Fig. 8.9 (a)).

A hump placed on the bed of the channel results in a local increase in the velocity of flow and a corresponding reduction in the elevation of the water surface.

A sufficient height is given to hump, critical flow will be produced in the flow over the hump. **There is then a direct relationship between Q and H,** i.e. the flow is modular. Long based weirs are designed for modular flow.

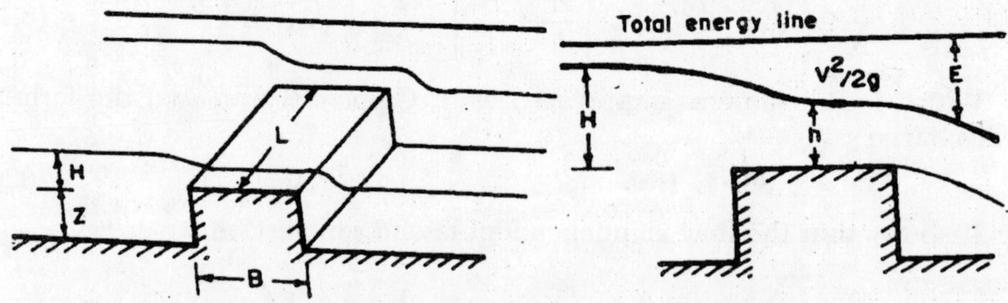

Fig. 8.9. (a) Broad crested weir (b) Flow over broad crested weir.

By definition, a rectangular weir is not streamlined. This, in turn, implies that the streamlines at the upstream end of the weir will not be parallel, since the flow is accelerating. If frictional resistance is ignored, then the streamlines will become parallel and the flow become critical given a sufficient length of crest. If is possible to derive a simple relation for discharge.

In Fig. 8.9 (b), the water is flowing over a weir. Let H is the head at, the upstream of the weir and, stream of the crest or sill. Assuming that the crest is a sufficient length to allow the velocity of the water to be uniform throughout its depth at the downstream edge. Then, neglecting losses

$$H = h + \frac{v^2}{2g}$$

$$v = \sqrt{2g\,(H - h)}$$

Let L is the length of the crest, then the discharge is given by

$$Q_{th} = (L\,h)\,(v)$$

$$Q_{th} = (L\,h)\,\sqrt{2g\,(H - h)}$$

Therefore, the actual discharge is given by

$$Q = C_d\ L\,h\,\sqrt{2g\ (H - h)} \qquad (8.19)$$

The condition for the discharge Q over a broad crested weir to be maximum for constant H may be obtained by differentiating equation (8.19) with respect to h, and equating to zero. Thus.

$$\frac{d}{dh}\left(Hh^2 - h^3\right) = 0$$

or, $$2\,Hh - 3\,h^2 = 0$$

or, $$h = \frac{2}{3}\,H \qquad (8.20)$$

Substituting the value of h equation (8.19)

$$Q_{\max} = C_d\ \sqrt{2g}\ L\,\frac{2}{3}\,H\,\sqrt{\frac{1}{3}\,H}$$

$$= C_d\ \sqrt{2g}\ L\ H^{3/2}\left(\frac{2}{3}\times\frac{1}{\sqrt{3}}\right) = 1.705\ C_d\ L\ H^{3/2} \qquad (8.21)$$

For large range of head H the weir seems to set itself automatically to this condition that the stable condition of maximum discharge; it thus appears that the stable condition of equilibrium is when the discharge is a maximum.

In reality the flow over a long crested rectangular weir is more complex than ideal flow frictional shearing action promotes the growth of a boundary layer. Additionally, as the flow approaches the critical condition, there is a tendency for ripples to form on the surface.

The value of c_d has to derived empirically. Singer proposed an equation for c_d in the form

$$C_d = 0.848\ C_F \qquad (8.22)$$

The correction factor, C_F is a function of H/B and $\dfrac{H}{H+Z}$ and is presented in graphical form in the book Ackers *et al.*

8.5.2 Round nosed weirs

These are so named because of the radius formed on the leading corner. In other respects they resemble the broad crested weir. The incorporation of the radius makes it the more robust of the two, and also means that the discharge coefficient is insensitive to minor damage. The magnitude of C_d is higher than that for a rectangular weir. However, very little is known about the performance of such weirs under submerged conditions.

It is worth noting that the region just downstream of any weir can be liable to scouring, especially in natural channels. For this reason some form of stilling arrangement may have to be incorporated.

8.5.3 Crump weirs

In 1952, E.S. Crump published details of a weir with a triangular profile, which had been developed at the Hydraulic research station. This was claimed to give a wide modular range and also to give a more predictable performance under submerged condition than other long based weirs (Fig 8.10).

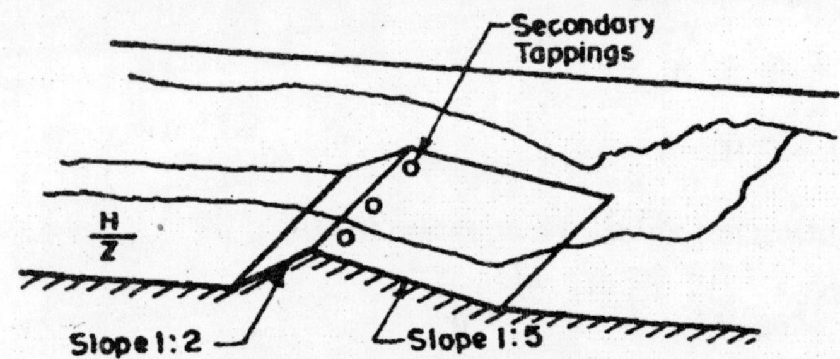

Fig. 8.10. Crump weir.

Crump proposed upstream and downstream slopes of 1:2 and 1:5 respectively which were based on sound principles. The upstream slope was designed so that sediment build up would not reach the crest. The downstream slope was shallow enough to permit a hydraulic jump to form on the weir under modular flow conditions, thus providing an integral energy dissipater. Also, under submerged conditions, losses are not too high and the afflux is minimised. The accuracy of the weir depends on the sharpness of the crest, so some weirs incorporate a metal insert at the crest.

The discharge is given by

$$Q = C_d \, C_v \, L \sqrt{g} \, H^{3/2} \tag{8.23}$$

The value of c_d = 0.63. The value of C_v varies with the ratio of $\left(\dfrac{H}{H+Z}\right)$.

For submerged flow,

$$\frac{Q_s}{Q} = 1.04\left[0.945 - \left(\frac{H_p}{H_1}\right)^{3/2}\right]^{0.256} \tag{8.24}$$

Where H_p is the head at the second gauging point. The onset of submergence occurs when

Downstream depth ≥ 0.75 × (upstream depth).

8.6 SUBMERGED WEIRS

When the downstream water level of nappe is higher than the crest, the weir is said to be submerged. The weir constructed across rivers become submerged during floods. Submerged weirs have larger discharging capacity and thus flooding can be prevented.

The discharge for submerged weir is calculated by dividing it into two horizontal section as shown in Fig. 8.11. The upper portion between the upstream and downstream water surface may be treated as free weir with a head of $(H_1 - H_2)$; the portion between the downstream water surface and the top of the crest may be treated as drowned orifice under a head H_2 the height of downstream water level above crest.

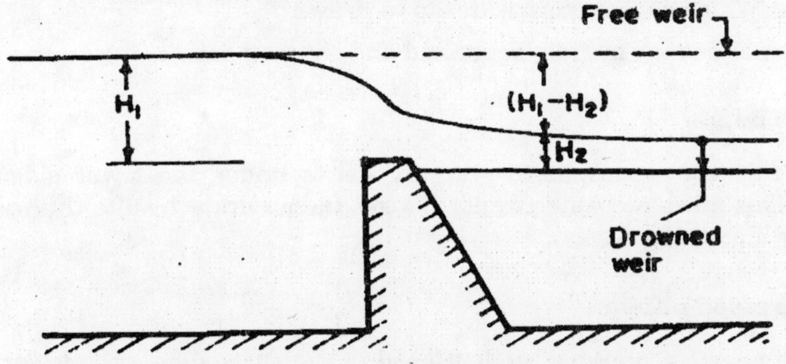

Fig. 8.11. Flow over a submerged weir.

Let Q_1 and Q_2 be the discharges for free weir and drowned weir respectively.

$$Q_1 = \frac{2}{3} C_{d_1} \sqrt{2g}\, (H_1 - H_2)^{3/2} \tag{8.25}$$

$$Q_2 = C_{d_2} (L \times H_2) \sqrt{2g\,(H_1 - H_2)} \tag{8.26}$$

Then the total discharge

$$Q = Q_1 + Q_2$$

The values of C_{d_1} and C_{d_2} are found experimentally.

8.7 VENTILATION OF WEIR

For accurate measurement of flow the stream passing over the weir should spring clear Fig. 8.12 (a), and the space under the nappe should be at atmospheric pressure and fully ventilated.

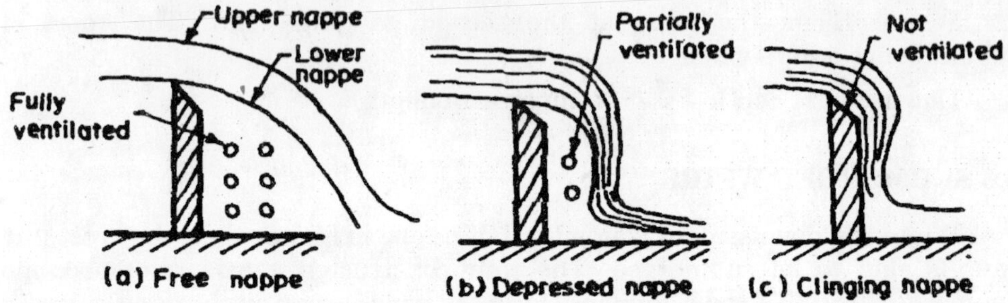

Fig. 8.12. Ventilation of suppressed weir.

In the suppressed weir, the nappe emerging form the weir touches the side walls of the channel. On this account air is trapped in the space, between the side walls of the channel, the falling nappe, the weir and the bottom of the channel.

This air is gradually carried away with the flowing water, thereby reducing the pressure in the space below the nappe, which may even become negative. Which causes more water to be drawn.

The nappe may be free, depressed and clinging one.

(a) Free nappe

In this the atmospheric pressure exists under the lower side of the nappe. This gives accurate results. To get the accurate results the weirs are fully ventilated Fig 8.12(a).

(b) Depressed nappe

The negative pressure so developed in the space draws the lower nappe more and more towards the downstream surface of the weir as shown in Fig. 8.12(b). Such a nappe is known as depressed nappe, which results in drawing more water thereby increasing the actual discharge over the weir. Discharge is about 8 to 10 percent greater than with a free nappe. Hence incorrect measurement discharge.

(c) Clinging nappe

The air is further withdraw form the space, no air is left below the nappe and water stream adheres to the downstream face of weir it is known as adhering or clinging nappe Fig 8.12 (a). Discharge in this more is as much as 25 to 30 percent more than that obtained by free nappe. Hence incorrect measurement of discharge.

ILLUSTRATIVE EXAMPLES

8.1 The discharge over a rectangular notch is to be 0.14 m³/s when the water level is 230 mm above the crest. If the coefficient of discharge is 0.6 calculate the width of notch or length of the crest required.

Solution

$$Q = \frac{2}{3} C_d \, L \sqrt{2g} \, H^{3/2}$$

Given, $Q = 0.14 \; m^3/s, \quad H = 0.23 \; m, \quad C_d = 0.6$

Therefore, $0.14 = \frac{2}{3} \times 0.6 \times L \times \sqrt{2 \times 9.81} \times (0.23)^{3/2}$

or, $0.195 \text{ L} = 0.14$

or, $L = \frac{0.14}{0.195} = 0.716 \; m$ Ans

8.2 State the Francis formula for a rectangular notch or weir, and explain why it is preferable to the simple theoretical formula.

A weir is 6 m long and is situated centrally in a channel 9 m wide. If the head above the crest is 250 mm, calculate the discharge.

Solution: For theory see text

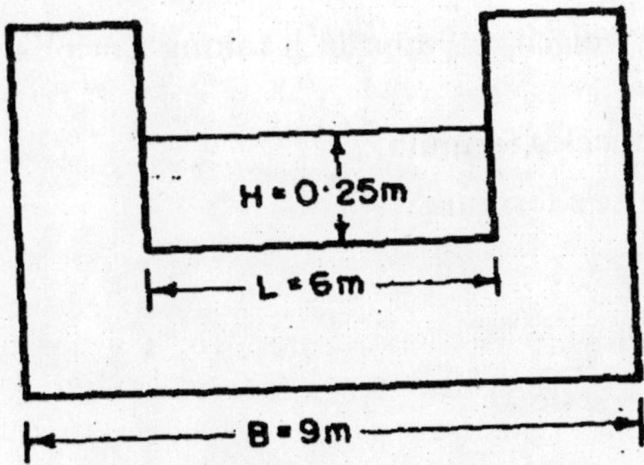

Fig. 8.13. Rectangular weir. P. 8.2.

The Francis formula is given by

$$Q = 1.84 \, (L - 0.1 \, n \, H) \, H^{3/2}$$

Given; L= 6 m, H = 0.25 m and n = 2

Therefore, $\text{Q} = 1.84 \, (6 - 0.1 \times 2 \times 0.25)\,(0.25)^{3/2}$

$= 1.84 \times (6 - 0.05) \times (0.125)$

$$= 1.84 \times 5.95 \times 0.125 = 1.3685\ m^3/s.$$ Ans

8.3 **Using the Francis formula, determine the length of the sill of the waste weir of a storage reservoir if the maximum daily flow over the weir is 125000 m³ and the level of the water must not rise more than 600 mm above the sill. Assume that the weir is fully contracted.**

Solution: Maximum daily flow = 125000 m³

$$\text{Therefore flow per second} = \frac{125000}{24 \times 3600} = 1.447\ m^3/s.$$

$$Q = 1.84\,(L - 0.1\,n\,H)\,H^{3/2}$$

$$1.447 = 1.84\,(L - 0.1 \times 2 \times 0.6)\,(0.6)^{3/2}$$

$$(L - 0.12)\,(0.4647) \times\ 1.447 = 1.84$$

$$1.447 = 0.855\,L - 0.1025$$

$$0.855\ \text{L} = 1.549598$$

$$\therefore\ \text{L} = 1.8124\ \text{m}$$ Ans

8.4 **A long channel 1.5 m wide terminates in a full width rectangular weir whose crest height is 300 mm above the stream bed. If the measuring station is recording a depth of 400 mm above the weir crest, estimate the discharge:**

(a) assuming a flow with a negligible velocity of approach. Take C_d as 0.7;

(b) including velocity of approach taking Coriolis coefficient, α, as 1.1 and C_d as 0.7.

(c) using Rehbock's formula.

(d) using White's formula.

Solution

(a) $$Q = \frac{2}{3} C_d\ L\sqrt{2g}\ H^{3/2}$$

$$= \frac{2}{3} \times 0.7 \times 1.5 \times \sqrt{2 \times 9.81} \times (0.4)^{3/2} = 0.784\ \text{m}^3/\text{s}$$ Ans

(b) If velocity of approach is to be incorporated. The formula is used

$$Q = \frac{2}{3} C_d\ L\sqrt{2g}\left[(H + \alpha\ ha_1)^{3/2} - (\alpha\ ha_1)^{3/2}\right]$$

But in this formula, initially the velocity is not known. The equation is therefore solved for successive approximations. For the sake of simplicity, it is usual to assume that the velocity at station 1 is uniform, i.e.

First trial If $Q = Q_1$ from (a), then

$$v_{a_1} = \frac{Q_1}{B(Z+H)} = \frac{0.784}{1.5(0.3+0.4)} = 0.747 \text{ m/s}$$

$$h_{a_1} = \frac{v_{a_1}^2}{2g} = \frac{(0.747)^2}{2\times 9.81} = 0.0284,\ \alpha\, ha_1 = 0.03125 \text{ and}$$

$$H + \alpha\, ha = 0.4 + 0.03125 = 0.4312 \text{ m}$$

$$Q_2 = \frac{2}{3}\times 0.7\times 1.5\times\sqrt{2\times 9.81}\left[(0.4312)^{3/2} - (0.037125)^{3/2}\right]$$

$$= 3.101 \times 0.27768 = 0.861 \text{ m}^3/\text{s}$$

Second trial

$$v_{a_2} = \frac{Q_2}{B(Z+H)} = \frac{0.861}{1.5(0.3+0.4)} = 0.82\ m/s$$

$$h_{a_2} = \frac{v_{a_2}^2}{2g} = \frac{(0.82)^2}{2\times 9.81} = 0.0342,\ \alpha\, ha_2 = 1.1\times 0.0342 = 0.0377$$

and $$H + \alpha\, h_{a_2} = 0.4 + 0.0377 = 0.4377$$

Therefore, $$Q_3 = \frac{2}{3}\times 0.7\times 1.5\times\sqrt{2\times 9.81}\left[(0.4377)^{3/2} - (0.377)^{3/2}\right]$$

$$= 3.101 \times 0.2822 = 0.8753 \text{ m}^3/\text{s}$$

which is sufficiently accurate for practical purposes.

(c) Rehbock's formula

$$Q = \left(1.786 + 0.236\frac{H}{Z}\right) LH^{3/2}$$

$$= \left(1.786 + 0.236\frac{0.4}{0.3}\right)\times 1.5\times(0.4)^{3/2} = 0.797\ m^3/s \qquad \text{Ans}$$

(d) White's formula

$$Q = 0.562\left(1 + 0.153\frac{H}{Z} L\sqrt{g}\right)(H + 0.001)^{3/2}$$

$$= 0.562\left[1 + \left(0.153\times\frac{0.4}{0.3}\right)\right]\times 1.5\sqrt{9.81}\,(0.4 + 0.001)^{3/2}$$

$$= 0.807\ m^3/s$$

8.5 A rectangular weir has a width of 900 mm the head over the weir is 200 mm and the height of the sill above the stream bed level is 150 mm. If the breadth of the channel is 1.2 m, calculate the discharge taking into account the velocity of approach. Use the Francis formula suitably modified.

Solution: Francis formula

$$Q = 1.84\left(L - 0.1\,nH\right)\,H^{3/2}$$

$$= 1.84\,\left(0.9\ - 0.1\times 2\ \times 0.2\right)\left(0.2\right)^{3/2} = 0.1415\ m^3/s = Q_1$$

First trial

Velocity of approach $v_{a_1} = \dfrac{Q_1}{B\left(H + Z\right)}$

$$= \frac{0.1415}{1.2\left(0.2 + 0.15\right)} = 0.3369\ \text{m/s}$$

$$h_{a_1} = \frac{v_{a_1}^2}{2g} = \frac{(0.3369)^2}{2\times 9.81} = 5878\times 10^{-3}\ \text{m}$$

and $\quad H + ha_1 = 0.2 + 5.878\times 10^{-3} = 0.205878$

Therefore, $Q_2 = 1.84\left[L - 0.1\,n\,\left(H + ha_1\right)\right]\left[\left(H + ha_1\right)^{3/2} - \left(ha_1\right)^{3/2}\right]$

$$= 1.84\left[0.9 - 0.1\times 2\times 0.2058\right]\left[\left(0.2058\right)^{3/2} - \left(5.878\ \times 10^{-3}\right)^{3/2}\right]$$

$$= 1.84\ (0.85884)\ (0.0929) = 0.1468\ \text{m}^3/\text{s}$$

Second trial

$$v_{a_2} = \frac{Q_2}{B\ (H + Z)} = \frac{0.1468}{1.2\left(0.2 + 0.15\right)} = 0.3495\ m/s$$

$$h_{a_2} = \frac{v_{a_2}^2}{2g} = \frac{(0.3495)^2}{2\times 9.81} = 6.2258\ \times 10^{-3}\ m$$

and $\quad H + ha_2 = 0.2 + 6.2258\times 10^{-3} = 0.20622$

$$Q_3 = 1.84\left[0.9\ - 0.1\times 2\ \left(0.20622\right)\right]$$

$$\left[\left(0.20622\right)^{3/2}\ -\ \left(6.2258\times 10^{-3}\right)^{3/2}\right]$$

$$= 1.58\times 0.09316 = 0.1472\ \text{m}^3/\text{s} = 147.2\ \text{liters/s}$$ Ans

8.6 A suppressed weir 0.9 m high extends fully across a rectangular channel 6m wide in which there is 2.83 m³/s flowing. Determine the depth of water upstream from the weir by the Francis formula.

Solution Francis formula for suppressed weir is given by

$$Q = 1.84\ L\ H^{3/2}$$

$$2.83\ = 1.84\times 6\times H^{3/2}$$

$$H^{3/2} = 0.256$$

$$H = (0.256)^{2/3} = 0.4035 \text{ m}$$

Depth of water upstream from the weir

$= Z + H = 0.9 + 0.4035 = 1.3035$ m Ans

8.7 A suppressed weir, 7.62 m long is to discharge 10.6 m³/s into a channel. The weir factor is 1.85. To what height may the weir be built, if the water behind the weir must not exceed 1.83 m in depth?

Solution

Given,

Z + H =1.83 m,

Weir factor = k = 1.85,

Q = 10.6 m³/s ,

L = 7.62 m,

$$v_a = \frac{Q}{B(Z+H)} = \frac{10.6}{7.62(1.83)} = 0.76 \ m/s$$

$$ha = \frac{v_a^2}{2g} = \frac{(0.76)^2}{2 \times 9.81} = 0.029 \ m$$

$$Q = k \ L \left|(H + ha)^{3/2} - (ha)^{3/2}\right|$$

$$10.6 = 1.85 \times 7.62 \left|(H + 0.029)^{3/2} - (0.029)^{3/2}\right|$$

$$0.75193 = (H + 0.029)^{3/2} - 4.938 \times 10^{-3}$$

$$(H + 0.029)^{3/2} = 0.75687$$

$$H + 0.029 = 0.8305$$

$$H = 0.8015 \text{ m}$$

$$Z + H = 1.83 \text{ m}$$

$Z = 1.83 - H = 1.83 - 0.8015 = 1.02848$ m Ans

8.8 A contracted weir, 1.22 m high, is to be installed in a channel 2.44 m wide. The maximum flow over the weir 1.7 m³/s when the total depth back of the weir is 2.13 m. If the weir factor is 1.85, what length of weir should be installed?

Solution Given,

Z = 1.22 m, B = 2.44 m , Q = 1.7 m³/s

K = 1.85, Z + H = 2.13

$\therefore$ H = 2.13 – 1.22 = 0.91 m

$$v_a = \frac{Q}{B(Z+H)} = \frac{1.7}{2.44\,(2.13)} = 0.327 \text{ m/s}$$

$$h_a = \frac{v_a^2}{2g} = \frac{(0.327)^2}{2 \times 9.81} = 5.453 \times 10^{-3} \text{ m}$$

Therefore, H + ha =0.91 + 5.453 × 10³

$$Q = K\left[L - 0.1n\,(H + ha)\right]\left[(H - ha)^{3/2} - ha^{3/2}\right]$$

$$1.7 = 1.85\left[L - 0.1 \times 2\,(0.91543)\right]$$

$$\left[(0.91543)^{3/2} - \left(5.453 \times 10^{-3}\right)\right]^{3/2}$$

$$1.7 \;\; 1.85\left[L - 0.183\right]\left[0.87549\right]$$

$$L - 0.183086 = 1.04959$$

$$L = 0.104959 + 0.183 = 1.23 \text{ m}$$ Ans

8.9 **The discharge in a channel varies from 0.16 m³/s to 0.74 m³/s. A contracted rectangular weir is to be constructed across this channel to measure the discharge. Determine the length of the weir such that the measured head will never be less than 0.16m or greater than one third of the length of the weir.**

Solution : I Case

$$Q_{min} = 1.84\,(L - 0.1\,nH)\,H^{3/2}$$

or, $$0.16 = 1.84\,(L - 0.1 \times 2 \times 0.16)\,(0.16)^{3/2}$$

or, $$0.16 = 1.84\,(L - 0.032)\,(0.064)$$

or, $$0.16 = 0.11776\,L - 3.768 \times 10^{-3}$$

$$0.11776\,L = 0.1637$$

$$L = 1.3907 \text{ m}$$

II Case, $H = \dfrac{L}{3}$

$$Q_{max} = 1.84\left(L - 0.1\,n\frac{L}{3}\right)\left(\frac{L}{3}\right)^{3/2}$$

$$0.74 = 1.84\left(L - \frac{0.2}{3}L\right)(L)^{3/2}\left(\frac{1}{3}\right)^{3/2}$$

$$= L^{5/2} \times \frac{1.84 \times 0.933}{3^{3/2}} = L^{5/2}\,\frac{1.84 \times 0.933}{5.196}$$

$$0.74 = L^{5/2}\,(0.3305)$$

or, $L^{5/2} = 2.239$

$$L = (2.239)^{2/5} = 1.3804\ m$$

Thus we would use L = 1.3804 m Ans

8.10 Develop an expression for the quantity of liquid flowing over a sharp edged V notch of total angle 2 θ, in terms of the head H above the bottom of the notch, the angle θ, and the coefficient of discharge C_d, assuming the velocity of approach to be small.

If the rate of flow of water over a v notch having θ = 35° is 42.5 liters/s, find the head in mm. Take C_d as 0.62.

Solution For theory see text.

$$Q = \frac{8}{15} C_d \sqrt{2g}\ \tan\theta\ H^{5/2}$$

$$42.5 \times 10^{-3} = \frac{8}{15} \times 0.62 \times \sqrt{2 \times 9.81}\ \tan 35^\circ\ H^{5/2}$$

$$H^{5/2} = 0.041435$$

$$H = (0.041435)^{2/5} = 0.2798 \approx 0.28\ m \approx 280\ mm$$ Ans

8.11 Working from first principles obtain an expression for the rate of flow over a V notch in terms of the angle θ between the sides and the head it above the bottom of the notch.

Calculate the value of the angle θ for a v notch which is to discharge 0.41 m³/s under a head of 0.6m, assuming a coefficient of discharge of 0.6.

Solution For theory see text.

$$Q = \frac{8}{15} C_d \sqrt{2g}\ \tan\theta\ H^{5/2}$$

$$0.41 = \frac{8}{15} \times 0.6 \times \sqrt{2 \times 9.81} \times (0.6)^{5/2}\ \tan\frac{\theta}{2}$$

$$\tan\frac{\theta}{2} = 1.0373$$

$$\frac{\theta}{2} = \tan^{-1}(1.0373) = 46.048$$

$$\theta = 92.097^\circ = 92^\circ\ 5'\ 52''$$ Ans

8.12 During a test in a laboratory the water which has passed through a venturimeter flows over a right angled V notch, the head at the V notch being registered. The inlet diameter of the venturimeter is 250 mm and the throat diameter is 100 mm. When a steady head over the V notch of 181 mm is maintained, the difference of pressure head at the venturimeter is found to be 0.25 m of water. Determine the

coefficient of V-notch on the assumption that the venturimeter results are correct, the coefficient being 0.99.

Solution The flow over right angled V-Notch is given by

$$Q = \frac{8}{15} C_{d_1} \sqrt{2g} \tan\frac{\theta}{2} H^{5/2}$$

$$= \frac{8}{15} \times C_{d_1} \times \sqrt{2 \times 9.81} \times 1 \times (0.181)^{5/2} = 0.0329\ C_{d_1} \quad (1)$$

The flow through the venturimeter is given by

$$Q = \frac{C_{d_2}\ a_1 \sqrt{2gh}}{\sqrt{a_1^2 / a_2^2 - 1}}$$

Here, $a_1 = \frac{\pi}{4}(0.25)^2 = 0.049\ m^2$

$$a_2 = \frac{\pi}{4}(0.1)^2 = 7.85 \times 10^{-3}\ m^2$$

Therefore, $\left(\frac{a_1}{a_2}\right)^2 = \left(\frac{0.049}{7.85 \times 10^{-3}}\right)^2 = 38.963$

Therefore, $Q = \frac{0.99 \times 0.049 \sqrt{2 \times 9.81 \times 0.25}}{\sqrt{38.963 - 1}}$

$$= \frac{0.16825}{6.164} = 0.0273\ m^3/s$$

Substituting in equation (1)

$$0.0273 = 0.0329\ C_{d_1}$$

$$\therefore\ C_{d_1} = \frac{0.0273}{0.0329} = 0.8296$$ Ans

8.13 Develop a formula for the theoretical discharge over a V notch.

Explain the way in which this may be modified to give the actual discharge, giving reasons why the theoretical discharge does not agree with the actual discharge.

A 90° V notch has a coefficient of discharge of 0.6. Calculate the discharge when the observed head is 0.65 m

Solution: For theory see text.

$$Q = \frac{8}{15} C_d \sqrt{2g} \tan\theta/2\ H^{5/2}$$

Here, $\theta = 90°$, $\therefore \theta/2 = 45°$, $C_d = 0.6$, $H = 0.65\ m$

$$Q = \frac{8}{15} \times 0.6 \times \sqrt{2 \times 9.81} \times \tan 45° \times (0.65)^{5/2}$$

$$= 0.4828\ m^3/s$$ Ans

8.14 **A 90° V-notch and a rectangular weir are placed in series. The length of the rectangular weir is 600 mm and its coefficient is 1.81. If the coefficient of contraction of the V-notch is 0.61, what will be its working head when the head on the weir is 150 mm?**

Solution

Assuming the coefficient of velocity for V notch is 0.985. The value of C_d is calculated with the formula

$$C_d = C_c\ C_v = 0.61 \times 0.985 = 0.6$$

The rectangular Weir and V-notch both are in series, therefore,

Discharge of rectangular weir = Discharge of triangular weir.

$$k\ LH^{3/2} = \frac{8}{15} c_d \sqrt{2g}\ \tan \theta/2\ H^{5/2}$$

$$181 \times 0.6\ (0.15)^{3/2} = \frac{8}{15} \times 0.6 \times \sqrt{2 \times 9.81} \times 1 \times H^{5/2}$$

$$0.063 = 1.4174\ H^{5/2}$$

$$H^{5/2} = 0.0444$$

$$H = (0.0444)^{2/5} = 0.288\ \text{m}$$ Ans

8.15 **Develop, from first principles, a discharge formula of the form Q = kH^n, giving the numerical values of k and n, for a V notch having a total angle (2 θ) to 60° and C_d = 0.6, if Q in m^3/s and H is in m.**

If an error of 1.8 mm is made in reading a head of 180 mm, what percentage error will result in the value of Q ?

Solution: For theory see text

$$Q = \frac{8}{15} C_d \sqrt{2g}\ \tan \theta\ H^{5/2}$$

$$Q = \frac{8}{15} \times 0.6 \times \sqrt{2 \times 9.81}\ \tan 30°\ H^{5/2}$$

$$Q = 0.818\ H^{5/2}$$

Comparing with the formulae

Thus k = 0.818 and n = 5/2 = 2.5

Now, $Q = 0.818\, H^{5/2}$

$$dQ = 2.5 \times 0.818\, H^{3/2}\, dH$$

Therefore, $\dfrac{dQ}{Q} = 2.5\, \dfrac{dH}{H}$

Putting dH = 1.8 mm when H = 180 mm

$$\frac{dQ}{Q} = 2.5 \times \frac{1.8}{180} = \frac{2.5}{100}$$

Therefore, percentage error = 2.5% **Ans**

8.16 Develop the formula $\frac{8}{15} C_d \sqrt{2g}\, \tan\theta/2\, H^{5/2}$ for the flow in m³/s over a V notch having an included angle of θ.

Find the flow if the measured head is 0.375 m, θ/2 being 45° and C_d being 0.6. If this flow is wanted within an accuracy of 2% up or down, what are the limiting values of head?

Solution: For theory see text

$$Q = \frac{8}{15} C_d \sqrt{2g}\, \tan\theta/2\; H^{5/2}$$

$$Q = \frac{8}{15} \times 0.6 \times \sqrt{2 \times 9.81}\, \tan 45^\circ \times (0.375)^{5/2}$$

$$= 0.122\ m^3/s$$ **Ans**

Now, $Q = \dfrac{8}{15} C_d \sqrt{2g}\, \tan\theta/2\; H^{5/2}$

$$= \frac{8}{15} \times 0.6 \times \sqrt{2 \times 9.81} \times 1 \times H^{5/2} = 1.4174\, H^{5/2}$$

when Q increases by 2% i.e. = 1.02 × 0.122. Therefore, H is also increases

Therefore, $1.02 \times 0.122 = 1.4174\, H^{5/2}$

$$H = \left(\frac{1.02 \times 0.122}{1.4174}\right)^{2/5} = (0.08779)^{2/5}$$

$$= 0.3779 = 0.378 \text{ m}$$ **Ans**

when Q decreases 2% i.e. Q = 0.98 × 0.122 = 0.11956 m³/s

Therefore, H is also decreases.

Therefore, $0.11956 = 1.4174\, H^{5/2}$

$$H = (0.08435)^{2/5} = 0.3719 \approx 0.372\ m$$ **Ans**

8.17 Water flows over a rectangular weir 1 m wide a depth of 0.15 m and afterwards passes through a triangular right angled weir. Find the depth of water through the triangular weir. Take Cd for rectangular and triangular weirs as 0.62 and 0.59 respectively

Solution: As per question, rectangular weir and right angled V notch are in series. Therefore

Discharge of rectangular weir = Discharge of right angled V notch.

$$\frac{2}{3} C_{d_1} L \sqrt{2g} H_1^{3/2} = \frac{8}{15} C_{d_2} \sqrt{2g} \tan\theta/2 \ H_2^{5/2}$$

$$\frac{2}{3} \times 0.62 \times (0.15)^{3/2} = \frac{8}{15} \times 0.59 \ H_2^{3/2}$$

$$0.024 = 0.3146 \ H_2^{5/2}$$

$$H_2^{5/2} = 0.076$$

$H_2 = (0.076)^{2/5} = 0.357$ m Ans

8.18 A flow from a channel is controlled by a trapezoidal notch so that the full supply discharge of 2 m³/s flows over the notch at a head of 1.2 m measured over the crest. At half this head, a discharge of 0.6 m³/s passes over the notch. Assuming Cd = 0.62, calculate the base width and side slopes of the notch.

Solution

$$Q = \frac{2}{3} C_d \sqrt{2g} \ H^{3/2} \left[L + \frac{4}{5} H \tan\theta/2 \right]$$

$$= \frac{2}{3} \times 0.62 \times \sqrt{2 \times 9.81} \ H^{3/2} \left[L + 0.8 \ H \tan\theta/2 \right]$$

$$= 1.8308 \ H^{3/2} \left[L + 0.8 \ H \tan\theta/2 \right] \quad (1)$$

when H = 1.2 m, and Q = 2 m³/s

$$2 = 1.8308 \ (1.2)^{3/2} \left[L + 0.8 \times 1.2 \tan\theta/2 \right]$$

$$L + 0.96 \tan\theta/2 = 0.831 \quad (2)$$

when H = 0.6 m and Q = 0.6 m³/s

Putting in equation (1)

$$0.6 = 1.8308 \ (0.6)^{3/2} \left[L + 0.8 \times 0.6 \tan\theta/2 \right]$$

$$L + 0.48 \tan\theta/2 = 0.7051 \quad (3)$$

Subtracting equation (3) from (2)

$$0.48 \tan\theta/2 = 0.1259$$

$$\tan\theta/2 = 0.26229$$

$$\theta/2 = \tan^{-1}(0.26229) = 14.697° = 14° \, 41' 14''$$ Ans

Substituting in equation (3)

$$L + 0.48 \times 0.26229 = 0.7051$$

$$\therefore L = 0.5792 \text{ m}$$ Ans

8.19 (a) What is a Cippoletti weir and what are its advantages.

(b) Prove that in a Cippoletti weir the sides have a slope of 1 : 4.

(c) The breadth of a Cippoletti weir is 0.52 m, the measured head is 0.21m and the velocity of approach is 1.5 m/s. Calculate the discharge.

Solution: For part (a) and (b) see text

(c) $Q = CLH^{3/2}$

For Cippoletti Weir C = 1.86.

$$Q = 1.86 \times 0.52 \, (0.21)^{3/2} = 0.093 \, m^3/s$$

Velocity of approach being taken into account, therefore

$$h_a = \frac{v_a^2}{2g} = \frac{(1.5)^2}{2 \times 9.81} = 0.1146 \, m$$

Therefore,
$$Q = CL\left[(H + h_a)^{3/2} - h_a^{3/2}\right]$$

$$= 1.86 \times 0.52 \left[(0.21 + .1146)^{3/2} - (0.1146)^{3/2}\right]$$

$$= 0.9672 \times [0.1849 - 0.0388] = 0.9672 \times 0.146$$

$$= 0.141 \, m^3/s$$ Ans

8.20 Prove that in the case of broad crested weir the maximum discharge is obtained when the flow depth over the weir is equal to two-third of the upstream depth neglecting the velocity of approach.

A broad crested weir, with flat top, is constructed across the entire 2.7m width of a rectangular channel. If the head on the weir crest is 0.4, find the discharge passing over the weir.

Solution: For theory see text

$$Q = C_d \, L \, h \sqrt{2g \, (H - h)}$$

For maximum discharge

$$h = \frac{2}{3} H$$

$$Q = C_d \, L \, h \left(\frac{2}{3} H\right) \sqrt{2g \, (H - 2/3 \, H)} = C_d \, L \, h \frac{2}{3} H \sqrt{\frac{2}{3} \, g H}$$

$$= 0.97 \times 2.7 \times \frac{2}{3} \times 0.4 \sqrt{\frac{2 \times 9.81 \times 0.4}{3}}$$

$$= 1.12959 \approx 1.3\ m^3/s$$ Ans

8.21 Water flow at the rate of 1 m^3/s perimeter width in a rectangular channel 2 m deep. A vertical weir is to be installed at the end of the channel to bring the free surface, 0.15 m from the top of the channel wall. Estimate the required height of weir, assuming full ventilation. $C_d = 0.611 + 0.75\ \frac{h}{z}$, where h = head over the weir, and z = height of weir from channel bottom.

Solution: The discharge over a broad crested weir per meter width is given by

$$q = C_d\ h \sqrt{2g\,(H - h)}$$

Here $\quad H - h = 0.15$ m

$q = m^3/s$ per meter width.

Therefore, $1 = C_d\ h \sqrt{2 \times 9.81 \times 0.15}$

$$C_d = \frac{1}{1.7155\ h}$$

As per question $C_d = 0.611 + 0.75\ \frac{h}{Z}$

and $\quad Z = 1.85 - h$

Therefore, $\quad \frac{1}{1.7155\ h} = 0.611 + 0.75 \left[\frac{h}{1.85 - h}\right]$

or, $\quad h^2 - 3.19\ h + 2.01 = 0$

Solving the above equation

h = 0.85

∴ Required height of weir = 1.85 – 0.87 = 0.98 m Ans

8.22 A broad crested weir spanning the full width of a 2 m wide channel is 1.5 m high and has a square entrance. What head is required to pass a discharge of 3 m^3/s? Take $C_d = 0.85$.

Solution

$$Q = 1.705\ C_d\ L\ H^{3/2}$$

$$3 = 1.705 \times 0.85 \times 2 \times H^{3/2}$$

$$H^{3/2} = 1.035$$

$$H = (1.035)^{2/3} = 1.0232 \text{ m}$$

Therefore, $Q = v_a \, B \, (H + Z)$

$$3 = v_a \times 2 \, (1.0232 + 1.5)$$

$$v_a = \frac{3}{2 \times 2.523} = 0.594 \text{ m/s}$$

$$h_a = \frac{v_a^2}{2g} = \frac{(0.594)^2}{2 \times 9.81} = 0.018$$

Now, required head = H – ha = 1.0232 – 0.018 = 1.005 m Ans

8.23 The height of water on the upstream and downstream side of a submerged weir of 3.5 length are 300 mm and 150 mm respectively. If C_d for free and drowned portions are 0.62 and 0.82 respectively, find the discharge over the weir.

Solution

Q = Free portion discharge, Q_1 + Drowned portion discharge, Q_2

$$Q = \frac{2}{3} C_{d_1} \sqrt{2g} \, (H_1 - H_2)^{3/2}$$

$$+ C_{d_2} (L \, H_2) \times \sqrt{2g \, (H_1 - H_2)}$$

$$= \frac{2}{3} \times 0.62 \times \sqrt{2 \times 9.81} \, (0.3 - 0.15)^{3/2}$$

$$+ 0.82 \, (3.5 \times 0.15) \times \sqrt{2 \times 9.81 \times 0.15}$$

$= 0.1063 + 0.7385 = 0.84483$ m³/s $= 844.83$ liters/s Ans

EXERCISES

A. THEORY

8.1 Define a notch or weir.

8.2 Define crest or sill of a weir.

8.3 Define upper nappe lower nappe, free, depressed and clinging, with a neat sketch.

8.4 How are the weir classified?

8.5 What is a suppressed weir?

8.6 What are end contractions? Why do they occur?

8.7 What is suppressed weir and weir with end contractions? where is it necessary to use a weir with end contractions?

8.8 Differentiate between sharp crested weir and, broad crested weir which is preferable for the measurement of discharge and why?

8.9 Why is triangular weir more suitable than a rectangular weir?

8.10 Explain why ventilation of suppressed rectangular weir necessary, whereas for V-notch it is not.

8.11 Where do you recommend the use of a triangular notch in practice?

8.12 How will you avoid the effect of end contractions in case of trapezoidal weir?

8.13 What is Cippoletti weir? Prove that the angle $\tan \theta / 2 = 1 : 4$.

8.14 What do you understand by velocity of approach? Explain it is generally used in rectangular weir.

8.15 Draw a neat sketch of ogee weir and explain its characteristic.

8.16 Draw a neat sketch of proportional or SUTRO weir. Write a formula for the same.

8.17 Draw a neat sketch of submerged weir and write the formula of discharge.

8.18 Prove a formula for maximum discharge for broad crested weir.

8.19 Derive a formula for discharge for rectangular weir.

8.20 Derive a discharge formula for V-notch.

8.21 Derive a discharge formula for right angled V-notch

8.22 Derive an expression for the discharge over a Cippoletti weir.

8.23 Derive the condition for trapezoidal weir to function as a Cippoletti weir.

8.24 Derive an expression for emptying a reservoir by using a rectangular weir.

8.25 Derive an expression for emptying a reservoir by using triangular notch.

8.26 Derive an expression for emptying a reservoir using a Cippoletti weir.

B. UNSOLVED PROBLEMS

8.1 A suppressed weir having a sill 4 m long discharge under a head of 200 mm. If the depth of the channel of approach is 70 m find the rate of flow. [644 liters/s]

8.2 Water flow from a reservoir through a rectangular opening 2 m high and 1.2 m wide in the vertical face of a dam. Calculate the discharge in m³/s when the free surface in the reservoir is 0.5 m above the top of the opening assuming a coefficient of the discharge of 0.64 [8.16 m³/s]

8.3 A rectangular weir is 2m long and has a head of 675 mm. Find the discharge taking into account the two end contractions [1.9 m^3/s]

8.4 A rectangular notch has a crest length 1m, the head over the notch is 200 mm and the height of the sill above the bed level is 150mm. If the width of the channel is 1.2 m, calculate the discharge in m^3/s, taking into account the velocity of approach [166 liters/s].

8.5 Determine the discharge over a sharp crested weir is 4.5 long with no end contractions, the measured head over the crest being 450 mm. The width of the approach channel is 7.5 m and its depth below the crest of the weir is 1 m [2.521 m^3/s].

8.6 A rectangular weir is to be made to discharge 18.2 m^3/min with a head over the crest equal to 0.5 of the width of the weir. Neglecting the velocity of approach and allowing for two end contractions determine the width of the sill and the head of water above the crest [768.7 mm, 384.4 mm].

8.7 The discharge over a rectangular notch is to be 0.14 m^3/s when the water level is 230mm above the sill. Assuming a coefficient of discharge of 0.6 and deducing the formula for theoretical discharge from first principles, calculate the width of notch required [720 mm].

8.8 In an experiment on a right angled V notch the flow is collected in a 0.9 m diameter vertical cylindrical tank. It is found that the depth of water increases by 0.685 m in 16.8 s when the head over the notch is 0.2m. Determine the coefficient of discharge of the notch [0.615].

8.9 Find the discharge through a triangular notch under a constant head of 250 mm if the angle of the notch is 120°. Take $C_d = 0.62$ [0.079 m^3/s]

8.10 In an experiment an a right angled V notch the flow is collected in a vertical cylindrical tank 900 mm diameter. It is found that the depth of water in the tank increases by 650 mm in 16.8 s when the head over the notch is 200 mm. Determine the coefficient of discharge of the notch [0.582]

8.11 A rectangular weir 400 mm long is used for measuring a discharge of 30 lps. An error of 1.5mm was made, while measuring the head over the weir. Calculate the percentage error in the discharge. Take C_d = 0.6. [1.85%]

8.12 A suppressed rectangular weir is constructed across a channel of 0.77 m width with head of 0.39m and the crest 0.6 m above the bed of the channel. Estimate the discharge over it. Consider velocity of approach and assume $C_d = 0.623$. [0.3573 m^3/s]

8.13 A sharp crested rectangular weir of 1 m height extends across a rectangular channel of 3 m width. If the head of water over the weir is 0.45m, calculate the discharge. Consider velocity of approach and assume $C_d = 0.623$. [1.703 m^3/s]

8.14 A weir 36 meters long is divided into 12 equal bays by vertical posts, each 600 mm wide. Determine the discharge over the weir if the head over the weir is 1.2 m and velocity of approach is 2m/s [75.246 m^3/s]

8.15 A reservoir has an area of 8.5×10^4 m^2 and is provided with a weir 4.5 m long. Find how long will it take for level at the sill to fall from 600 mm to 300 mm [3.05 hrs]

8.16 A Cippoletti weir has a sill of 250 mm. If the head on the sill is 150 mm, calculate the discharge flowing over it. Take C_d=0.64 [0.0275 m^3/s]

8.17 Determine the discharge over a broad crested weir 30 m long with rounded entrance, the measured upstream level over the crest being 650 mm. The width of channel of approach is 40 m and its depth below the crest of weir is 600 mm.

8.18 A submerged weir at the edge of a pond is 2 m wide. The crest of the weir is 150 mm below the water level of the pond and 75 mm blow the tail water level. Compute the discharge over this weir in liters per hour, if the coefficients of discharge for free and drowned portions are 0.58 and 0.8 respectively. [780×10^3 liters/hour]

8.19 A right angled V notch is inserted in the side of a tank of length 4m and width 2.5 m. Initial height of water above the apex of the notch is 300 m. Find the height of water above the apex if the time required to lower the head in tank from 300 mm to final height in 3 minutes. Take C_d = 0.6. [82.2 mm].

9

Flow through Pipes

9.1 INTRODUCTION

A study of pipe flow is useful in many ways besides giving information about a common case. Pipes convey fluids from place to place under pressure or vacuum, where it is necessary to prevent contamination or loss by evaporation and in heat exchangers. Industry is employing increasing numbers of closed conduits to transport fluids used in processing and fluids being processed. Air conditioning (heating and ventilating) systems use conduits of various shapes. Water works involve extensive piping systems, both large and small. Oil and gas pipe lines supply users thousands of kilometers away from the well. Various other industries are using pipe lines to transport their product as raw material, during processing, and as the finished product. We shall restrict the discussion here to incompressible and homogenous fluids of constant viscosity.

9.2 REYNOLDS EXPERIMENTS

Osborne Reynolds in his classic experiments at Manchester University demonstrated most effectively the characteristics of laminar and turbulent flow. He showed that, under suitable conditions, the two types of flow could be made to occur in the one pipe. His apparatus was extremely simple and consisted essentially of a glass tube through which water could be passed at varying velocities. Provision was made for the insertion of a thin jet aniline dye (do not mix with water) into the stream of water at the upstream end. A diagrammatic sketch of the layout of the apparatus is given in Fig. 9.1.

The whole apparatus was filled and allowed to stand for some 24 hours to ensure that the water was completely static before the tests were started.

(i) The control valve opens gradually to allow a very slow flow (small enough flow rate) through the glass pipe. The dye streak will remain as a well-defined line as it flows along, with only slight blurring due to molecular diffusion of the dye into the surrounding water. Since the fluid may therefore be considered as moving in the layers or parallel to the axis of the glass tube, this kind of flow is now usually called Laminar flow [Fig. 9.2(a)].

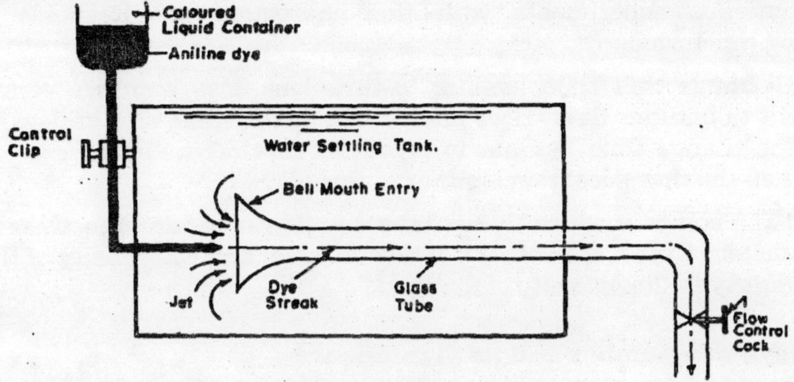

Fig. 9.1. Reynolds apparatus

(ii) With the increase of velocity of flow (intermediate flow rate) the dye streak fluctuates in time and space, and intermittent bursts of irregular behaviour appear along the streak. This type of flow is now called the transitional flow [Fig. 9.2 (b)].

(iii) With a further increase in the velocity of flow (large enough flow rates) the dye streak almost immediately becomes blurred and spreads across the entire pipe in a random fashion. This type of flow is usually called the turbulent flow [Fig. 9.2 (c)].

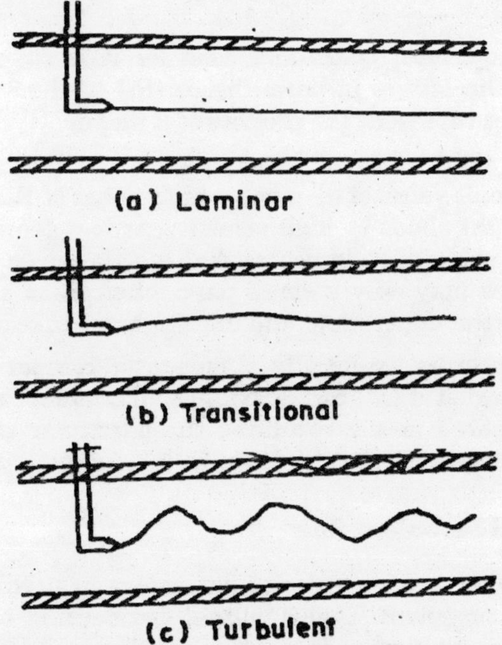

Fig. 9.2. Typical dye streaks illustrating type of flow.

Subsequently, when valve was gradually closed, a point would again be reached when the dye stream returned to its original line, indicating that the flow in glass tube had returned to the laminar condition.

Numerous experiments with this equipment led Reynolds to the following conclusions

(1) Change over from laminar to turbulent flow- and vice versa, from turbulent to laminar flow - does not happen at the same rate of flow in glass tube. The change from laminar to turbulent flow occurs at a lesser velocity than when the flow goes from laminar to turbulent.

(2) The points at which these changes in flow occur are directly related to the diameter of tube, the velocity of flow in tube, and the density of the fluid and inversely to the viscosity of the fluid.

9.2.1 Reynolds number and its Significance

The Reynolds number is of great importance in fluid mechanics. It is a dimension less parameter, established by Osborne Reynolds, which combines the properties of the fluid, the dimension of the surrounding solid boundary and the relative velocity between the fluid and the boundary. It may be considered as the ratio of inertia forces of the fluid to the viscous forces present when a body is completely immersed in the fluid, or the fluid completely fills a pipe line.

The Reynolds number is written as

$$\text{Reynolds number} = \frac{\rho \, v \, d}{\mu} \tag{9.1}$$

The symbol now most commonly used for it is R_e, and it is often an advantage to put the letters in parantheses (R_e) to show that it is a single symbol and does not represent the product of R and e.

It is evident from the form of the expression (ρ v d /μ) > that a high value of ρ, d or v, or a small value of μ, gives a high value of R_e. Conversely, a low value of R_e is brought about by high viscosity, or low density, low velocity or small size. A high value of R_e indicates that inertia forces dominate the flow white viscous forces play only a small part; when R_e is small in value, the viscous forces have the upper hand and inertia forces take second place.

Reynolds number is essentially a means of comparing one flow with another, and provided that corresponding diameters and corresponding velocities are compared in the two flows the particular choices of diameter and velocity do not matter.

9.2.2 Laminar and turbulent flow

Reynolds number is a very useful parameter in predicting whether the flow is laminar or turbulent. Under normal engineering condition, however, where disturbances are always present, transition occurs at values between 2000 and 4000.

There is, then apparently no upper limit to the value of R_e at which the changes from laminar flow to turbulent flow occurs. There is, however, a definite lower limit. When R_e is less than this lower limit any disturbances in the flow are damped out by viscous forces, which are then predominant.

The technique using a stream of dye is not suitable for demonstrating the change from turbulent to laminar flow. Reynolds therefore employed another method to study the changes which occurs as the velocity is reduced.

Reynolds set up an apparatus of which the essential part is indicated in Fig. 9.3. The velocity of water through a uniform horizontal pipe was controlled by a valve at its downstream end, and for various values of the mean velocity the drop in pressure over a given length of pipe was measured by differential manometer. The first manometer connection was situated several meters from the inlet to the pipe so that any disturbances from the inlet could be damped out and the flow pattern could settle down to its final form before measuring section was reached.

If the head loss in a given length of uniform pipe is measured at different values of the velocity, it will be found that, as long as the velocity is low enough to secure laminar flow, the head loss, due to friction, will be directly proportional to the velocity, as shown in Fig. 9.4. But with increasing velocity, at some point B, where visual observation in a transparent tube would show that the flow changes from laminar to turbulent, there will be an abrupt increase in the rate at which the head loss varies. If the logarithms of these two variables are plotted on linear scale or if the values are plotted directly on log - log paper, it will be found that, after a certain transition region has been passed, lines will be obtained with scopes ranging from about 1.75-2.

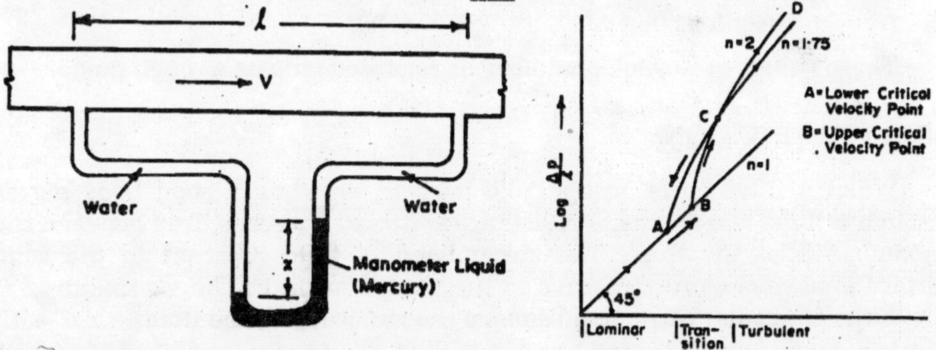

Fig. 9.3. Apparatus for measuring loss of head in pipe (Reynolds second experiment)

Fig. 9.4. Log-Log plot for flow in a uniform pipe

It is thus seen that for laminar flow the drop in energy due to friction varies as V, while for turbulent flow the friction varies as V^n, where n ranges from about 1.75 to 2. The lower value of 1.75 for turbulent flow is found for pipes with very smooth walls, as the wall roughness increases, the value of n increases up to its maximum value of 2. In the transition zone, the values of n are even greater than 2.

If, however, the velocity is carefully reduced, the previous route is not exactly retraced. The line DC is followed but, instead of path CB being taken, the line CA is traced out. Then the point A, the line corresponding to laminar flow is followed. The point B is known as the higher critical point, and A as the lower critical point.

9.2.3 Critical Reynolds number

The lower critical value is at the point A, where the Reynolds number is 2000. This lower value is definite and is the dividing point between the laminar and turbulent flow. The turbulent flow below the Reynolds number 2000 is practically impossible, because any turbulence that is set up will be damped out by viscous friction. Therefore, the flow is laminar upto point A and this is known as the lower critical point. The upper critical Reynolds number, corresponding to point B of Fig. 9.4, is really indeterminate and depends upon the care taken to present any initial disturbance from affecting the flow. In normal practice, the usual value of the upper critical Reynolds number is between 2500 and 4000.

It is the lower critical value which is of much greater interest and importance because it is at this point that laminar flow becomes unstable. It is usual to refer to this lower critical value simply as the critical Reynolds number.

The flows in commercial pipe lines normally conform to the following pattern.

Laminar flow	$R_e < 2000$
Transition region	$2000 < R_e < 4000$
Turbulent flow	$R_e > 4000$.

These values of Reynolds number be regarded only as a rough guide.

9.3 FLUID FRICTION

When a fluid flows past a solid surface, or when a solid body moves through a fluid, the frictional resistance is due to a viscous drag between the streambands of the fluid. The streamband of fluid adjacent to the solid surface is always at rest relative to the wetted surface. The viscous drag is due to the molecular attraction between the molecules of the fluid.

The frictional resistance of surfaces moving in water were investigated by Froude. The concluded results of the Froude's experiment are as follows.

(1) The frictional resistance varies approximately with the square of the velocity.

(2) The frictional resistance varies with the nature of the surface.

(3) The frictional resistance per unit area of surface decreases as the length of the board increases but is constant for long lengths. The total frictional resistance F is given by

$$F = f' A v^n = f' (\text{Wetted Perimeter} \times \text{length})\, v^n$$

Assuming n = 2

$$F = f' P l v^2 \tag{9.2}$$

when f' be an experimental coefficient.

9.4 HEAD LOST DUE TO FRICTION IN PIPES-DARCY'S WEISBACH EQUATION

Henri P.G. Darcy (1803-58), who was Inspector-General of the Paris water works in the first half of the nineteenth Century, first proposed the friction loss equation after he made very comprehensive and careful tests on the flow of water in 21 pipes constructed with different diameters and of different materials. He introduced the coefficient f which is referreds as Darcy's coefficient *f*.

Julius Weisbach (1806-71), a German hydraulician, wrote a text book of Engineering Mechanics in 1845, which even now can be considered a text book on hydraulics. Weisbach not only incorporated the best available experimental information, but in many cases supplemented it with the results of his own experiments. He advocated the use of the non dimensional coefficient and was first to express the resistance equation in the pipes in the form $h_f = K \dfrac{v^2}{2g}$.

Where K is equal to 4fl/d. Therefore, the equation is generally called the Darcy's Weisbach equation. The equation is written as

$$h_f = \frac{4 f l v^2}{2g\, d} \tag{9.3}$$

9.4.1 Proof of the Darcy's Weisbach equation

Let a fluid flow through a circular pipe of diameter *d* and length *l*, and let the pressure difference necessary to maintain a steady flow be $(p_1 - p_2)$ when the mean velocity in the pipe *v*.

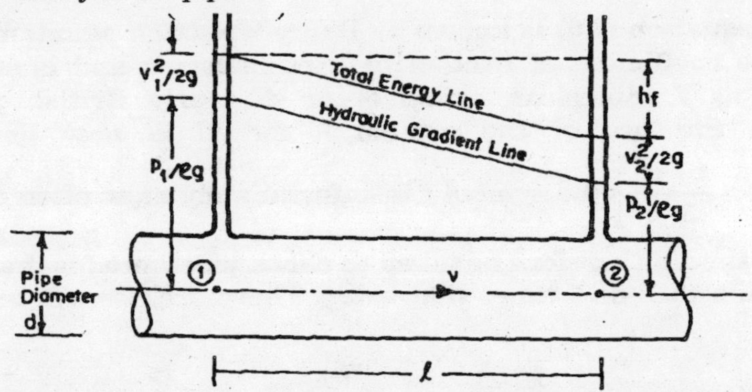

Fig. 9.5. Horizontal uniform pipe line

Therefore, the pressure difference acting on the fluid is equal to the frictional force and therefore

$$(p_1 - p_2) A = f' \, p \, l \, v^2$$

or, $$p_1 - p_2 = f' \frac{p}{A} l \, v^2 \qquad (9.4)$$

The ratio of the flow cross sectional area A and the wetted perimeter P appearing in the above expression is called the hydraulic radius or hydraulic mean depth and is denoted by m or R_h.

Therefore, $$m = \frac{A}{P}$$

For a circular pipe running full

$$m = \frac{A}{P} = \frac{\pi d^2}{4 \pi d} = \frac{d}{4}$$

and Let h_f = head lost due to friction

$$= \frac{p_1 - p_2}{\rho g}$$

Putting in equation (9.4)

$$h_f = \frac{f' 4}{\rho g \, d} l \, v^2 \qquad (9.5)$$

Since the factor f′ is determined experimentally it may, for convenience to incorporate other constants, and in fact it was found useful to make the substitution.

$$\frac{f'}{\rho} = \frac{f}{2}$$

With this modification the equation (9.5) becomes

$$h_f = \frac{4 f \, l \, v^2}{2g \, d} \qquad (9.6)$$

The equation (9.6) is known as Darcy-Weisbach equation and f is called the coefficient of friction or friction factor and is sometimes referred as f (American practice) or 4f (early British practice). Although the use of the symbol λ for 4f is now encouraged $\left(\text{giving } h_f = \frac{\lambda \, l \, v^2}{2g \, d}\right)$ the symbol *f* is unfortunately most often employed and care should therefore be taken to check whethere f is that defined by equation (9.6) or 4 times that value. Therefore, we will use in this book,

$$h_f = \frac{\lambda \, l \, v^2}{2g \, d} \qquad (9.7)$$

The original investigators persumed that the friction factor was constant. This was subsequently found to be incorrect. Equations relating λ to both the

Reynolds number and the pipe roughness were developed later. We will discuss in a separate chapter of turbulent flow.

9.4.2 Chezy's formula

The Darcy's-Weisbach equation is given as

$$h_f = \frac{f' l v^2}{m(\rho g)}$$

or,
$$\frac{hf}{l} = \frac{f' v^2}{m(\rho g)}$$

The ratio h_f/l is known as hydraulic gradient and is denoted by i.

Thus,
$$i = \frac{hf}{l} = f' \frac{v^2}{m(\rho g)}$$

or,
$$v^2 = \frac{\rho gm}{f'} i \quad v = \sqrt{\frac{\rho g m}{f'}\, i} = \sqrt{\frac{\rho g}{f'}} \sqrt{mi} = C\sqrt{m\, i} \tag{9.8}$$

The equation (9.8) is known as Chezy's equation in which C is the Chezy's constant determined experimentally. However, Chezy's formula is not commonly used in the case of flow through pipes.

9.4.3 Manning's formula

Manning's formula is very popular and simple. It states that Chezy's coefficient is given by

$$C = \frac{m^{1/6}}{n}$$

Here, the mean velocity is now given by

$$V = \frac{1}{n} m^{2/3}\, t^{1/2} \tag{9.9}$$

Where n is a roughness coefficient or Manning's roughness (or rugosity). The formula is very appropriate for rough surfaces and turbulent flow, i.e. the conditions in which most channels operate.

9.4.4 Hazen William's formula

Hazen William's is a simple empirical formula. The formula is given by

$$= 0.85\, C\, m^{0.63}\, i^{0.54}$$

$$v = 0.355\, C\, d^{0.63} \left(\frac{h_f}{l}\right)^{0.54}$$

or, alternatively

$$h_f = \frac{6.78\, l}{d^{1.165}} \left(\frac{v}{C}\right)^{1.85} \tag{9.10}$$

where, C is a coefficient. The value of C varies from about 70 to 150, depending on pipe diameter, material and age.

This formula is applicable only to the flow of water in pipes larger than 50 mm and at velocities less than 3m/s, but widely used in the waterworks pipe net work analysis.

9.5 MINOR LOSSES

A part from viscous and friction energy losses (Mazer losses), there are other losses-known as minor losses associated with pipe line problems. These losses occur due to any disturbance of the flow, e.g. changes in section, at bends and all pipe fittings. They are termed minor losses because they can often be neglected, particularly in long pipelines where the losses due to friction are high in proportion to the minor losses. However, in short pipelines, particularly those with a large number of fittings, the effect of the minor loss will be proportionally greater, and should be taken into considerations such losses are considered to occur locally at the disturbance to the flow.

Minor losses are usually associated with a change in velocity, either in magnitude or direction. It has been shown by experiment that the magnitude of the loss is approximately proportional to the square of the average velocity. Hence it is common practice to express minor losses as a function of the velocity head $v^2 / 2g$ in the pipe. Therefore,

$$h_l \propto \frac{v^2}{2g}$$

or,
$$h_f = K \frac{v^2}{2g} \tag{9.11}$$

where K is known as loss coefficient.

9.5.1 Head loss due to sudden enlargement

Let us consider fluid flowing in a pipe having sudden enlargement. Taking two sections at (1) and (2) as shown in the Fig. 9.6. A sudden enlargement in the pipe wall causes an increase in pressure from P_1 to P_2 and a decrease in the velocity from v_1/v_2 . The separation and large scale turbulence occur just after the change of section takes place and they are responsible for the loss of head. The normal conditions are not restored until some distance downstream. The pressure P_o acts in the eddying zone, and experimental work has shown that $P_o = P_1$. Isolating the body of fluid between sections (1) and (2) the forces acting on the fluid are shown in Fig. 9.7. Applying the impulse momentum theorem.

Forces acting on fluid in the direction of flow = Rate of change of momentum

$$p_1 a_1 + p_o (a_2 - a_1) - p_2 a_2 = \rho Q (v_2 - v_1)$$

but $\quad p_o = p_1 \ and \ Q = a_1 v_1 = a_2 v_2$

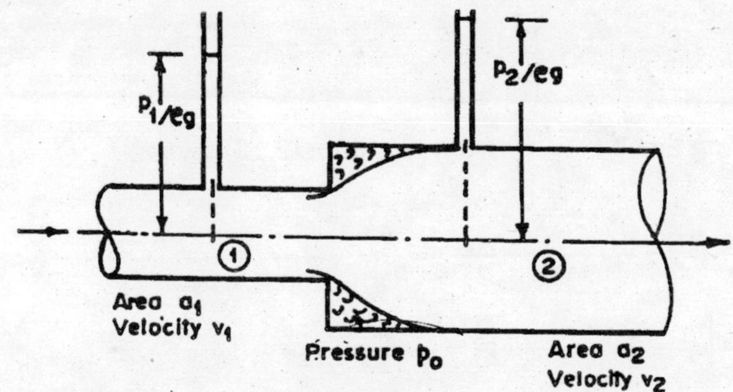

Fig. 9.6. Sudden enlargement

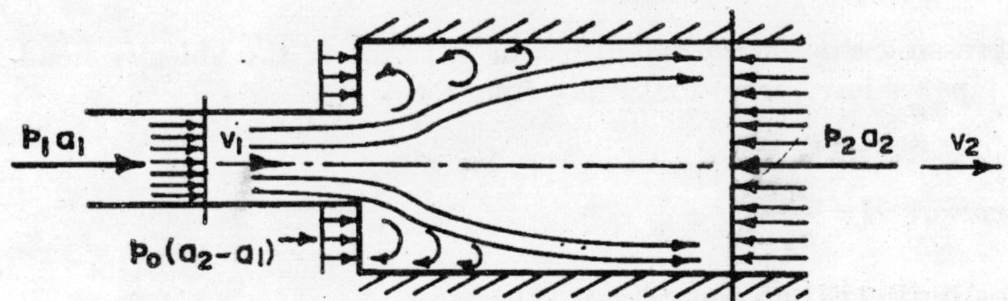

Fig. 9.7. Force diagram for sudden enlargement of flow.

therefore

$$p_1 a_1 + p_1 a_2 - p_1 a_1 - p_2 a_2 = \rho a_2 v_2 (v_2 - v_1)$$

or, $$p_1 a_2 - p_2 a_2 = \rho a_2 v_2 (v_2 - v_1)$$

or, $$(p_1 - p_2) a_2 = \rho a_2 v_2 (v_2 - v_1)$$

or, $$(p_1 - p_2) = \rho v_2 (v_2 - v_1)$$

or, $$\left(\frac{p_1 - p_2}{\rho}\right) = v_2 (v_2 - v_1)$$

or, $$\left(\frac{p_1 - p_2}{\rho g}\right) = \frac{v_2}{g} (v_2 - v_1) \tag{9.12}$$

Applying the Bernoulli's equation between section (1) and section (2) gives

$$\frac{p_1}{\rho g} + \frac{v_1^2}{2g} + z_1 = \frac{p_2}{\rho g} + \frac{v_2^2}{2g} + z_2$$

+ head loss h_l due to sudden enlargement.

If the pipe is in horizontal plane then $z_1 = z_2$

therefore,

$$\frac{p_1}{\rho g} + \frac{v_1^2}{2g} = \frac{p_2}{\rho g} + \frac{v_2^2}{2g} + h_l$$

or, $$h_l = \frac{p_1 - p_2}{\rho g} + \frac{v_1^2}{2g} - \frac{v_2^2}{2g}$$

Substituting for $\frac{(p_1 - p_2)}{\rho g}$ from equation (9.12) gives

$$h_l = \frac{v_2}{g}(v_2 - v_1) + \frac{v_1^2}{2g} - \frac{v_2^2}{2g} = \frac{v_2^2}{g} - \frac{v_1 v_2}{g} + \frac{v_1^2}{2g} - \frac{v_2^2}{2g}$$

$$= \frac{1}{2g}\left[v_1^2 - 2 v_1 v_2 + v_2^2\right] = \frac{1}{2g}(v_1 - v_2)^2$$

$$= \frac{(v_1 - v_2)^2}{2g} \qquad (9.13)$$

Expressing the energy head loss h_l in terms of the velocity head in the smaller pipe we have, from continuity equation

$$a_1 v_1 = a_2 v_2$$

Therefore $v_2 = \frac{a_1}{a_2} v_1$

Substituting v_2 into equation (9.13) gives

$$h_l = \frac{\left(v_1 - \frac{a_1 v_1}{a_2}\right)^2}{2g}$$

$$h_l = \left(1 - \frac{a_1}{a_2}\right)^2 \frac{v_1^2}{2g} \qquad (9.14)$$

$$h_l = K \frac{v_1^2}{2g}$$

where, K = loss coefficient = $\left[1 - \frac{a_1}{a_2}\right]^2$

9.5.2 Head loss due to sudden contraction

The loss of head due to sudden contraction (Fig. 9.8) is not due to the contraction it self but to the sudden enlargement which follows the contraction. Consider a pipe whose cross sectional area is reduced from a_1 to a_2. When the fluid is forced to pass through a contracted section, it contracts further to form vena contracta at (c), after which its cross section is enlarged again upto point (2) to fill the pipe. The total energy head loss at a sudden contraction is constituted by two separate minor losses. These are caused by :

(i) the convergence of the streamlines from the upstream pipe to the section of the vena contracta : and

(ii) the divergence of the streamlines from the section of the vena contracta to the pipe downstream.

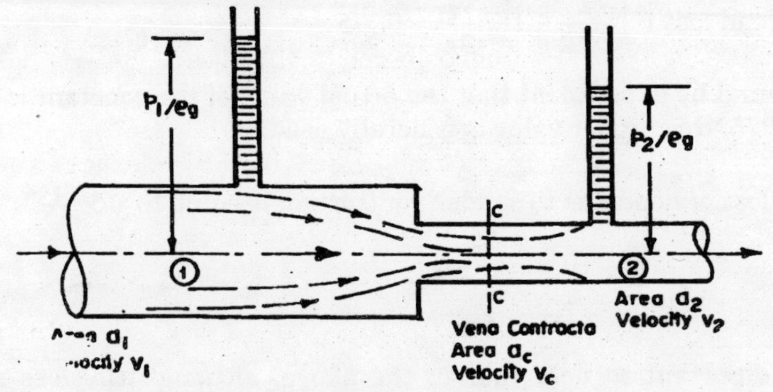

Fig. 9.8. Sudden contraction of flow.

Experimentation has shown that the loss due to the contraction of the streamlines is very small and a satisfactory estimate of the total head loss h_l may be established only by considering the head loss due to the expansion of the streamlines. Therefore from equation (9.13).

$$h_l = \left(\frac{v_c - v_2}{2g}\right)^2$$

For the vena contracta

$$a_c = c_c\, a_2$$

and from continuity

$$a_c\, v_c = a_2\, v_2$$

Therefore,

$$v_c = \frac{a_2}{a_c} v_2$$

Substituting for a_c gives

$$= \frac{a_2\, v_2}{c_c\, a_2} = \frac{v_2}{c_c}$$

Therefore

$$h_l = \frac{\left(\dfrac{v_2}{c_c} - v_2\right)^2}{2g} = \left(\frac{1}{c_c} - 1\right)^2 \frac{v_2^2}{2g} \tag{9.15}$$

$$= K\frac{v_2^2}{2g}$$

Where K = loss coefficient $= \left(\dfrac{1}{c_c} - 1\right)^2$

Assuming c_c to be 0.62 for a circular orifice

$$\text{Head loss} = \left(\frac{1}{0.62} - 1\right)^2 \frac{v_2^2}{2g} = 0.375 \frac{v_2^2}{2g}$$

It is found by experiment that the actual value of the constant is nearer to 0.5 than 0.375; this higher value is generally used.

Then, loss of head due to sudden contraction is equal to $0.5 \dfrac{v_2^2}{2g}$

$$h_l = 0.5 \frac{v_2^2}{2g} \qquad (9.16)$$

It is important to note that in the above, although the loss of energy occurs downstream of vena contracta, it very much depends upon the area ratio a_2/a_1, Since the smaller it is, the smaller is the cross sectional area at the vena contracta and thus smaller is the value of c_c.

Note : If the value of c_c is given in the numerical problem the equation (9.15) is used. If the value of c_c is not given in the numerical problem in equation (9.16) is used.

9.5.3 Head loss at entrance to pipe

The loss of head due to the fluid entering a pipe from a large reservoir is actually a loss due to sudden contraction. The equation (9.15) or (9.16) is used as per the condition.

In case of fluid flowing along long pipes, this loss of head is very small compared with the frictional loss and may be neglected.

9.5.4 Exit loss

From the equation (9.14)

$$h_l = \left(1 - \frac{a_1}{a_2}\right)^2 \frac{v_1^2}{2g}$$

The area a_1 of the pipe is negligibly small compared with the area a_2 of the reservoir, hence

$$\left(1 - \frac{a_1}{a_2}\right)^2 = 1$$

Therefore,

$$h_l = \frac{v_1^2}{2g} \qquad (9.17)$$

This happens at the outlet of a submerged pipe discharging into a large reservoir, for example (Fig. 9.9). The velocity head in the pipe, corresponding to the kinetic energy of the fluid per unit weight, is thus lost in turbulence in the reservoir. In such circumstances the loss is usually termed the exit loss for the pipe.

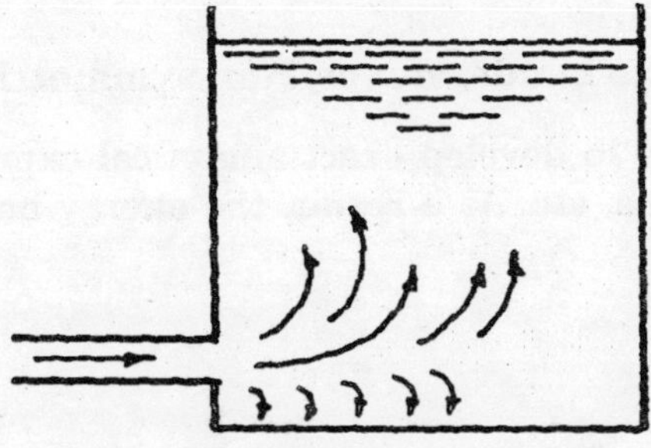

Fig. 9.9. Exit loss

9.5.5 Head loss due to obstruction

Any obstruction in the passage, such as a diaphragm of a projection from the passage walls, will interfere with the steady flow of the fluid and from eddies, the energy of which will be ultimately lost in friction.

An obstruction will cause a contraction of the area of flow which will be followed by an enlargement when obstruction is passed. The loss of head will be due to this sudden enlargement.

Consider a pipe of cross sectional area A, and let an obstruction of area a be placed in the pipe (Fig. 9.10). The fluid will flow in streamlines by the obstruction, the vena contracta occuring just beyond at the section c-c.

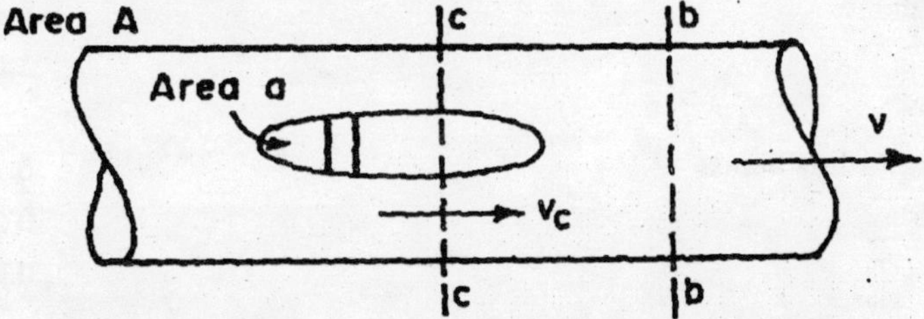

Fig. 9.10. Flow through a pipe with an obstruction

There will be a loss of head due to the enlargement between the section c-c and b-b equals to

$$h_l = \frac{(v_c - v)^2}{2g}$$

Area of section of flow at c – c = C_c (A – a)

Also,

$$v_c \, c_c \, (A - a) = v \, A$$

$$v_c = \frac{v \, A}{c_c \, (A - a)}$$

Then loss of head $= \left[\frac{A}{c_c\,(A-a)}\right]^2 \frac{v^2}{2g}$ (9.18)

9.5.6 Head loss due to bends, valves, Non symmetrical sections, etc.

It is not possible to develop exact analytical expressions for the energy losses at bends, valves, etc. As a result, the energy head loss h_l is expressed simply in the form

$$h_l = K\frac{v^2}{2g} \quad (9.19)$$

Where K = loss coefficient

v = velocity of flow in pipe downstream of the disturbance.

For such fittings, it is necessary to resort to experimentation to determine the value of K.

Values of the factor K critically depend upon the specific geometry and relative roughness of pipe fitting material. A few typical loss coefficients for various pipe fittings values are given in Table 9.1 but should be regarded as very approximate.

TABLE 9.1 Approximate loss coefficient K for commercial pipe fittings :

S.N.	Fitting	K
1	Elbows	0.2
	22.5°	0.4
	45°	1.0
	90°	
2	Long radius bends	0.1
	22.5°	0.2
	45°	0.4
	90°	
3	Gate Value	0.12
	Fully open	1.0
	Quarter closed	6.0
	Half closed	24.0
	Three quarter closed	1.5
4	Pump foot Valve	0.9
	90° elbow (threaded)	0.4
	45° elbow (threaded)	1.8
	Side outlet at T junction	
5	Globe Valve, wide open	9
	Screwed	12

	Flanged	0.8
6	Foot Valve, wide open	
7	Swing check valve, wide open	3.0
	Screwed	2.0
	Flanged	–
8	Angle Valve, wide open	2.1
	Screwed	2.4
	Flanged	

The pressure drop, or head loss owing to energy dissipation in an elbow is largely due to secondary flow superimposed on the main flow. As a result of centrifugal effects, the fluid in the faster core is directed outward and, in order to satisfy continuity (the outward flowing fluid must be replaced), the slower fluid near the boundaries is directed in ward.

9.6 TOTAL ENERGY LINE AND HYDRAULIC GRADIENT

In the study of fluid flow in pipes the concepts of total energy line and pressure line or piezometric line and more usually in America hydraulic grade line or hydraulic gradient line are often useful.

9.6.1 Total energy line (TEL)

The quantity $\left[\dfrac{p}{\rho g} + \dfrac{v^2}{2g}\right]$ at any point in the pipe may be represented by a vertical ordinate above the centre line of the pipe. The line linking the tops of such ordinates is termed the total energy line; and its height above a horizontal datum represents the total head of the fluid relative to the datum.

9.6.2 Hydraulic gradient line (HGL)

The pressure line is obtained by plotting values of $\left(\dfrac{p}{\rho g}\right)$ vertically above the pipe centre line. It is therefore, a distance $\dfrac{v^2}{2g}$ below the total energy line. If the pressure line and axis of the pipe itself coincide at any point then the gauge pressure in the pipe is zero, i.e. atmospheric. A pipe which rises above its pressure line has a negative pressure within it and is known as a siphon.

Fig. 9.11 shows two reservoirs connected by three pipes of different diameters. Due to the introduction of the pipe of smaller diameter in the centre, there will be contraction at the entrance of the smaller pipe and an enlargement at its exit. As such in addition to the friction losses, enlargement loss and exit loss. It should be noted that the energy gradient or total energy line will not rise in the direction of flow, since there is always a loss of energy in the direction of flow.

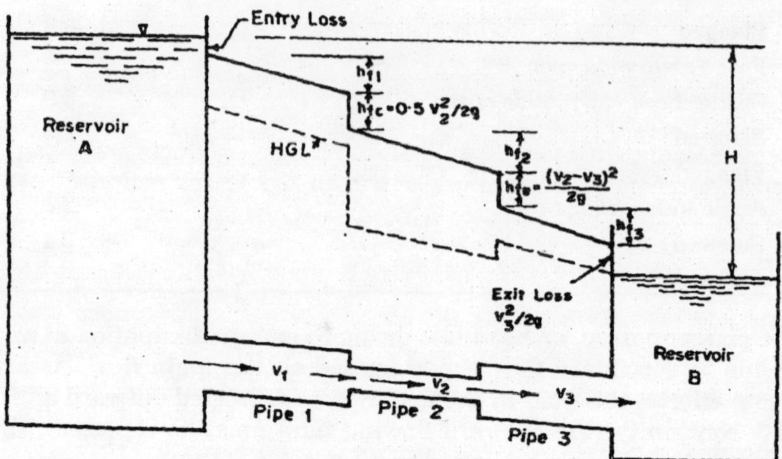

Fig. 9.11. Total energy line and hydraulic gradient line for pipes of different diameters connected in series.

If values of total head $\left(\frac{p}{\rho g}+\frac{v^2}{2g}+z\right)$ are computed at successive points along a pipe and plotted as shown in Fig. 9.11, the resulting line is known as energy gradient (total energy line). It is located initially at the liquid surface in the reservoir, but it decreases almost instantaneously by an amount equal to the entrance head loss as the liquid moves into pipe 1. The energy gradient then drops or decreases in the direction of flow to just before the sudden contraction by an amount equal to the head loss due to friction in pipe 1. At the sudden contraction, the energy gradient decreases almost instantaneously by an amount equal to the sudden contraction head loss. It then decreases, increases, drops again in the direction of flow to the sudden enlargement by an amount equal to the head loss due to friction in pipe 2. At the sudden enlargement, the energy gradient increases almost instantaneously by an amount equal to the sudden enlargement head loss. It then decreases or drops again in the direction of flow to the reservoir. B by an amount equal to the head loss due to friction in pipe 3. At this point an entrance loss. If values of potential, or piezometric, head $\left(\frac{p}{\rho g}+Z\right)$ are computed at successive points along a pipe and plotted as shown by the dashed line in Fig. 9.11., the resulting line is known as the hydraulic gradient. The distance between the energy gradient and hydraulic gradient at any point along the pipe line is equal to the velocity head at that point. For pipe segments with constant area and therefore constant velocity, the HGL and TEL are parallel.

9.7 FLOW THROUGH PIPE LINES

The losses of energy which occur in a pipe line are due to friction and separation (minor losses). In order that the flow may exist in the pipeline an amount of energy equal to the losses must be supplied to the fluid. As the losses vary with velocity and hence discharge, the larger is the required flow the more energy must be supplied to produce it. Reversing the argument it will be seen that for a given amount of energy available in the fluid the resulting flow in the pipeline will be such that all the losses are equal to the energy available. In fact the problems dealing with flow of fluid through pipes or ducts are solved by the application of Bernoulli's equation, which with the losses included represents the energy balance between any two selected sections of flow.

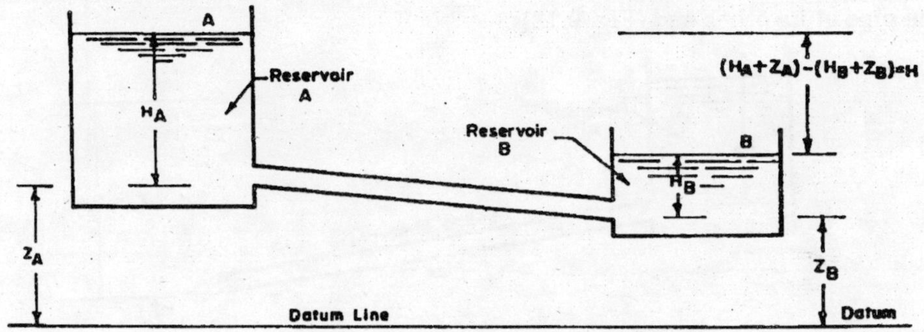

Fig. 9.12. Flow through pipes

Consider fluid flowing from a higher reservoir to lower reservoir through a pipe. Let the water levels in the two reservoir be kept constant at H_A and H_B above the centre line of the pipe. [Fig. 9.12].

Applying Bernoulli's equation to Reservoirs A and B.

$$\frac{p_A}{\rho g}+\frac{v_A^2}{2g}+z_A=\frac{p_B}{\rho g}+\frac{v_B^2}{2g}+Z_B+\sum_A^B \text{ losses} \tag{9.20}$$

but $\frac{p_A}{\rho g}=H_A,\ \frac{p_B}{\rho g}=H_B$ *and* $v_A=0$ *and* $v_B=0$

So that above equation becomes

$$H_A+Z_A=H_B+Z_B+\sum_A^B \text{ losses}$$

or,

$$(H_A+Z_A)-(H_B+Z_B)=\sum_A^B \text{ losses}$$

but $(H_A+Z_A)-(H_B+Z_B)=H=$ Difference between fluid levels in the two reservoirs.

Therefore,

$$H=\sum_A^B \text{ losses} \tag{9.21}$$

If the above condition is modified to give the pipe discharges freely into atmosphere, the above equation (9.20).

Modifies to

$$H_A + Z_A = Z_B + \sum losses$$

or, $$H_A + Z_A - Z_B = \sum losses$$

or, $$H = \sum losses$$

Where H is the height of fluid level in the reservoir A above the outlet of the pipe at its other end (Fig. 9.13).

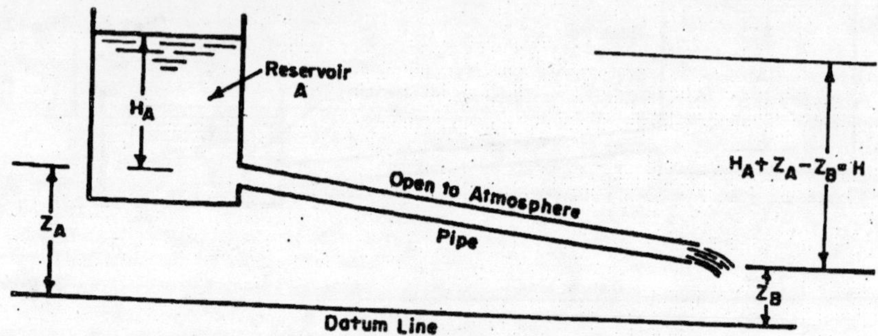

Fig. 9.13. Flow through pipe discharges free.

9.8 FLOW THROUGH PIPES CONNECTED IN SERIES

The discussion in article 9.7 was restricted to the case of a single pipe. If the pipe is made of sections of different diameters, and different lengths, are connected in series, that is end to end, the same flow passes through each in turn and, the head loss is the sum of the losses in all the individual pipes and fittings, the continuity and energy equations establish the following two simple relations which must be satisfied.

$$Q = Q_1 = Q_2 = Q_3 = Q_4 \ldots\ldots \qquad (9.22)$$

$$h_l = h_{l_1} + h_{l_2} + h_{l_3} + h_{l_4} + \ \ldots\ldots \qquad (9.23)$$

Consider the two tanks are joined by three pipes in series the pipes being of different length and different diameter. Fig. 9.14 shows the arrangement and also indicates all the losses occurring in the pipeline.

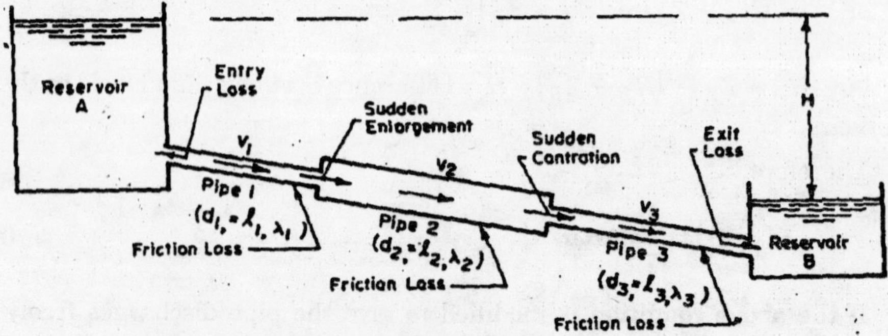

Fig. 9.14. Pipe line in series indicates all the losses

Thus the following equation can be written for the Fig. 9.14

H = Entry loss (A to 1) + Friction loss in pipe 1 + Sudden enlargement (1 to 2) + Friction loss in pipe 2 + Sudden Contraction (2 to 3) + Friction loss in pipe 3 + Exit loss 3 to B

$$H = \left(\frac{1}{c_c} - 1\right)^2 \frac{v_1^2}{2g} \text{ or } 0.5\frac{v_1^2}{2g} + \frac{\lambda_1 l_1 v_1^2}{2gd_1} + \frac{(v_1 - v_2)^2}{2g}$$

or $$+\frac{\lambda_2 l_2 v_2^2}{2gd_2} + \left(\frac{1}{c_c} - 1\right)^2 \frac{v_3^2}{2g} \text{ or } 0.5\frac{v_3^2}{2g} + \frac{\lambda_3 l_3 v_3^2}{2gd_3} + \frac{v_3^2}{2g} \tag{9.24}$$

Also by the continuity equation

$$Q = a_1 v_1 = a_2 v_2 = a_3 v_3$$

Thus two of the unknown velocities can be expressed in terms of the first velocity, for example

$$v_2 = \frac{a_1}{a_2} v_1 \text{ and } v_3 = \frac{a_1}{a_3} v_1$$

By Substituting for v_2 and v_3 into equation (9.25) it can be solved for V_1 for a given level difference H. The discharge can then be calculated from

$$Q = a_1 v_1$$

The alternative way of solving equation (9.24) is by expressing each velocity in terms of Q and then solving for it

Thus $v_1 = \frac{Q}{a_1}$, $v_2 = \frac{Q}{a_2}$ and $v_3 = \frac{Q}{a_3}$

Equation (9.25) then takes the general form

$$H = k_1 Q^2 + k_2 Q^2 + k_3 Q^2$$

$$+ k_4 Q^2 + k_5 Q^2 + k_6 Q^2 + k_7 Q^2 \tag{9.25}$$

$$H = (k_1 + k_2 + k_3 + k_4 + k_5 + k_6 + k_7) Q^2$$

in which

$$k_1 = \frac{\left(\frac{1}{c_c} - 1\right)^2}{a_1^2 \times 2g} \text{ or } \frac{0.5}{a_1^2 \times 2g}$$

$$k_2 = \frac{\lambda_1 l_1}{d_1^2 a_1^2 \times 2g}, \text{ etc.}$$

If now $k = k_1 + k_2 + k_3 + k_4 + k_5 + k_6 + k_7$, then

$$H = k Q^2 \tag{9.26}$$

This equation demonstrates the fact stated earlier that for a given amount of energy available, there is one value of discharge possible and given by the above equation. This equation, although obtained for a specific case, is in fact quite general and applicable to any system of pipes in series.

The frictional losses in terms of Q can be written as

$$h_f = \frac{\lambda l v^2}{2gd}$$

$$Q = av = \frac{\pi}{4} d^2 v$$

So, $$v = \frac{Q}{\frac{\pi}{4} d^2} = \frac{4Q}{\pi d^2}$$

and substituting into the equation

$$h_f = \frac{\lambda\, l}{2gd} \times \frac{16\, Q^2}{\pi^2 d^4} = \frac{\lambda\, l\, Q^2}{d^5} \times \left[\frac{16}{2\pi^2 g}\right] = \frac{\lambda\, l\, Q^2}{12.1\, d^5} \tag{9.27}$$

9.9 METHOD OF EQUIVALENT LENGTHS

An inspection of equation (9.27) will indicate that if the diameter were to be solved for a problem in which minor losses are included, a fifth order polynomial involving the unknown diameter must be solved at each iteration level, and this results in a very tedious computation procedure. The method of equivalent lengths avoids this problem and often makes for a more straight forward solution to the discharge problem as well. However, there is no advantage to be gained for the direct computation of head loss.

Equivalent pipe is defined as the pipe of uniform diameter having loss of head and discharge equal to the loss of head and discharge of a compound pipe consisting of several pipes of different lengths and diameters. The uniform diameter of the equivalent pipe is called equivalent size of pipe. The length of equivalent pipe is equal to sum of lengths of the compound pipe consisting of different pipes.

Assuming friction factors are same for all pipes. The equivalent head loss for the compound pipe neglecting minor losses.

$$h_{fe} = h_{f1} + h_{f2} + h_{f3} + \text{- - -}$$

$$\frac{\lambda\, L_e\, Q^2}{12.1\, d_e^{\ 5}} = \frac{\lambda_1\, L_1\, Q^2}{12.1\, d_1^{\ 5}} + \frac{\lambda^2\, L_2\, Q^2}{12.1\, d_2^{\ 5}} + \frac{\lambda_3\, L_3\, Q^2}{12.1\, d_3^{\ 5}} + \ldots\ldots\ldots\ldots$$

$$\lambda_1 = \lambda_2 = \lambda_3 = \ldots\ldots\ldots\ldots$$

$$\therefore \quad \frac{L_e}{d_e^5} = \frac{L_1}{d_1^5} + \frac{L_2}{d_2^5} + \frac{L_3}{d_3^5} + \ldots\ldots\ldots \tag{9.28}$$

If the diameter of the equivalent pipe de is fixed up, its length Le can be determined from the equation (9.28).

In Principle, all loss terms could be converted to a equivalent length of a pipe that has nothing to do with the problem but there is no practical advantage to so doing. Generally, it is most reasonable to convert to equivalent lengths of the pipe in which most of the losses occur.

9.10 FLOW THROUGH PIPES CONNECTED IN PARALLEL

When two or more pipes are connected as shown in Fig. 9.15, so that the flow divides and subsequently comes together again, the pipes are said to be in parallel. Using the continuity equation, the rate of discharge in the main pipe line is equal to the sum of the discharges in each of the parallel pipes.

$$Q = Q_A + Q_B \tag{9.29}$$

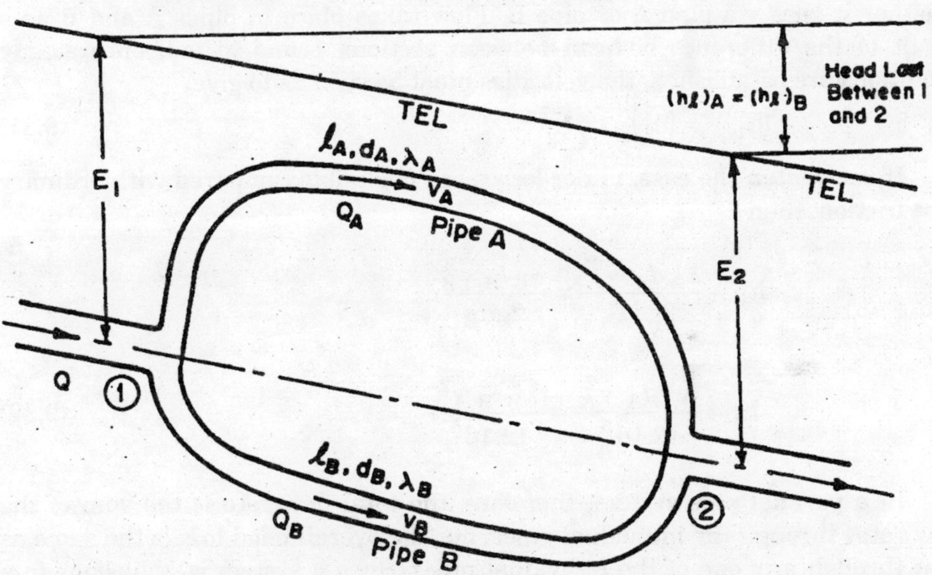

Fig. 9.15. Pipe lines in parallel

Now at any point there can be only one value of the total head. So all the fluid passing point 1 has the same total head $\left(\frac{p_1}{\rho g} + \frac{v_1^2}{2g} + z_1\right)$. Similarly, at point 2, after passing through the parallel pipes, all the fluid must have the same total head $\left(\frac{p_2}{\rho g} + \frac{v_2^2}{2g} + z_2\right)$. No matter which path it took between points 1 and 2. Therefore, whatever changes occur in the total head lines for pipes A and B, these lines must join common ones at sections 1 and 2 because there cannot be more than one total head line for the single pipe upstream or downstream of the functions.

The steady flow energy equation may be written

$$\frac{p_1}{\rho g} + \frac{v_1^2}{2g} + z_1 = \frac{p_2}{\rho g} + \frac{v_2^2}{2g} + z_2 + h_l$$

or, $\quad E_1 = E_2 + h_l$

$$h_l = E_1 - E_2 \tag{9.30}$$

The term h_l represents the loss of head between points 1 and 2. In Fig. 9.15, each unit of fluid has the same total head at the point where the flow branches. As the fluid flows through the branches, same energy is lost. But at the point where the flow rejoins, the total head of each unit of fluid must again be the same. Consequently all the fluid must suffer the same loss of head h_l whether it goes via pipe A or pipe B. Flow takes place in pipes A and B as a result of the difference of head between sections 1 and 2 and, once steady conditions are established, the velocities must be such as to give

$$(h_l)_A = (h_l)_B \tag{9.31}$$

If, as is often the case, minor losses are negligible compared with ordinary pipe friction, then

$$h_l = \frac{\lambda_A\, l_A\, v_A^2}{2gd_A} = \frac{\lambda_B\, l_B\, v_B^2}{2gd_B}$$

$$h_l = \frac{\lambda_A\, l_A\, v_A^2}{12.1 d_A^5} = \frac{\lambda_B\, l_B\, v_B^2}{12.1 d_B^5} \tag{9.32}$$

In a parallel pipe system, therefore, the total flow rate is the sum of the flow rates through the individual pipes, but the overall head loss is the same as that through any one of the individual pipes. Such a system is analogous to a parallel electric circuit in which the drop in potential and flow of electric current can be compared to head loss and rate of discharge in a fluid flow respectively.

9.11 POWER TRANSMISSION THROUGH PIPES

A pipeline to be able to deliver power at exit, only a fraction of the total head H at the inlet can be used to overcome the loss of head due to friction and remainder will then be available as the power head, which is the energy per unit weight available at the outlet.

Fig. 9.16 shows a pipe AB connected to a high level storage tank.

Let $\quad$ L = Length of the pipe

D = Diameter of the pipe

V = Velocity of flow in pipe.

$$h_f = \frac{4f\,LV^2}{2gd} = \frac{\lambda\,LV^2}{2g\,D}$$

where $\lambda = 4f$

H = Head available at the inlet to the pipe

h_p = Total head available at the outlet of the pipe.

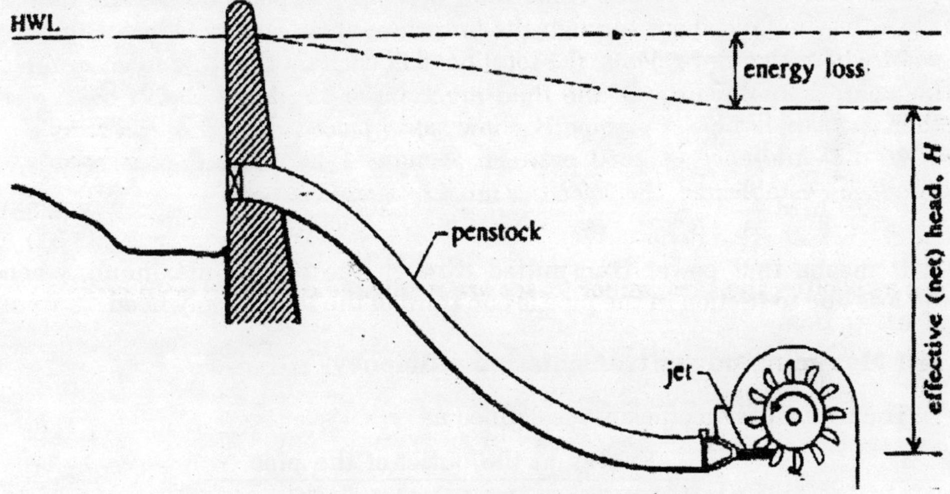

Fig. 9.16. Power transmission through pipe.

$$h_p = H - h_f$$

$$= H - \frac{\lambda\,LV^2}{2g\,D} \tag{9.33}$$

Power available at outlet

$$P = \rho Q\, g\, h_p$$

$$= \rho\, g\, AV\left(H - \frac{\lambda\,LV^2}{2g\,D}\right)$$

$$= \rho\, g\, A\left(HV - \frac{\lambda\,LV^3}{2g\,D}\right) \tag{9.34}$$

Power transmitted will be maximum when

$$\frac{dp}{dV} = 0$$

From equation (9.34)

$$P = \rho g A\left(HV - \frac{\lambda L V^3}{2g\ D}\right)$$

Differentiating with respect to V and equating to zero.

$$\frac{dP}{dV} = \rho g A\left(H - \frac{3\lambda L V^2}{2g\ D}\right) = 0$$

or $$\rho g A\left(H - \frac{3\lambda L V^2}{2g\ D}\right) = 0$$

or $$H - \frac{3\lambda L V^2}{2g\ D} = 0$$

or $H - 3h_f = 0$

or, $H = 3h_f$

$$\therefore\ h_f = \frac{H}{3} \tag{9.35}$$

It means that power transmitted through the pipe is maximum, when head loss due to friction in the pipe is one third of the total supply head.

9.11.1 Maximum power transmission efficiency

The transmission efficiency is defined as

$$\eta_{trans} = \frac{\text{Energy at the outlet of the pipe}}{\text{Energy available at the inlet of the pipe}}$$

$$= \frac{h_p}{H} = \frac{H - h_f}{H} \tag{9.36}$$

From Equation (9.35)

$$\text{Maximum } \eta_{trans} = \frac{H - H/3}{H} = \frac{2}{3}$$

$$= 66.67\% \tag{9.37}$$

9.12 FLOW THROUGH NOZZLE AT THE END OF A PIPE

Fig. 9.17 shows a nozzle fitted at the end of a pipe connected to a reservoir.

Let V = Velocity of flow in pipe

D = Diameter of the pipe

v = Velocity of flow at the outlet of the nozzle

d = Diameter of the nozzle.

H = Height of water level in the reservoir above the centreline of the nozzle.

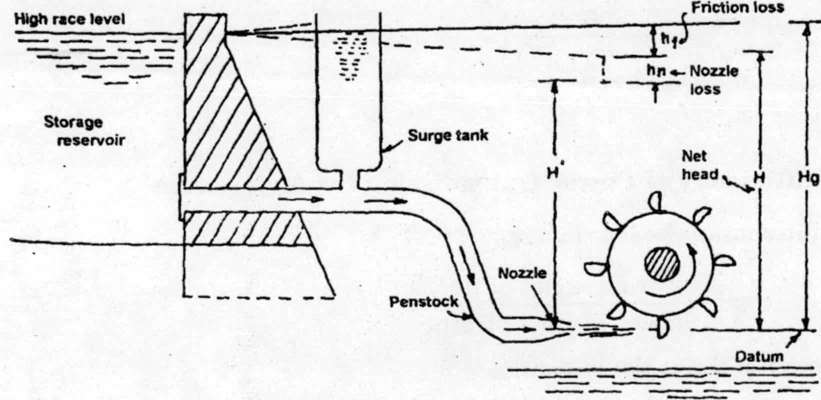

Fig. 9.17. Flow through nozzle.

Head available at the base of the nozzle

= Head at the inlet of pipe – Head loss due to friction.

$$= H - h_f$$

$$= H - \frac{\lambda L V^2}{2g\ D} \quad (9.38)$$

Total head at inlet of pipe

= Total head at the outlet of nozzle + Losses

But total head at outlet of nozzle = kinetic head = $\frac{v^2}{2g}$

Therefore,

$$H = \frac{v^2}{2g} + h_f$$

$$= \frac{v^2}{2g} + \frac{\lambda L V^2}{2g\ D} \quad (9.39)$$

Using continuity equation

$$AV = av \qquad V = \frac{a}{A} v$$

Substituting in equation (9.39), we get

$$H = \frac{v^2}{2g} + \frac{\lambda L}{2g\ D}\left(\frac{a\ V}{A}\right)^2 = \frac{v^2}{2g} + \frac{\lambda L a^2 v^2}{2g\ D\ A^2}$$

$$= \frac{v^2}{2g}\left(1 + \frac{\lambda L a^2}{DA^2}\right) \quad (9.40)$$

$$v = \frac{\sqrt{2g\,H}}{\left(1 + \dfrac{\lambda L a^2}{D\,A^2}\right)} \tag{9.41}$$

9.12.1 Efficiency of Power transmission through nozzle

Efficiency of the nozzle = η_n

$$= \frac{\text{Power at outlet of nozzle}}{\text{Power in inlet of pipe}}$$

Power of the outlet of nozzle

= Kinetic energy of the jet at the outlet

$$= \frac{m\,V^2}{2}$$

$$= \frac{\rho\,a\,v}{2}\,v^2 \qquad = \frac{\rho\,a\,v^3}{2}$$

Power at the inlet of pipe

$$= \rho\,g\,Q\,H$$

$$\therefore \eta_n = \frac{\frac{1}{2}\rho\,a\,v^3}{\rho\,g\,Q\,H} = \frac{\frac{1}{2}\rho\,a\,v^3}{\rho\,g\,a\,v\,H}$$

$$= \frac{v^2}{2\,g\,H} \tag{9.42}$$

Using equation (9.43), we get

$$n_n = \frac{1}{\left(1 + \dfrac{\lambda\,L\,a^2}{D\,A^2}\right)} \tag{9.43}$$

9.12.2 Condition for maximum power transmission through nozzle

From equation (9.40)

$$\frac{v^2}{2g} = H - \frac{\lambda\,L\,V^2}{2gD}$$

$$\therefore v^2 = 2g\left(H - \frac{\lambda\,L\,V^2}{2gD}\right) \tag{9.44}$$

But power transmitted through nozzle

$$= \frac{1}{2}\rho\,a\,v^3 \qquad = \frac{1}{2}\rho\,a\,v\,v^2$$

$$= \frac{1}{2} \rho a v \left[2g \left(H - \frac{\lambda L V^2}{2g D} \right) \right]$$ [Using equation (9.44)]

$$= \rho g a v \left(H - \frac{\lambda L V^2}{2g D} \right)$$

$$= \rho g a v \left(H - \frac{\lambda L a^2 v^2}{2g D A^2} \right) \quad (9.45)$$

The power will be maximum, when

$$\frac{dP}{dv} = 0$$

$$\rho g a \left[H - \left(\frac{3 \lambda L a^2 v^2}{2g D A^2} \right) \right] = 0$$

or, $$H - \frac{3 \lambda L V^2}{2g D} = 0$$

or, $$H - 3h_f = 0$$

$$\therefore h_f = \frac{H}{3} \quad (9.46)$$

It means that power transmitted through nozzle is maximum when the head loss due to friction in pipe is one-third of the total head supplied at the inlet of pipe.

9.12.3 Diameter of nozzle for maximum transmission of power through nozzle

From equation (9.46)

$$H = 3 h_f$$

and $$H = \frac{v^2}{2g} + h_f \quad (9.47)$$

Equating equations (9.46) and (9.47)

$$3 h_f = \frac{v^2}{2g} + h_f$$

$$2 h_f = \frac{v^2}{2g} \qquad \frac{2 \lambda L V^2}{2g D} = \frac{v^2}{2g}$$

$$\frac{2 \lambda L}{D} \left(\frac{V^2}{v^2} \right) = 1$$

or, $$\frac{2\lambda L}{D} \times \frac{a^2}{A^2} = 1 \qquad (9.48)$$

or, $$\frac{2\lambda L}{D} \times \frac{d^4}{D^4} = 1 \qquad d^4 = \frac{D^5}{2\lambda L}$$

$$\therefore\ d = \left(\frac{D^5}{2\lambda L}\right)^{1/4} \qquad (9.49)$$

From equation (9.48)

$$\frac{A^2}{a^2} = \frac{2\lambda L}{D}$$

$$\frac{A}{a} = \sqrt{\frac{2\lambda L}{D}} \qquad (9.50)$$

Putting $\lambda = 4f$ in equations (9.49) and (9.50), we get

$$d = \left(\frac{D^5}{8fL}\right)^{1/4} \qquad (9.51)$$

and $$\frac{A}{a} = \sqrt{\frac{8fL}{D}} \qquad (9.52)$$

9.13 WATER HAMMER

In a long pipe, when the flowing water is suddenly brought to rest by closing the valve, there will be a sudden rise in pressure due to the momentum of water being destroyed. A pressure wave is transmitted along the pipe. A sudden rise in pressure has the effect of hammering action on the walls of the pipe. This phenomenon is knwon as water hammer or hammer blow.

The magnitude of the rise in pressure depends on :

(1) Speed at which the valve is closed.

(2) The length of pipe

(3) Elasticity of the pipe material as well as of the flowing fluid.

9.14 Water hammer analysis

There are broadly two ways available to predict the behaviour of fluid column when it is under the action of a force.

(1) Rigid column theory which considers the entire fluid column to be accelerated at the same value throughout its length, the wave speed being infinitely large.

(2) Elastic theory which considers any pressure change to be transmitted through the fluid column at a large but finite wave speed.

9.14.1 Rigid Water Column theory

The theory was presented initially in simplest form. The following assumption were made.

(1) The water in the pipe line is incompressible.

(2) The pipe walls do not stretch regardless of the pressure inside the pipe.

(3) The pipe line remains full of water at all times and the minimum pressure inside the pipe is in excess of the vapour pressure of water.

(4) The hydraulic losses and velocity head are negligible when compared with the pressure changes.

(5) The velocity of water in the direction of the axis of the pipe is uniform over any cross section of pipe.

(6) The pressure is uniform over a transverse cross section of the pipe and is equal to the pressure at the centre line of the pipe.

(7) The reservoir level remains constant during the gate movement time.

In this theory, situation arise in which pressure transients are not of great interest but in which fluid movements are important and capable of describing such motion moderately accurately.

(a) Gradual Closure of Valve

Consider a long pipe of length l and of cross sectional area A carrying fluid flowing with a velocity V. The valve is closed gradually in time t seconds. It is assumed that the rate of closure of the valve is so adjusted that the liquid column in the pipe is brought to rest with a uniform retardation from an initial velocity V to zero velocity in time t seconds (Fig. 9.18).

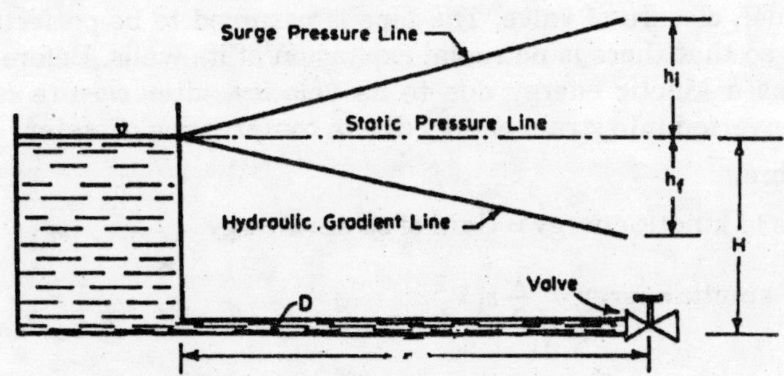

Fig. 9.18. Incompressible surge in a pipe line system

Mass of fluid in pipe line = ρ A L

$$\text{Retardation} = \frac{V}{t}$$

Axial force available for producing retardation

= Inertia pressure × Area

$F = p_i A$

But Retarding force = Mass × Retardation

$$p_i\ A = \rho\ A\ L\frac{V}{t}$$

$$p_i = \rho\ L\frac{V}{t} \tag{9.53}$$

But

$$P_i = \rho\ g\ h_i$$

Therefore, $\rho\ gh_i = \rho L\frac{V}{t}$

$$h_i = \frac{L}{g}\frac{V}{t} \tag{9.54}$$

The inertia head given by equation (9.54) is developed at the end of pipe. At any other point on the pipe, the pressure rise can be determined easily. The maximum pressure head from all causes is represented as

$(h_i)_{\max}$ = Static pressure head + Inertia pressure head – Friction head (9.55)

(1) The closure of valve is said to be gradual when $T > \frac{2L}{C}$.

(2) The closure of valve is said to be instantaneous when $T < \frac{2L}{C}$.

where C = velocity of pressure wave.

(b) Sudden closure of valve

This is the simplest case to treat. Let the fluid flow in a pipe is brought to rest by sudden closure of valve. The pipe is assumed to be perfectly rigid and non elastic, so that there is no radial expansion of its walls. Before the closure the fluid has a kinetic energy due to its velocity, after closure of valve this energy is converted into strain energy due to compression of water.

Therefore,

Loss of kinetic energy = Gain of strain energy (9.56)

Loss of kinetic energy = $\frac{1}{2}m\,V^2$

$$= \frac{1}{2}\rho A\ L\,V^2$$

Gain of strain energy = Average pressure × Volumetric strain

$$= \frac{p_i}{2} \times \text{ Volumetric strain}$$

$$= \left(\frac{p_i}{2}\right) \times \left(\frac{p_i}{K}\right) \times Volume = \frac{1}{2}\left(\frac{p_i^2}{K}\right) \times Volume \qquad \left[Because\ K = \frac{p_i}{dV/V}\right]$$

$$= \frac{1}{2}\frac{p_i^2}{K} \times Volume = \frac{1}{2}\frac{p_i^2}{K} \times AL$$

putting in equation (9.56)

$$\frac{1}{2}\rho\, A\, L\, V^2 = \frac{1}{2}\frac{p_i^2}{K}\, AL$$

$$p_i^2 = \rho\, K\, V^2$$

$$p_i = \sqrt{\rho K V^2} = V\,\sqrt{\rho K} = \rho V\,\sqrt{\frac{K}{\rho}} = \rho C V \qquad (9.57)$$

Where C = velocity of pressure wave

$$= \sqrt{\frac{K}{\rho}}$$

9.14.2 Elastic pipe theory

The rigid pipe theory is often applicable in circumstances that are of little interest to the engineer. It can be used only if valve clousers are slow and the pressure surges so caused are small. Usually it is necessary to know the worst conditions under which the pipe may have to operate, and it is precisely these conditions that the rigid pipe theory cannot describe. It is thus essential to be able to predict the pressure rises generated when velocity fluctuations are rapid and the consequent pressure surges large.

It is helpful to consider an infinitely rapid, that is, instantaneously valve closure although in practice rapidly varying pipe velocities are produced by valve closures that are only moderately rapid.

When a valve at the end of a long pipe line is shut instantaneously, the fluid upstream of valve will impinge upon the closed valve and its pressure will immediately rise. The force on the due to sudden closure will produce a longitudinal stress f_l and a circumferential stress f_c in the pipe wall. The original kinetic energy of the fluid will be converted partly into strain energy of the fluid and partly into strain energy of the pipe.

Kinetic energy given up by fluid or Loss of kinetic energy

= Strain energy stored in fluid + strain energy stored in pipe walls (9.58)

Loss of kinetic energy $= \frac{1}{2}\, mV^2$

$$= \frac{1}{2}\rho\, A\, L\, V^2$$

$$f_l = \frac{pD}{4t} \quad and \quad f_c = \frac{pD}{2t}$$

Therefore, strain energy stored in pipe walls per unit volume

$$= \frac{1}{2E}\left[f_l^2 + f_c^2 - \frac{2 f_l\, f_c}{m}\right]$$

or, $$= \frac{1}{2E}\left[f_l^2 + f_c^2 - 2 f_l f_c \sigma\right]$$

Where $\frac{1}{m} = \sigma$ = Poission's ratio of pipe material

$$= \frac{1}{2E}\left[\left(\frac{pD}{4t}\right)^2 + \left(\frac{pD}{2t}\right)^2 - \frac{2pD}{4t} \times \frac{pD}{2t} \times \sigma\right]$$

$$= \frac{1}{2E}\left[\frac{p^2D^2}{16t^2} + \frac{p^2D^2}{4t^2} - \frac{2p^2D^2}{8t^2} \times \sigma\right]$$

$$= \frac{1}{2E} \times \frac{p^2D^2}{t^2}\left[\frac{1}{16} + \frac{1}{4} - \frac{1}{16}\right] \qquad \left\{\text{Assuming } \sigma = \frac{1}{4}\right\}$$

$$= \frac{p^2D^2}{8Et^2}$$

Total volume of pipe = π D t L

Therefore, strain energy stored in pipe

$$= \frac{p^2D^2}{8Et^2}(\pi D t L) = \frac{p^2}{2}\frac{\pi D^2}{4}\frac{Dl}{Et}$$

$$= \frac{p^2 ADL}{2Et}$$

Gain in strain energy in water

$$= \frac{1}{2}\left(\frac{p^2}{K}\right) \times Volume = \frac{1}{2}\left(\frac{p^2}{K}\right)(AL)$$

Substituting in equation (9.58)

$$\frac{1}{2}\rho A LV^2 = \frac{1}{2}\left(\frac{p^2}{K}\right)(AL) + \frac{p^2 ADL}{2Et}$$

$$\frac{\rho V^2}{2} = \frac{p^2}{2K} + \frac{p^2 D}{2Et}$$

or, $$\frac{\rho V^2}{2} = \frac{p^2}{2}\left[\frac{1}{K} + \frac{D}{Et}\right]$$

or, $$\frac{p^2}{\rho V^2} = \frac{1}{\left(\frac{1}{K} + \frac{D}{Et}\right)}$$

or, $$p_i^2 = \frac{\rho V^2}{\left(\frac{1}{K} + \frac{D}{Et}\right)}$$

$$\text{or,}\quad p_i = \frac{V\sqrt{\rho}}{\sqrt{\frac{1}{K}+\frac{D}{Et}}} \tag{9.59}$$

$$p_i = \frac{V\rho}{\sqrt{\rho\left(\frac{1}{K}+\frac{D}{Et}\right)}} \tag{9.60}$$

The value of term $\left(\frac{D}{Et}\right)$ is usually small as compared with the value of $\left(\frac{1}{K}\right)$, hence neglected, therefore,

$$p_i = \frac{V\rho}{\sqrt{\rho\left(\frac{1}{K}\right)}} \tag{9.61}$$

$$\text{or,}\quad p_i = \rho V\sqrt{\frac{K}{\rho}}$$

$$p_i = \rho V C \qquad \left[C = \sqrt{K/\rho}\right] \tag{9.62}$$

This is often known as **Allievi formula**

Using the equation (9.62), in the equation (9.59), we get

$$C = \frac{1}{\sqrt{\rho\left(\frac{1}{K}+\frac{D}{Et}\right)}}$$

$$\text{or,}\quad C = \sqrt{\frac{1}{\rho\left(\frac{1}{K}+\frac{D}{Et}\right)}} \tag{9.63}$$

ILLUSTRATIVE EXAMPLES

9.1 Water flows vertically downwards through a 150 mm diameter pipe with a velocity of 2.4 m/s. The pipe suddenly enlarges to 300 mm in diameter. Find the loss of head and also the loss of head if the flow is reversed, the coefficient of contraction being 0.62.

Solution

$$a_1 = \frac{\pi}{4}d_1^2 = \frac{\pi}{4}(0.15)^2 = 0.01767\ m^2$$

$$a_2 = \frac{\pi}{4}d_2^2 = \frac{\pi}{4}(0.3)^2 = 0.07068\ m^2$$

$$Q = a_1\, v_1 = 0.01767 \times 2.4 = 0.0424\ m^3/s$$

$$Q = a_2\, v_2$$

$$0.0424 = 0.07068\ v_2$$

$$v_2 = 0.6\ m/s$$

$$h_e = \frac{(v_1 - v_2)^2}{2g} = \frac{(2.4 - 0.6)^2}{2 \times 9.81} = 0.165\ m$$ Ans

$$h_c = \left(\frac{1}{C_c} - 1\right)^2 \frac{v_2^2}{2g} \qquad [Here\ v_1 = v_2]$$

$$= \left(\frac{1}{0.62} - 1\right)^2 \times \frac{(2.4)^2}{2 \times 9.81} = 0.11\ m$$ Ans

9.2 A 300 mm diameter pipe through which water flows at the rate 283 liters/s suddenly enlarges to 600 mm diameter. If the axis of the pipe is horizontal and the water in a vertical tube connected to the large pipe stands 0.36 m higher than the level in a tube connected to the smaller pipe, determine the coefficient K if the shock loss is expressed as $K\frac{v_1^2}{2g}$, Where v_1 is the velocity in the smaller pipe.

Solution

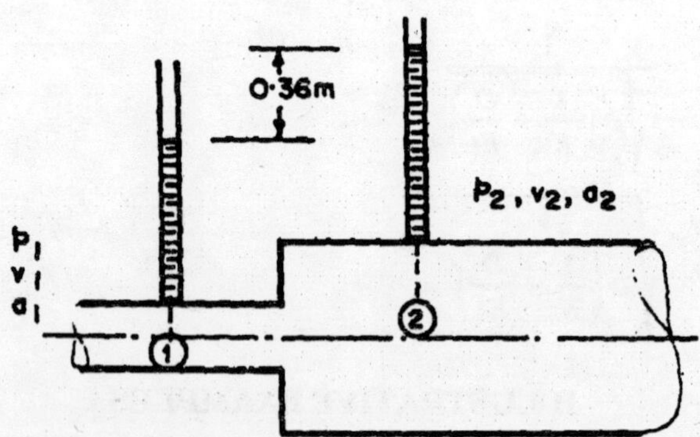

Fig. 9.19. Sudden enlargement of flow with Piezo meter tubes. P. 9.2

$$\frac{p_2 - p_1}{\rho g} = 0.36\ m$$

$$a_1 = \frac{\pi}{4}(0.3)^2 = 0.07068\ m^2$$

$$a_2 = \frac{\pi}{4}(0.6)^2 = 0.2827\ m^2$$

$$Q = a_1 v_1 = a_2 v_2$$

$$v_1 = \frac{Q}{a_1} = \frac{0.283}{0.07068} = 4\ m/s$$

$$v_2 = \frac{Q}{a_2} = \frac{0.283}{0.2827} = 1\ m/s$$

Applying the Bernoulli's equation between point (1) and (2)

$$\frac{p_1}{\rho g} + \frac{v_1^2}{2g} = \frac{p_2}{\rho g} + \frac{v_2^2}{2g} + K\frac{v_1^2}{2g}$$

$$\frac{p_2}{\rho g} - \frac{p_1}{\rho g} = \frac{v_1^2}{2g} - K\frac{v_1^2}{2g} - \frac{v_2^2}{2g}$$

$$0.36 = \frac{4^2}{2 \times 9.81} - K\frac{(4)^2}{2 \times 9.81} - \frac{1^2}{2 \times 9.81}$$

$$0.36 = \frac{16}{19.62} - K\frac{16}{19.62} - \frac{1}{19.62}$$

$$0.36 \times 19.62 = 16 - 1 - 16\,K$$

$$7.0632 = 15 - 16\,K$$

$$16\,K = 7.9368$$

$$K = 0.496$$ **Ans**

9.3 **A pipe, 0.093 m² in area, carries a discharge of 283 liters/s of water. If it enlarges suddenly to 0.372 m² and if the pressure in the smaller section is 4.8 kPa, find (a) the head lost, (b) the pressure in the larger part, (c) the power needed to force the water through the enlargement.**

Solution

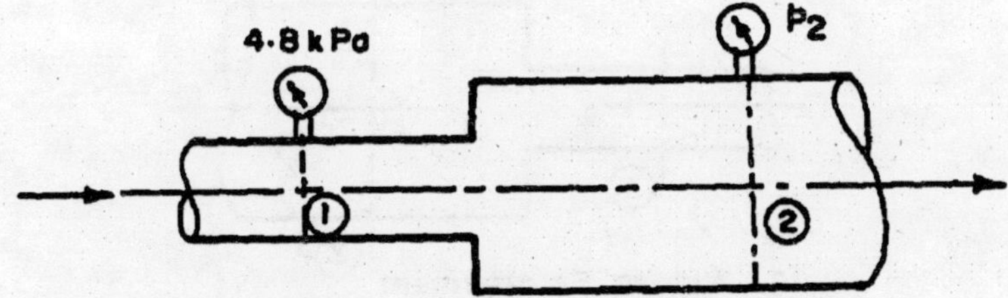

Fig. 9.20. Sudden enlargement of flow with pressure gauge P. 9.3.

$$Q = a_1 v_1 = a_2 v_2$$

$$283 \times 10^{-3} = 0.093\ v_1$$

$$v_1 = 3.043\ m/s$$

$$Q = a_2\ v_2$$

$$283 \times 10^{-3} = 0.372\ v_2$$

$$v_2 = 0.76\ m/s$$

(a) $$h_l = \frac{(V_1 - v_2)^2}{2g} = \frac{(3.043 - 0.76)^2}{2 \times 9.81}$$

$$= \frac{(2.2822)^2}{19.62} = 0.2654\ m$$ Ans

(b) Applying the Bernoulli's theorem between points (1) and (2)

$$\frac{p_1}{\rho g} + \frac{v_1^2}{2g} = \frac{p_2}{\rho g} + \frac{v_2^2}{2g} + h_l$$

$$\frac{4.8}{9.81} + \frac{(3.043)^2}{2 \times 9.81} = \frac{p_2}{\rho g} + \frac{(0.76)^2}{2 \times 9.81} + 0.2654$$

$$0.48929 + 0.47195 = \frac{p_2}{\rho g} + 0.02944 + 0.2654$$

$$\frac{p_2}{\rho g} = 0.96125 - 0.29484 = 0.6664$$

$p_2 = 0.6664 \times 1000 \times 9.81 = 6537.39$ Pa

$= 6.5374$ kPa Ans

(c) $P = \rho g\ Q\ h_l$

$= 1000 \times 9.81 \times 283 \times 10^{-3} \times 0.2654$

$= 737$ Watts Ans

9.4 A pipeline conveying a given quantity of water per second is suddenly changed in section. Find the ratio of the pipe diameter if the loss of head at the change in section is independent of the direction of flow. Assume a value of 0.61 for the coefficient of contraction.

(a) Sudden enlargement (b) Sudden contraction

Solution

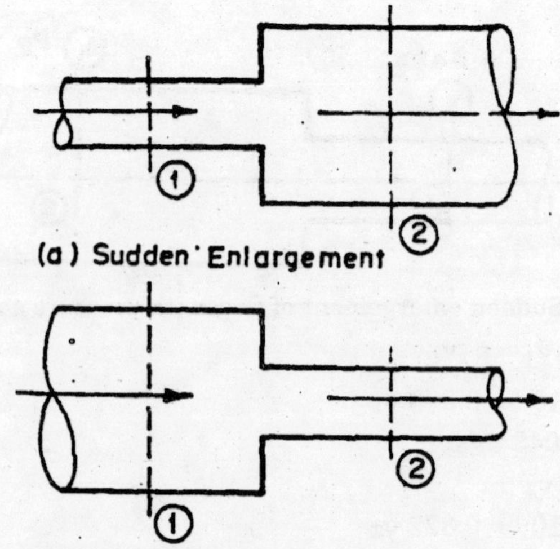

Fig. 9.21. Sudden enlargement and contraction P. 9.4.

Head loss due to sudden enlargement

$$h_e = \frac{(v_1 - v_2)^2}{2g} = \frac{v_1^2}{2g}\left(1 - \frac{v_2}{v_1}\right)^2$$

Head loss due to sudden contraction

$$h_c = \left(\frac{1}{c_c} - 1\right)^2 \frac{v_2^2}{2g}$$

As per the question, both the head losses are equal. The velocity v_1 in Fig. 9.27(a) is equal to velocity v_2 in Fig. 9.27(b). Therefore,

$$\left(1 - \frac{v_2}{v_1}\right)^2 = \left(\frac{1}{c_c} - 1\right)^2$$

or, $$1 - \frac{v_2}{v_1} = \frac{1}{c_c} - 1 = \frac{1}{0.61} - 1 = 0.6394$$

or, $$1 - 0.6394 = \frac{v_2}{v_1}$$

$$0.36065 = \frac{v_2}{v_1}$$

$$\frac{v_2}{v_1} = 0.36065 \qquad (3)$$

Using continuity equation

$$Q = a_1 v_1 = a_2 v_2$$

$$d_1^2 v_1 = d_2^2 v_2$$

or, $$\frac{v_2}{v_1} = \frac{d_1^2}{d_2^2}$$

equating (3) and (4)

$$\frac{d_1^2}{d_2^2} = 0.36065$$

or, $$\frac{d_1}{d_2} = 0.6$$

Therefore,

$$d_1 : d_2 = 1 : 1.665$$ Ans

9.5 A pipeline connecting two reservoirs having a difference of level of 6 m is 720 m long, and rises to a height of 3 m above the upper reservoir at a distance of 240 m from the entrance before falling to the lower reservoir. If the pipe is 1.2 m in diameter and the frictional coefficient λ = 0.04, what will be the discharge and the pressure at the highest point of the pipeline? Minor loss are neglected.

Solution

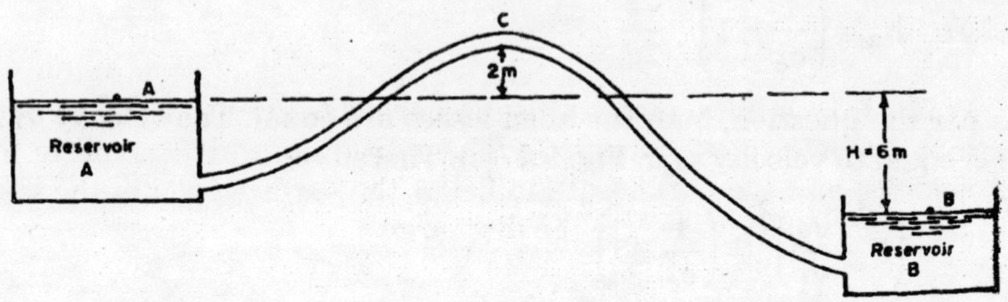

Fig. 9.22. Pipe line connecting two reservoirs P. 9.5.

Applying Bernoulli's equation between A and B on the free surface of the reservoirs where the velcities are zero and the pressures atmospheric. Take B as datum level, therefore, $Z_B = 0.$, and $Z_A = 6$ m.

$$\frac{p_A}{\rho g} + \frac{v_A^2}{2g} + Z_A = \frac{p_B}{\rho g} + \frac{v_B^2}{2g} + Z_B + \frac{\lambda l v^2}{2gd}$$

$$p_A = p_B \textit{ and } v_A = v_B = 0, \textit{ so that}$$

$$z_A - z_B = \frac{\lambda l v^2}{2gd} = \frac{\lambda l Q^2}{12.1\, d^5}$$

$$6 = \frac{0.04 \times 720\, Q^2}{12.1 \times (1.2)^5} = 0.9565\, Q^2$$

or, $$Q^2 = \frac{6}{0.9565} = 6.2726$$

$Q = 2.5045$ m³/s Ans

$$Q = \frac{\pi}{4}(1.2)^2 \times v \quad \therefore\ v = 2.214\ m/s$$ Ans

Apply Bernoulli's equation between A and C, taking datum level at A (i-e $z_4 = 0$) and $v_A = 0$

$$\frac{p_A}{\rho g} + \frac{v_A^2}{2g} + Z_A = \frac{p_c}{\rho g} + \frac{v_c^2}{2g} + Z_c + \frac{\lambda l_1 v^2}{2gd}$$

$$v_c = v, \quad v_A = 0, \; p_A = \textit{zero gauge pressure}$$

or, $$\frac{p_c}{\rho g} = -z_c - \frac{v^2}{2g}\left[\frac{1 + \lambda l_1}{d}\right]$$

$$= -2 - \frac{(2.214)^2}{2 \times 9.81}\left[1 + \frac{0.04 \times 240}{1.2}\right]$$

$$= -2 - 0.2499\,[1 + 8] = -2 - 2.295 = -\ \mathbf{4.249\ m\ of\ water}$$

$$p_c = -4.249 \times 9810 = -41682.69\ N/m^2$$

$$= -41.68\ kPa$$ **Ans**

9.6 **Water from a large reservoir is discharged to atmosphere through a 100 mm diameter pipe 450 m long. The entry from the reservoir to the pipe is sharp and the outlet is 12 m below the surface level in the reservoir. Taking $\lambda = 0.04$, calculate the discharge.**

Solution

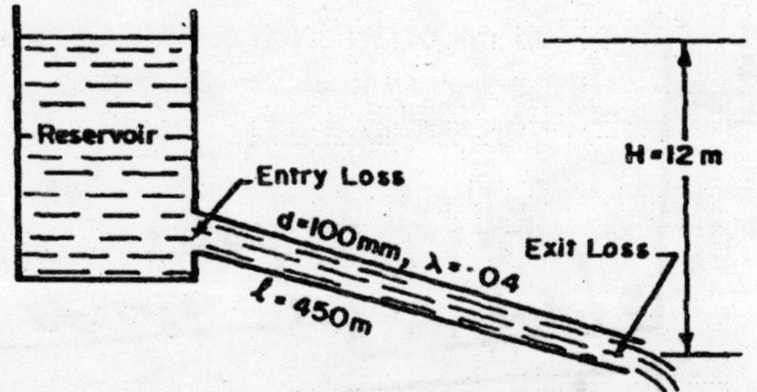

Fig. 9.23. Large reservoir discharge to atmosphere P. 9.6.

H = Entry loss + Friction in pipe + Exit loss (1)

$$\text{Entry loss} = 0.5\frac{v^2}{2g}$$

$$\text{Exit loss} = \frac{v^2}{2g}$$

$$\text{Friction loss in pipe} = \frac{\lambda l v_1^2}{2gd} = \frac{004 \times 450}{0.1} \times \frac{v^2}{2g} = 180\frac{v^2}{2g}$$

Therefore,

$$H = 12 = 0.5\frac{v^2}{2g} + 180\frac{v^2}{2g} + \frac{v^2}{2g}$$ [Using equation (1)]

$$12 = (0.5 + 180 + 1)\frac{v^2}{2g}$$

$$v^2 = \frac{12 \times 2 \times 9.81}{181.5} = 1.2972$$

$$v_1 = 1.1389 \approx 1.14\ m/s$$

$$Q = av = \frac{\pi}{4}(0.1)^2 \times 1.14 = 8.95 \times 10^{-3}\ m^3/s$$

$= 8.95$ liters/s Ans

9.7 Water is discharged from a reservoir into the atmosphere through a 40 m long. There is a sharp entrance to the pipe and the diameter is 50 mm for 15 m from the entrance. The pipe them enlarges suddenly to 75 mm diameter for the remainder of its length. Taking into account the loss of head at entry at the enlargement, and at the exit, calculate the difference of level between the surface of the reservoir and the pipe exit which will maintain a flow of 10 liters/s. Take λ as 0.02 for the 50 mm pipe and 0.025 for the 75 mm diameter pipe and $C_c = 0.58$. Calculate also the difference of level when minor losses are neglected.

Solution

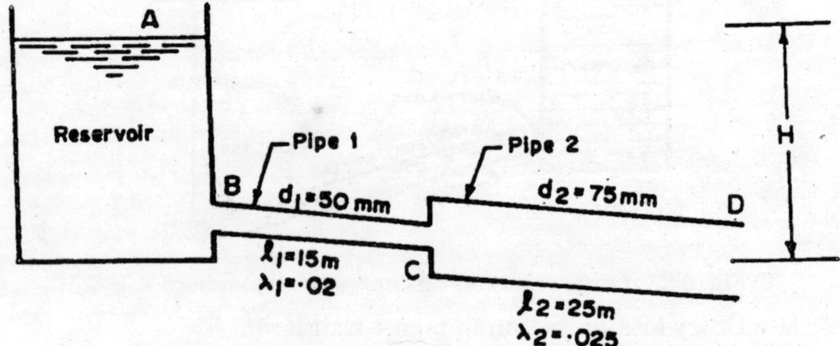

Fig. 9.24. Pipe in series discharge to atmosphere P. 9.7.

$$a_1 = \frac{\pi}{4}d_1^2 = \frac{\pi}{4}(0.05)^2 = 1.963 \times 10^{-3}\ m^2$$

$$a_2 = \frac{\pi}{4}d_2^2 = \frac{\pi}{4}(0.075)^2 = 4.41786 \times 10^{-3}\ m^2$$

$$Q\ a_1 v_1 = a_2 v_2$$

$$v_1 = \frac{Q}{a_1} = \frac{0.01}{1.963 \times 10^{-3}} = 5.094\ m/s$$

$$v_2 = \frac{Q}{a_2} = \frac{0.01}{4.41786 \times 10^{-3}} = 2.26\ m/s$$

H = Entry loss at B + Friction loss in pipe (1) + Sudden enlargement loss at C + Friction loss in pipe (2) + Exist loss at D. (1)

$$\text{Entry loss at B} = \left(\frac{1}{c_c} - 1\right)^2 \frac{v_1^2}{2g} = \left(\frac{1}{0.58} - 1\right)^2 \frac{(5.094)^2}{2 \times 9.81}$$

$$= 0.6935 \text{ m}$$

Friction loss in pipe (1) = $\frac{\lambda_1 l_1 v_1^2}{2gd_1} = \frac{0.02 \times 15 \times (5.094)^2}{2 \times 9.81 \times 0.05}$

$= 7.935$ m

Sudden enlargement loss at C $= \frac{(v_1 - v_2)^2}{2g}$

$= \frac{(5.094 - 2.26)^2}{2 \times 9.81} = 0.409$ m

Friction loss in pipe (2) $= \frac{\lambda_2 l_2 v_2^2}{2gd_2}$

$= \frac{0.025 \times 25 \times (2.26)^2}{2 \times 9.81 \times .075} = 2.17$ m

Exit loss at D $= \frac{v_2^2}{2g} = \frac{(2.26)^2}{2 \times 9.81} = 0.26$

Substituting in equation (1)

H = 0.6935 + 7.935 + 0.409 + 2.17 + 0.26 = 11.4675 m Ans

Neglecting minor losses

H = 7.935 + 2.17 = 10.105 m Ans

9.8 Two reservoirs are connected by a pipe line which is 150 mm in diameter for the first 6 m and 225 mm in diameter for the remaining 15 m. The water surface in the upper reservoirs is 6 m above that in the lower. The entrance and exit are sharp and the change of section is sudden. Tabulate the losses of head which occur and calculate the rate of flow. Friction coefficient λ is 0.04 for both pipes. Draw also the hydraulic gradient and total energy gradient.

Solution

Using continuity equation

$$Q = a_1 v_1 = a_2 v_2$$

$$v_1 = \frac{a_2}{a_1} v_2 = \left(\frac{225}{150}\right)^2 v_2 = 2.25\ v_2 \quad (1)$$

H = Entry loss at C + Friction losses in pipe (1) + Sudden enlargement loss at D + Friction losses in pipe (2) + Exit loss at E

Entry loss at C $= 0.5 \frac{v_1^2}{2g} = 0.5\,(2.25)^2 \frac{v_2^2}{2g}$

$$= 2.53 \frac{v_2^2}{2g}$$ [Using equation (1)]

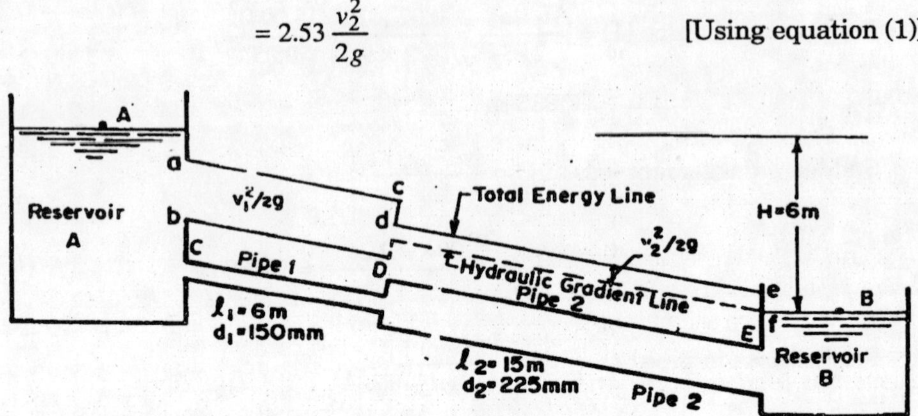

Fig. 9.25. Pipes in series P. 9.8.

$$\text{Friction loss in pipe (1)} = \frac{\lambda\, l_1\, v_1^2}{2gd_1} = \frac{0.04 \times 6}{15} \times \frac{v_1^2}{2g}$$

$$= 1.6 \frac{v_1^2}{2g} = 8.1 \frac{v_2^2}{2g}$$

$$\text{Sudden enlargement loss at D} = \frac{(v_1 - v_2)^2}{2g}$$

$$= \frac{(2.25\ v_2 - v_2)^2}{2g}$$ [Using equation (1)]

$$= 1.5625 \frac{v_2^2}{2g}$$

Friction loss in pipe (2)

$$= \frac{\lambda\, l_2\, v_2^2}{2gd_2} = \frac{0.04 \times 15}{0.225} \times \frac{v_2^2}{2g} = 2.67 \frac{v_2^2}{2g}$$

$$\text{Exit loss at E} = \frac{v_2^2}{2g}$$

Substituting in equation (2)

$$H = 6 = [2.53 + 8.1 + 1.5625 + 2.67 + 1] \frac{v_2^2}{2g}$$

$$= 15.859 \frac{v_2^2}{2g}$$

$$6 = \frac{15.859\, v_2^2}{2 \times 9.81}$$

or, $$v_2^2 = \frac{6}{0.808} = 7.422$$

$$v_2 = 2.724\ m/s$$

$$Q = a_2 v_2 = \frac{\pi}{4} d_2^2 v_2 = \frac{\pi}{4}(0.225)^2 \times 2.724 = 0.1083 \text{ m}^3/\text{s}$$

$$= 6.4985 \text{ m}^3/\text{min}$$ Ans

The TEL can be presented graphically, as shown in Fig. 9.37. At the entry to the pipeline there will be an entry loss ab, next follows the friction loss in the first pipe, which gives the sloping straight line bc. The loss at the sudden enlargement is shown as Cd. From d to e the friction loss in the larger pipe causes the total energy to fall slowly and the shock loss at exit ef brings the total energy gradient to the level of the surface in the lower reservoir.

The HGL is obtained by plotting the sum of $\left(\frac{P}{w} + z\right)$ and will therefore be a distance equal to the velocity head in the pipe below the TEL.

9.9 Two reservoirs are connected by three pipes laid in parallel, their diameters are respectively d, 2d and 3d and they are all the same length l. Assuming λ to be the same for all the pipes, what will be the discharge through the larger pipes if that through the smallest is 50 liters/s.

Solution

If Q_1, Q_2, and Q_3 are the discharges in the pipes of diameter d, 2d and 3d respectively and H is the difference of level of the reservoirs,

$$H = \frac{\lambda l Q_1^2}{12.1\, d_1^5} = \frac{\lambda l Q_2^2}{12.1\, d_2^5} = \frac{\lambda l Q_3^2}{12.1\, d_3^5}$$

or, $$H = \frac{\lambda l Q_1^2}{12.1\,(d)^5} = \frac{\lambda l Q_2^2}{12.1\,(2d)^5} = \frac{\lambda l Q_3^2}{12.1\,(3d)^5}$$

Putting $Q_1 = 0.05 \text{ m}^3/\text{s}$

$$\frac{Q_1^2}{d^5} = \frac{Q_2^2}{32\, d^5}$$

$$Q_2^2 = (0.05)^2 \times 32 = 0.014142$$

$Q_2 = 0.11892 \text{ m}^3/\text{s}$ Ans.

and $$\frac{Q_1^2}{d^5} = \frac{Q_3^2}{243\, d^5}$$

$$Q_3^2 = (0.05)^2 \times 243 = 0.6075$$

$Q_3 = 0.7794 \text{ m}^3/\text{s}$ Ans

9.10 (a) Two pipes, of 0.3 and 1.2 m diameters respectively and each 1940 m long, are arranged in parallel, they meet in a common junction at their upper and lower ends. With a discharge of 93.5 lit/s is available at their upstream junction, find the discharge passing through each parallel pipe of the loop.

(b) The above said pipes are now arranged in series and the water flows from 0.3 diameter pipe to 1.2 m diameter pipe. There is a reservoir A at the upstream end of the compound pipe line and the reservoir B at the downstream end. What should be the available head between the reservoirs so that discharge of the pipe system should be same viz 93.5 lit/s. Take $\lambda = 0.04$

Solution (a)

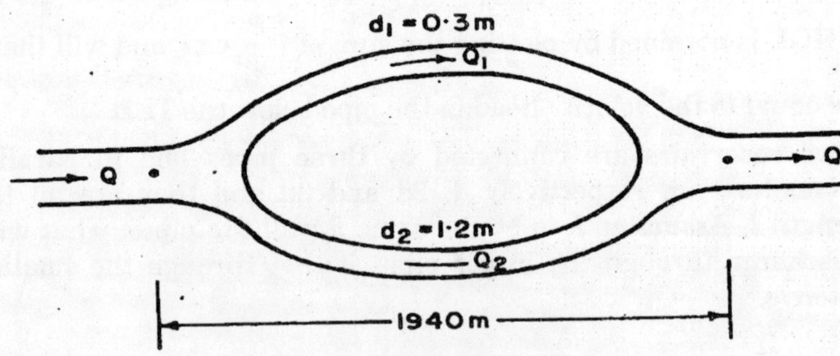

Fig. 9.26. Pipes in series P. 9.10.

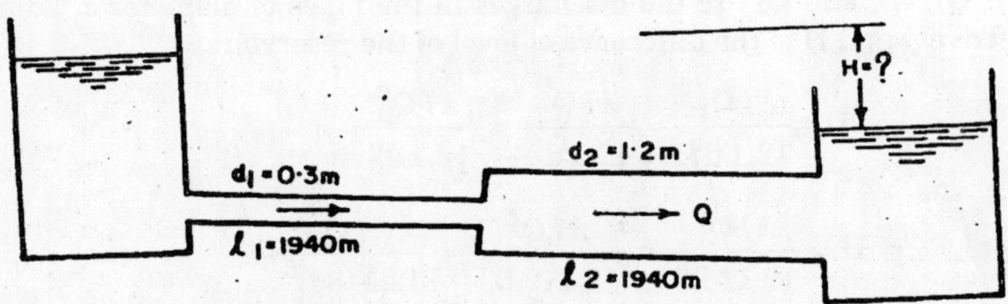

Fig. 9.27. Pipes in sereis P. 9.10.

$$h_{f_1} = h_{f_a}$$

$$= \frac{\lambda l Q_1^2}{12.1\, d_1^5} = \frac{\lambda l Q_2^2}{12.1\, d_2^5}$$

$$\frac{Q_1^2}{Q_2^2} = \left(\frac{d_1}{d_2}\right)^5$$

or, $$\frac{Q_2^2}{Q_1^2} = \left(\frac{d_2}{d_1}\right)^5 = \left(\frac{1.2}{0.3}\right)^5 = 1024$$

$$\frac{Q_2}{Q_1} = 32$$

$$Q_2 = 32\,Q_1 \quad (1)$$

Here, $Q = Q_1 + Q_2$

$93.5 = Q_1 + 32\,Q_1$ [Using equation (1)]

$93.5 = 33Q_1$

$Q_1 = 2.833$ liters/s

Therefore,

$Q_2 = 90.66$ liters/s [Using equation (1)]

(b) Fig. 9.49 Pipes in series

$$Q = Q_1 = Q_2$$

$$H = h_f + h_{f_2} = \frac{\lambda l_1 Q_1^2}{12.1\, d_1^5} = \frac{\lambda l_2 Q_2^2}{12.1\, d_2^5} = \frac{\lambda l Q^2}{12.1}\left[\frac{1}{d_1^5} + \frac{1}{d_2^5}\right]$$

$$= \frac{0.04 \times 1940 \times (0.0935)^2}{12.1}\left[\frac{1}{(0.3)^5} + \frac{1}{(1.2)^5}\right]$$

$$H = 0.056\ [411.52 + 0.4018] = 0.056\ [411.9245]$$

$= 23.067$ m Ans

9.11 A pipeline, with diameter 0.9 m and length 4000 m, connects two fresh water open reservoirs which have their water surfaces at elevations 75 m and 52 m above ordnance datum. In order to increase the rate of flow between the reservoirs by 30% it is decided to lay an additional 0.9 m diameter pipe line from the upper reservoir. The second pipe line is to be parallel to the original pipe line and is to be connected to the latter at same suitable point. Determine the point of connection, assuming that the friction factor $\lambda = 0.04$ for each pipe line. Neglect minor losses.

Solution:

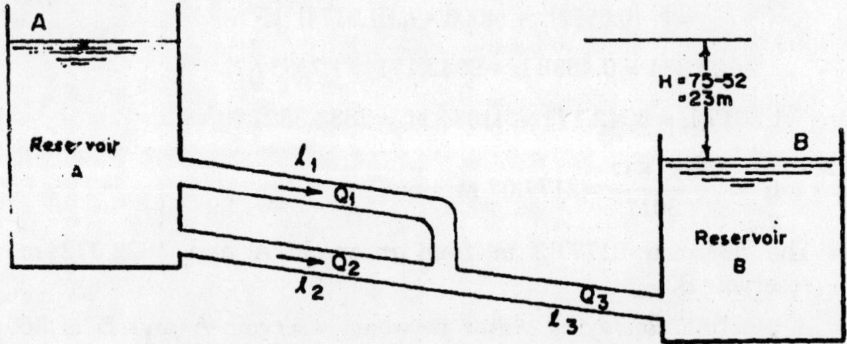

Fig. 9.28. Pipes in parallel upto centre line P. 9.11.

Case I

$$H = h_f = 23 = \frac{\lambda l Q^2}{12.1\, d^5}$$

$$23 = \frac{0.04 \times 4000 \times Q^2}{12.1 \times (0.9)^5}$$

$$Q^2 = 1.027$$

$$Q = 1.01345 \text{ m}^3/\text{s}$$

As per question $Q_3 = 1.3 \times 1.01345 = 1.3174 \text{m}^3/\text{s}$

Case II

$$Q = Q_1 + Q_2 = Q_3$$

Here, $Q_1 = Q_2$

Therefore,

$$Q_3 = 2\, Q_1$$

$$1.3174 = 2\, Q_1$$

$$Q_1 = \frac{1.3174}{2} = 0.6587\, m^3/s$$

For combined pipe the total head loss

$$H = h_{f_1} + h_{f_3}$$

$$23 = \frac{\lambda\, l_1\, Q_1^2}{12.1\, d_1^5} + \frac{\lambda\, l_3\, Q_3^2}{12.1\, d_3^5}$$

$$= \frac{0.04}{12.1\,(0.9)^5}\left[l_1\, Q_1^2 + l_3\, Q_3^2\right] \qquad [but\ d_1 = d_3,\ l_1 = l_2]$$

$$\frac{23 \times 12.1 \times (0.9)^5}{0.04} = \left[l_1\, Q_1^2 + (4000 - l_1)\, Q_3^2\right]$$

$$= \left[l_1\,(0.6587)^2 + (4000 - l_1)(1.3174)^2\right]$$

$$4108.334 = 0.4338\, l_1 + 6942.171 - 1.7355\, l_1$$

or, $1.3017\, l_1 = 6942.171 - 4108.334 = 2833.837$

$$\therefore \quad l_1 = \frac{2833.837}{1.3017} = 2177.02 \text{ m}$$

The distance 2177.02 m from reservoir A and 1822.972 m from reservoir B.

Ans

9.12 A pipeline conveying water between reservoir A and B is 300 mm diameter and 400 m long. The difference of head between the two surfaces is 5 m. Determine the flow rate if $\lambda = 0.02$.

It is required to increase the flow by 50 percent by duplicating a portion of the pipe. If the head and friction factor are unchanged and minor losses are ignored, find the length of the second pipe which is of the same diameter as the first.

Solution: Case I

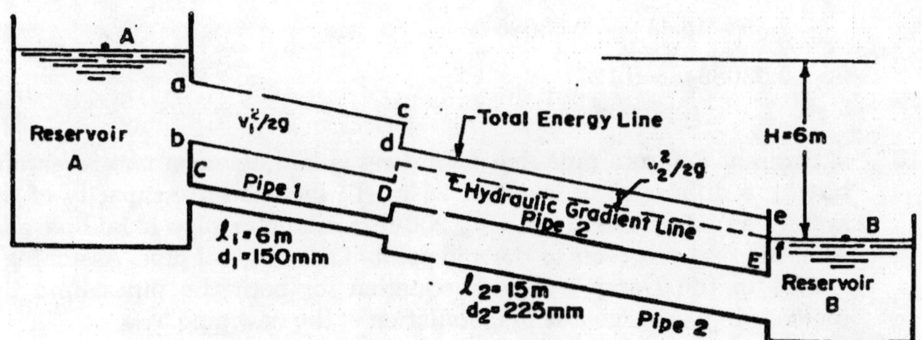

Fig. 9.29. Pipes in parallel with a unknown distance P. 9.12.

$$h_f = H = 5 = \frac{\lambda\, l\, v^2}{2gd}$$

$$5 = \frac{0.02 \times 400 \times v^2}{2 \times 9.81 \times 0.3}$$

$v^2 = 3.67875$

$v = 1.918$ m /s

$Q = a\,v = \frac{\pi}{4}(0.3)^2 \times 1.918 = 0.1355$ m³/s

$Q_1 = Q + 50\%$ of $Q = 1.5\,Q = 1.5 \times 0.1355$

$= 0.203$ m³/s (1)

Case II

$Q_1 = Q_2 + Q_3$

$Q_1 = Q_2 + Q_2$ [because $a_2 = a_3$]

$0.203 = 2Q_2$

$$Q_2 = \frac{0.203}{2} = 0.10168\ m^3/s \tag{2}$$

Here $l_1 + l_2 = 400$ (3)

$H = h_{f_1} + h_{f_2}$

$$5 = \frac{\lambda\, l_1\, Q_1^2}{12.1\, d_1^5} + \frac{\lambda\, l_2\, Q_2^2}{12.1\, d_2^5} = \frac{0.02}{12.1\, d_1^5}\left[l_1\, Q_1^2 + l_2\, Q_2^2\right]$$

$$= \frac{0.02}{12.1 \times (0.3)^5} \left[(400 - l_2)(0.203)^2 + l_2\,(0.10168)^2\right]$$

$$5 = 0.68\,[400 \times 0.0412 - 0.0410\,l_2 + 0.01033\,l_2]$$

$$\frac{5}{0.68} = 16.48 - 0.03086\,l_2$$

$$7.35 - 16.48 = -\ 0.03086\,l_2$$

$$0.03086\,l_2 = 9.127$$

$$l_2 = 295.75 \text{ m}$$ Ans

9.13 A Straight 900 mm pipe line 4 km long is laid between two reservoirs having a difference of levels of 23 m. To increase the capacity of the system an additional 2 km long 900 mm diameter pipe is laid parallel from the first reservoir to the mid point of the original pipe. Assuming λ = 0.04 in the Darcy-Weisbach equation for both the pipes, find the increase in discharge due to installation of the new pipe.

Solution

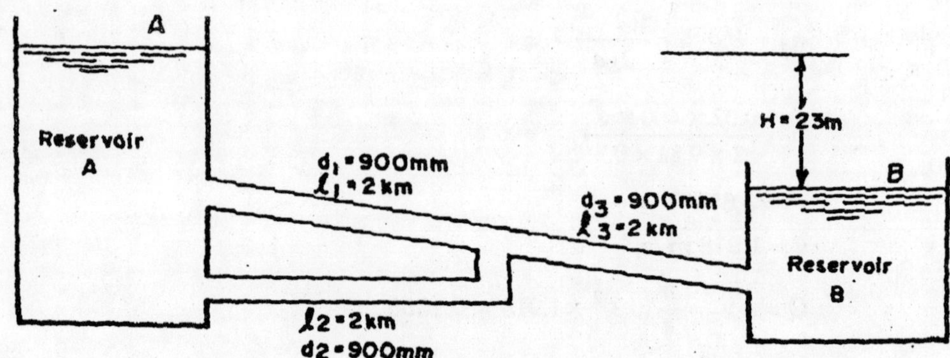

Fig. 9.30. Pipes in parallel upto centre line P. 9.13.

Case I

$$H = h_f = \frac{\lambda\, l\, Q^2}{12.1\, d^5}$$

$$23 = \frac{0.04 \times 4000 \times Q^2}{12.1 \times (0.9)^5}$$

$$Q^2 = 1.02708$$

$$Q = 1.01345 \text{ m}^3/\text{s}$$

Case II

$$h_{f_1} = h_{f_2}$$

$$\frac{\lambda\, l\, Q_1^2}{12.1\, d_1^5} = \frac{\lambda\, l\, Q_2^2}{12.1\, d_2^5}$$

but length and diameter are equal, therefore

$$Q_1 = Q_2$$

$$Q = Q_1 + Q_2 = 2\, Q_1 = 2\, Q_2 = Q_3$$

or, $Q_1 = 0.5\, Q$

and $Q_2 = 0.5\, Q$

$$H = 23 = h_{f_1} + h_{f_3}$$

$$= \frac{\lambda\, l_1\, Q_1^2}{12.1\, d_1^5} + \frac{\lambda\, l_3\, Q_3^2}{12.1\, d_3^5}$$

$$\frac{0.04 \times 2000 \times (.5Q)^2}{12.1 \times (0.9)^5} + \frac{0.04 \times 2000 \times Q^2}{12.1 \times (0.9)^5}$$

$$= 2.79918\, Q^2 + 11.1967\, Q^2$$

$$23 = 13.9959\, Q^2$$

$$Q^2 = 1.6433$$

$$Q = 1.2819 \text{ m}^3/\text{s}$$

Increase in discharge = 1.2819 – 1.01345 = 0.2684 m³/s **Ans**

9.14 Two pipe lines are connected to a large water reservoir, one is 150 mm in diameter and 300 m long with an outlet 4 m below the reservoir water level. The other is 200 mm in diameter and 600 m long. Both pipes have free discharge at outlet end. The total discharge is 50 liters/s. Assume λ = 0.08 for each pipe, find the difference of elevation between the reservoir water level and the outlet of 200 mm pipe. Neglect entrance loss.

Solution

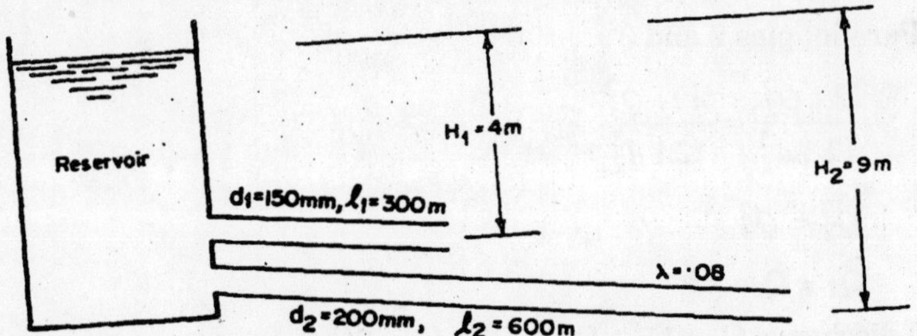

Fig. 9.31. Two pipes connected to a upstream reservoir P. 9.14.

The two pipes are in parallel condition the total head loss is equal to the loss of head in one pipe, therefore

$$H_1 = 4 = \frac{\lambda_1 l_1 Q_1^2}{12.1\, d_1^5} = \frac{0.08 \times 300 \times Q_1^2}{12.1 \times (0.15)^5}$$

$$Q_1^2 = \frac{4 \times 12.1 \times (0.15)^2}{0.08 \times 300} = 1.5314 \times 10^{-4}$$

$Q_1 = 0.012375\ m^3/s$

$Q = Q_1 + Q_2$

$0.05 = 0.0123 + Q_2$

$\therefore \quad Q_2 = 0.0376\ m^3/s$

Therefore,

$$H_2 = \frac{\lambda_2 l_2 Q_2^2}{12.1\, d_2^5} = \frac{0.08 \times 600 \times (0.0376)^2}{12.1 \times (0.2)^5}$$

$= 17.549 \approx 17.55$ m Ans

9.15 Two reservoir have a difference of level of 6 m and are connected by a pipe line which consists of a single 600 mm diameter pipe 3000 m long feeding a junction from which two pipes, each of 300 mm diameter and 3000 m long, lead in parallel to the lower reservoirs. If $\lambda = 0.04$ what will be the total discharge ?

Solution

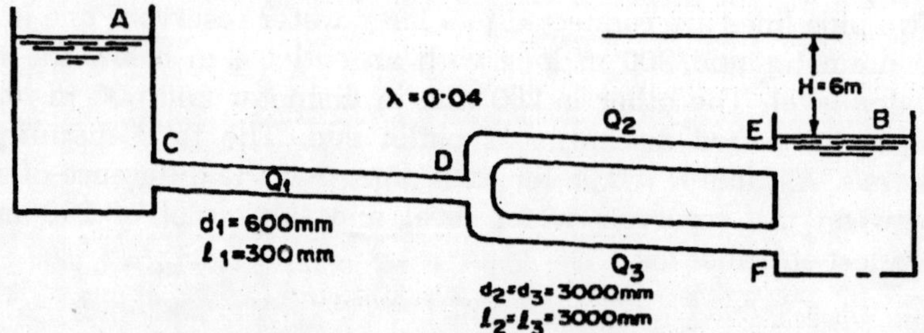

Fig. 9.32. Pipes in parallel after the centre point P. 9.15.

For Parallel pies 2 and 3

$$\frac{\lambda\, l\, Q_2^2}{12.1\, d_2^5} = \frac{\lambda\, l_2\, Q_3^2}{12.1\, d_2^5}$$

$$Q_2^2 = Q_3^2$$

$Q_2 = Q_3$

Total discharge $Q_1 = Q_2 + Q_3 = Q_2 + Q_2 = 2Q_2$

or, $\quad Q_2 = 0.5\, Q_1$

and $\quad Q_3 = 0.5\, Q_1$

Therefore,

$$H = 6 = \frac{\lambda l_1 Q_1^2}{12.1\, d_1^5} + \frac{\lambda l_2 Q_2^2}{12.1\, d_2^5}$$

$$6 = \frac{0.04 \times 3000 \times Q_1^2}{12.1 \times (0.6)^5} + \frac{0.04 \times 3000 \times (0.5Q_1)^2}{12.1 \times (0.3)^5}$$

$$6 = 127.538\, Q_1^2 + 1020.30\, Q_1^2$$

$$6 = 1147.842\, Q_1^2$$

$$Q_1^2 = 5.2272 \times 10^{-3}$$

$$Q_1 = 0.072299$$

$Q_1 = 0.0723$ m³/s **Ans**

9.16 A pumping plant forces water through a 600 mm diameter main, the friction head being 30 m. In order to reduce the power consumption, it is proposed to lay another main of appropriate diameter along side the existing one, so that the two pipes may work in parallel for the entire length and reduce the friction head of 9 m only. Find the diameter of the new main if, with the exception of the new main if, with the exception of the diameter it is similar to the existing one in every respect.

Solution

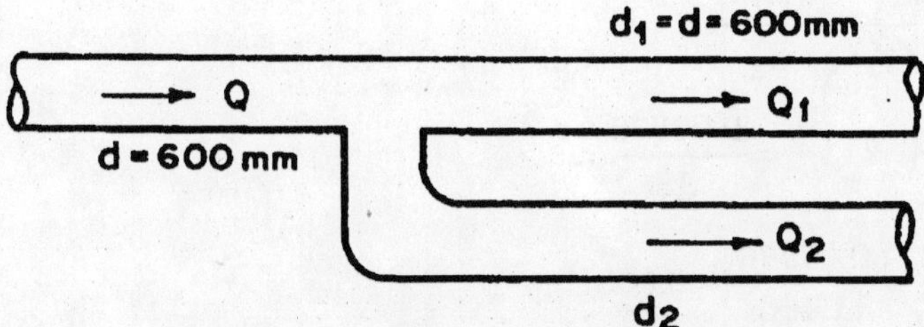

Fig. 9.33. Two pipes in parallel P. 9.16.

First case

Single pipe conveying the discharge

Head loss $h_f = \frac{\lambda\, l\, Q^2}{12.1\, d^5} = 30$ (1)

Second case

When the same total discharge is conveyed by two pipes in parallel

Let Q_1 be the discharge in the 600 mm pipe

Let Q_2 be the discharge in the second pipe

whose diameter d_2 is unknown.

In a parallel pipes flow

$$h_{f_1} = h_{f_2} = H$$

$$h_{f_1} = \frac{\lambda\, l\, Q_1^2}{12.1\, d_1^5} = \frac{\lambda\, l\, Q_1^2}{12.1\, d^5} = 9 \qquad (2)$$

and $$h_{f_2} = \frac{\lambda\, l\, Q_2^2}{12.1\, d_1^5} = 9 \qquad (3)$$

From equations (2) and (3)

$$\frac{Q_1^2}{Q^2} = \frac{9}{30} = 0.3$$

$$\frac{Q_1}{Q} = 0.5477$$

or, $$Q_1 = 0.5477\, Q \qquad (4)$$

From the figure $Q = Q_1 + Q_2$

$$Q_2 = Q - Q_1 = Q - 0.5477\, Q = 0.45227\, Q \qquad (5)$$

Therefore from equation (4) and (5)

$$\frac{Q_1}{Q_2} = \frac{0.5477\, Q}{0.45227\, Q} = 1.211$$

From equations (2) and (3)

$$\frac{Q_1^2}{Q_2^2} = \frac{(0.6)^5}{d_2^5}$$

$$(1.211)^2 = \frac{0.07776}{d_2^5}$$

or, $$d_2^5 = \frac{0.07776}{1.466} = 0.053$$

$$d_2 = (0.053)^{0.2} = 0.5558 = 555.8 \text{ mm}$$ Ans

9.17 Two pipes have a length *l* each. One of them has a diameter D, and the other a diameter d. If the pipes are arranged in parallel the loss of head when a total quantity of water Q flows through them is h, but, if the pipes are arranged in series and the same quantity Q flows through them, the loss of head is H; If D = 2.25 d, find the total ratio of H to h, neglecting secondary losses and assuming the pipe coefficient λ has a constant value. (b) Find the ratio of H to h for D = 2d, (c) Find the ratio of H to h for D = 2.5 d,(d) Find the ratio of H to h for D = 3 d

Solution

First case

When pipes are arranged in parallel

$$h = \frac{\lambda\, l\, Q_1^2}{12.1\, D^5} = \frac{\lambda\, l\, Q_2^2}{12.1\, d^5} \quad (1)$$

Therefore, $\dfrac{Q_1^2}{D^5} = \dfrac{Q_2^2}{d^5}$

or, $\left(\dfrac{Q_1}{Q_2}\right)^2 = \left(\dfrac{D}{d}\right)^5 = (2.25)^5 = 57.665$

Therefore,

$$\frac{Q_1}{Q_2} = 7.593 \approx 7.6$$

$$Q_1 = 7.6\, Q_2 \quad (2)$$

Total discharge for parallel flow pipes

$$Q = Q_1 + Q_2 = 7.6\, Q_2 + Q_2 = 8.6\, Q_2$$

or, $Q_2 = 0.11628\, Q$ (3)

and $Q_1 = 7.6 \times 0.11628\, Q = 0.88372\, Q$ (4)

Second case

When pipes are arranged in series

Total discharge $= Q = Q_1 = Q_2$

Total head loss $H = h_{f_1} + h_{f_2} = \dfrac{\lambda\, l\, Q^2}{12.1\, D^5} + \dfrac{\lambda\, l\, Q^2}{12.1\, d^5}$

$$= \frac{\lambda\, l\, Q^2}{12.1}\left[\frac{1}{D^5} + \frac{1}{d^5}\right] = \frac{\lambda\, l\, Q^2}{12.1}\left[\frac{1}{D^5} + \frac{(2.25)^2}{D^5}\right]$$

$$= \frac{\lambda\, l\, Q^2}{12.1\, D^5}[1 + 57.665]$$

$$H = \frac{58.665\, \lambda\, l\, Q^2}{12.1\, D} \quad (5)$$

From equations (1) and (5)

$$\frac{H}{h} = \frac{58.665\, \lambda\, l\, Q^2}{12.1\, D^5} \Big/ \frac{\lambda\, l\, Q_1^2}{12.1\, D^5} = 58.665\left(\frac{Q}{Q_1}\right)^2$$

$$= 58.665\left(\frac{Q}{0.88372\, Q}\right)^2 \quad \text{[Form equation (4)]}$$

$= 75.11898 \approx 75.12$ Ans

(b) Same procedure, $\frac{H}{h} = 45.7069$ Ans.

(c) Same procedure , $\frac{H}{h} = 119.6346$ Ans.

(d) Same procedure, $\frac{H}{h} = 277.085$ Ans.

9.18 **A single jet pelton wheel is fed by three equal parallel pipes each of length l, diameter D and having a coefficient of friction , connected to the machine through a junction box. Determine the diameter of the theoretically best nozzle through which the joint flow in the pipes should be discharged from the junction box to give maximum power output. Calculate the nozzle diameter and power of the jet if for each pipe D = 200 mm, l = 1500 m, λ = 0.032 and the gross head available is 225 m.**

Solution (a)

$$h_f = \frac{H}{3} \text{ and } h_p = \frac{2H}{3}$$

$$h_f = \frac{3h_p}{2} \times \frac{1}{3} = \frac{h_p}{2} \quad (1)$$

$$h_f = \frac{\lambda l V^2}{2gD}$$

also $h_p = \frac{v^2}{2g}$

Putting in equation (1)

$$\frac{\lambda l V^2}{2gD} = \frac{1}{2}\left(\frac{v^2}{2g}\right)$$

$$\frac{2\lambda l}{D} = \frac{v^2}{V^2} \quad (2)$$

For continuity of flow, since there are three supply pipes

$$3\left(\frac{\pi}{4}D^2\right)V = \frac{\pi}{4}d^2 v$$

$$\frac{v}{V} = 3\frac{D^2}{d^2}$$

Substituting in equation (2)

$$\frac{2\lambda l}{D} = \frac{9\,D^4}{d^4}$$

$$d = \left(\frac{9\,D^5}{2\,\lambda\,l}\right)^{1/4}$$ Ans

(b) Putting D = 200 mm, l = 1500 m, λ = 0.032

$$d = \left(\frac{9\,(2)^5}{2 \times 0.032 \times 1500}\right)^{0.25} = \left(3 \times 10^{-5}\right)^{0.25}$$

$= 0.074 = 74$ mm Ans

(c) Power head $h_p = \frac{2}{3}H$

$$= \frac{2}{3} \times 225 = 150 \text{ m}$$

$$v = \sqrt{2gh_p}$$

$$= \sqrt{2 \times 9.81 \times 150}$$

$$= \sqrt{2943} = 54.25 \text{ m/s}$$

Power $P = \rho\, g\, Q\, h_p$

$$s = 1000 \times 9.81 \times \frac{\pi}{4}(0.074)^2 \times 54.25 \times 150$$

$= 343.327$ kW Ans

9.19 A pipeline is 1800 m long and 375 mm in diameter and the supply head at inlet is 240 m. A nozzle with an effective diameter of 50 mm is fitted at the discharge end and has a coefficient of velocity of 0.972. If λ for the pipe is 0.02, calculate (a) the velocity of the jet, (b) the discharge, (c) the power of the jet

Solution (a)

Let D = Pipe diameter

v = Pipe velocity

V = Jet velocity

d = Jet diameter

Using continuity equation

$$Q = \frac{\pi}{4}d^2 V = \frac{\pi}{4}D^2 v$$

$$v = \frac{d^2}{D^2} V = 0.0178\, V \tag{1}$$

Head loss in pipe friction

$$h_f = \frac{\lambda\, l\, v^2}{2gD} = \frac{0.02 \times 1800 \times (0.0178\, V)^2}{2 \times 9.81 \times 0.375}$$

$$= 1.5464 \times 10^{-3}\, V^2$$

Head at nozzle = H – h_f

$$= 240 - 1.5464 \times 10^{-3}\, V^2$$

Jet velocity $V = c_v \sqrt{2g \times Head\ at\ nozzle}$

$$V^2 = c_v^2\, 2g\left(240 - 1.5464 \times 10^{-3}\, v^2\right)$$

$$V^2 = (0.972)^2 \times 2 \times 9.81 \left[240 - 1.5464 \times 10^{-3}\, V^2\right]$$

$$V^2 = 4448.798 - 0.02865\, V^2$$

$$1.02865\, V^2 = 4448.798$$

$$V^2 = 4324.88$$

V = 65.76 m/s Ans

(b) Discharge $= \frac{\pi}{4} d^2 V$

$$= \frac{\pi}{4}(0.05)^2 \times 65.76$$

$$= 0.1291\ \text{m}^3/\text{s}$$

(c) Power of jet $= \frac{\rho\, g\, Q\, V^2}{2g}$

$$= \frac{1000 \times 0.1291 \times (65.76)^2}{2}$$

= 279230.2589 watt

= 279.23 kW Ans

9.20 Determine the conditions for maximum transmission of power through a pipe assuming loss of head by friction only. Calculate the maximum power available at the far end of a hydraulic pipeline 5000 m long and 200 mm in diameter when water at 70 bar pressure is fed in at the near end. Take the friction coefficient $\lambda = 0.032$

Solution: For theory see text

$$p = \rho g\, H$$

$$70 \times 10^5 = 1000 \times 9.81 \times H$$

$$H = \frac{7000}{9.81} = 713.5\ m \text{ of water}$$

For maximum power

$$h_f = \frac{H}{3} = \frac{713.55}{3} = 237.85\ m$$

Power head $h_f = \frac{2}{3} H = \frac{2}{3} \times 713.55 = 475.7\ m$

$$h_f = \frac{\lambda\, l\, v^2}{2gd}$$

$$237.85 = \frac{0.032 \times 5000 \times v^2}{2 \times 9.81 \times 0.2}$$

$$v^2 = 5.833$$

$$v = 2.4152 \text{ m/s}$$

Maximum power = $\rho\, g\, Q\, h_p$

$$= 1000 \times 9.81 \times \frac{\pi}{4} (0.2)^2 \times 2.4152 \times 4750.7$$

$$= 354083.04 \text{ Watt} = 354.08 \text{ kW}$$ Ans

9.21 Find the least number of 150 mm diameter pipes required to transmit 170 kW to a machine 3200 m from the power station, if the efficiency of transmission is to be 92 percent and λ = 0.03. The feed pressure is 48 bar

Solution

If P_1 = input pressure; P_2 = pressure at machine

Therefore, Input head H = $\frac{p_1}{\rho g}$

Power head $h_p = \frac{p_2}{\rho g}$

$$\eta = \frac{h_p}{H} = \frac{p_2}{p_1} = 0.92$$

$$p_2 = 0.92\ p_1 = 0.92 \times 48 = 44.2\ bar$$

Loss of head due to friction $h_f = \frac{\Delta p}{\rho g} = \frac{\lambda\, l\, v^2}{2gd}$

$$\frac{\Delta p}{\rho g} = \frac{\lambda \, l \, v^2}{2gd}$$

$$\frac{(4800 - 4420)}{1000 \times 9.81} = \frac{0.03 \times 3200 \times v^2}{2 \times 9.81 \times (0.15)}$$

$$v^2 = \frac{2 \times 0.15 \times 380 \times 1000}{0.03 \times 3200 \times 1000} = 1.1875$$

v = 1.08972 m/s

Power transmitted by one 150 mm pipe

$$= p_2 Q = 4420 \times \frac{\pi}{4}(0.15)^2 \times 1.089 = 85.116 \, kW$$

Total power required = 170 kW

Number of pipes required $= \dfrac{170}{85.116} = 1.9972 \approx 2$ Ans

EXERCISES

A. THEORY

9.1 What is `Reynold's number' ? What value of it indicates a stream line flow?

9.2 Describe Reynold's experiment to demonstrate the existence and the characteristics of laminar and turbulent flows in real fluids.

9.3 What is a Reynold's number? Comment on the criterion to establish the prevalent type of flow and thus define critical velocity.

9.4 What are upper and lower critical velocities? Describe the upper and lower critical Reynolds number.

9.5 Describe Reynold's experiment for pipe flow and spell out the characteristics of laminar flow and turbulent flow as demonstrated by the experiment. What is the significance of lower critical velocity here?

9.6 Describe Reynold's experiment to show the existence of Laminar and Turbulent flows.

9.7 Derive Darcy-Weisbach expression for loss of head due to friction in a pipe.

9.8 Derive an expression for loss of head due to sudden expansion in a pipe.

9.9 Distinguish between Laminar and turbulent flow in pipes.

9.10 Explain the terms hydraulic gradient and total energy line.

9.11 How will you determine the loss of head due to friction in pipes by using (a) chezy's formula (b) Manning's formula, and (c) Hazen William's formula.

9.12 Obtain the expression for the head loss due to sudden enlargement and sudden contraction of pipes.

9.13 Derive an expression for the loss of head due to friction in a pipeline in terms of the velocity head, assuming that the frictional resistance per unit area of pipe wall is proportional to the square of the mean velocity of flow.

9.14 Define the following: (i) Hydraulic gradient, (ii) Energy gradient, (iii) Hydraulic mean depth, and (iv) critical velocity.

9.15 Derive on expression for the head loss due to sudden contraction in a pipe flow.

9.16 What are the effect of contractions and expansion in a flow through pipes? Explain briefly.

9.17 Discuss various minor losses involved in a pipe flow system. Derive an expression for head loss due to sudden enlargement.

9.18 Explain the hydraulic gradient line with the help of diagram and explain its significance.

9.19 Show that the loss of head due to sudden expansion in pipe line is a function of velocity head.

9.20 Differentiate clearly between majar and minor losses. Why the minor losses in a parallel pipes flow are neglected ?

9.21 Explain hydraulic gradeline and its utility in solving pipe flow problems. Sketch the hydraulic grade line for a pipe system in which water is pumped from lower level sump to an elevated service reservoir.

9.22 Obtain an expression for the loss of energy head in a sudden expansion of a conduit carrying a fluid. State clearly the assumptions involved. Sketch the layout and mark on it the energy line hydraulic grade line.

9.23 For the two cases of flow in a sudden contraction and sudden enlargement in a pipe line, draw the flow pattern, piezometric grade line and total energy line.

9.24 Derive the equation for discharge and total head loss for a compound pipe.

9.25 What is a compound pipe? How would you determine the equivalent size of a compound pipe ?

9.26 Define pipe, pipes lines and long pipe lines. Where are they used.

9.27 Why are the pipes connected in parallel ?

9.28 Explain the terms (a) Pipes in parallel; (b) Equivalent pipe (c) Equivalent size of the pipe.

9.29 For a pipe flow. show that Total head = Summation of total losses

9.30 What are branching pipes ? What is three reservoir system ?

9.31 What are pipe networks ? How will you solve the pipe network problem ?

9.32 What is a loop method ? Explain briefly.

9.33 What is a nodal method ? Explain briefly.

9.34 Explain pipe network methods which method is preferred generally ?

9.35 Explain with reasons the choice of loop or nodal method for the analysis of complex networks.

9.36 Explain Hardy Cross method used in solving problems on pipe network. What is its modified form suggested by Cornish.

9.37 Obtain the condition for maximum efficiency in transmission of power through a pipeline.

9.38 A pipe line of a given length and diameter has a difference of head H between the input and output ends. Show by means of sketch diagrams how the head lost in friction h_f, the power head h_p, the power P and efficiency of power transmission η vary with the discharge Q.

9.39 Determine the conditions for maximum transmission of power through a pipe assuming loss of head by friction only.

9.40 A hose pipe of diameter D and length l is fitted with a nozzle of diameter d at the end. If the losses in the nozzle and pipe entry are 10 per cent of the friction loss in the pipe, show that, for any fixed head and pipe, the force of the jet from the nozzle is a maximum when

$$\frac{d}{D} = \left(\frac{D}{1.1\lambda l}\right)^{1/4}$$

9.41 A turbine is fed by four pipes laid in parallel and connected at the discharge end by a short connecting pipe to the turbine nozzle. Neglecting losses in the connecting pipe, prove that the power of the jet is a maximum when

$$d = \left(\frac{D^5}{\lambda}\right)^{1/4}$$

9.42 Show that if ηp is the efficiency of power transmission through a pipeline conveying water to a reaction turbine having efficiency ηt and H is the head from reservoir level to the turbine tailrace, then the power developed by the turbine will be proportional to

$$Ht\,\eta p\,\sqrt{1-\lambda_p}\,.\,H\sqrt{H}$$

9.43 Prove that the head loss due to friction is equal to one third of the total head at inlet for maximum power transmission through pipes or nozzles.

9.44 Prove that in case of transmission of power by means of water under pressure, the maximum power is transmitted when the frictional loss of head in the pipe is one third of the total head supplied.

9.45 Draw the curves for power transmission through pipes and discuss the curves.

9.46 Prove the formula

$$v = \frac{\sqrt{2gH}}{\sqrt{1 + \frac{\lambda l a^2}{DA^2}}}$$

where symbols has a usual meanings.

9.47 Prove that the efficiency of power transmission through nozzle is given by

$$h = \frac{1}{\frac{1 + \lambda 1_2^a}{DA^2}}$$

9.48 Prove the formula for maximum power transmitted through nozzle is given by

$$h_f = \frac{h_P}{2}$$

9.49 Prove that the maximum efficiency of transmission through the pipes is given by

$$Ht = 66.67\ \% = \frac{h_P}{H}$$

9.50 What is meant by water hammer ?

9.51 How the water hammer classified ?

9.52 What are causes of water hammer ?

9.53 What is meant by oil hammer or steam hammer in rigid pipes

9.54 Describe the rigid water column theory. what are the assumptions made ?

9.55 Describe the elastic pipe theory.

9.56 Describe the pressure wave theory.

9.57 What are the water hammer controls.

9.58 What is meant by hydraulic transient ?

B.UNSOLVED PROBLEMS

9.1 At a sudden enlargement of flow of a water main from 240 mm to 480 mm diameter, the hydraulic gradient rises by 10 mm. Estimate the rate of flow [32.75 liters/s].

9.2 A horizontal pipe of 100 mm diameter and carrying 45 liters of water per second suddenly enlarges to one of 150 mm diameter. If the intensity of pressure just before sudden change in section is 68.6 kPa, find the intensity of pressure, immediately after the junction of the two portions, at a point where the flow is steady [76.81 kPa].

9.3 Water flows vertically downward through a 150 mm diameter pipe with a velocity of 2.4 m/s. The pipe suddenly enlarges to 300 mm in diameter. Find the loss of head and also the loss of head if the flow is reversed, the coefficient of contraction being 0.62 [0.165 m, 0.110 m].

9.4 A pipeline carrying 0.236 m^3/s is reduced suddenly from 450 mm to 300 mm diameter. Calculate the change in (a) total energy head; (b) pressure energy head. Take $c_c = 0.67$. [0.135 m fall; 0.589 m fall].

9.5 A pipe carrying 0.056 m^3/s suddenly changes diameter (a) 200 mm to 150 mm, (b) 300 mm to 150 mm, (c) 450 mm to 150 mm. Find the loss of head and the pressure difference across the contraction in each case, $c_c = 0.62$. [0.19 m, 0.54 m; 0.19 m, 0.673 m, 0.19 m, 0.699 m].

9.6 A diaphragm with a 150 mm diameter opening is increased in a 300 mm diameter pipe. If the velocity of flow is 0.6 m/s, calculate the loss of head due to the diaphragm assuming that $c_c = 0.64$. [0.507 m].

9.7 Water flows down a straight inclined pipe. At a point A, 45 m above datum, the section of the pipe suddenly doubles in area. The pressure in the smaller pipe at A is 860 kPa and the velocity of flow in the larger pipe is 2.4 m/s. Find the pressure in kPa at a point B which is 2.5 m above datum.

9.8 Water flows at the rate of 0.028 m^3 through 150 mm diameter pipe which reduces suddenly to 100 mm diameter. Find the loss of head and the difference of pressure head between tapping points either side of the contraction if the coefficient of contraction is 0.62. Assume the full loss due to the enlargement of the streamlines to be included, but make no allowance for pipe friction [0.258 m, 0.812 m].

9.9 Using the Chezy formula, find the loss of head in a circular pipe 120 m long and 75 mm diameter when the velocity of flow is 4.8 m/s. Take c = 54. 6 [49.5 m].

9.10 Water is discharged from a reservoir through a pipe 1200 m long which is 400 mm diameter for 600 m and 250 mm diameter for the rest. Calculate the flow, taking only friction into account, if the far end of the pipe is 30 m below the reservoir level; λ for the 400 mm pipe is 0.016 and for the 250 mm pipe is 0.024. [0.151 m^3/s]

9.11 Two vessels in which the difference of surface levels is maintained constant at 2.4 m are connected by a 75 mm diameter pipeline 15 m long. If the frictional coefficient λ may be taken as 0.032, determine the volume rate of flow through the pipe. [11.98 liters/s].

9.12 The difference in surface levels in two reservoirs connected by a syphon is 7.5 m. The diameter of the syphon is 300 mm and its length 750 m. The friction coefficient λ is 0.0256. If air is liberated from solution when the absolute pressure is less than 1.2 m of water, what will be the maximum length of the inlet leg of the syphon to run full, if the highest point is 5.4 m above the surface level in the upper reservoir. What will be the discharge [373 m, 105.5 liters/s].

9.13 A pipeline 30 m long connects two tanks which have a difference of water level of 12 m. The first 10 m of pipeline from the upper tank is 40 mm diameter and next 20 m in 60 mm diameter. At the change in section a valve is fitted. Calculate the rate of flow when the valve is fully opened assuming that its resistance is negligible and that λ for both pipes is 0.0216. In order to restrict the valve the valve is then partially closed. If K for the valve is now 5.6 find the percentage reduction in flow. [0.00738 m^3/s, 25.8%]

9.14 A pipe, 50 mm in diameter, is 6 m long and the velocity of water in the pipe is 2.4 m/s. What loss of head would be saved if the central 1.8 m of pipe were replaced by a 75 mm diameter pipe, the change of section being sudden? Take $\lambda = 0.04$, $C_c = 0.62$. [0.164 m]

9.15 A pipe of 50 mm diameter and 45 m long is connected to a large tank, the entrance to the pipe being 3 m below the surface. The lower end of the pipe which is 6 m below the upper end is joined to a horizontal pipe of 100 mm diameter and 75 m long which discharges to atmosphere. Calculate the discharge taking into account the sudden enlargement and entry losses. Take $\lambda = 0.032$ for both pipes. [4.66 liters/s]

9.16 Two reservoirs are connected by two pipelines in parallel. Their diameters are 300 mm and 600 mm and their respective lengths are 1500 m and 3000 m. If the value of λ for the smaller pipe is 0.032 and for the larger is 0.024, what will be the discharge from the larger when the smaller pipe is delivering 0.056 m^3/s. [0.256 m^3/s]

9.17 A pipe line conveying water consists of 750 m of an 200 mm diameter pipe with another 200 mm pipe running in parallel for the last 300 m. A gate valve is installed in one of the branches. If the total flow is reduced by 5 percent by partially closing the valve, find the loss of head across the valve in terms of the new velocity in that pipe, taking $\lambda = 0.028$ [6.43 v^2]

9.18 Two pipelines of equal length, 3000 m, are laid in parallel between two reservoirs whose difference of level is 15 m. If their diameters are respectively 300 mm and 600 mm and if the frictional resistance is given by

$$h_f = \frac{\lambda \, l \, v^{1.8}}{d^{1.2}}$$

Where $\lambda = 0.04$ what will be the total discharge ? [0.158 m^3/s]

9.19 Two reservoirs having a difference of level of 18 m are connected by a straight pipe, 450 mm in diameter and 900 m long, which connects to the upper reservoir 2.4 m below the surface, and to the lower one 1.5 m below the surface. If $\lambda = 0.032$, calculate the flow in the pipe and the pressure at a point 300 m from the discharge end.

If a 300 mm main is connected in parallel with 450 mm main over the last 300 m, what will be the flow from the upper reservoir and the pressure at the branch point? [0.369 m³/s; 15.75 kPa, 0.4 m³/s, - 5.9 kPa].

9.20 Two open water reservoirs are connected by 800 m of 0.6 m internal diameter pipeline and the resulting gravitational flow, caused by the difference in levels of the reservoirs, is 0.5 m³/s. For the pipeline, λ = 0.04 in the Darcy's equation.

Determine the total flow between the reservoirs when an additional pipeline, 0.6 m internal diameter and 550 m in length, has been laid from the higher reservoir, parallel to the first pipeline. For the second pipeline λ = 0.02 in the Darcy's equation. Losses at entry to and exit from the pipelines may be assumed to be negligible. [0.762 m³/s]

9.21 Two reservoirs having a difference of surface level of 30 m are connected by two parallel pipes each 1 km long and diameter 300 mm and 400 mm. A length of 200 m of 300 mm pipe is removed for repair and the total flow is diverted over this length through the 400 mm pipe. Calculate the percentage reduction in discharge resulting from the diversion. Consider friction losses only and take λ = 0.032. [27%]

9.22 Three pipes : 300 m long 300 mm diameter, 150 m long 200 mm diameter and 200 m long 250 mm diameter are connected in series in the same order between two reservoirs. The Darcy's-Weisbach friction factor (λ) for the three pipes are 0.018, 0.02 and 0.019 respectively. Determine the rate of flow, if the difference in elevation of water levels in the two reservoirs is 15 m. Account for all losses. Contractions and expansions are sudden. Determine also all the losses in flow between the reservoirs. [0.1059 m³/s,

$[h_{ent} = 5.1\,Q^2 = 0.0572\,m \quad h_{f_1} = 183.6\,Q^2 = 2.0590\,m$

$h_{cont} = 25.82\,Q^2 = 0.2896\,m \quad h_f = 774.6\ \ Q^2 = 8.6870\,m$

$h_{exp} = 6.69\,Q^2 = 0.0750\,m \quad h_{f_3} = 321.51\ Q^2 = 3.6057\,m$

$h_{exit} = 21.15\,Q^2 = 0.2372]$

9.23 A pipeline of 600 mm diameter is 1500 m long. To augment the discharge another line of the same diameter is introduced parallel to the first in the second half of this length. Neglect minor losses, find the increase in discharge if λ = 0.04 and the head above the inlet is 30 m. [26.6 %]

9.24 What is the maximum rate in power at which energy can be transmitted through a 150 mm pipe line 3000 m long supplied with water at 84 bar pressure if λ = 0.04 ? [260 kW]

9.25 Water is supplied through a pipe of 50 mm bore 60 m long under a head of 16.5 m and is discharged through a nozzle which has a coefficient of velocity of 0.98. Calculate the bore of the nozzle to give discharge of 220 liters/min if $\lambda = 0.024$. [17.9 mm]

9.26 The head of water at the inlet to a pipe 180 m long and 75 mm diameter is 30 m and λ for the main is 0.04. What diameter of nozzle fitted to the end of the pipe will give maximum power and what power will be available? [20.2 mm, 1.26 kW]

9.27 The gross head from reservoir level to the four nozzles of a twin runner Pelton wheel is 300 m. The length of the supply pipeline is 3000 m with $\lambda = 0.020$. The turbine develops 10.5 MW with an efficiency of 85 percent and the efficiency of power transmission in the pipeline is 90 percent. If the coefficient of velocity for the nozzle is 0.98, calculate the diameters of (a) the pipeline, and (b) of each jet [1.31 m, 0.146 m]

9.28 A pump feeds two hoses each of which 45 m long and is fitted with a nozzle. Each nozzle has a coefficient of velocity of 0.97 and discharges a 37.5 mm diameter jet of water at 24 m/s when the nozzle is at the same level as the pump. If the power lost in overcoming friction in the hoses is not to exceed 20 percent of the hydraulic power available at the inlet end of the hoses, calculate (a) the diameter of the hoses, taking $\lambda = 0.028$, (b) the power required to drive the pump if its efficiency is 70 percent and its draws it water supply from a level 3 m below the nozzle. [98.5 mm, 31.1 kW]

9.29 The power of a jet driving a pelton wheel is 1350 kW. The gross head is 270 m and the water is conveyed from the reservoir to the nozzle through a pipe 1500 m long, for which the friction coefficient is 0.02. The velocity coefficient for the nozzle is 0.97. Losses in the pipeline, including the nozzle, are a percent of the gross head: Find (a) the diameter of the pipe, and (b) the diameter of the jet [0.62 m, 101 mm].

10

Laminar Viscous Flow

10.1 INTRODUCTION

When the Euler equations were developed (Chapter 6) it was assumed that there were no frictional forces acting upon the fluid element. In any fluid such forces must exist and the Euler equations cannot give an accurate description of the behaviour of the fluid especially if its viscosity is large.

There are many problems of fluid flow where the viscosity should be taken into account. Even a fluid with small viscosity may cause considerable shear stress on an immersed body.

More complex laminar viscous flows are multidimensional and multidirectional. The normal stresses are no longer simply the pressure but also involve the linear strain rates and viscosity. The normal stresses are different in different directions. More than one component of the shear stress is present, and the expressions for the shear stresses involve sums of two velocity gradients. These more complex viscous flows are analyzed using a set of general momentum differential equations called the Navier-Stoke equations, which contain complete expressions for the normal and shear stresses. Except for the simplest flow situations, the exact solution of the Navier-Stokes equations is so formidable that numerical integration using a high speed digital computer become necessary. Before the advent of computers, however, many approximate analytical solutions of the Navier-Stokes equations were developed. They involved some very sophisticated mathematics coupled with great insight into the physics of the flow.

This chapter briefly introduces the general differential equations of motion for a laminar viscous flow, the Navier-Stokes equations.

10.2. HAGEN-POISEUILLE FLOW

Probably the best known exact solution to the Navier-Stokes equation is straight circular tube of constant cross section. This type of flow is commonly called Hagen-Poiseuille flow. It is named in honour of J.L. Poiseuille (1799-1869), a French Physician and G.H.L. Hagen (1797-1884), a German hydraulic engineer. Poiseuille was interested in blood flow through capillaries and deduced experimentally the resistance laws, for laminar flow through circular tubes. Hagen's investigation of flow in tubes was also experimental. It was actually after the work of Hagen and Poiseuille that the theoretical results presented in this section were determined, but their names are commonly associated with the solution of this problem.

The derivation of the Hagen-Poiseuille equation for laminar flow in straight round pipes is based on the following two assumptions :

(1) the viscous property of fluid following Newton's law of viscosity, that is $\tau = \mu \dfrac{du}{dr}$ and

(2) there is no relative motion between fluid particles and solid boundaries, that is, no slip of fluid particles at the solid boundary.

Consider the flow through a horizontal circular tube of radius R as is shown in Fig. 10.1.

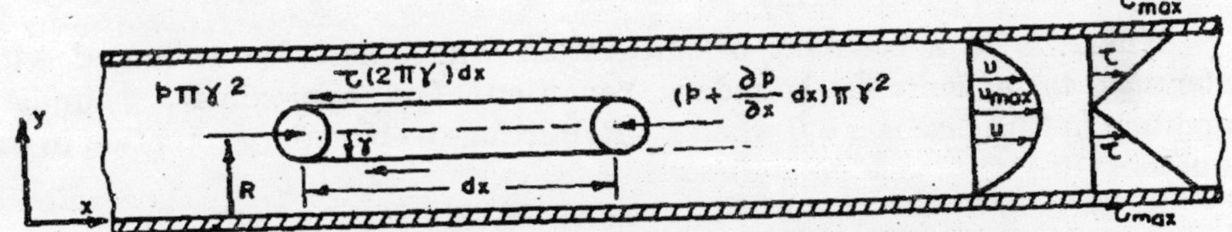

Fig. 10.1. Laminar flow in circular pipe-Hagen Poiseuille flow.

The momentum equation for the flow of fluid through the chosen free body (in the absence of grawitational force) is obtained from the conditions of equilibrium in the x direction which requires that shearing force acting on the circumferential area of the cylinder be equal to the pressure force acting on the cross section of the cylinder. At any radial position, the equilibrium condition can be expressed by

Pressure force = Shear force

$$p \pi r^2 - \left(p + \frac{\partial p}{\partial x} dx\right) \pi r^2 = \tau 2\pi r\, dx$$

or, $$p \pi r^2 - p \pi r^2 - \frac{\partial p}{\partial x} dx\ \pi r^2 = \tau 2\pi r\, dx$$

or, $$-\frac{\partial p}{\partial x} dx\ \pi r^2 = \tau\, 2\, \pi\, r\, dx$$

or, $$\left(-\frac{\partial p}{\partial x}\right) \frac{r}{2} = \tau$$

or, $$\tau = \left(-\frac{\partial p}{\partial x}\right) \frac{r}{2} \tag{10.1}$$

In accordance with the first assumption $\tau = \mu \frac{du}{dy}$. Since $y = R - r$ it follows that $dy = -dr$, Newton's law of viscosity then becomes

$$\tau = -\mu \frac{du}{dr} \tag{10.2}$$

By Combining equation (10.1) and (10.2)

$$-\mu \frac{du}{dr} = -\frac{\partial p}{\partial x}\frac{r}{2}$$

$$du = \frac{r}{2\mu}\left(\frac{\partial p}{\partial x}\right) dr \tag{10.3}$$

Integrating with respect to r

$$u = \left(\frac{\partial p}{\partial x}\right)\frac{r^2}{4\mu} + A \tag{10.4}$$

Where A is a constant of integration which may be evaluated with reference to a suitable boundary condition. The approximate boundary condition in this case is u = 0 when r = R. Putting the boundary condition in the equation (10.4).

$$0 = \left(\frac{\partial p}{\partial x}\right)\frac{R^2}{4\mu} + A$$

$$A = \left(-\frac{\partial p}{\partial x}\right)\frac{R^2}{4\mu} \tag{10.5}$$

Substituting into equation (10.4)

$$u = \left(\frac{\partial p}{\partial x}\right)\frac{r^2}{4\mu} - \left(\frac{\partial p}{\partial x}\right)\frac{R^2}{4\mu}$$

$$\frac{1}{4\mu}\left(-\frac{\partial p}{\partial x}\right)\left(R^2 - r^2\right) \tag{10.6}$$

Thus, at any cross section the velocity distribution is parabolic.

The maximum velocity u_{max} occurs at the centre of the tube. The maximum velocity is obtained by putting r = o in equation (10.6)

$$u_{\max} = \left(-\frac{\partial p}{\partial x}\right)\frac{R^2}{4\mu} \tag{10.7}$$

To obtain a relationship between the volume rate of flow, Q, passing through the tube and the pressure gradient, we consider the flow through the differential, washer shaped ring of Fig. 10.2. Since u is constant on this ring the volume rate of flow through the differential area is given by

$$dQ = u\left(2\pi r\right) dr$$

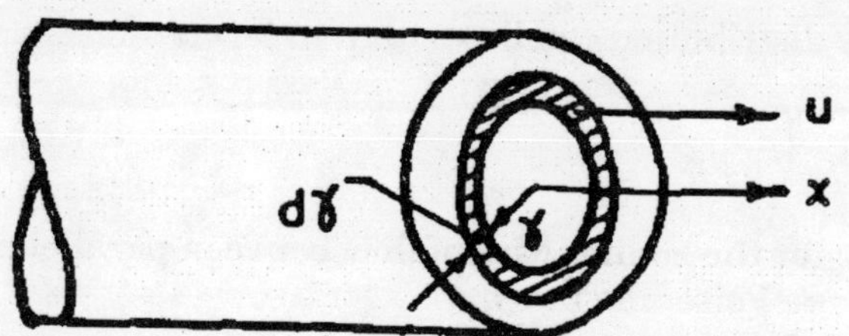

Fig. 10.2. Flow through differential annular ring.

and therefore

$$Q = 2\pi \int_0^R u\, r\, dr$$

Substituting the value of u from equation (10.7)

$$Q = 2\,\pi \int_o^R \frac{1}{4\mu}\left(-\frac{\partial p}{\partial x}\right)\left(R^2 - r^2\right) r\, dr$$

$$= \frac{2\,\pi}{4\,\mu}\left(-\frac{\partial p}{\partial x}\right) \int_o^R \left(R^2\, r - r^3\right) dr$$

$$= \frac{\pi}{2\,\mu}\left(-\frac{\partial p}{\partial x}\right)\left[\frac{R^2\, r^2}{2} - \frac{r^4}{4}\right]_o^R = \frac{\pi}{2\,\mu}\left(-\frac{\partial p}{\partial x}\right)\left[\frac{R^4}{2} - \frac{R^4}{4}\right]$$

$$= \frac{\pi\, R^4}{2\,\mu}\left(-\frac{\partial p}{\partial x}\right)\left[\frac{1}{2} - \frac{1}{4}\right] = \frac{\pi\, R^4}{8\,\mu}\left(-\frac{\partial p}{\partial x}\right) \qquad (10.8)$$

The average velocity can be obtained by using the relation

Q = Area × Average velocity = $\pi\ R^2\ u_{av}$

or, $$u_{av} = \frac{Q}{\pi\, R^2}$$

Substituting the value of Q from equation (10.8)

$$u_{av} = \frac{1}{\pi\, R^2}\left[\frac{\pi\, R^4}{8\,\mu}\left(-\frac{\partial p}{\partial x}\right)\right]$$

$$u_{av} = \frac{R^2}{8\,\mu}\left(-\frac{\partial p}{\partial x}\right) \qquad (10.9)$$

or, $$u_{av} = \frac{1}{2}\left[\frac{R^2}{4\,\mu}\left(-\frac{\partial p}{\partial x}\right)\right] = \frac{1}{2}\left(u_{max}\right)$$ [From equation (10.7)]

$$u_{av} = \frac{u_{max}}{2}$$

or, $$u_{max} = 2\ u_{av} \qquad (10.10)$$

The velocity distribution can be written in terms of u_{max} as

$$\frac{u_{av}}{u_{max}} = 1 - \left(\frac{r}{R}\right)^2 \tag{10.11}$$

As was true for the similar case of flow between parallel plates (some times referred to as plane Poiseuille flow).

The relationship can be expressed in terms of pressure drop, dp, which occurs over a length $l = x_2 - x_1$, along the tube, since

$$\frac{dp}{l} = -\frac{\partial p}{\partial x} \tag{10.12}$$

Substituting in equation (10.7)

$$Q = \frac{\pi R^4 \, dp}{8 \mu l} \tag{10.13}$$

For a given pressure drop per unit length, the volume rate of flow is inversely proportional to the viscosity and proportional to the tube radius to the fourth power. Equation (10.12) is commonly called Poiseuille's law.

Substituting the equation (10.11) into equation (10.8) gives

$$u_{av} = \frac{R^2}{8 \mu} \frac{dp}{l}$$

or, $$dp = \frac{8 \mu u_{av} l}{R^2}$$

$$dp = \frac{32 \mu u_{av} l}{D^2} \tag{10.14}$$

Dividing by ρg in the equation (10.14) gives

$$\frac{dp}{\rho g} = \frac{32 \mu u_{av} l}{\rho g D^2}$$

or, $$h_f = \frac{32 \mu u_{av} l}{\rho g D^2} \tag{10.15}$$

We know that the head loss due to friction is given by

$$h_f = \frac{\lambda l u_{av}^2}{2g D}$$

therefore,

$$\frac{\lambda l u_{av}^2}{2g D} = \frac{32 \mu u_{av} l}{\rho g D^2}$$

or, $$\lambda = \frac{64 \mu}{\rho u_{av} D}$$

$$\lambda = \frac{64}{R_e} \tag{10.16}$$

The shear stress distribution can be obtained very easily by using either equation (10.1) or equation (10.2)

$$\tau = \left(-\frac{\partial p}{\partial x}\right)\frac{r}{2} = \frac{dp}{l}\left(\frac{r}{2}\right) = \frac{8\,\mu\,u_{av}\,l}{R^2\,l}\left(\frac{r}{2}\right)$$

$$= \frac{4\,\mu\,u_{av}\,r}{R^2} \tag{10.17}$$

The generalized shear distribution is obtained by using equation (10.17). The maximum shear stress at the wall of the pipe is obtained by putting r = R, therefore,

$$\tau = \tau_{max} = \frac{4\,\mu\,u_{av}\,R}{R^2} = \frac{4\mu\,u_{av}}{R} = \frac{8\,\mu\,u_{av}}{D} \tag{10.18}$$

$$= \frac{8\mu}{D}\left(\frac{u_{max}}{2}\right) = \frac{4\mu u_{max}}{D}$$

or, $$\frac{\tau_{max}}{\tau_{max}} = \frac{4\,\mu}{D} \tag{10.19}$$

Numerous experiments performed to substantiate the theoretical results show that the theory and experiment are in agreement for the laminar flow of Newtonian fluids in circular tube or pipes. As long as the Reynold's number has a value less than the critical Reynold's number, the laminar flow can be maintained. According to experiments the critical Reynold's number is approximately equal to 2300. For $R_e > R_{e_{critical}}$ the flow pattern is entirely different and becomes turbulent.

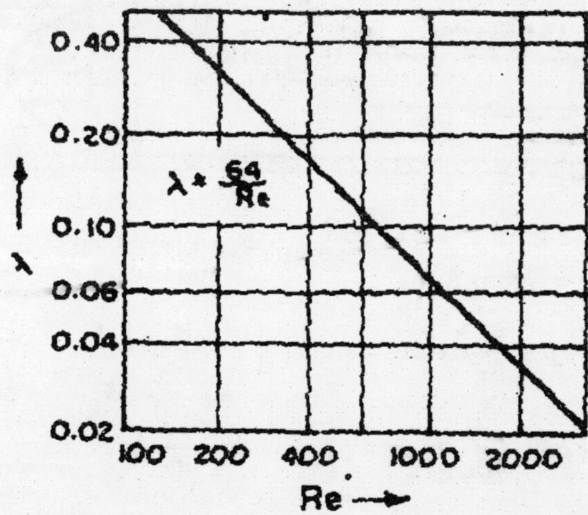

Fig. 10.3. Resistance coefficient λ v/s Reynold's number Re.

The laminar equation for pressure loss in pipes, equation (10.18), is in excellent agreement with experimental results for the laminar range, as seen from Fig. 10.3 which reproduces experimental points measured by G. Hagen.

The Navier-Stokes equation (10) in cylindrical coordinates subject to the following assumptions steady flow, axially symmetric flow, uniform flow, flow parallel to pipe wall and continuity equation are used to simplify the above derivations for velocity distribution

$$v_z = \frac{1}{4\,\mu}\left(\frac{\partial p}{\partial z}\right) r^2 + A \ln\, r + B$$

with the proper boundary conditions at r = 0, it follows that A = 0 at the centre of the tube. At the wall (r = R) the velocity must be zero so that

$$B = \frac{1}{4\,\mu}\left(-\frac{\partial p}{\partial z}\right) R^2$$

and the velocity distribution becomes

$$v_z = \frac{1}{4\,\mu}\left(-\frac{\partial p}{\partial z}\right)\left(R^2 - r^2\right)$$

This is same equation (10.5) which we obtained earlier.

10.3. PLANE POISEUILLE FLOW

Consider the steady flow as shown in Fig. 10.4. which is assumed to occur between parallel fixed plates. The coordinate system is defined such that the x direction is parallel to the plate orientation and y direction is perpendicular to it. The only forces acting on a thin lamina (dx x dy) of unit width are presure force and the shear force. The equation of motion for the lamina is the x direction is as follows:

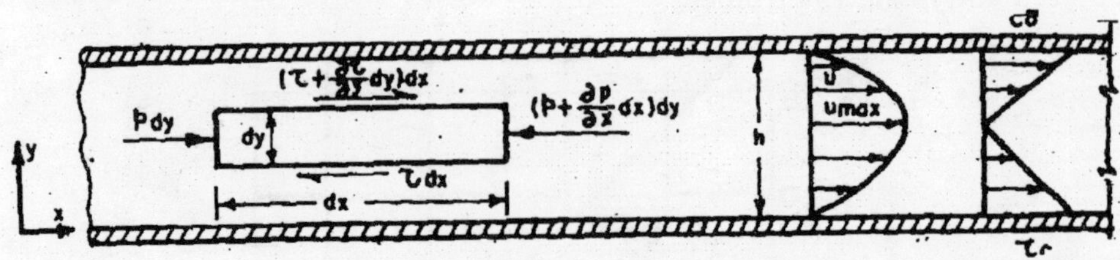

Fig. 10.4. Flow between fixed parallel plates.

Pressure force = Shear force

$$pdy - \left(p + \frac{\partial p}{\partial x}\,dx\right) dy = \tau\, d\, x - \left(\tau + \frac{\partial \tau}{\partial y} dy\right) dx$$

or, $$pdy - pdy - \frac{\partial p}{\partial x}\,dx\,dy = \tau\,dx - \tau\,dx - \frac{\partial \tau}{\partial y}\,dx\,dy$$

or, $$-\frac{\partial p}{\partial x}\,dx\,dy = -\frac{\partial \tau}{\partial y}\,dx\,dy$$

or, $$\frac{\partial \tau}{\partial y} = \frac{\partial p}{\partial x} \quad (10.20)$$

When Newton's law of viscosity $\tau = \mu \dfrac{du}{dy}$ is substituted in the equation (10.20)

$$\frac{\partial}{\partial y}\left(\mu \frac{\partial u}{\partial y}\right) = \frac{\partial p}{\partial x}$$

or, $$\mu \frac{\partial^2 u}{\partial y^2} = \frac{\partial p}{\partial x} \quad (10.21)$$

Integrating twice the equation (10.21) with respect to y we get

$$u = \frac{1}{2\,\mu}\left(\frac{\partial p}{\partial x}\right) y^2 + Ay + B \quad (10.22)$$

The two constants A and B can be determined from the boundary conditions. In this case

$$u = 0 \;\; at \;\; y = 0 \quad (i)$$

and $$u = 0 \;\; at \;\; y = h$$

The boundary conditions are used in the basic governing equation (10.22)

$$u = \frac{1}{2\,\mu}\left(\frac{\partial p}{\partial x}\right) y^2 + Ay + B$$

From first boundary condition

$$0 = 0 + 0 + B$$

Therefore $\quad B = 0$

From second boundary condition

$$0 = \frac{1}{2\,\mu}\left(\frac{\partial p}{\partial x}\right) h^2 + Ah + 0$$

$$Ah = \frac{1}{2\,\mu}\left(-\frac{\partial p}{\partial x}\right) h^2$$

$$A = \frac{h}{2\,\mu}\left(-\frac{\partial p}{\partial x}\right)$$

Substituting the values of the constant A and B in equation (10.22)

$$u = \frac{1}{2\mu}\left(\frac{\partial p}{\partial x}\right) y^2 - \frac{hy}{2\mu}\left(\frac{\partial p}{\partial x}\right)$$

$$= \frac{1}{2\mu}\left(\frac{\partial p}{\partial x}\right)\left[y^2 - hy\right]$$

$$= \frac{1}{2\mu}\left(-\frac{\partial p}{\partial x}\right)\left[hy - y^2\right] \tag{10.23}$$

The velocity distribution curve for the laminar flow between parallel flat plates is a parabola. The negative sign for the pressure gradient indicates that there is a drop in pressure in the direction of flow.

The maximum velocity which occur at the mid point between the plates, can be obtained by putting $y = \frac{h}{2}$ in equation (10.23).

$$u_{max} = \frac{h^2}{8\mu}\left(-\frac{\partial p}{\partial x}\right) \tag{10.24}$$

The average velocity can be calculated as follows

$$u_{av} = \frac{1}{h}\int_o^h u\, dy \qquad = \frac{1}{h}\int_o^h \frac{1}{2\mu}\left(-\frac{\partial p}{\partial x}\right)\left(hy - y^2\right)dy$$

$$= \frac{1}{2\mu h}\left(-\frac{\partial p}{\partial x}\right)\int_o^h \left(hy - y^2\right)dy = \frac{1}{2\,\mu h}\left(-\frac{\partial p}{\partial x}\right)\left[\frac{hy^2}{2} - \frac{y^3}{3}\right]_o^h$$

$$= \frac{1}{2\,\mu h}\left(-\frac{\partial p}{\partial x}\right)\left[\frac{h^3}{2} - \frac{h^3}{3}\right] = \frac{h^3}{2\,\mu h}\left(-\frac{\partial p}{\partial x}\right)\left[\frac{1}{2} - \frac{1}{3}\right]$$

$$= \frac{h^2}{12\,\mu}\left(-\frac{\partial p}{\partial x}\right) \tag{10.25}$$

The discharge Q of the laminar flow can be obtained very easily

$$Q = h \times \text{width of the plate} \times u_{av}$$

$$q = \frac{Q}{\text{width of the plate}} = h\, u_{av} = \frac{h^3}{12\,\mu}\left(-\frac{\partial p}{\partial x}\right)$$

or, $$q = \frac{h^3}{12\mu}\left(-\frac{\partial p}{\partial x}\right) \tag{10.26}$$

The relation between the average velocity and maximum velocity can be obtained by combining the equation (10.24) and (10.25)

$$u_{av} = \frac{8}{12} \times \frac{h^2}{8\mu}\left(-\frac{\partial p}{\partial x}\right)$$

or, $$u_{av} = \frac{2}{3}u_{max}$$

or, $$u_{max} = \frac{3}{2}u_{av} \tag{10.27}$$

The pressure drop can be obtained from the equation (10.25)

$$\left(-\frac{\partial p}{\partial x}\right) = \frac{12\mu\, u_{av}}{h^2}$$

or, $$\partial p = -\frac{12\,\mu\, u_{av}}{h^2}$$

Integrating between the points 1 and 2

$$p_1 - p_2 = \frac{-12\mu\, u_{av}}{h^2}\,(x_1 - x_2)$$

$$\Delta p = \frac{12\,\mu\, u_{av}}{h^2}\,(x_2 - x_1)$$

$$\Delta p = \frac{12\,\mu\, u_{av}\, l}{h^2} \tag{10.28}$$

The loss of head can be written as

or, $$h_f = \frac{\Delta p}{\rho g} = \frac{12\,\mu\, u_{av}\, l}{\rho\, g\, h^2} \tag{10.29}$$

The shear stress distribution in the flowing fluid across any section of the parallel plates may be determined by using the Newton's law of viscosity

$$\tau = \mu\frac{du}{dy} = \mu\left[\frac{1}{2\mu}\left(-\frac{\partial p}{\partial x}\right)[h-2y]\right] = \frac{1}{2}\left(-\frac{\partial p}{\partial x}\right)(h-2y)$$

$$= (-\partial p/\partial x)\,[h/2 - y] \tag{10.30}$$

The maximum shear stress or at the wall can be determined by putting y = 0 in the equation (10.30)

$$\tau_o = \tau_{\max} = \left(-\frac{\partial p}{\partial x}\right)\frac{h}{2} \tag{10.31}$$

or, $$\tau_o = \frac{12\,\mu\, u_{av}}{h^2}\times\frac{h}{2} = \frac{6\,\mu\, u_{av}}{h} \tag{10.32}$$

$$= \frac{6\,\mu}{h}\times\frac{2}{3}\,u_{\max} = \frac{4\,\mu\, u_{\max}}{h} \tag{10.33}$$

The shearing stress is zero at the centre line of the parallel flat plates $\left(y = \frac{h}{2}\right)$ and increases linearly with the distance from the centre line, attaining a maximum value at the boundary or at plate wall (y = o and y = h). The linear variation of shearing stress is shown in Fig. 10.4.

The local friction coefficient c_f is defined by

$$(\tau)_{y=0} = \frac{1}{2}\rho\, u_{av}^2\; C_f$$

$$C_f = \frac{(\tau)_{y=0}}{\frac{1}{2}\rho u_{av}^2} = \frac{6\,\mu\, u_{av}}{h} / \frac{1}{2}\rho\, u_{av}^2 \qquad \text{[Using equation 11.30]}$$

$$= \frac{12\,\nu}{u_{av}\,h} = \frac{12}{R_e}$$

Where $R_e = \frac{u_{av}\,h}{\nu}$ is the Reynold's number of the flow based on the average velocity and the channel height.

The flow will be laminar if the Reynold's number, $R_e = \frac{\rho\, u\, h}{\mu}$ remains below about 1400. For flow with larger Reynold's number the flow becomes turbulent and the preceding analysis is not valid since the flow field is complex, three dimensional and unsteady.

10.4 COUTTE FLOW

Another simple parallel plate flow can be developed by fixing one plate and letting the other plate move with a constant velocity U, as is illustrated in Fig. 10.6.

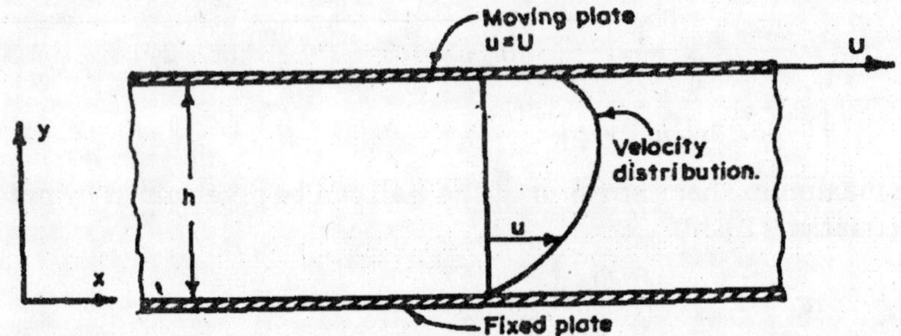

Fig. 10.5. Coutte flow

The Navier-Stoke's equations reduce to the same form as those in the preceding section and the solution for the velocity distribution is given by the equation (10.36). However, for the moving plate problem the boundary conditions are different. The origin of the coordinate system at the bottom plate and designate the distance between the two plates as h (Fig. 10.5). The two constants A and B in equation (10.36) can be determined from the boundary conditions

u = 0 at y = 0

and u = U at y = h

From first boundary condition

0 = 0 + 0 + B

therefore, B = 0

From second boundary condition

$$U = \frac{1}{2\mu}\left(\frac{\partial p}{\partial x}\right)h^2 + Ah$$

$$A = \frac{U}{h} - \frac{h}{2\mu}\left(\frac{\partial p}{\partial x}\right) \tag{10.34}$$

Substituting in equation (10.36)

$$u = \frac{1}{2\mu}\left(\frac{\partial p}{\partial x}\right)y^2 + U\frac{y}{h} - \frac{hy}{2\mu}\left(\frac{\partial p}{\partial x}\right)$$

$$u = \frac{Uy}{h} - \frac{h^2}{2\mu}\left(\frac{\partial p}{\partial x}\right)\frac{y}{h}\left(1 - \frac{y}{h}\right) \tag{10.35}$$

Equation (10.33) is the equation for the generalized Coutte flow.

In non dimensional form the velocity distribution in equation (10.35) becomes

$$\frac{u}{U} = \frac{y}{h} - \frac{h^2}{2\mu U}\left(\frac{\partial p}{\partial x}\right)\frac{y}{h}\left(1 - \frac{y}{h}\right) \tag{10.36}$$

The actual velocity profile will depend on the dimensionless parameter or dimensionless pressure gradient.

$$\alpha = \frac{h^2}{2\mu U}\left(-\frac{\partial p}{\partial x}\right) \tag{10.37}$$

The dimensionless velocity is plotted in Fig. 10.6 as a function of the distance from the stationary plate for various dimensionless pressure gradient α. In particular, one can observe that, when the pressure gradient along the passage is zero, the velocity distribution between the two plate is linear. From the velocity distribution it can also be seen that, for positive dimensionless pressure gradient $\alpha > 1$, there is a reverse flow in the vicinity of the stationary wall.

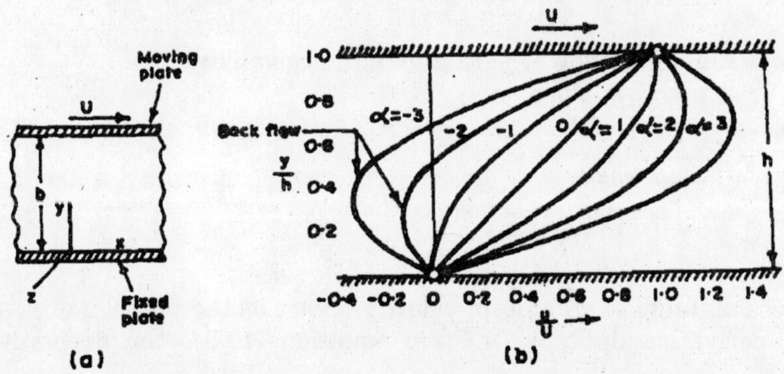

Fig. 10.6. Velocity distribution as a function of parameter α

The simplest type of Coutte flow is one for which the pressure gradient is zero, that is, the fluid motion is caused by the fluid being dragged along by the moving boundary. In this case, with $\frac{\partial p}{\partial x} = 0$, equation (10.38) simply reduces to

$$u = U\frac{y}{h} \tag{10.38}$$

Which indicates that the velocity varies linearly between the two plates as shown in Fig. 10.7 for $\alpha = 0$. This situation would be approximated by the flow between closely spaced concentric cylinder in which one cylinder is fixed and the other cylinder rotates with a constant angular velocity w.

The average velocity distribution of the Coutte flow between two parallel straight walls can be calculated by the expression.

$$u_{av} = \frac{1}{h}\int_o^h u\,dy$$

$$u_{av} = \frac{1}{h}\int_o^h \frac{Uy}{h} + \alpha U\left(\frac{y}{h} - \frac{y^2}{h^2}\right)dy \quad = \frac{1}{h}\left[\frac{Uy^2}{2h} + \alpha U\left[\frac{y^2}{2\,h} - \frac{y^3}{3\,h^2}\right]\right]_o^h$$

$$= \frac{Uh^2}{2h^2} + \frac{\alpha U}{h}\left[\frac{h^2}{2h} - \frac{h^3}{3h^2}\right] = \frac{U}{2} + \frac{\alpha U}{h}\left[\frac{h}{2} - \frac{h}{3}\right]$$

$$= \frac{U}{2} + \frac{\alpha u}{6} = U\left[\frac{\alpha}{6} + 1/2\right] \tag{10.39}$$

In the case of simple Coutte flow, $\alpha = 0$, therefore

$$u_{av} = \frac{U}{2} \tag{10.40}$$

When $\alpha = -3$ the average velocity becomes zero. This means that there is actually no flow across the passage between plates because the influence of the adverse pressure gradient in balance with the dragging action of the viscous force.

The volumentric flow Q per unit width is given by

$$Q = h \times \text{unit width} \times u_{av}$$

$$q = \frac{Q}{unit\ width} = h\,u_{av}$$

$$q = \left(\frac{\alpha}{6} + \frac{1}{2}\right)U\,h \tag{10.41}$$

The maximum or minimum velocity occurs at the position of y where the velocity derivative du/dy = 0. From equation (10.34) the derivative of the velocity is

$$\frac{du}{dy} = \frac{U}{h} + \frac{\alpha U}{h}\left(1 - \frac{2y}{h}\right) \tag{10.42}$$

The position of the maximum or minimum velocity is

$$0 = \frac{U}{h} + \frac{\alpha U}{h}\left(1 - \frac{2y}{h}\right)$$

or, $$\frac{U}{h} + \frac{\alpha U}{h} = \frac{\alpha U}{h}\left(\frac{2y}{h}\right)$$

$$1 + \alpha = \alpha\left(\frac{2y}{h}\right)$$

or, $$\frac{y}{h} = \frac{1}{2\alpha} + \frac{1}{2} \tag{10.43}$$

It is interesting to note that the maximum velocity for α = 1 occurs at $\frac{y}{h} = 1$. The minimum velocity for α = –1 at $\frac{y}{h} = 0$.

This implies that for α = –1 the velocity gradient at the stationary wall is zero and it becomes negative (reverse flow) for any value of α less than –1.

Equation (10.57) breaks down when $-1 < \alpha < 1$ because the maximum and minimum values of y/h have already been reached at α = 1 and α = –1, respectively.

The values of minimum and maximum velocities can be obtained by substituting the value of y from equation (10.43) into equation (10.38).

$$u_{\min} = \frac{U}{4}(1+\alpha)^2 / \alpha \quad for\ \alpha \le -1 \tag{10.44}$$

$$u_{\max} = \frac{U}{4}\frac{(1+\alpha)^2}{\alpha} \quad for\ \alpha \ge +1$$

For the purpose of comparison, this equation is plotted in Fig. 10.7.

From equation (10.56) the shearing stress distribution in a Coutte flow can be obtained, and it is

$$\tau = \mu\frac{du}{dy} = \mu\frac{U}{h} + \frac{\mu U \alpha}{h}\left(1 - \frac{2y}{h}\right) \tag{10.45}$$

In dimensionless form, this shear stress distribution becomes.

$$\frac{\tau h}{\mu U} = 1 + \alpha(1 - 2y/h)$$

For a simple Coutte flow when $\alpha = 0$ the shearing stress remains constant across passage between plates.

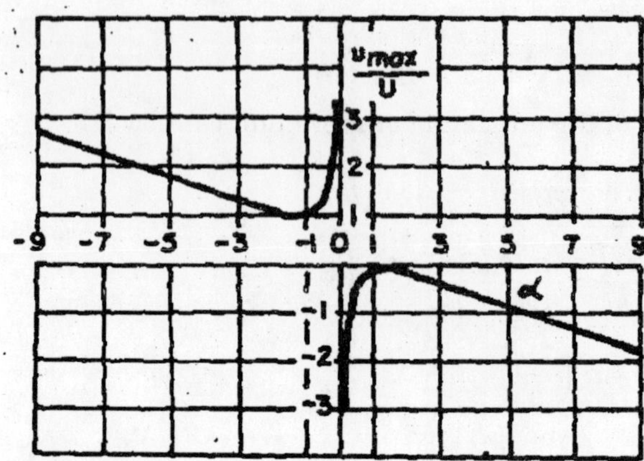

Fig. 10.7. Value of maximum velocity as a function of the dimensionless pressure gradient.

At the position $\frac{y}{h} = \frac{1}{2}$, the second term on the right hand side of equation (10.59) becomes zero; hence the shearing stress is independent of α at the centre of the passage. The shearing stress at the stationary wall is positive for $\alpha > -1$ and negative for $\alpha < -1$. Since the velocity gradient at the stationary wall is zero for $\alpha = -1$, the value of the shearing stress is zero for this condition.

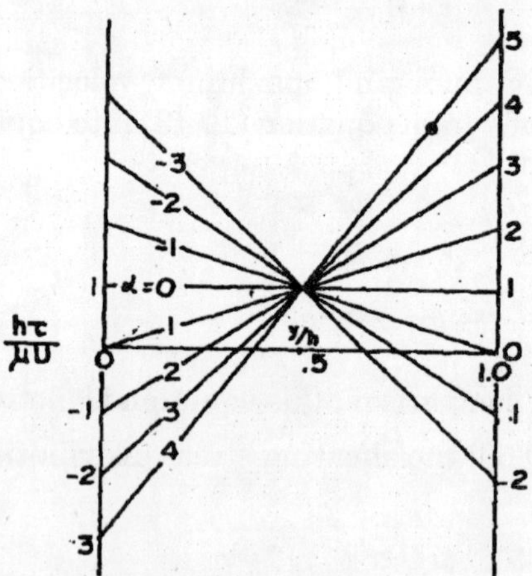

Fig. 10.8. Pressure distribution of the dimensionless pressure gradient.

ILLUSTRATIVE EXAMPLES

10.1. Oil of relative density 0.9 and kinematic viscosity of 330 cS. is pumped over a distance of 1500 meter through a 75 mm diameter tube at a rate of 25000 kg/hr. Determine whether the flow is laminar and calculate the pumping power required, assuming 70 percent mechanical efficiency. Also calculate the shear stress at the tube walls.

Solution

$$\nu = \frac{\mu}{S\rho}$$

or, $\mu = \nu\, S\, \rho = 330 \times 10^{-6} \times 0.9 \times 10^{3} = 0.297$ kg/ms

m = s ρ Q

$$\frac{25000}{3600} = 0.9 \times 1000 \times Q$$

$Q = 7.716 \times 10^{-3}$ m³/s

$$Q = \frac{\pi}{4} D^2\, u_{av}$$

$$7.716 \times 10^{-3} = \frac{\pi}{4} (0.075)^2\, u_{av}$$

$$u_{av} = 1.7465\ m/s$$

$$R_e = \frac{S\, \rho\, u_{av}\, D}{\mu} = \frac{0.9 \times 1000 \times 1.7465 \times .075}{0.297}$$

= 396.9318 Less than 2000

Therefore the flow is laminar

$$h_f = \frac{32\, \mu\, u_{av}\, l}{s\rho\, g\, D^2} = \frac{32 \times 0.29 \times 1.7465 \times 1500}{0.9 \times 1000 \times 9.81 \times (0.075)^2}$$

= 501.3398 m

$$P = \frac{S\, \rho\, g\, Q\, h_f}{\eta}$$

$$= \frac{0.9 \times 1000 \times 9.81 \times 7.716 \times 10^{-3} \times 501.3398}{0.7}$$

= 48790.798 = 48.79 kW

$$\tau = \frac{4\, \mu\, u_{av}\, r}{R^2}$$

The maximum shear stress or at wall shear stress is obtained by putting r = R

$$\tau_o = \frac{4\, \mu\, u_{av}\, R}{R^2} = \frac{4\, \mu\, u_{av}}{R} = \frac{8\, \mu\, u_{av}}{D}$$

$$= \frac{8 \times 0.297 \times 1.7465}{0.075} = 55.329\ \text{N/m}^2$$ Ans

10.2. **900 liters/min of oil are pumped through a pipe 600 m long and 100 mm diameter. The oil has an absolute viscosity of 1 Poise and a relative density of 0.928. Find the Reynold's number and friction head.**

If a pump having an efficiency of 65 percent is used to overcome the friction, also to elevate the oil through 20 m, determine the power required to drive the pump.

Solution

Discharge through pipe,

$$Q = a \times u$$

$$\frac{900 \times 10^{-3}}{60} = \frac{\pi}{4}(0.1)^2 \times u$$

$$u = 1.9098 \text{ m/s}$$

$$S = \frac{\rho_{oil}}{1000}$$

$$\therefore \rho_{oil} = 0.928 \times 1000 = 928 \ kg/m^3$$

$$\therefore R_e = \frac{\rho\, ud}{\mu} = \frac{928 \times 1.9098 \times 0.1}{1/10} = 1772.29$$

The flow is laminar

Therefore, $$\lambda = \frac{64}{R_e} = \frac{64}{1772.29} = 0.036$$

Friction head, $$h_f = \frac{\lambda\, l\, u^2}{2g\, d} = \frac{0.036 \times 600 \times (1.9098)^2}{2 \times 9.81 \times 0.1} = 40.278 \text{ m} \quad \text{Ans}$$

Total Pumping head $H = 40.278 + 20 = 60.278$ m

Therefore,

Power to drive pump $= \dfrac{S\rho Q g H}{\eta_p}$ $\quad \eta_p$ is the efficiency of pump.

$$= \frac{0.928 \times 10^3 \times 15 \times 10^{-3} \times 9.81 \times 60.278}{0.65} = \frac{8231.328}{0.65}$$

$= 12663.58$ Watt $= 12.663$ kW Ans

10.3. **A 100 mm diameter smooth pipe carries Olive oil horizontally at an average velocity of 1.4 m/s. What is the pressure gradient when the oil temperature is (a) 20°C and (b) 40°C (c) Explain why the pressure gradient is changed so little for the warmer, less viscous oil.**

Take $v_{20°c} = 8.3 \times 10^{-5} \ m^2/s$ *and* $v_{40°c} = 3.5 \times 10^{-5} \ m^2/s$

Density is 918 kg/m³

Solution

(a) $R_e = \dfrac{uD}{\nu} = \dfrac{1.4 \times 0.1}{8.3 \times 10^{-5}} = 1686.7469 \approx 1687$ (Laminar flow)

$$\lambda = \frac{64}{R_e} = \frac{64}{1687} = 0.03793$$

$$\frac{dp}{l} = \frac{\lambda}{D} \times \frac{\rho u^2}{2} = \frac{0.03793}{0.1} \times \frac{918 \times (1.4)^2}{2} = 341.2979 \text{ Pa/m}$$

(b) $R_e = \dfrac{uD}{\nu} = \dfrac{1.4 \times 0.1}{3.5 \times 10^{-5}} = 4000$ (Trubulent flow)

From the chart λ = 0.04

$$\frac{dp}{l} = \frac{\lambda}{D} \times \frac{\rho u^2}{2} = \frac{0.04}{0.1} \times \frac{918 \times (1.4)^2}{2} = 359.856 \text{ Pa/m}$$

(c) The pressure gradient is changed so little because the flow changes from laminar to turbulent for the warmer, less viscous oil.

10.4. Crude oil of dynamic viscosity 1.5 poise and relative density 0.9 flows through a 20 mm diameter vertical pipe. Two pressure gauges have been fixed at 20 m apart. The pressure gauges fixed at higher level reads 200 kPa and that at the lower level reads 6 bar. Find the direction and the rate of flow through the pipe.

Solution

$$\rho = 0.9 \times 1000 = 900 \; kg/m^3$$

$$\mu = 1.5 \; poise = \frac{1.5}{10} \; Ns/m^2 = 0.15 \frac{Ns}{m^2}$$

$$p_1 = 6 \times 10^5 \; N/m^2 = 6 \times 10^5 \; N/m^2$$

$$p_2 = 200 \times 10^3 \; N/m^2 = 6 \times 10^5 \; N/m^2$$

$$Z = 20 \text{ m}$$

Now,

$$p_1 - \rho g z - p_2 = \frac{32 \, \mu \, u_{av} \, l}{D^2}$$

$$(p_1 - p_2) - \rho g z = \frac{32 \, \mu \, u_{av} \, l}{D^2}$$

$$(6-2) \times 10^5 - 900 \times 9.81 \times 20 = \frac{32 \times 0.15 \, u_{av} \times 20}{(0.02)^2}$$

$$400000 - 176580 = 240000 \, u_{av}$$

$$223420 = 240000 \, u_{av}$$

$u_{av} = 0.9309$ m/s

$$Q = \frac{\pi}{4} D^2 u_{av} \qquad = \frac{\pi}{4}(0.02)^2 \times 0.9309 = 2.9245 \times 10^{-4}\ m^3/s$$

$= 0.29245$ liters/s Ans

10.5. **Glycerin (density is 1250 kg/m³ and dynamic viscosity is 1 kg/ms) flows through a 100 mm diameter pipe 60 m long and at an upward slope of 10° at a flow rate of 25 liters per second. At the lower inlet the pressure is 8 bar gauge.**

(a) What is the Reynold's number of the flow ?

(b) Why is the pressure drop from viscous shear? Also calculate head loss due to friction.

(c) What is the average wall shear stress?

(d) What is the pressure at the higher, downstream end of the pipe?

Solution

$$Q = \frac{\pi}{4} D^2 \times u$$

$$0.025 = \frac{\pi}{4}(0.1)^2 \times u$$

u = 3.188 m/s

$$R_e = \frac{\rho\, u\, D}{\mu} = \frac{1250 \times 3.183 \times 0.1}{1}$$

$= 397.875 \approx 398$ Ans

(b) $$dp = \lambda \frac{l}{D}\frac{\rho u^2}{2} = \frac{64}{R_e} \times \frac{60}{0.1} \times 1250 \times \frac{(3.183)^2}{2}$$

$= 610944.0603$ Pa ≈ 6.1 bar Ans

$$h_f = \frac{dp}{\rho g} = 6.1 \times 10^5 / 1250 \times 9.81 = 49.745\ m$$ Ans

(c) $$\tau_o = \frac{dp\, R}{2\, l} = \frac{61 \times 10^5 \times 0.05}{2 \times 60} = 254.167\ N/m^2$$

(d) $$\frac{p_1}{\rho g} = \frac{p_2}{\rho g} + dz + h_f$$

or, $$\frac{p_2}{\rho g} = \frac{P_1}{\rho g} - dz - h_f$$

$$\frac{P_2}{\rho g} = \frac{8 \times 10^5}{1250 \times 9.81} - 60 \sin 10^\circ - 49.745$$

$= 65.2395 - 10.418 - 49.745 = 5.075$ m

Therefore,

$P_2 = 5.075 \times 1250 \times 9.81 = 62238.35\ Pa = 62.2$ kPa Ans

10.6. An oil (relative density 0.8) a kinematic viscosity of $6 \times 10^{-5}\ m^2/s$ flows through a 50 mm diameter tube at an average velocity of 0.12 m/s. Calculate the pressure drop in 65 m of tube and the velocity 10 mm from the tube centre.

Solution

$$\mu = \nu\,\rho = 6 \times 10^{-5} \times 800 = 0.048 \text{ kg/ms}$$

$$R_e = \frac{u_{av}\,D}{\nu} = \frac{0.12 \times 0.05}{6 \times 10^5} = 100$$

Hence the flow is laminar

Now pressure drop,

$$\Delta p = \frac{32\,\mu\,u_{av}\,l}{D^2} = \frac{32 \times 0.048 \times 0.12 \times 65}{(0.05)^5}$$

$= 4792.32$ N/m² $= 4.792$ k Pa Ans

The velocity at 10 mm from the tube centre i.e. r = 10 mm

$$u = \left(-\frac{\partial p}{\partial x}\right)\frac{1}{4\mu}\left(R_e - r^2\right)$$

$$= \frac{4792.32}{65} \times \frac{1}{4 \times 0.048}\left[(0.025)^2 - (.01)^2\right]$$

$$= \frac{4792.32 \times 5.25 \times 10^{-4}}{65 \times 4 \times 0.048}$$

$= 0.2016$ m/s Ans

10.7. Linseed oil at 25°C flows in a 25 mm diameter horizontal pipe. (a) What is the maximum average velocity for which the flow may be considered laminar? (b) What is the pressure drop in 50 m of pipe at that flow ? (c) What is the wall shear stress? Take S = 0.93, $\mu = 3.31 \times 10^{-2}$ kg/ms.

Solution

$$\rho = 0.93 \times 1000 = 930\ kg/m^3$$

$$\nu = \mu/\rho = \frac{3.31 \times 10^{-2}}{930} = 3.559 \times 10^{-5}\ m^2/s$$

The Reynold's number is 2000 for calculating maximum average velocity for laminar flow,

$$R_e = \frac{u_{av}\ D}{\nu}$$

$$2000 = \frac{u_{av} \times 0.025}{3.559 \times 10^{-5}}$$

$u_{av} = 2.8473$ m/s **Ans**

(b) $$dp = \frac{32\ \mu\ u_{av}\ l}{D^2} = \frac{32 \times 3.31 \times 10^{-2} \times 2.8473 \times 50}{(0.025)^2}$$

= 241268.8 N/m² = 2.412 bar **Ans**

(c) $$\tau_o = \frac{8\ \mu\ u_{av}}{D} = \frac{8 \times 3.31 \times 10^{-2} \times 2.8473}{0.025}$$

= 30.1586 N/m² **Ans**

10.8. **Fluid of density 1260 kg/m³ and viscosity 0.9 × 10^{-3} Ns/m² passes between two infinite plates 20 mm separation. If the flow rate is 0.5 liters/s unit width calculate the pressure drop/unit length if both plates are stationary.**

Solution

$Q = A \times u_{av}$

Q / unit width = $h\ u_{av}$

$0.5 = 0.02 \times u_{av}$

$u_{av} = 0.025$ m/s

$$R_e = \frac{\rho\ u_{av}\ h}{\mu} = \frac{1260 \times 0.025 \times 0.02}{0.9 \times 10^{-3}} = 700$$

The flow is laminar

$$\frac{dp}{dx} = \frac{12\ \mu\ u_{av}}{h^2}$$

$$= \frac{12 \times 0.9 \times 10^{-3} \times 0.025}{(0.02)^2} = 0.675 \text{ N/m}^2\text{/m}$$ **Ans**

10.9. **Oil of specific gravity 0.90 and coefficient of viscosity 100 cp (centi poise) flows between two fixed parallel plates kept 12 mm apart. If the average velocity is 1.5 m/s Calculate, (a) the maximum velocity, (b) plate shear stress, (c) velocity and shear stress at a distance of 3 mm from one of the plates and (d) head loss in a distance of 30 m.**

Solution

For two fixed parallel plates

(a) $$u_{max} = \frac{3}{2} u_{av} = \frac{3}{2} \times 1.5 = 2.25 \text{ m/s}$$ Ans

(b) $$u_{av} = \left(-\frac{dp}{dx}\right) \frac{h^2}{12\mu}$$

$$1.5 = \left(-\frac{dp}{dx}\right) \times \frac{\left(\frac{12}{1000}\right)^2}{12 \times 0.1}$$

$$\left(-\frac{dp}{dx}\right) = \frac{1.5 \times 12 \times 0.1 \times (1000)^2}{(12)^2} = 12500$$

$$\tau_o = \left(-\frac{dp}{dx}\right) \frac{h}{2} = 12500 \times \frac{12}{1000 \times 2} = 75 \text{ N/m}^2$$ Ans

(c) $$(u)_y = \frac{1}{2\mu}\left(-\frac{dp}{dx}\right)\left(hy - y^2\right)$$

$$= \frac{1}{2 \times 0.1}(12500)\left[\frac{12}{1000} \times \frac{3}{1000} - \left(\frac{3}{1000}\right)^2\right]$$

$$= \frac{12500 \times 2.7 \times 10^{-5}}{2 \times 0.1} = 1.69 \text{ m/s}$$ Ans

$$(\tau)_y = \left(-\frac{dp}{dx}\right)\left(\frac{h}{2} - y\right)$$

$$= 12500\left(\frac{12}{2 \times 1000} - \frac{3}{1000}\right) = 37.5 \text{ N/m}^2$$ Ans

(d) head loss

$$h_f = \frac{12\, \mu\, u_{av}\, l}{\rho\, g\, h^2}$$

$$= \frac{12 \times 0.1 \times 1.5 \times 30}{(0.9 \times 1000 \times 9.81)\left(\frac{12}{1000}\right)^2} = 42.47\ m$$ Ans

10.10. Kerosene oil flows upwards through inclined parallel plates at the rate of 2 liters/s per meter width. The distance between the parallel plates is 10 mm and inclination of plates is 30° with the horizontal. Determine the difference of pressure between two sections 10 m apart. Take S = 0.82 and $\mu = 1.92 \times 10^{-3}$ Ns/m^2 for kerosene oil.

Solution

$$u = \frac{Q}{h} = \frac{2 \times 10^{-3}}{10 \times 10^{-3}} = 0.2 m/s$$

$$h_f = \frac{12 \mu u l}{\rho g h^2} = \frac{12 \times 1.92 \times 10^{-3} \times 0.2 \times 10}{820 \times 9.81 \times (10 \times 10^{-3})^2}$$

$$= \frac{0.04608}{0.80442} = 0.05728 \ m$$

But $$h_f = \left(\frac{p_1}{\rho g} + z_1\right) - \left(\frac{p_2}{\rho g} + z_2\right)$$

$$0.05728 = \frac{p_1 - p_2}{\rho g} + (z_1 + z_2)$$

$$\frac{p_1 - p_2}{\rho g} = 0.05728 + 10 \sin 30° = 0.05728 + 5 = 5.05728$$

$$p_1 \ p_2 = (\rho g)(5.05728) = 40681.77 \text{ N/m}^2 = 40.68 \text{ k Pa}$$ Ans

10.11. Show that the rate of change of pressure at any point along laminar flow between parallel surfaces is given by

$$\frac{dp}{dx} = \mu \left(\frac{d^2 u}{dy^2}\right)$$

Where μ is the viscosity of the fluid and u is viscosity at the distance y from one of the surface.

Water at 20°C flows between two large parallel plates at a distance 2 mm apart. If the average velocity is 0.15 m/s. Find the maximum velocity, pressure drop, wall shearing stress and the frictional coefficient. The dynamic viscosity at 20°C is given by 1.02×10^{-3} kg/ms.

Solution: For theory see text.

The maximum velocity is given by

$$u_{\max} = \frac{3}{2} u_{average} = \frac{3}{2} \times 0.15 = 0.225 \text{ m/s}$$ Ans

The pressure drop

$$\frac{dp}{dx} = 12 \mu \frac{u_{av}}{h^2}$$

$$= \frac{12 \times 1.02 \times 10^{-3} \times 0.15}{(2 \times 10^{-3})^2} = 459 \text{ N/m}^2$$ Ans

The wall shearing stress

$$\tau_o = -\frac{h}{2}\left(\frac{dp}{dx}\right) = -\frac{2}{2} \times 10^{-3}\,(-459) = -0.459 \text{ N/m}^2 \qquad \text{Ans}$$

$$R_e = \frac{\rho\, u\, h}{\mu} = \frac{1000 \times .15 \times 2 \times 10^{-3}}{1.02 \times 10^{-3}} = 294.11$$

The frictional coefficient

$$\lambda = 12 / R_e = \frac{12}{294.11} = 0.408 \qquad \text{Ans}$$

EXERCISES

A. THEORY

10.1. Show that for laminar flow in a circular pipe head loss is given by

$$h_f = \frac{32 \mu\, u\, l}{S \rho\, g\, D^2}$$

10.2. Prove that for a steady laminar flow through a circular pipe the velocity distribution across the section is parabolic and the average velocity is half the maximum local velocity.

10.3. Show that the volumetric flow through a circular pipe is given by the expression $\frac{\pi}{8\,\mu} R^4 \left(-\frac{dp}{dx}\right)$ where R is the radius of the pipe, μ = dynamic velocity.

10.4. Derive an expression for velocity distribution and pressure drop in the fully developed flows region of the flow through a circular pipe.

10.5. Derive an expression for velocity distribution for laminar flow in a circular pipe. Show that the maximum velocity is twice the average velocity.

10.6. For a laminar flow in a circular pipe, find out the relation between friction coefficient, λ and the Reynold's number of flow.

10.7. Show from first principles that the loss of pressure due to laminar flow in a horizontal circular pipe is given by

$$dp = \frac{32\,\mu\, u\, l}{D^2}$$

and hence that under these conditions the friction coefficient λ in the expression.

$$\frac{\lambda\, l\, u^2}{2g\, D} \text{ is } \frac{64}{\text{Reynold's number}}$$

10.8. Derive an expression for velocity distribution for laminar flow occuring between two parallel plates.

10.9. For laminar flow in a pipe, show that the shear stress at a radial distance r from the centre line of the pipe is given by

$$\tau = \left(-\frac{dp}{dx}\right)\frac{r}{2}$$

10.10. Using Hagen-Poiseuille equation show that

$$\frac{\tau_{max}}{u_{max}} = \frac{4\mu}{D}$$

10.11. Using Hagen-Poiseuille equation show that

$$u_{max} = \left(-\frac{dp}{dx}\right)\frac{R^2}{4\,\mu}$$

10.12. Using Hagen-Poiseuille equation show that

$$\frac{u_{av}}{u_{max}} = 1 - \left(\frac{r}{R}\right)^2$$

10.13. Using Hagen-Poiseuille equation show that

$$Q = \frac{\pi\,R^4\;dp}{8\,\mu l}$$

10.14. Laminar flow takes place between two stationary plates parallel to each other at a distance h apart from each other where h/2 is the distance from the centre line to the plate.

Show that the velocity profile for the flowing fluid is given by

$$u = \frac{h^2}{8\,\mu}\left(-\frac{\partial p}{\partial x}\right)\left[1 - 4\,(y/h)^2\right]$$

Also show that

$$\tau = 4\,\mu\,\frac{u_{max}}{h}$$

10.15. For a plane Poiseuille flow, when both the plates are stationary, show that, the velocity u is given by

$$u = \frac{h^2}{8\,\mu}\left(-\frac{\partial p}{\partial x}\right)\left[1 - 4\,(y/h)^2\right]$$

10.16. Show that for a plane Poiseuille flow the maximum velocity is given by

$$u_{max} = \frac{h^2}{8\,\mu}\left(\frac{\partial p}{\partial x}\right)$$

where h/2 is the distance from the centre line to the plate.

Also show that the friction coefficient

$$\lambda = \frac{12}{R_e}$$

where R_e is the Reynold's number

10.17. Show that for a plane Poiseuille flow that pressure drop across length l is given by

$$p_2 - p_1 = \frac{12\mu \, u_{av} \, l}{h^2}$$

B. UNSOLVED PROBLEMS

10.1. Estimate the power required in kW per 1 km length of a pipe line to over come the viscous resistance to the flow of crude oil (sp. gravity = 0.85) having dynamic viscosity equal to 0.1 Ns/m^2 through horizontal 75 mm diameter pipe at the rate of 5 liters/s. [3.21 kW]

10.2. Calculate the loss of head due to friction and the power required to maintain flow in a horizontal circular pipe of 40 mm diameter and 750 m long when water ($\mu = 1.14 \times 10^{-3}$ Ns/m^2) flow at a rate of 4 liter/min. [R_e = 1856, $h_f = 92.4 \times 10^{-3}$ m, P = 0.0605 Watts]

10.3. The centre line velocity is measured as 3m/s in a 0.1 m diameter tube. If the fluid flowing has a density of 1260 kg/m^3 and a viscosity of 0.9 Ns/m^2 determine whether the flow is laminar and calculate the pressure gradient necessary. [R_e = 210, 4.3 $kN/m^2/m$]

10.4. Oil of specific gravity 0.9 and viscosity 1 poise pumped through a 50 mm diameter pipe at the rate of 280 liters/min. Show that the flow is stream line and hence estimate the power required to pump the oil up through a pipe line 75 m long which rises 7.5 m over this length. [R_e = 1071, P = 1.404 kW]

10.5. What power will be required per one kilometer length of pipe line to overcome the viscous resistance to the flow of crude oil (S.G. 0.85) of viscosity 1.9 poises through a horizontal 100 mm diameter pipe at the rate of 650 liters/hr. [2.5 Watt]

10.6. Oil of viscosity 0.5 poise flows through a 20 mm diameters pipe. The velocity along the centre line of the pipe is 0.4 m/s.

(a) The pressure drop per unit length of pipe

(b) The velocity at a distance d/4 from the wall of the pipe.

(c) The discharge

(d) The shear stress in the oil at the pipe wall.

[7.85 kPa, 0.3 m/s, 62.8 cm^3/s.3.92 kPa]

10.7. Oil of dynamic viscosity 0.07 kg/ms flows through a 20 mm diameter pipe with a mean velocity of 0.5 m/s. Calculate the pressure drop which occurs over a length of 50 m of pipe. [1.4 bar]

10.8. The velocity along the centre line of the Hagen-Poiseuille flow in a 100 mm diameter pipe is 2 m/s. If the viscosity of the fluid is 0.07 kg/ms and its relative density is 0.92, Calculate (a) the volumetric flow in m^3/s, (b) the shearing stress of the fluid at the pipe wall, (c) the local frictional coefficient. [7.85×10^{-3} m^3/s, 2.2 Pa, 0.0487]

10.9. Calculate the diameter of a vertical pipe needed for flow of a liquid at a Reynold's number of 1200 when the pressure remains constant throughout the pipe. Take kinematic viscosity of the fluid as 1.92×10^{-3} m^2/s. [D = 243 mm]

10.10. A lubricating oil of relative density 0.92 and dynamic viscosity 0.085 Ns/m^2 is to be pumped at a rate of 2400 liters/min with energy gradient in laminar flow not exceeding 0.04. Determine the least diameter of the pipe to satisfy the above requirements. [275 mm]

10.11. An oil of coefficient of viscosity 1.5 poise and specific gravity 0.9 flows through a 30 mm vertical pipe. Two pressure gauges are fixed 20 m apart. The gauge A fixed at the top records 200 kPa and the gauge B fixed at the bottom records 500 kPa. Find the direction of flow and the rate of flow. [Upwards, R_e = 139, Q = 0.5468 liters/s]

10.12. An oil of density 917 kg/m^3 is being pumped in a 150 mm diameter pipe. The discharge is measured as 14.2 liters/s. The drop in pressure in a stretch of 800 mm of pipe line, both ends of which are at th same elevation, is measured as 95 kPa. Estimate the absolute viscosity of the oil. [R_e = 1059, μ = 0.1041 kg/ms]

10.13. In a steady laminar flow between parallel plates, h = 2 mm, $\mu = 5 \times 10^{-2}$ Ns/m^2 and $\frac{\partial p}{\partial x} = -10$ kPa. Determine q, maximum shear stress (τ_o) and maximum velocity (u_{max}). [q = 0.1333 liter/s m, τ_o = 10 Pa, u_{max} =0.1 m/s]

10.14. Kerosene oil flows upwards through inclined parallel plates at the rate of 2 liters/s per meter width. The distance between the parallel plates is 10 mm and the inclination of plates is 20° with the horizontal. Determine the difference of pressure between two sections 10 m apart. Take density and viscosity of kerosene oil as 800 kg/m^3 and 2×10^{-3} Ns/m^2

10.15. A flow of 60 liters/s per meter width of glycerine of relative density 1.25 and dynamic viscosity 1.5 kg/ms takes place between two parallel plates having a gap of 25 mm between them. Calculate the (a) maximum velocity (b) boundary shear stress and (c) energy gradient. [3.6 m/s, 864 Pa, h_f/l = 5.648]

10.16. Water flows between two very large parallel flat plates 20 mm apart. If the average velocity of the water is 0.15 m/s, what is the shearing stress at 5 mm and 10 mm from the lower plate. [$\tau_{5\,mm}$ = 0.0225 Pa, $\tau_{10\,mm}$ = 0]

10.17. Oil flows between parallel plates, one of which is at rest and other moves with a velocity U. If the pressure is decreasing in the direction of flow at a rate of 5 N/m^3, dynamic viscosity is 0.05 Ns/m^2. The space between two plates is 40 mm and volumetric flow per unit width is 0.02 m^2/s. What is the velocity. [u_{av} = 0.5 m/s, U = 1.16 m/s].

11

Turbulent flow through Pipes

11.1 INTRODUCTION

In the chapter eleven i.e. in the last chapter various properties of fully developed laminar pipe flow were discussed. Since turbulent pipe flow is actually more likely to occur than laminar flow in practical situations, it is necessary to obtain similar information for turbulent pipe flow. However, turbulent flow is a very complex process.

The turbulent flow is characterized by rapid and continuous mixing of the fluid in a rather chaotic manner throughout the pipe. In the turbulent flow the Reynold's number is large (more than 4000), the inertia forces, which dominate in the flow, tend to diffuse the fluid particles, hence causing the mixing of the fluid.

The discussion so far has hinged upon only the physical phenomenon of the disorderlines and random irregularities of turbulent flow; yet we do not have a clear cut definition of turbulence.

In his book Turbulence, Hinze suggests, "Turbulent fluid motion is an irregular" condition of flow in which the various quantities show a random variation with time and space coordinates, so that statistically distinct average values can be discerned.

According to the definition suggested by Taylor and von Karman, turbulence can be generated by fluid flow past solid surfaces or by the flow of layers of fluids at different velocities past or over one another. The definition above indicates that there are two distinct types of turbulence.

(1) Turbulence generated by the various effect due to the pressure of a solid wall is designated by wall turbulence;

(2) Turbulence, in the absence of a wall, generated by the flow of layers of fluids at different velocities is called free turbulence.

Turbulent flow through pipes and past bodies are examples of wall turbulence, and turbulent jet mixing regions and wakes fall into the category of free turbulence.

Numerous persons have devoted considerable effort in attempting to understand the variety of baffling aspects of turbulence. Although a considerable amount of knowledge about the topic has been developed, the field of turbulent flow still remains the least understood area of fluid mechanics.

Fortunately, engineers can evade such troublesome details by time averaging values of the flow. Time averaging produces statistical correlations that involve fluctuating velocities, pressure and temperatures. These fluctuating properties are the unknown variable that appear in the governing equation of turbulent motion. Since no general solution of these fluctuating properties are known to exist, we are forced to model them in terms of properties we do know.

In this chapter, we shall be investigating internal turbulent flows or elementary topics of wall turbulence. The simplest example is a fully developed turbulent flow in a circular pipe. This is a flow that is independent, in the time mean, of the stream wise position along the pipe, that is independent of either an inlet or outlet state and that has conditions statistically identical at each axial position.

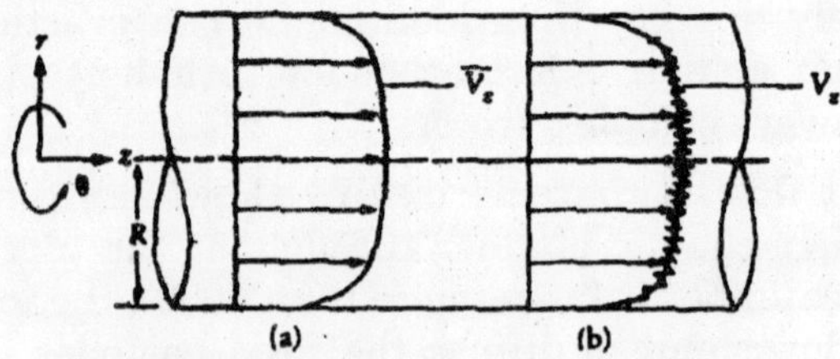

Fig. 11.1. Turbulent flow in a tube.

Fig. 11.1 shows turbulent flow in a tube. If the velocity at each point in the cross section is measured with the help of pitot static tube, the velocity profile in Fig. 11.1(a) results. If the velocity is instead measured with a more sensitive device, such as hot wise anemometer, the profile in Fig. 11.1(b) is obtained. Transition from laminar shown in Fig. 11.2, Fig. 11.3 shows shear stress and velocity distribution in a pipe.

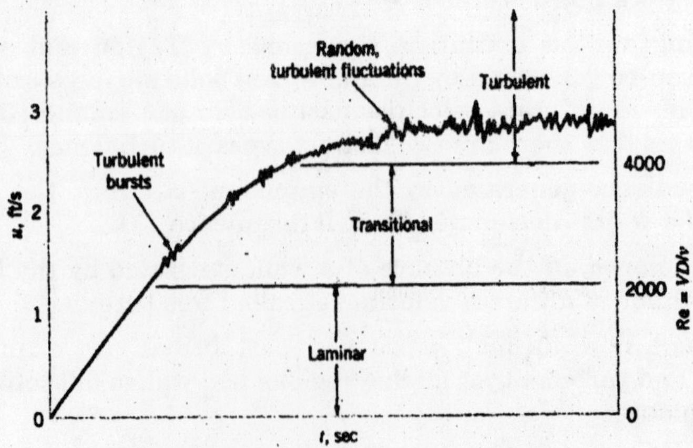

Fig. 11.2. Transition from Laminar flow.

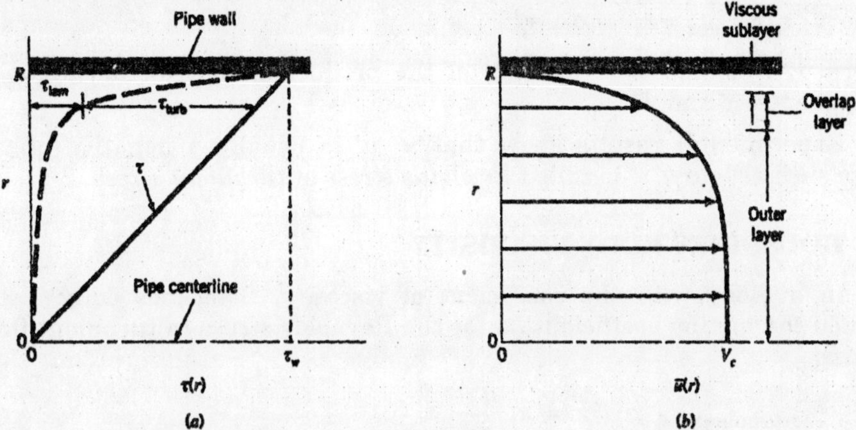

Fig. 11.3. Velocity and shear stress distribution in a turbulent flow near a wall.

11.2 TURBULENT SHEAR STRESS

Turbulent flow is characterized by random and rapid fluctuations of swirling regions of fluid, called eddies, throughout the flow. These fluctuations provide an additional mechanism for momentum and energy transfer (Fig. 11.4).

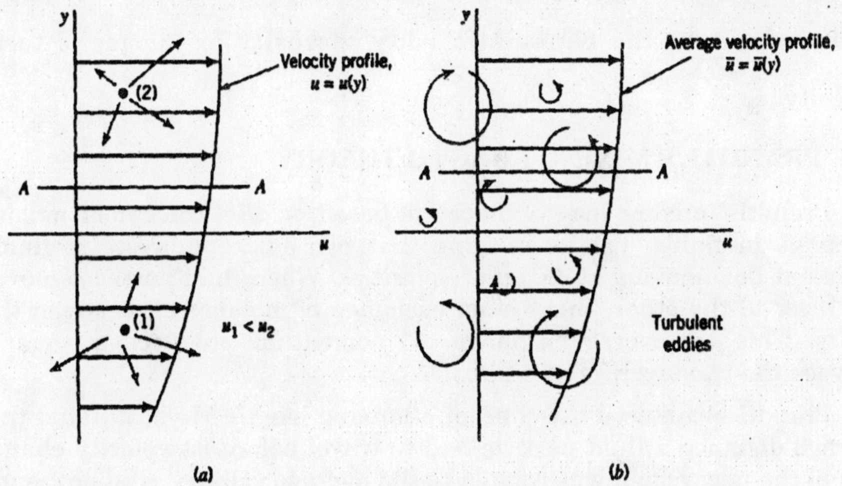

Fig. 11.4. Turbulent eddies

The total shear stress in turbulent flow can be expressed as

$$\tau_{total} = \tau_{laminar} + \tau_{turbulent} \tag{11.1}$$

Laminar shear stress is given by

$$\tau_{laminar} = \tau_l = \mu \frac{du}{dy} \tag{11.2}$$

Turbulent shear stress is expressed as

$$\tau_{turbulent} = \tau_t = -\rho \overline{u'v'} \quad (11.3)$$

where $\overline{u'v'}$ is the time average of the product of the fluctuating velocity components, u' and v'.

Experimental results show that $u'v'$ is usually a negative quantity. Term such as $-\rho \mu' v'$ is called *Reynolds stress* or *turbulent stress.*

11.3 BOUSSINES EDDY VISCOSITY

In analogy with the coefficient of viscosity, Boussines eddy viscosity defined the mixing coefficient, μ_t for the Reynolds stress in turbulent flow by putting

$$\tau_{turbulent} = -\rho \overline{u'v'} = \mu_t \frac{d\bar{u}}{dy} \quad (11.4)$$

where, μ_t is the eddy viscosity or turbulent viscosity or apparent or virtual viscosity, which accounts for momentum transport by turbulent eddies. Then the total shear stress can be expressed as

$$\tau_{total} = (\mu + \mu_t) \frac{d\bar{u}}{dy} \quad (11.5(a))$$

$$= \rho(v + v_t) \frac{d\bar{u}}{dy} \quad (11.5(b))$$

where $v_t = \frac{\mu_t}{\rho}$ is the **Kinematic eddy viscosity** or kinematic turbulent viscosity.

11.4. PRANDTL'S MIXING LENGTH THEORY

Prandtls' mixing length theory is based on the concept of momentum exchange in fluids. Let us imagine that two adjacent layers of fluid in a turbulent flow moving at different velocities. When fluid particles move from one layer to the other, there is an exchange of momentum between the two layers. This transverse exchange of momentum produces a shear force between the two layers (Fig. 11.5).

Prandtl introduced the concept of mixing length. He postulated that the average distance a fluid particle had to travel before its velocity changed to that of the new region was related to the average velocity gradient, and so in a shear flow

$$u' = l \frac{d\bar{u}}{dy}$$

Prandtl assumed v' to be of the some order of magnitude as u' that is u' = v'.

$$v' = l \frac{d\bar{u}}{dy}$$

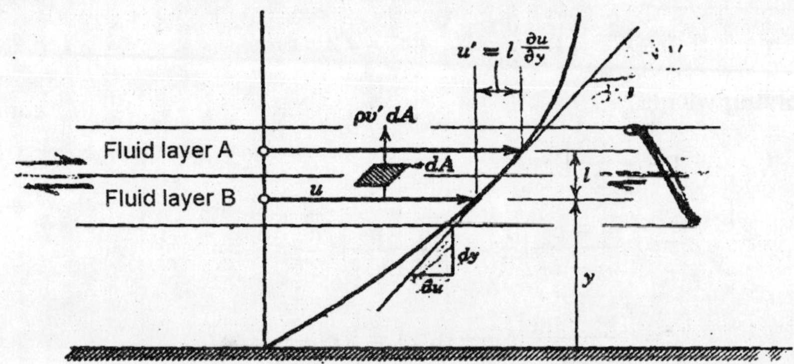

Fig. 11.5. Prandtl conceptional model of a two dimensional parallel turbulent flow.

Substituting the values u' and v' in equation (11.3)

$$\rho u' v' = \rho l^2 \left(\frac{d\bar{u}}{dy}\right)^2 \tag{11.6}$$

The equation (11.6) is known as **Prandtl's mixing length** hypothesis.

The l can be calculated from the experimental data. For pipes, l to be a linear function of the distance y from the pipe wall.

$$l \;\alpha\; y$$

or $$l = ky \tag{11.7}$$

where K is the constant of proportionality and is called the universal constant.

Nikurdse confirmed Prandtl hypothesis experimentally l = ky, where, k = 0.4 for those regions very close to the pipe wall (Fig. 11.6).

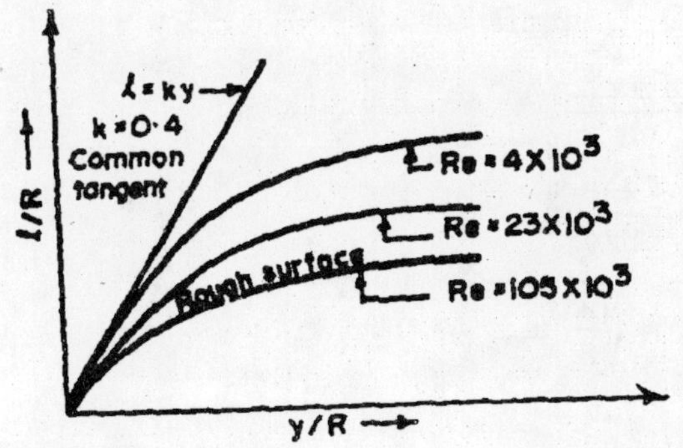

Fig. 11.6. Curve showing l/R versus v/R for turbulent pipe flow (not to scale)

11.5 SHEAR VELOCITY OR FRICTION VELOCITY

At any radial position, the equilibrium condition for pipe flow (Hagen-Poiseuille flow) can be expressed by

Pressure force = Shear force

$$p\,\pi r^2 - \left(p + \frac{\partial p}{\partial x}dx\right)\pi r^2 = \tau\, 2\pi r\, dx$$

Solving we get,

$$\tau = \left(-\frac{\partial p}{\partial x}\right)\frac{r}{2} \tag{11.8}$$

$$\tau = \left(\frac{p_1 - p_2}{x_2 - x_1}\right)\frac{r}{2} = \left(\frac{p_1 - p_2}{L}\right)\frac{r}{2}$$

$$= \left(\frac{p_1 - p_2}{\rho g}\right)\left(\frac{r}{2L}\right)(\rho g) = h_f\left(\frac{r}{2L}\right)(\rho g)$$

when r = R, τ = τo

$$\tau_o = h_f\left(\frac{\rho g R}{2L}\right)$$

or, $$\tau_o = h_f\left(\frac{\rho g D}{4\,L}\right)$$

or, $$h_f = \frac{4\,L\,\tau_o}{\rho g D} \tag{11.9}$$

We know that, head lost due to friction can be written as

$$h_f = \frac{\lambda\, L\, u_{av}^2}{2 g D} \tag{11.10}$$

where $\lambda = 4f$

Equating (11.9) and (11.10)

$$\frac{4\,L\,\tau_o}{\rho g D} = \frac{\lambda\, L\, u_{av}^2}{2 g D}$$

or, $$\tau_o = \frac{\lambda\,\rho\, u_{av}^2}{8}$$

or, $$\frac{\tau_o}{\rho} = \left(\frac{\lambda}{8}\right)u_{av}^2$$

or, $$\sqrt{\frac{\tau_o}{\rho}} = \sqrt{\frac{\lambda}{8}}\; u_{av}$$

or, $$u_* = \sqrt{\frac{\lambda}{8}}\; u_{av} \tag{11.10}$$

$u_* = \sqrt{\dfrac{\tau_o}{\rho}}$ is termed as **shear velocity or friction velocity**. It is to be noted that u_* is not an actual velocity of the fluid-it is merely a quantity that has dimensions of velocity. u_* is constant for any given case of turbulent pipe flow.

11.6 PRANDTL'S UNIVERSAL VELOCITY DISTRIBUTION EQUATION FOR TURBULENT PIPE FLOW

The following derivation is for a general case of turbulent pipe flow and therefore apply equally well to turbulent flow in both smooth and rough pipes. The mechanism of turbulent motion in the central region of the pipe must necessarily be the same, regardless of the surface conditions at the pipe wall.

According to Prandtl's mixing length theory, the turbulent shearing stress between any two layers of fluid is given by

$$\tau_t = \rho \left(l \frac{du}{dy} \right)^2$$

Prandtl further assumed that near the pipe wall.

(a) $\tau_t = \tau_o$, that is, the turbulent shearing stress remaining approximately constant.

(b) The mixing length l has a linear relationship with y, the distance from the boundary (i.e. l = ky).

Thus

$$\tau_o = \rho k^2 y^2 \left(\frac{du}{dy} \right)^2$$

or, $$\left(\frac{du}{dy} \right)^2 = \frac{\tau_o}{\rho K^2 y^2}$$

or, $$\frac{du}{dy} = \sqrt{\frac{\tau_o}{\rho}} \times \frac{1}{k\,y}$$

or, $$\frac{du}{dy} = \frac{u_*}{k\,y}$$

or, $$du = \frac{u_*}{k\,y} dy$$

By Integrating,

$$\int du = \int \frac{u_*}{k\,y} dy$$

or, $$u = \frac{u_*}{k} \ln y + C \qquad (11.12)$$

The constant of integration C is determined from the boundary conditions at the centre of the pipe.

At the pipe axis (i.e. at centre of the pipe)

$u = u_{max}$ and $y = R$, so that

$$u_{max} = \frac{u_*}{k} \ln R + C$$

$$C = u_{max} - \frac{u_*}{k} \ln R$$

By substituting the expression for C into equation (11.12) and noting that K = 0.4 (suggested by Nikuradse)

$$u = \frac{u_*}{k} \ln y + v_{max} - \frac{u_*}{k} \ln R$$

or, $$u = \frac{u_*}{k} \ln \frac{y}{R} + u_{max}$$

or, $$u = \frac{u_*}{0.4} \ln \frac{y}{R} + u_{max}$$

or, $$u = 2.5\, u_* \ln \frac{y}{R} + u_{max} \qquad (11.13)$$

The equation (11.13) is known as **Prandtl's universal velocity distribution equation for turbulent flow in both smooth and rough pipes.**

The equation (11.13) can also be expressed in a non dimensional form.

$$\frac{u_{max} - u}{u_*} = 2.5 \ln \frac{R}{y} \qquad (11.14)$$

and $$\frac{u_{max} - u}{u_*} = 5.75 \log_{10} \frac{R}{y} \qquad (11.15)$$

The term ($u_{max} - u$) is known as `velocity defect'. As depicted in Fig. 11.7 the equation governs the relationship between u_{max} and u. Unless either the maximum or wall velocity is known it does not of itself provide sufficient information to enable the actual velocities across the section to be determined. The logarthmic form of the equation does, however, indicate that the velocity gradient is much steeper adjacent to the walls than in the central portion of the pipe.

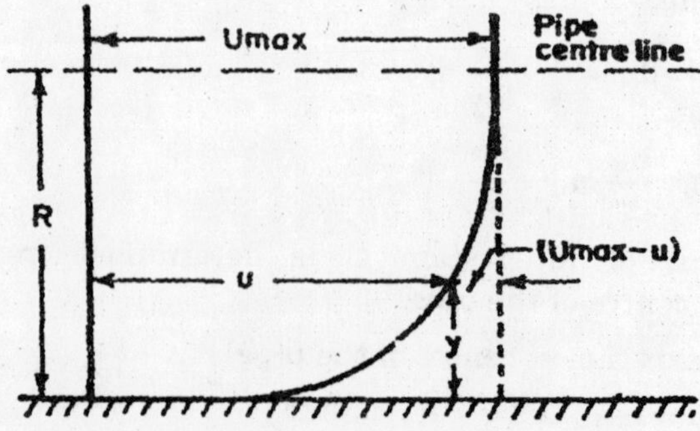

Fig. 11.7. Velocity deficiency distribution

11.7 HYDRODYNAMICALLY SMOOTH AND ROUGH BOUNDARIES

The flow conditions in the vicinity of the pipe wall depend on the relative magnitude of the laminar sublayer thickness δ' and the pipe wall roughness Ks as shown in Fig. 11.8. If the Reynolds number of flow is of such a numerical value that Ks is smaller than δ', the wall roughness are submerged within the laminar sublayer and, therefore, do not interfere with the formation of the laminar sublayer and buffer zone. The velocity distribution would be exactly the same as that in a smooth pipe. The pipe is in fact considered to be hydraulically smooth. When Ks is larger than δ', the wall roughness protrude beyond the laminar sub-layer and create additional turbulence in the flow. As a result, the relative roughness R/Ks may be regarded as a similarity parameter for turbulent flow in rough pipes because geometrically similar rough surfaces will result in dynamically similar turbulent flow patterns. Hence the resulting velocity distribution equation for turbulent flow in rough pipes is a function of the relative roughness R/Ks and the Reynolds number Re of flow.

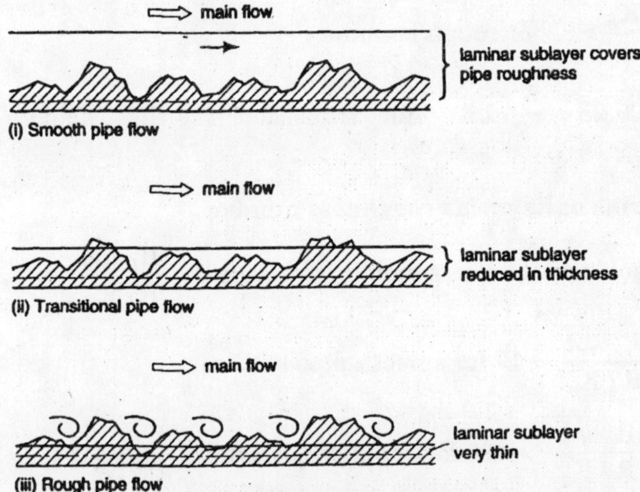

Fig. 11.8(a). Smooth and Rough boundaries with reference to Laminar sub layer thickness

Fig. 11.8(a) shows with reference to laminar sublayer, Fig. 11.8(b) shows the structure of the turbulent boundary layer and 12.8(c) shows the difference between rough and smooth boundary.

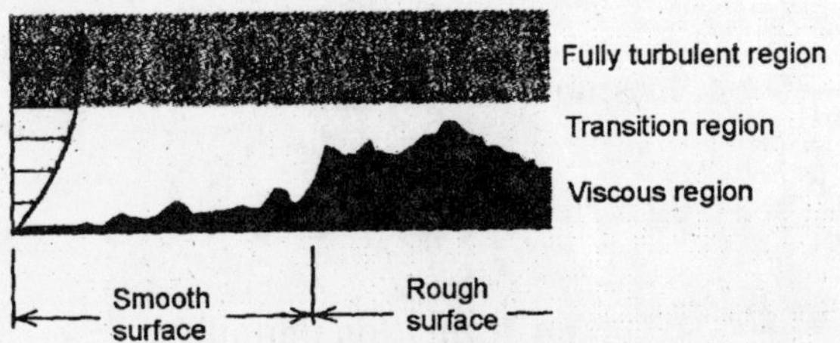

Fig. 11.8(b). Structure of turbulent boundary layer.

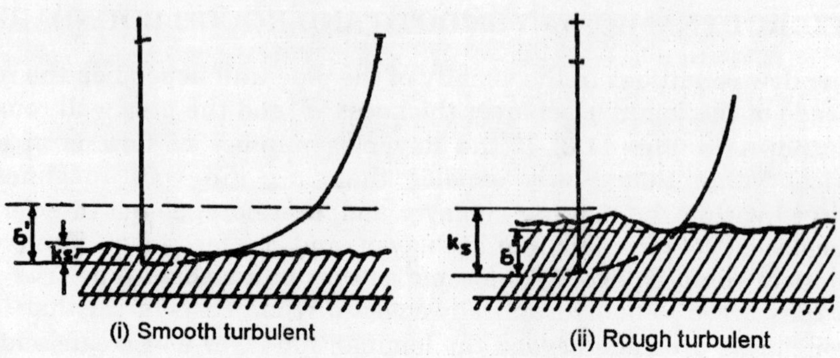

Fig. 11.8c. Turbulent flow near boundary

(a) Nikuradse's experiment results for hydraulic smooth and rough boundaries are given below.

(1) $\frac{K_s}{\delta'} < 0.25$ for smooth boundary

(2) $\frac{K_s}{\delta'} > 6$ for rough boundary

(3) $0.25 < \frac{K_s}{\delta'} < 6$ for transition (neither smooth nor rough) boundary

(b) In terms of Reynolds roughness number.

Reynolds roughness number is written as $\frac{R_e \sqrt{\lambda}}{R / K_s}$

(1) $\frac{R_e \sqrt{\lambda}}{R / K_s} < 17$ for smooth pipe surface.

(2) $\frac{R_e \sqrt{\lambda}}{R / K_s} > 400$ for rough pipe surface.

(3) $17 < \frac{R_e \sqrt{\lambda}}{R / K_s} < 400$ the pipe surface in transition stage.

(c) In terms of Reynolds number with reference to roughness.

$R_e = \frac{u_* K_s}{v}$

(1) $\frac{u_* K_s}{v} < 4$ for smooth boundary

(2) $\frac{u_* K_s}{v} > 100$ for rough boundary

(3) $4 < \frac{u_* K_s}{v} < 100$ for transition boundary.

11.8 VELOCITY DISTRIBUTION FOR TURBULENT FLOW IN SMOOTH AND ROUGH PIPES-PRANDTL KARMAN VELOCITY DISTRIBUTION EQUATION

Prandtl's equation (equation 11.14) gives a general non-dimensional form for the logarthmic velocity distribution in the central region of the pipe where the turbulent flow is completely developed. This equation, however, is not applicable to the points near the pipe wall because the velocity distribution in the laminar sub layer is entirely different. At y = 0, equation (11.13) gives the value of u equal to – ∞ as (minus infinity). This is absurd and is due to the fact that equation that equation (11.13) is not applicable at y = 0, At a certain distance y′ from the wall, the velocity is zero (Fig. 11.9).

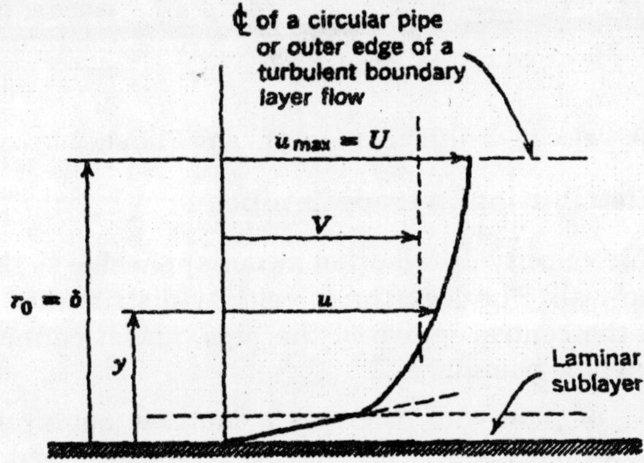

Fig. 11.9. Velocity profile near a pipe wall (vertical scale greatly exaggerated)

The constant C can be obtained by putting y = y′ and u = 0 in the equation (11.12).

$$0 = \frac{u_*}{k} \ln y' + C$$

$$C = -\frac{u_*}{k} \ln y'$$

Substituting the value of C in equation (11.12)

$$u = \frac{u_*}{k} \ln y - \frac{u_*}{k} \ln y' = \frac{u_*}{k} \ln \frac{y}{y'}$$

Putting k = 0.4

$$u = 2.5\, u_* \ln \frac{y}{y'}$$

or, $$u = 2.5 \times 2.3\, u_* \log_{10} \frac{y}{y'}$$

$$= 5.75\, u_* \log_{10} \left(\frac{y}{y'}\right) \qquad (11.16)$$

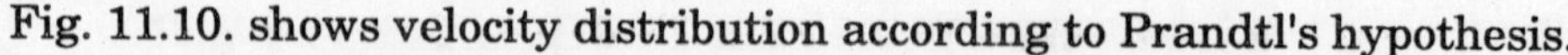

Fig. 11.10. shows velocity distribution according to Prandtl's hypothesis.

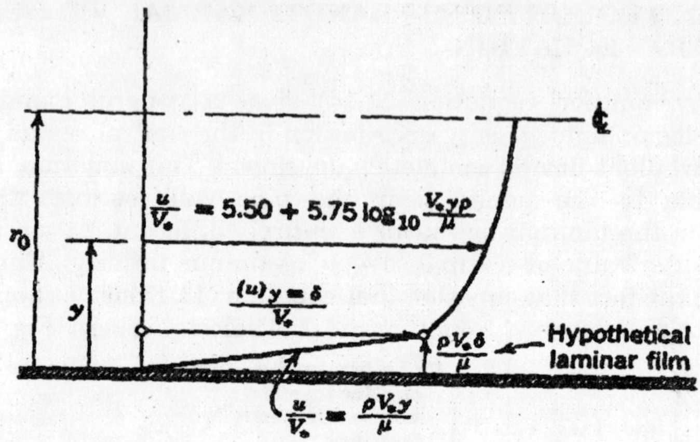

Fig. 11.10. Velocity distribution according to Prandtl's hypothesis.

11.8.1 Velocity distribution in smooth pipes

The logarthmic velocity distribution in not applicable to the region in the vicnity of the pipe wall. The logarthmic velocity distribution applies only to turbulent flow in the central region of the pipe, and it cannot be applied to regions close to the pipe boundary.

In the vicnity of pipe wall, there is a laminar sublayer in which the velocity distribution is parabolic. Therefore, it can be assumed that the flow is laminar with a constant velocity gradient. Fig. 11.11 and velocity distribution in a turbulent boundary layer is shown in Fig. 11.12.

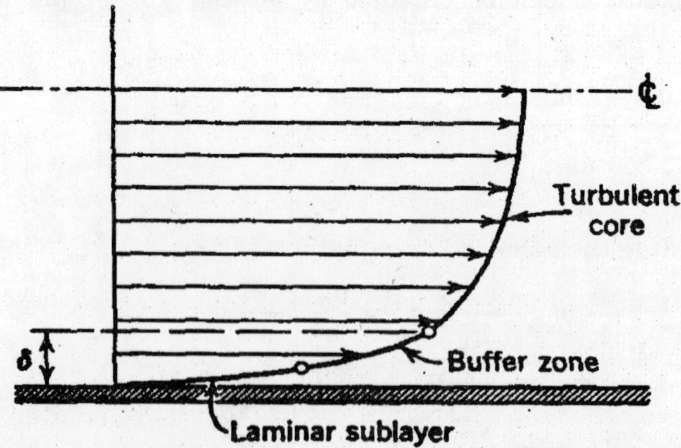

Fig. 11.11. Velocity distribution in a turbulent boundary layer.

That is,

$$\left(\frac{du}{dy}\right)_{\delta' - 0} = \frac{y}{f'} = \frac{\tau_o}{\mu}$$

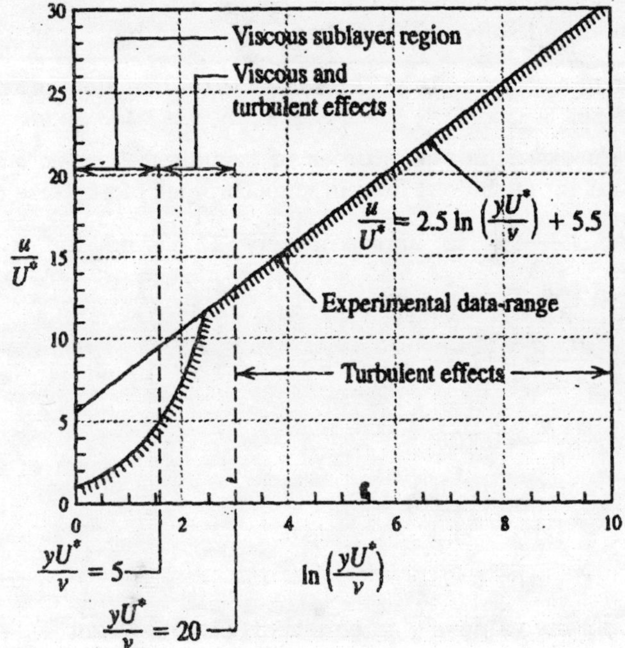

Fig. 11.12. Graph of experimental data between verses $ln\left(\frac{yu}{\nu}\right)\frac{u}{u_*}$

or, $$\frac{u}{y'} = \frac{\tau_o}{\rho} \times \frac{\rho}{\mu}$$

or, $$\frac{u}{y'} = \frac{u_*^2}{v}$$

or, $$\frac{u}{u_*} = \frac{u_* y'}{v} \tag{11.17}$$

or, expressed in a more general non dimensional form,

$$\frac{u}{u_*} = \frac{u_* \delta'}{\nu} = \text{Constant} \tag{11.18}$$

The terms $\frac{u_* y'}{\nu}$ and $\frac{u_* \delta'}{\nu}$ on the right hand side of equations (11.17) and (11.18) are dimensionally equivalent to a form of Reynold's number.

Since the velocity distribution is assumed to be linear within the hypothetical laminar film of thickness δ', the non dimensional equations (11.17) and (11.18) are significant in that

(1) the $\frac{u}{u_*}$ is a linear function of same modified Reynold's number $u_* y'/\nu$;

(2) the modified Reynold's number $\frac{u_* \delta}{\nu}$ and has a constant numerical value of approximately 11.6 for all turbulent flow in smooth pipes of any size.

Nikuradse's experimental studies of turbulent flow in a smooth pipes have shown that for turbulent flow in smooth pipes of any size of the value of the parameter $\left(\frac{u_* y}{\nu}\right)$ is approximately 11.6 and for y = y′ it is approximately 0.108. That means

$$\frac{u_* \delta}{\nu} = 11.6 \tag{11.19}$$

and
$$\frac{u_* y'}{\nu} = 0.108 \tag{11.20}$$

Solving (11.19) and (11.20), we get

$$y' = \frac{\delta}{107} \tag{11.21}$$

Substituting the value of y' in equation (11.16), we get

$$\frac{u}{u_*} = 5.75 \quad \log_{10} \frac{y}{\frac{0.108\,\nu}{u_*}}$$

$$\frac{u}{u_*} = 5.75 \ \log_{10} \left(\frac{u_* y}{0.108\,\nu}\right)$$

$$= 5.75 \log_{10} \frac{u_* y}{\nu} + 5.75 \log_{10} \left(\frac{1}{0.108}\right)$$

$$= 5.75 \log_{10} \frac{u_* y}{\nu} + 5.75 \log_{10} (9.259)$$

$$= 5.75 \ \log_{10} \frac{u_* y}{\nu} + 5.5$$

$$\frac{u}{u_*} = 5.5 + 5.75 \ \log_{10} \frac{u_* y}{\nu} \tag{11.22}$$

The equation (11.22) is known as the **Karman-Prandtl equation** for velocity distribution in turbulent flow near hydrodynamically **smooth boundaries.**

11.8.2 Velocity distribution in rough pipes

In the case of hydrodynamically rough boundaries, the height Ks, of the roughness elements is large in comparison with the laminar sublayer (Fig. 11.8c), create additional turbulence in the flow. In this case laminar sub layer is destroyed and hence one would except y' to be proportional to the average height of protrusions Ks.

Nikuradses experimental data on velocity distribution measurements for turbulent flow in artifically roughned pipes with Ks >> δ, gave

$$y' = \frac{K_s}{30} \tag{11.23}$$

Substituting the value of y′ in equation (11.20), we get.

$$\frac{u}{u_*} = 5.75 \log_{10} \frac{y}{K_s/30} = 5.75 \log_{10}\left(\frac{y}{K_s}\right) + 5.75 \ \log_{10} 30$$

$$= 5.75 \ \log_{10} y/K_s + 8.5 \tag{11.24}$$

This is the **Karman-Prandtl equation** for velocity distribution in turbulent flow near **rough boundaries**.

11.9 VELOCITY DISTRIBUTION FOR TURBULENT FLOW IN TERMS OF AVERAGE VELOCITY

In the last article (11.8), velocity distribution equation for turbulent flow in smooth (equation 11.22) and rough (equation 11.24) pipes were derived separately. However, these two equations can be transformed to identical form if both are expressed in terms of the average velocity of flow in pipe.

By definition, the average velocity u_{av}, is obtained by

$$u_{av} = \frac{Q}{A} = \frac{\text{Total discharge}}{\text{Area of the pipe}}$$

$$= \frac{Q}{\pi R^2} = \frac{\int_0^R u(2\pi r)\,dr}{\pi R^2} \tag{11.25}$$

Consider an elementary ring of thickness dr at a radial distance *r* from the centre of the pipe at a distance *y* from the pipe boundary as shown in Fig. 11.13.

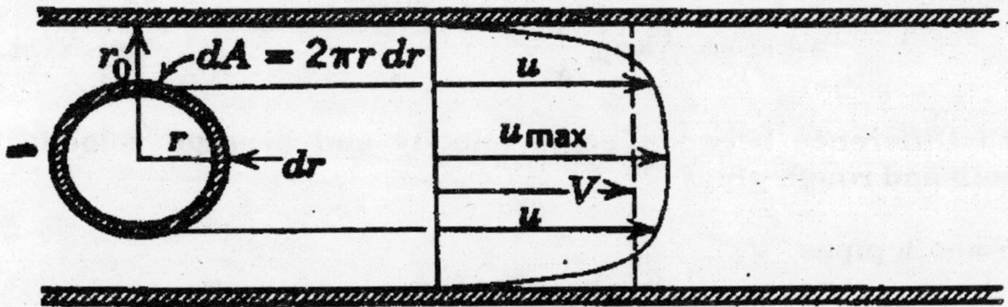

Fig. 11.13. Circular pipe with elemental ring.

Thus $\quad y = R - r$

Where R is the radius of the pipe.

11.9.1 Turbulent flow in smooth pipes

In a turbulent flow in smooth pipes, equation (11.22) gives the velocity u as

$$u = u_* \left(5.5 + 7.5 \log_{10} \frac{u_* y}{\nu}\right)$$

$$u = u_* \left(5.5 + 7.5 \log_{10} \frac{u_* (R - r)}{\nu}\right)$$

Substituting the above value of u in equation (11.25), we obtain

$$u_{av} = \frac{1}{\pi R^2} \int_0^R u_* \left(5.5 + 5.75 \log_{10} \frac{u_* (R - r)}{\upsilon}\right) 2 \pi r \, dr$$

Integrating and simplifying the above expression

$$\frac{u_{av}}{u_*} = 1.75 + 5.75 \log_{10} \frac{u_* R}{\nu} \quad (11.26)$$

11.9.2 Turbulent flow in rough pipes

In a turbulent flow in rough pipes, equation (11.24) gives the velocity u as

$$u = u_* \left(8.5 + 5.75 \log_{10} \frac{y}{K_s}\right)$$

$$= u_* \left(8.5 + 5.75 \log_{10} \frac{(R - r)}{K_s}\right)$$

Substituting the above value of u in equation (11.25), we obtain

$$u_{av} = \frac{1}{\pi R^2} \int_o^R u_* \left(8.5 + 5.75 \log_{10} \frac{(R - r)}{K_s}\right) 2 \pi r \, dr$$

Integrating and simplifying the above expression

$$\frac{u_{av}}{u_*} = 4.75 + 5.75 \log_{10} \frac{R}{K_s} \quad (11.27)$$

11.9.3 Difference between point velocity and average velocity for smooth and rough pipes

(a) Smooth pipes

The point velocity is given by the equation (11.22)

$$\frac{u}{u_*} = 5.5 + 5.75 \log_{10} \frac{u_* y}{\nu}$$

or, $$\frac{u}{u_*} = 5.5 + 5.75 \log_{10} \frac{u_* (R - r)}{\upsilon}$$

and the average velocity is given by the equation (11.26)

$$\frac{u_{av}}{u_*} = 1.75 + 5.75 \log_{10} \frac{u_* R}{\upsilon}$$

Difference of the above two velocities is given by

$$\frac{u}{u_*} - \frac{u_{av}}{u_*} = 5.5 + 5.75 \log_{10} \frac{u_* (R - r)}{\upsilon} - \left[1.75 + 5.75 \log_{10} \frac{u_* R}{\upsilon}\right]$$

$$\frac{u - u_{av}}{u_*} = (5.5 - 1.75) + 5.75 \left[\log_{10} \frac{u_* (R - r)}{\upsilon} - \log_{10} \frac{u_* R}{\upsilon}\right]$$

$$= 3.75 + 5.75 \log_{10} \frac{u_* (R - r)}{\upsilon} \times \frac{\upsilon}{u_* R}$$

$$= 3.75 + 5.75 \log_{10} \left(\frac{R - r}{R}\right) = 3.75 + 5.75 \log_{10} \frac{y}{R} \qquad (11.28)$$

(b) Rough pipes

The point velocity is given by the equation

$$\frac{u}{u_*} = 8.5 + 5.75 \log_{10} \left(\frac{y}{K_s}\right)$$

and the average velocity is given by the equation

$$\frac{u_{av}}{u_*} = 4.75 + 5.75 \log_{10} \frac{R}{K_s}$$

Difference of the above two velocities is given by

$$\frac{u - u_{av}}{u_*} = (8.5 - 4.75) + 5.75 \left[\log_{10} \frac{y}{K_s} - \log_{10} \frac{R}{K_s}\right]$$

$$= 3.75 + 5.75 \log_{10} \frac{y}{K_s} \times \frac{k_s}{R} = 3.75 + 5.75 \log_{10} \frac{y}{R} \qquad (11.29)$$

Equation (11.28) and (11.29) are identical and independent of kinematic viscosity υ and roughness factor K_s.

Thus the velocity distribution for the turbulent core is seen to be identical for both smooth and rough pipes because the mechanism of turbulent motion of fluid in the region outside the boundary layer is essentially independent of the conditions at the boundary surface. Therefore, when equation (11.28) and (11.29) are used to plot a velocity distribution curve in rough pipes, the pipe wall roughness k need not be known. Thus the two equations (11.28) and (11.29) have a definite advantage over the special equation (11.24), which are for rough pipe only.

11.10 TURBULENT PIPE COEFFICIENT

The ratio of average velocity to maximum velocity (u_{av}/u_{max}) which occurs at the centre of the pipe is sometimes called the pipe coefficient or pipe factor. For laminar flow in round pipes the pipe coefficient is always equal to 0.5. The expression for the turbulent pipe coefficient may be derived from equation (11.28) or (11.29)

$$\frac{u - u_{av}}{u_*} = 3.75 + 5.75 \log_{10} \frac{y}{R}$$

From equation (11.11)

$$u_* = u_{av} \sqrt{\frac{\lambda}{8}} \qquad \text{[where } \lambda = 4f]$$

Substituting in above equation

$$\frac{u - u_{av}}{u_{av}\sqrt{\frac{\lambda}{8}}} = 3.75 + 5.75 \log_{10} \frac{y}{R}$$

or, $$\frac{u}{u_{av}} - 1 = \sqrt{\frac{\lambda}{8}} \left[3.75 + 5.75 \log_{10} \frac{y}{R}\right]$$

$$\frac{u}{u_{av}} = 1 + \sqrt{\lambda}\left(1.33 + 2 \log_{10} \frac{y}{R}\right) \qquad (11.30)$$

Which indicates the velocity distribution for turbulent flow to be dependent on λ, regardless of whether the pipe wall is smooth or rough. However, better agreement with Nikuradse's experimental data occurs when the constant 1.33 and 2 are changed to 1.43 and 2.15, respectively, equation (11.30) then becomes

$$\frac{u}{u_{av}} = 1 + \sqrt{\lambda}\left(1.43 + 2.15 \log_{10} \frac{y}{R}\right) \qquad (11.31)$$

By Substituting the conditions at the centre of the pipe, that is $u = u_{max}$ at $y = R$, into equation (11.31), the turbulent pipe coefficient

$$\frac{u_{max}}{u_{av}} = 1 + 1.43\sqrt{\lambda} \qquad (11.32)$$

or, $$\frac{u_{av}}{u_{max}} = \frac{1}{1 + 1.43\sqrt{\lambda}} \qquad (11.33)$$

The pipe coefficient equation (11.33) is seen to be a function of coefficient of friction or friction factor (λ = 4 f) alone.

11.11 THE CHRONOLOGICAL DEVELOPMENT OF TURBULENT PIPE FLOW THEORIES

Table 11.1 lists the names of the main contributors, and their contributions, to turbulent pipe flow theories in chronological order.

The Colebrook–White transition formula represent the culmination of all the previous work, and can be applied to any fluid in any pipe operating under turbulent flow conditions. The later contributions of Moody, Ackers and Barr etc are mainly concerned with the practical application of Colebrook–White equation.

In the initial research, the original investigators persumed that the friction factor (λ or 4f) was constant. This was subsequently found to be incorrect. Equations relating λ or 4f to both the Reynold's number and the pipe roughness were developed later.

Table 11.1 The chronological development of turbulent pipe flow theories

S.N.	Year	Name	Contribution.
1.	1913	Blasius	Friction factor equation for smooth pipes.
2.	1914	Stanton & Pannell	Experimental values of the friction factor for smooth pipes.
3.	1930	Nikuradse	Experimental values of the friction factor for artifically rough pipes.
4.	1930s	Prandtl & Von Karman	Equation for rough and smooth friction factors.
5.	1937-1939	Colebrook & White	Experimental values of the friction factor for commercial pipes and transition formula.
6.	1944	L.F. Moody	The Moody diagram for commercial pipes
7.	1958	Ackers	The hydraulic research station charts and tables for the design of pipes and charts.
8.	1975	Barr	Direct solution of the Colebrook-White equation.
9.	1976	J W Murdock	No iteration is needed for sufficient accurancy D in most problems.
10.	1976	Swamee & Jain	Simple and more accurate explicit equation for friction factor for flow in commercial pipes.
11.	1983	S.E. Haaland	The best yet produced. It combines reasonable simplicity with acceptable accuracy (with in about 1.5%) and is useful if an algebraic expression is required.

11.11.1 Smooth pipes and Blasius equation

At the beginning of the 20th century, Blasius came to the important conclusion that there were two pipes of pipe friction in turbulent flow. The first type be associated with smooth pipes where the viscosity effects predominate so that the friction factor is dependent solely on the Reynold's number [$\lambda = \Phi\left(R_e\right)$] the second type was relevant to rough pipes where the viscosity and roughness effects influence the flow and friction factor is dependent both on the Reynold's number and a parameter of relative roughness.

On the basis of his own researches and of the experimental data supplied by Saph and Schoder, Blasius deduced the following expression for the friction in smooth pipes.

$$\lambda = \frac{0.316}{(R_e)^{0.25}} \tag{11.34}$$

11.11.2 Stanton and Pannell

Stanton and Pannell working at the National Physical laboratory, investigated in considerable detail the head loss in smooth brass pipes conveying both air and water. On a dimensionless graph of λ (actually τ_0/u_{av}^2) against R_e the plotted points were found to be compactly situated along a single curve. The curve in the turbulent zone agreed quite well with the Blasius formula for values of R_e up to 10^5, but after this point there was increasing divergence, indicating that the power by which u should be raised was not constant but increased with R_e. Confirmation of this characteristic was provided by the experimental results of other investigators.

11.11.3 Nikuradse experimental results using artificially rough pipes

The next major experimental contributions were made by Nikuradse at Gottingen in the period (1933).

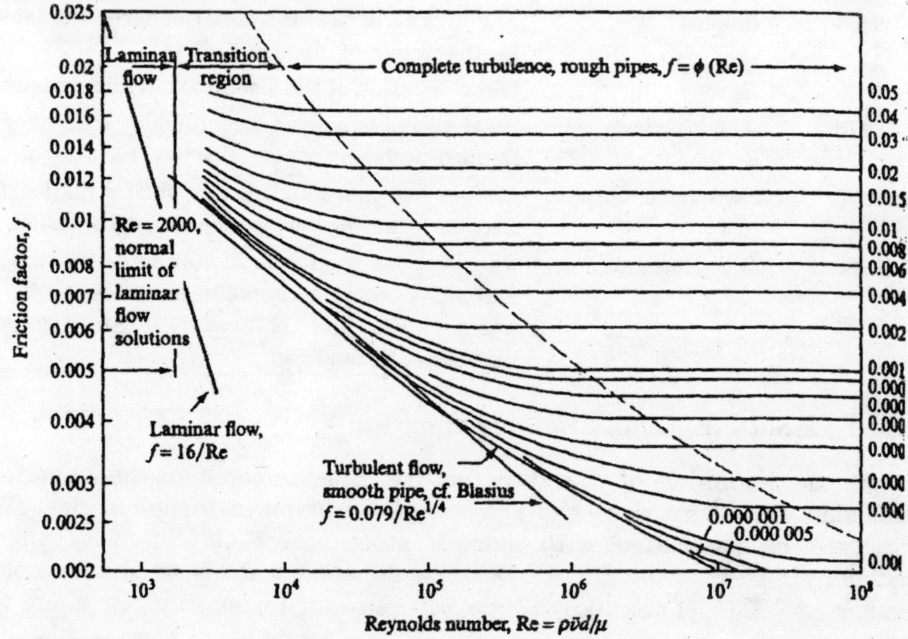

Fig. 11.14. Nikurdse experimental results.

Nikuradse made a major contribution to the theory of pipe flow by objectively differentiating between smooth and rough turbulence in pipes. He carried out a painstaking series of experiments to determine both the friction factor and the velocity distributions at various Reynold's numbers up to 3×10^6. In these experiments, pipes were artificially roughened by sticking uniform sand grains on to smooth pipes. He defined the relative roughness (K_s/D) as the ratio of the sand grain size to the pipe diameter. By using pipes of different diameter and sand grains of different size, he produced a set of experimental results of λ and R_e for a six values of K_s/D, ranging from $K_s/D = \frac{1}{30}$ to $K_s/D = 1/1014$. The test data produced the series of smooth curves shown in Fig. 11.14. These curves were found to be defined not by D but by the ratio K_s/D.

He plotted his results as log λ against log R_e for each value of K_s/D, as shown in Fig. 11.14. This figure shows that these are five regions of flow as follows.

(a) Laminar flow

The laminar portion is a straight line and is unaffected by the degree of roughness. In this region the relative roughness has no influence on the friction factor λ.

Hagen-Poiseuille equation equated with Darcy-Weisbach equation gives

$$\lambda = \frac{64}{R_e} \tag{11.35}$$

Darcy-Weisbach equation may also be used for laminar flow, provided that λ is evaluated by equation (11.35).

(b) Transition from Laminar to Turbulent flow

An unstable region between R_e = 2000 and 4000. Fortunately, pipe flow normally lies outside this region.

(c) Smooth turbulence

The limiting line of turbulent flow, approached by all values of relative roughness as R_e decreases.

(d) Transitional turbulence

The region in which λ varies with both R_e and K_s/D. The limit of this region varies with K_s/D. In practice, most of pipe flow lies within this region.

(e) Rough turbulence

The region in which λ remains constant for a given K_s/D, and is independent of R_e.

The artificial roughness used by Nikurdse does not simulate geometrically the rather irregular pattern of roughness normally found in commercial pipes. Nikurdses's curves in Fig. 11.14 should not be used directly in solving engineering problems of pipe flow in commercial pipes. However, the excellent results of Nikuradse's tests have been used as a basis for verifying theoretical studies such as those developed by Prandtl and von Karman.

11.11.4 The smooth and rough laws of Prandtl and von Karman

The publication of Nikuradse's experimental results (particularly his velocity distribution measurements) were used by Prandtl and Von Karman to support their own work on turbulent boundary layers. By combining their theories of turbulent boundary layer flows with the experimental results, they derived the semi empirical smooth and rough laws.

(a) Universal resistance law for turbulent pipe flow

The fact that both the velocity distribution and the friction loss in turbulent pipe flow are related to the shearing stress τo at the pipe wall as shown in preceding articles in the equation (11.11)

$$u_* = \sqrt{\frac{\lambda}{8}}\, u_{av}$$

or, $$\lambda = \frac{8\, u_*^2}{u_{av}^2} \tag{11.36}$$

Which may be used in the derivation of universal resistance laws for turbulent flow in smooth and rough pipes.

For turbulent flow in smooth pipes, the average velocity u_{av} of flow can be expressed in a non dimensional form shown in equation (11.26).

$$\frac{u_{av}}{u_*} = 1.75 + 5.75 \log_{10} \frac{u_* R}{\upsilon}$$

Combining equation (11.36) and (11.37) yields

$$\frac{u_{av}}{u_{av}\sqrt{\lambda/8}} = 1.75 + 5.75 \log_{10} \frac{u_{av}\sqrt{\lambda/8}\, R}{\upsilon}$$

Which may be simplified and rearranged in the following form

$$\frac{1}{\sqrt{\lambda}} = -0.91 + 2.03 \log_{10} \frac{u_{av}\, D\sqrt{\lambda}}{\upsilon}$$

or, $$\frac{1}{\sqrt{\lambda}} = -\,0.91 + 2.03 \log_{10} \left(R_e \sqrt{\lambda}\right) \tag{11.38}$$

By plotting Nikurdse's experimental data of turbulent flow in smooth pipes in Fig. 11.15 with $\frac{1}{\sqrt{\lambda}}$ *versus* $\log_{10}\left(R_e \sqrt{\lambda}\right)$ equation (11.38) is found to confirm with experimental results after the constants are changed slightly, the equation then becomes

$$\frac{1}{\sqrt{\lambda}} = -0.8 + 2\log_{10}\left(R_e\sqrt{\lambda}\right) \quad (11.39)$$

This is Prandtl's universal resistance equation for turbulent flow in smooth pipes. The universal equation (11.39) may be safely used to predict the value of λ for any arbitrarily large Reynold's number of flow beyond the range of available experimental data.

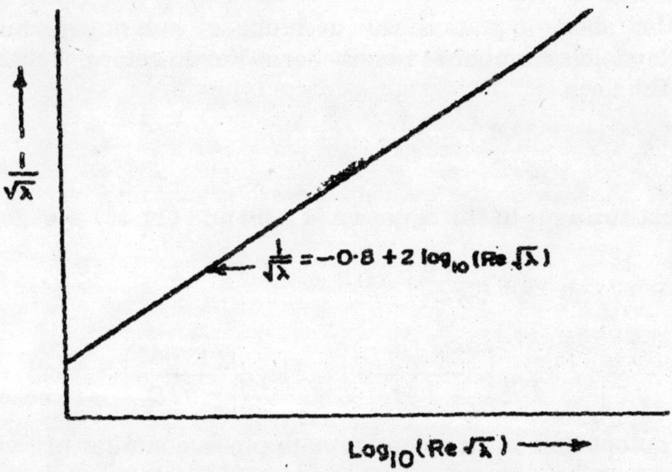

Fig. 11.15. Experimental verification of universal resistance law for turbulent flow in smooth pipe.

The equation (11.39) is simplified in a form given below.

$$\frac{1}{\sqrt{\lambda}} - \log\lambda = 2\log_{10} R_e - 0.8$$

With the help of above equation λ can be calculated by trial and error method.

The value of λ can alternatively be obtained by Nikurdse

$$\lambda = 0.0032 + \frac{0.221}{(R_e)^{0.237}} \quad (11.40)$$

and from Dr. A.K. Jain's equation.

$$\frac{1}{\sqrt{\lambda}} = 1.8\log_{10} R_e - 1.5186 \quad (11.41)$$

Alternatively the two equations 12.40 and 12.41 give approximately the same value of λ as obtained by the equation (11.39).

The equation (11.39) can also be written in a new form as

$$\frac{1}{\sqrt{\lambda}} = 2\log_{10}\left(\frac{R_e\sqrt{\lambda}}{2.51}\right) \quad (11.42)$$

Since the wall shear stress is governed by the viscous force, that is $\tau = \mu$ (du/dy) at the wall, the thickness of the viscous sublayer must be an important parameter in relation to friction effects. The dimensionless ratio δ/D from equation

$$\frac{\delta}{D} = \frac{11.6\,\upsilon}{u_* D} = \frac{11.6\,\upsilon}{u\sqrt{\lambda/8}\,D} = \frac{32.8}{R_e\sqrt{\lambda}}$$

From the above equation the decrease of sub layer thickness with increasing Reynold's number is readily seen. Furthermore, writing the above equation in the form

$$R_e\sqrt{\lambda} = \frac{32.8}{(\delta/D)}$$

and substituting it in the equation (11.39) and (11.42) we get

$$\frac{1}{\sqrt{\lambda}} = 2\log_{10}\left(\frac{32.8}{\delta/D}\right) - 0.8 \quad (11.43)$$

and
$$\frac{1}{\sqrt{\lambda}} = 2\log_{10}\left(\frac{32.8/(\delta/D)}{2.51}\right) = 2\log_{10}\left(\frac{13.067}{\delta/D}\right) \quad (11.44)$$

For turbulent flow in completely rough pipes, a similar procedure may be followed.

$$\frac{u_{av}}{u_*} = 4.75 + 5.75\log_{10}\frac{R}{K_s} \quad (11.45)$$

Substituting $u_* = \sqrt{\frac{\lambda}{8}}\,u_{av}$ in equation (11.45) we get

$$\frac{1}{\sqrt{\lambda}} = 1.68 + 2.03\log_{10}\frac{R}{K_s} \quad (11.46)$$

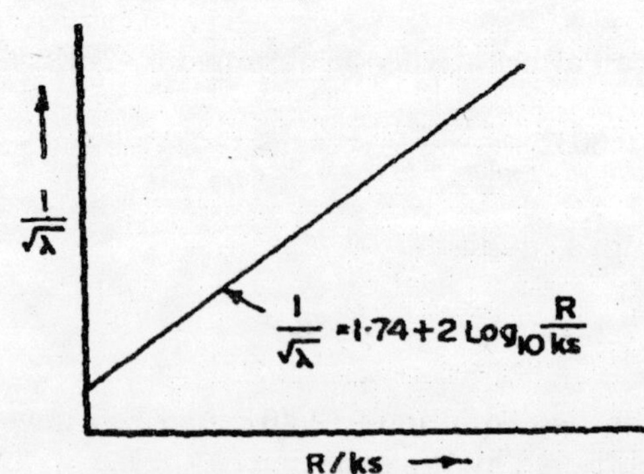

Fig. 11.16. Experimental verification of universal resistance law for turbulent flow in completely rough pipes.

Again a comparison with Nikurdse's experimental data on turbulent flow in completely rough pipes as plotted in Fig. 11.16 shows the slight modification of constant in equation (11.46) will improve the agreement with test results. The modified universal resistance equation for turbulent flow in completely rough pipes is

$$\frac{1}{\sqrt{\lambda}} = 1.74 + 2 \log_{10} \frac{R}{K_s} \tag{11.47}$$

The equation (11.47) can also be written in a new form as

$$\frac{1}{\sqrt{\lambda}} = 2 \log_{10} \frac{3.7\,D}{K_s} \tag{11.48}$$

11.10.5 The Colebrook-White transition formula

The experimental work of Nikurdse and the theoretical work of van Karman and Prandtl provided the frame work for a theory of pipe friction. However, these results were not to direct use to engineers because they applied only to artificially roughned pipes. Commercial pipes have roughness which is uneven both in size and spacing, and do not, therefore, necessarily correspond to the pipes used in Nikurdse's experiment.

Colebrook and White (1938 and 1939) therefore carried out tests at Imperial College with pipes that had been roughened with sand grains which were mixed in size. They discovered that in the turbulent transition region the $\lambda - R_e$ curves exhibited a gradual change from smooth to rough turbulent in constrast to Nikurdse's `S' shaped curves for uniform roughness, size and spacing. Colebrook then went on to determine the effective roughness size of many commercial pipes.

He was thus able to publish a list of K_s values applicable to commercial pipes.

Colebrook and White found that by judiciously combining the Prandtl and von Karman smooth and rough laws in a single expression which would represent the entire turbulent flow range of the λ - R_e diagram. The expression is

$$\frac{1}{\sqrt{\lambda}} = -2 \log_{10}\left(\frac{2.51}{R_e\sqrt{\lambda}} + \frac{K_s}{3.7\,D}\right) \tag{11.49}$$

For $K_s \to 0$ this equation (11.49) transform into equation (11.48), valid for hydraulically smooth pipes.

For $R_e \to \infty$, it transforms into equation (11.48) for the completely rough regime.

The Colebrook and White equation with its asymptotes, are shown in Figs. 11.17 & 11.18

The coordinates of Fig. 11.17 are chosen to be $1/\sqrt{\lambda} - 2\log_{10}\dfrac{3.7\,D}{K_s}$ and

Reynold's roughness number $R_e = \dfrac{u_* K_s}{\upsilon} = R_e\sqrt{\dfrac{\lambda}{8}}\dfrac{K_s}{D}$, so that the smooth pipe equation (11.42) appears as a straight line with a slope of 2, and the rough pipe equation (11.48) plots as a horizontal line.

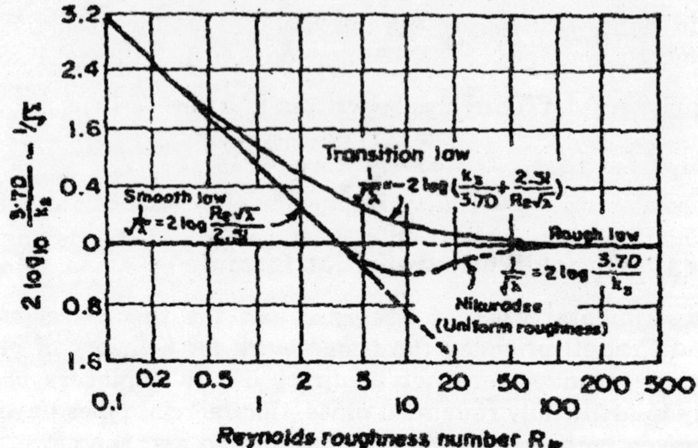

Fig. 11.17. Comparison of transitition zone curves for commercial and uniform roughness.

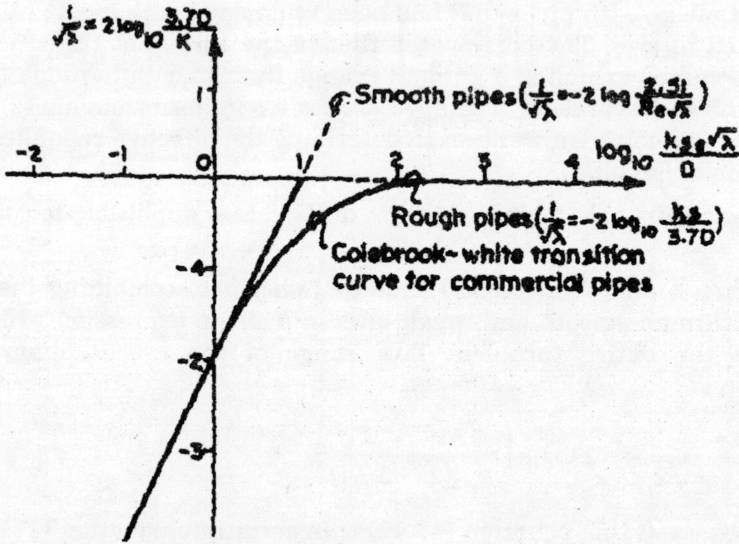

Fig. 11.18. Colebrook-White curve for commercial pipes.

The transition curve, shown in Figs. 12.17 and12.18, was found to be in reasonable agree-ment with the experimental data. An essential feature of the expression is that at low values of R_e, the first term in the brackets predominate and the smooth law is approached; whilst at high values of R_e the second term predominates and the rough law is approached. The curve is asymtotic at approximately those point which the experimental results indicate as being appropriate – that is, at about R* = 0.3 to the smooth law; the departure from smooth pipe law occurs at R* = 4, and the rough pipe law commences approximately where R_* = 60.

A unique curve would be obtained between $\frac{1}{\sqrt{\lambda}} - 2\log_{10}\frac{R}{K_2}$ *and* $\frac{R_e\sqrt{\lambda}}{R/K_s}$ which would define the nature of the boundary. Since the nature of the boundary smooth, rough or transition depends on the value of K_s/δ', it can be expected that a unique curve will define variation of $\frac{1}{\sqrt{\lambda}} - 2\log_{10}\frac{R}{K_s}$ *with* K_s/δ', *or* $\frac{R_e\sqrt{\lambda}}{R/K_s}$ for Nikurdse's data over the entire range of roughness. Such variation is shown in Fig. 11.25. It should be noted that the transition of the curve begins when the ratio K_s/δ' is approximately 0.25 or $\frac{R_e\sqrt{\lambda}}{R/K_s}$ is about 17 and it ends when K_s/δ' is about 6 or $\frac{R_e\sqrt{\lambda}}{R/K_s} \approx 400$. Therefore a pipe will act as a smooth pipe if K_s/δ' is less than 0.25 and will act as a rough when K_s/δ' is greater than 6.

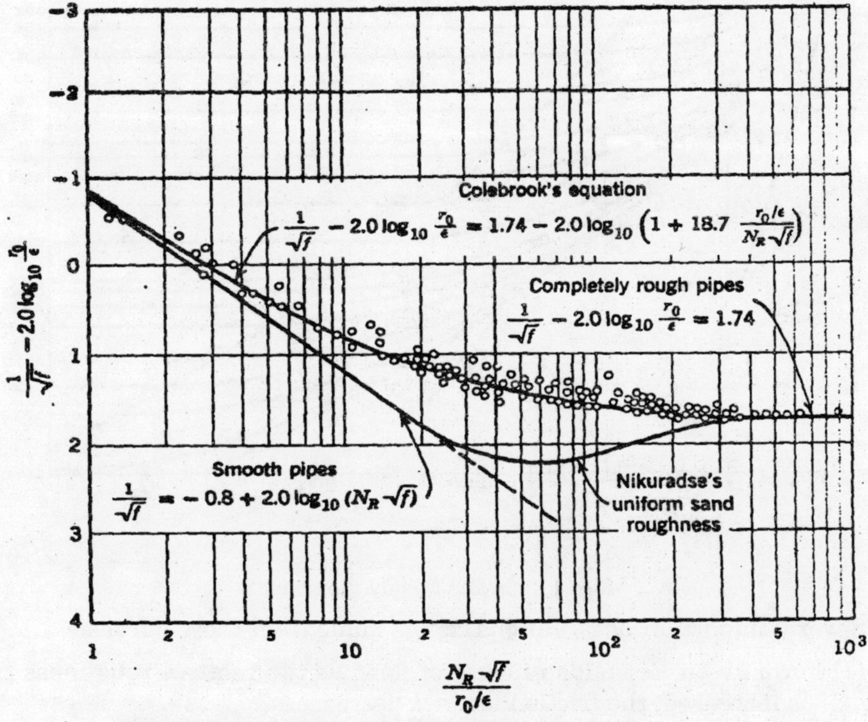

Fig. 11.19. Nikurdse data over the entire range of roughness

Equation (11.49) was not at first used very widely by engineers, mainly because it was not expressed directly in terms of the standard engineering variables of diameter, discharge and hydraulic gradient. In addition, the equation is implicit and requires a trial and error solution. In the 1940s, slide rules and logarithm tables were the main computational aids of the engineer, since pocket calculators and computers were not then available. So these objections to the use of the Colebrook-White equation were not unreasonable.

11.11.6 Moody diagram for commercial pipes

The first attempt to make engineering calculations easier was made by Moody. He produced a $\lambda - R_e$ plot based on equation (11.49) for commercial pipes, as shown in Fig. 11.20 which is now known as the Moody diagram. He also presented an explicit formula for λ.

$$\lambda = 0.0055\left[1+\left(\frac{2000\,K_s}{D}+\frac{10^6}{R_e}\right)^{1/3}\right] \tag{11.50}$$

Which gives λ correct to $\pm 5\%$ *for* $4\times10^3 < R_e < 1\times10^7$ and for $\frac{K_s}{D} <$ 0.01.

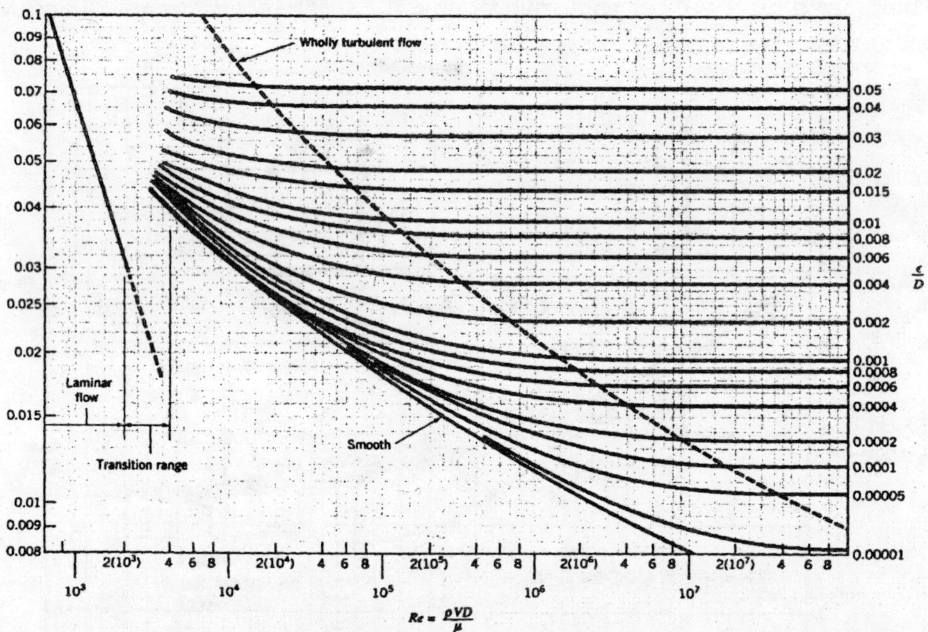

Fig. 11.20. The Moody diagram

Several important observations can be made from these curves.

(1) For a given Reynolds number of flow, as the relative roughness D/K_s is increased, the friction factor λ decreases.

(2) For a given relative roughness D/K_s, the friction factor λ decreases with increasing Reynold's number until the zone of complete turbulence is reached.

(3) Within the zone of complete turbulence, the Reynolds number has no effect on the friction factor λ.

(4) As the relative roughness D/K_s increases, the value of the R_e at which the zone of complete turbulence begins also increases.

Fig. 11.21 shows a chart for fully developed flow in circular pipes Fig. 11.22, 11.23 and 11.24 shows moody diagrams for type 1, type 2 and type 3 pipe flow problems.

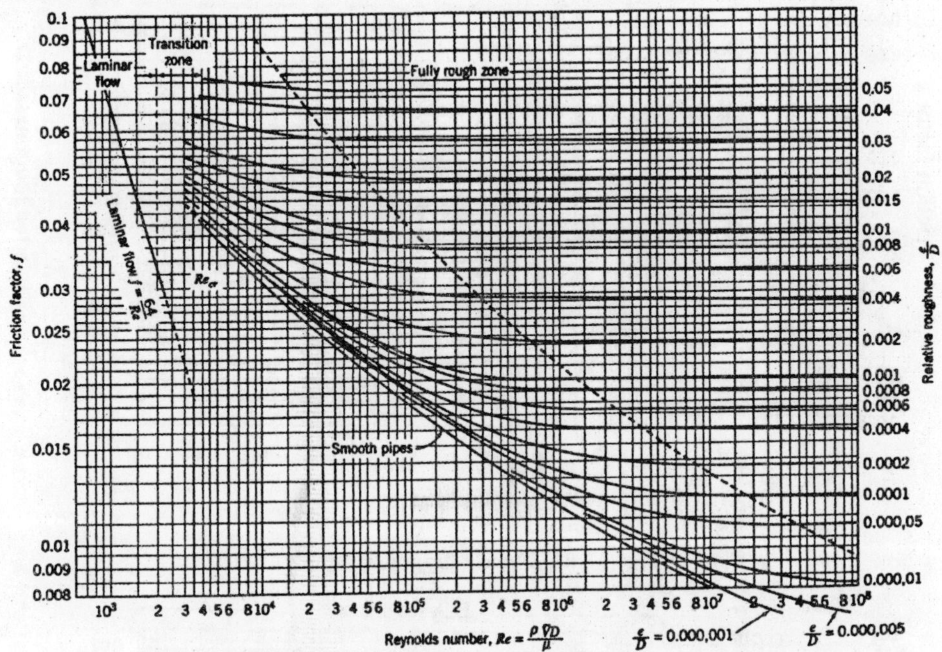

Fig. 11.21. Chart for fully developed flow in a circular pipe.

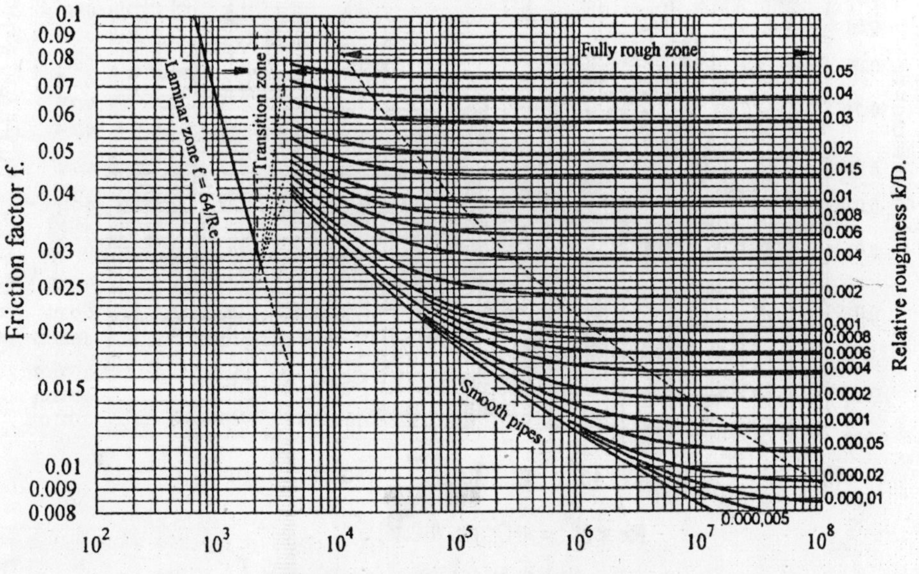

Fig. 11.22. Moody diagram type 1

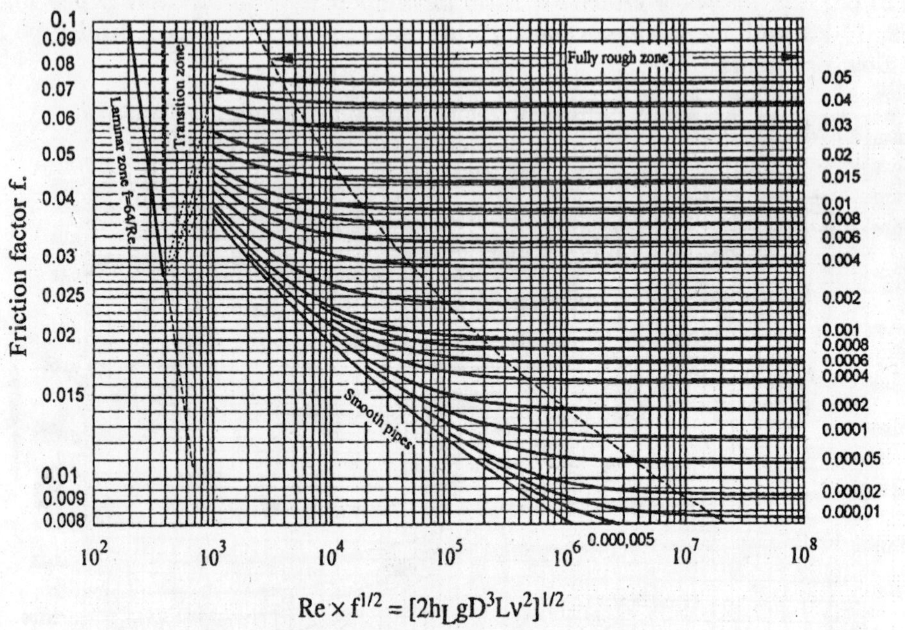

Fig. 11.23. Moody diagram type 2

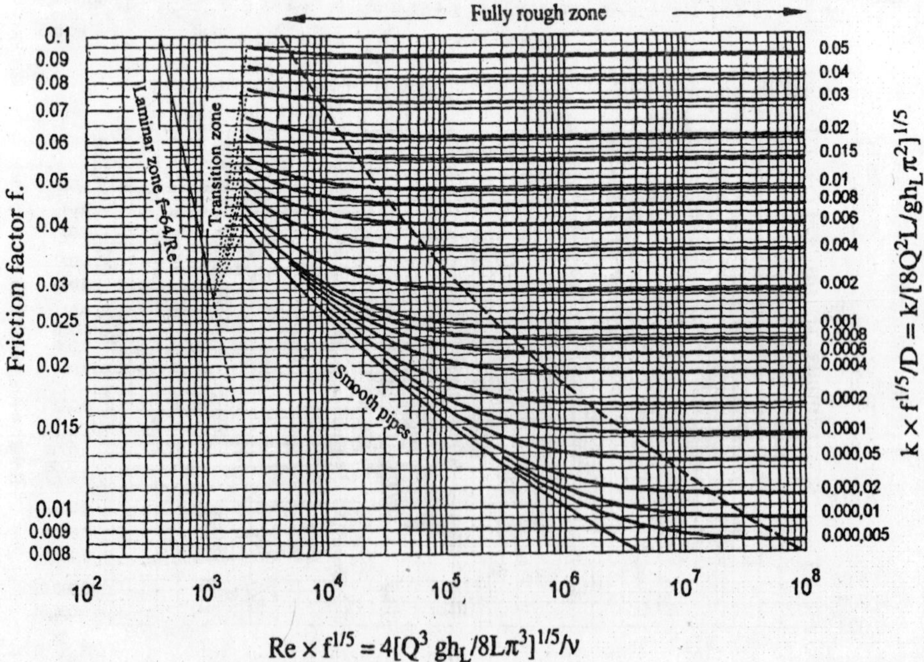

Fig. 11.24. Moody diagram type 3

Fig. 11.25 is a simplified sketch of Moody's diagram in which the various zones are identified.

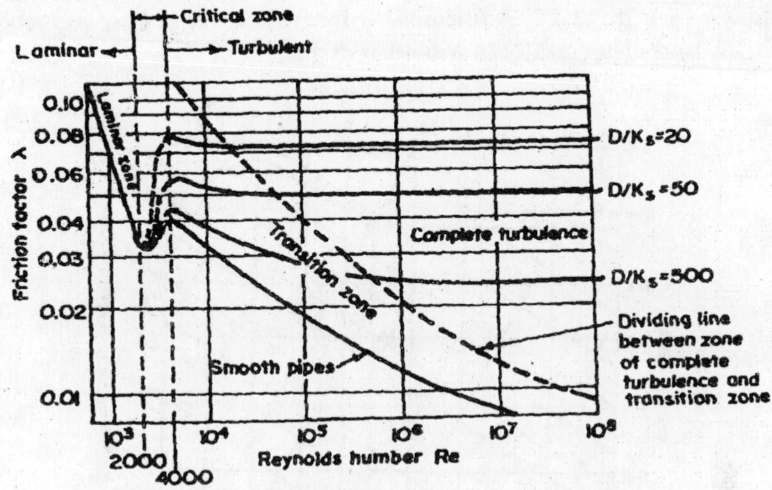

Fig. 11.25. Explanation of parts of Moody diagram.

Fig. 11.26 shows relative surface roughness for various pipes.

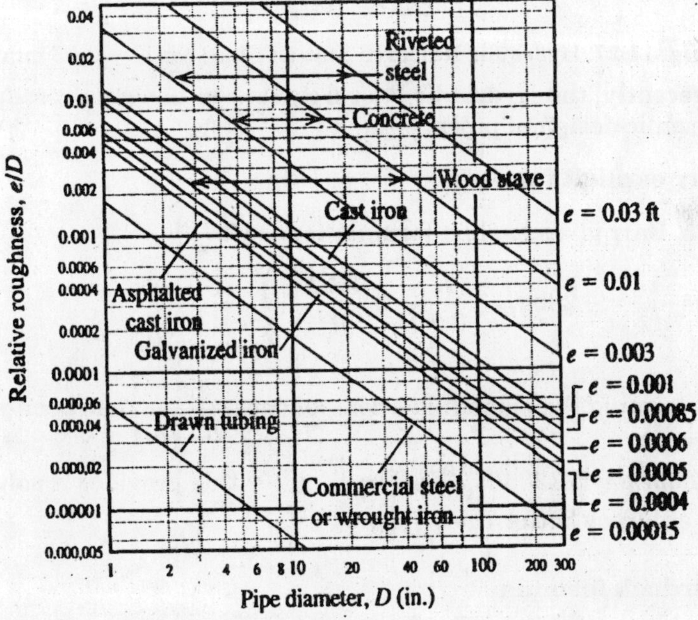

Fig. 11.26. Relative surface roughness for various pipes.

11.11.7 Hydraulic Research station charts (HRS) Acketes

However, the basic engineering objections to the use of the Colebrook White equation were not overcome until the publication of charts for the hydraulic design of channels and pipes in 1958 by the Hydraulic Research Station. In this publication, the three dependent engineering variables (Q, D and S_f) were presented in the form of a series of charts for various values of

K_s, as shown in Fig. 11.27. Additional information regarding suitable design values for K_s and other matters was also included.

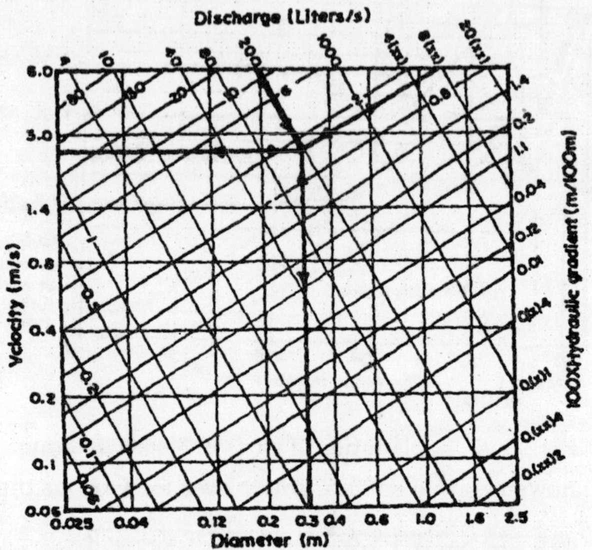

Fig. 11.27. Hydraulic Research Station Chart for K_s = 0.03 mm.

More recently, the Hydraulic Research Station have also produced tables for the hydraulic design of pipes.

11.11.8 Barr explicit formula

In 1975, Barr gives explicit formulation for λ

$$\frac{1}{\sqrt{\lambda}} = -2\log_{10}\left(\frac{K_s}{3.7\,D} + \frac{5.1286}{R_e^{0.99}}\right) \tag{11.51}$$

In this formula the smooth law component $\left(\frac{2.51}{R_e\sqrt{\lambda}}\right)$ has been replaced by an approx-imation $\left(5.1286/R_e^{0.99}\right)$. For $R_e > 10^5$ this provides a solution for S_f (h_f/L) to an accuracy better than ± 1 %.

11.11.9 Murdock formula

In 1976, J.W. Murdock solves the Colebrook-White equation. First, an approximate value for λ can be calculated from

$$\lambda_a = 0.0055\left[1 + \left(\frac{2\times10^4}{D/K_s} + \frac{10^6}{R_e}\right)^{1/3}\right] \tag{11.52}$$

Then, we can use a modified form of equation (11.49) to find the actual λ as follows

$$1/\sqrt{\lambda} = -2\log_{10}\left[\frac{1}{3.7\,D/K_s} + \frac{2.51}{R_e\sqrt{\lambda_a}}\right] \qquad (11.53)$$

Since λ_a is quite close to the actual value of λ, no iteration is needed for sufficient accuracy in most problems.

11.11.10 Swamee and Jain's explicit equation

Dr. A.K. Jain has obtained a simple and more accurate explicit equation for friction factor λ for flow in commercial pipes. The equation has a logarithmic form.

$$\frac{1}{\sqrt{\lambda}} = 1.14 - 2\log_{10}\left(\frac{K_s}{D} + \frac{21.25}{R_e^{0.9}}\right) \qquad (11.54)$$

It is valid for $5000 \le R_e \le 10^8$ and $10^{-6} < K_s / D < 10^{-2}$. This equation yields λ within 1% of the equation (11.49).

11.11.11 S.E. Haaland's formula

The best yet produced is probably that by S.E. Haalland.

$$\frac{1}{\sqrt{\lambda}} = -1.8\log_{10}\left[\frac{6.9}{R_e} + \left(\frac{K_s}{3.71\,D}\right)^{1.11}\right] \qquad (11.55)$$

It combines reasonable simplicity with acceptable (within about 1.5 %) and is useful if an algebraic expression is required.

11.12 NON CIRCULAR PIPES

For non circular pipes it is common practice to use an effective diameter termed the hydraulic diameter D_h which is defined as

$$D_h = 4\left(\frac{A}{P}\right) = 4 \times hydraulic\ radius \qquad (11.56)$$

For regular noncircular shapes of pipes, the friction factors for Turbulent flow can be predicted from Moody's diagram or, HRS chart. The equations already developed for circular pipes may also be used by applying the concept of hydraulic radius. D_h is used in Reynold's equation as $R_e = \dfrac{\rho u D_h}{\mu}$ and in the Darcy's-Weisbach equation as

$$h_f = \frac{\lambda l u^2}{2 g D_h} \qquad (11.57)$$

The above method using Moody's diagram and Reynolds number computed with the help of hydraulic radius gives satisfactory results.

11.13 ROUGHNESS OF PIPES WITH AGE (OLD PIPES)

When the pipes become old, the friction increases with the rust and sedimentation accumulated on the pipe wall.

The value of K_s given are for new and clean pipes. Colebrook-White gave the following equation for the determination of roughness of old pipes.

$$K_s = K_o + \propto t \tag{11.58}$$

where $\propto$ is a coefficient, t is number of years of use of the pipe and K_o is the initial value of K_s.

For pipelines in service for a long period, say 20 years or more, the absolute roughness of old concrete or cement lined steel pipes may increase three times over that of new pipes; for cast-iron pipes the increase may be upto twenty times and for wrought iron and steel pipes it may reach even forty times.

ILLUSTRATIVE EXAMPLES

11.1. Oil (specific gravity S = 0.85) is flowing in a 150 m long smooth pipe line at the rate of 80 liters per second. The loss of head in the pipe line is 67 m. If the kinematic viscosity of the oil is 0.1 stoke. Calculate diameter of the pipe.

Solution

$$Q = \frac{\pi}{4} d^2 u$$

$$80 \times 10^{-3} = \frac{\pi}{4} d^2 u$$

$$u = \frac{80 \times 10^{-3} \times 4}{\pi d^2} = \frac{101.859 \times 10^{-3}}{d^2}$$

$$R_e = \frac{ud}{v} = \frac{101.859 \times d^{1/4}}{0.1 \times 10^{-4} \times d^2} = \frac{10185.916}{d} \tag{1}$$

Using Blasius formula

$$\lambda = \frac{0.3164}{(R_e)^{1/4}} = \frac{0.3164 \times d^{1/4}}{(10185.916)^{0.25}} = 0.03149\ d^{1/4}$$

$$h_f = \frac{\lambda\, l\, u^2}{2g\, d}$$

$$67 = \frac{0.03149\, d^{1/4} \times 150 \times (101.859 \times 10^{-3})^2}{2 \times 9.81 \times d^5}$$

$$67 = \frac{2.4978 \times 10^{-3}}{d^{19/4}}$$

$$d^{19/4} = 3.728 \times 10^{-5}$$

$$d = \left(3.728 \times 10^{-5}\right)^{4/19} = 0.11687 \text{ m} = 116.8 \text{ mm} \qquad \text{Ans}$$

Check

$$R_e = \frac{10185.916}{0.11687} = 87154.5856 \quad \text{Less than } 10^5$$

11.2. **A smooth pipe carries 7 liters of water per second at 20°C (υ = 1 centistoke) with a head loss of 75 mm per 10 m of length. Determine the pipe diameter.**

Solution

$$Q = \frac{\pi}{4} d^2 u$$

$$7 \times 10^{-3} = \frac{\pi}{4} d^2 u$$

$$u = \frac{8.915 \times 10^{-3}}{d^2}$$

$$R_e = \frac{ud}{\upsilon} = \frac{8.915 \times 10^{-3} \times d}{d^2 \times 10^{-6}} = \frac{8.915 \times 10^{-3}}{d} \qquad (1)$$

$$h_f = \frac{\lambda l u^2}{2gd} = \frac{\lambda l}{d} \times \left(\frac{8.915 \times 10^{-3}}{d^2}\right)^2 \times \frac{1}{2 \times 9.81}$$

$$\frac{75}{1000} = \frac{\lambda}{d^5} \times 40.5 \times 10^{-6}$$

$$d^5 = \frac{40.5 \times 10^{-6} \times 1000}{75} \lambda = 54 \times 10^{-5} \ \lambda$$

or, $$\lambda = \frac{d^5}{54 \times 10^{-5}}$$

Using the Blasius formula

$$\lambda = \frac{0.3164}{(R_e)^{1/4}} = \frac{0.3164}{\left(\frac{8.915 \times 10^3}{d}\right)^{1/4}}$$

or, $$\frac{d^5}{54 \times 10^{-5}} = \frac{0.3164\, d^{1/4}}{\left(8.915 \times 10^3\right)^{1/4}}$$

or, $$d^{19/4} = \frac{54 \times 10^{-5} \times 0.3164}{9.716} = 1.7583 \times 10^{-5}$$

$$d = 0.100392 \text{ m} = 100.3 \text{ mm}$$

Therefore $$R_e = \frac{8.915 \times 10^{+3}}{0.100392} = 88801.89 \text{ Less than } 10^5$$

11.3. Determine (a) the shear stress at the walls of a 300 mm diameter pipe when water flowing causes a measured lost head of 15 mm in 300 m of pipe length, (b) the shear stress 50 mm from the centre line of the pipe, (c) shear velocity, (d) the average velocity for a value of 0.05 and (e) the ratio of u/u*.

Solution

(a) $$\tau_o = \frac{\rho g h_f R}{2l}$$

$$= \frac{1000 \times 9.81 \times 15 \times 0.15}{2 \times 300} = 36.7875 \text{ N/m}^2$$ **Ans**

(b) The shear stress varies linearly from centre line to wall

$$\frac{\tau}{r} = \frac{\tau_o}{R}$$

$$\tau = \frac{\tau_o r}{R} = \frac{36.7875 \times 50}{150} = 12.2625 \; N/m^2$$ **Ans**

(c) $$u_* = \sqrt{\frac{\tau_o}{\rho}} = \sqrt{\frac{36.7875}{1000}} = 0.1918 \; m/s$$ **Ans**

(d) $$h_f = \frac{\lambda l u^2}{2g\, d}$$

$$15 = \frac{0.05 \times 300 \times u_{av}^2}{2 \times 9.81 \times 0.3}$$

$$u_{av}^2 = 5.886$$

Therefore,

$$u_{av} = 2.426 \; m/s$$

The average velocity = 2.426 m/s **Ans**

(e) $$\tau_o = \frac{\mu u}{y}$$

or, $$\tau_o = \frac{\rho \upsilon u}{y}$$

or, $$\frac{\tau_o}{\rho} = \frac{\upsilon u}{y}$$

or, $$u_*^2 = \frac{\upsilon u}{y}$$

or, $$\frac{u}{u_*^2} = \frac{y}{\upsilon}$$

or, $$\frac{u}{u_*} = \left(\frac{y}{\upsilon}\right) u_*$$ **Ans**

11.4. **When 0.28 m³/s of water flow in a 300 mm pipeline, 63 kW are lost in friction in 300 mm of pipe. Calculate head loss, friction factor, friction velocity, and shear stress at the pipe wall.**

Solution

$$P = \rho\ g\ Q\ h_f$$

$$63 \times 1000 = 1000 \times 9.81 \times 0.28 \times h_f$$

$$h_f = \frac{63}{9.81 \times 0.28} = 22.93\ m$$ **Ans**

$$h_f = \frac{\lambda\ l\ u^2}{2g\ d}$$

$$22.93 = \frac{\lambda\ l\ Q^2}{12.1\ d^5} = \frac{\lambda \times 300 \times (0.28)^2}{12.1 \times (.3)^5}$$

$\lambda = 0.0286$ **Ans**

$$Q = \frac{\pi}{4}(d)^2\ u$$

$$0.28 = \frac{\pi}{4}(0.3)^2\ u$$

u = 3.96 m/s

$$u_* = u\sqrt{\frac{\lambda}{8}} = (3.96)\sqrt{\frac{0.0286}{8}} = 0.237$$ **Ans**

$$\tau_o = \rho\ u_*^2 = 1000 \times (0.237)^2 = 56.169 = 56.169\ \text{Pa}$$ **Ans**

11.5. **Fluid of density 1030 kg/m³ and kinematic viscosity 1.86 × 10⁻⁵ m²/s flows parallel to a very smooth plane surface. The velocities at 75 mm and 4.4 mm from the wall are measured and found to be 0.3 m/s and 0.08 m/s, respectively. Using both of these measurements calculate the shearing stress on the surface.**

Solution

The locations and magnitudes of velocities suggest that one is in the turbulent region, the other in the viscous sub layer.

We know that

$$\frac{u}{u_*} = 5.5 + 5.75 \log_{10}\left(\frac{u_* y}{\upsilon}\right)$$

$$\frac{0.3}{u_*} = 5.5 + 5.75 \log_{10}\left(\frac{u_* \times 0.075}{1.86 \times 10^{-5}}\right)$$

Solving u* by trial and error method.

u* = 0.018 m/s

In the laminar film

$$\tau_o = \rho u_*^2 = \frac{\upsilon \rho u}{y}$$

Therefore, $u_*^2 = \frac{1.86 \times 10^{-5} \times 0.08}{0.0044} = 3.38 \times 10^{-4}$

u* = 0.01838 as before

Which seems to confirm the original assumption.

For friction confirmation calculate δ

We know that

$$\frac{u_* \delta}{\upsilon} = 11.6$$

or, $\frac{0.018 \delta}{1.86 \times 10^{-5}} = 11.6$

$$\delta = 0.01198$$

Since $\delta > y$

or, $0.01198 > 0.0044$,

u = 0.08 m/s is the viscous sub layer

Therefore,

$$\tau_o = \rho u_*^2 = 1030 \times (0.018)^2 = 0.33372 \text{ N/m}^2$$ **Ans**

11.6. **A pipe 200 mm diameter conveys water. Measurements indicate that velocity at the pipe centre and that at a distance 50 mm from the pipe centre are 2.5 m/s and 2 m/s respectively. The flow in the pipe is turbulent, calculate the shear velocity and the wall shear stress.**

Solution

Using the velocity deficiency equation

$$\frac{u_{max} - u}{u_*} = 5.75 \log_{10}\left(\frac{R}{y}\right)$$

$$\frac{2.5 - 2}{u_*} = 5.75 \log_{10}\left(\frac{100}{50}\right)$$

$$\frac{0.5}{u_*} = 1.73092$$

$$u_* = 0.288\ m/s$$ Ans

$$u_* = \sqrt{\frac{\tau_o}{\rho}}$$

$$\tau_o = \rho u_*^2 = 1000 \times (0.288)^2 = 83.442 \text{ N/m}^2$$ Ans

11.7. **Crude oil of relative density 0.86 and kinematic viscosity 1.3 $\times 10^{-5}$ m^2/s is discharged through a 300 mm diameter pipe line at the rate of 100 liters/s. Determine (a) the centre line velocity and (b) the velocity just above the laminar sub layer.**

Solution

$$Q = \frac{\pi}{4} d^2 u$$

$$100 \times 10^{-3} = \frac{\pi}{4}(0.3)^2 \times u$$

u = 1.4147 m/s

$$R_e = \frac{ud}{\upsilon} = \frac{1.4147 \times 0.3}{1.3 \times 10^{-5}} = 32646.923$$

The value of R_e number comes out to be less than 10^5. The Blasius formula can be used for finding the friction factor

$$\lambda = \frac{0.3164}{(R_e)^{1/4}} = \frac{0.3164}{(32646.923)^{0.25}} = \frac{0.3164}{13.44} = 0.0235$$

$$\sqrt{\lambda/8} = \sqrt{\frac{0.0235}{8}} = 0.054198$$

$$u_* = u_{av}\sqrt{\lambda/8} = 1.4147 \times 0.054198 = 0.0766$$

For finding the value of u_{max}

$$\frac{u_{max}}{u_{av}} = 1 + 1.43\sqrt{\lambda} = 1 + 1.43\sqrt{0.0235}$$

$$u_{max} = 1.4147\,(1 + 0.2192)$$

$$u_{max} = 1.7248\ m/s$$ Ans

(b) Hence the velocity just above the laminar sub layer

$$\tau_o = \mu\frac{u}{\delta} = \frac{\upsilon \rho u}{\delta}$$

Therefore, $u = \frac{\tau_o \delta}{\rho \upsilon} = \frac{\tau_o}{\rho} \times \frac{1}{\upsilon} \times \frac{11.6\,\upsilon}{u_*} = \frac{\tau_o}{\rho} \times \frac{11.6}{\sqrt{\tau_o/\rho}} = \sqrt{\tau_o/\rho} \times 11.6$

$$= u_* \times 11.6 = 0.0766 \times 11.6 = 0.89\ m/s$$ Ans

11.8. **A 100 mm smooth pipe line carries 38 liters/s of water at 20°C. Calculate (a) the friction factor, (b) maximum velocity, (c) shear**

velocity, (d) shear stress at the wall (e) the thickness of the laminar sub layer. Take ν = 1 cs.

Solution

$$Q = \frac{\pi}{4} d^2\, u$$

$$38 \times 10^{-3} = \frac{\pi}{4} (0.1)^2 \times u$$

u = 4.838 m/s

$$R_e = \frac{ud}{v} = \frac{4.838 \times 0.1}{1 \times 10^{-6}} = 483831.027 = 4.838 \times 10^5$$

$$\log_{10} R_e = \log_{10}\left(4.838 \times 10^5\right) = 5.6846$$

The value of R_e comes out to greater than 10^5 therefore, Blausius formula cannot be used.

(a) For turbulent flow in smooth pipe

By Nikurdse

(1) $$\lambda = 0.0032 + \frac{0.221}{(R_e)^{0.237}} = 0.0032 + \frac{0.221}{\left(4.838 \times 10^5\right)^{0.237}}$$

$$= 0.0032 + 9.934 \times 10^{-3} = 0.013$$

(b) $$u_* = u_{av} \sqrt{\lambda/8} = 4.838 \sqrt{\frac{0.013}{8}} = 0.195 \text{ m/s}$$ **Ans**

(c) $$\frac{u_{max}}{u_{av}} = 1 + 1.43\sqrt{\lambda}$$

$$= 1 + 1.43\sqrt{0.013} = 1 + 0.163 = 1.163$$

$$u_{max} = 4.838 \times 1.163 = 5.6268 \text{ m/s}$$ **Ans**

(d) $$u_* = \sqrt{\frac{\tau_o}{\rho}}$$

$$\tau_o = \rho\, u_*^2 = 1000 \times (0.195)^2 = 38.025 \text{ N/m}^2$$ **Ans**

(e) $$\delta' = 11.6 \frac{v}{u_*} = \frac{11.6 \times 1 \times 10^{-6}}{0.195}$$

$$= 5.9487 \times 10^{-5} \text{ m} = 0.05948 \text{ mm}$$ **Ans**

11.9. **A 300 mm diameter pipe line carries 500 liters/s of water at 20°C. Determine the maximum permissible height of roughness so that the pipe may still be called smooth. Take ν = 1 cs.**

Solution

$$Q = \frac{\pi}{4} d^2 u$$

$$500 \times 10^{-3} = \frac{\pi}{4} (0.3)^2 \times u$$

u = 7.0753 m/s

$$R_e = \frac{ud}{\upsilon} = \frac{7.0735 \times 10^{-6}}{1 \times 10^{-6}} = 2.11 \times 10^6$$

The R_e comes out to be greater than 10^5 therefore, Blausius formula cannot be used.

For turbulent flow in smooth pipes

By Nikurdse

$$\lambda = 0.0032 + \frac{0.221}{R_e^{0.237}}$$

$$= 0.0032 + \frac{0.221}{2.11 \times 10^{6\ 0.237}} = 0.0105$$

$$\therefore u_* = u\sqrt{\lambda/8} = 7.0735\sqrt{\frac{0.0105}{8}} = 0.2562 \text{ m/s}$$

Thickness of the laminar sub layer

$$\delta' = \frac{11.6\ \upsilon}{u_*} = \frac{11.6 \times 1 \times 10^{-6}}{0.25626}$$

$= 4.526619 \times 10^{-5}$ m $= 0.04526$ mm

Limiting value of height of roughness so that the pipe behaves as smooth

$$K_s = \frac{\delta'}{4} = \frac{0.04526}{4} = 0.0113 \text{ mm}$$ Ans

11.10. Kerosene (S = 0.8 and $\mu = 1.90 \times 10^{-3}$ kg/ms) is to flow in a 200 mm diameter pipe at a velocity of 3 m/s with a minimum loss of energy. Determine the permissible height of surface roughness of the pipe wall.

Solution

$$R_e = \frac{\rho\, u\, d}{\nu} = \frac{800 \times 3 \times 0.2}{1.9 \times 10^{-3}} = 252631.5789 = 2.526 \times 10^5$$

The R_e number comes out to be greater than 10^5. Therefore, Blausius formula cannot be used. At the permissible height of roughness in the pipe the flow will be in smooth turbulent regime.

By Nikurdse method

$$\lambda = 0.0032 + \frac{0.221}{(R_e)^{0.237}} = 0.0032 + \frac{0.221}{(2.526 \times 10^5)^{0.237}}$$

$$= 0.0032 + \frac{0.221}{19.0717} = 0.0032 + 0.01158 \quad = 0.01478$$

Since the pipe wall behave as smooth, at the limiting value

$$\frac{R_e \sqrt{\lambda}}{(R_o / K_s)} = 17$$

$$\frac{2.56 \times 10^5 \sqrt{0.01478}}{(R_o / K_s)} = 17$$

$$\frac{R_o}{K_s} = \frac{30733.95}{17} = 1807.879$$

$$K_s = 5.5313 \times 10^{-5} = 0.00553 \text{ mm}$$ **Ans**

11.11. **The wall shear stress and hence the pressure gradient may be estimated for fully developed turbulent flow from velocity measurements. Water flows at $u_{av} = 0.86\ u_{max} = 2.1$ m/s in a 150 mm diameter pipe.**

(a) What is the wall shear stress?

(b) What is the pressure drop in 30 m of pipe?

Solution

We know that

$$\frac{u_{max}}{u_{av}} = 1 + 1.43\sqrt{\lambda}$$

$$\frac{u_{max}}{0.86 u_{max}} = 1 + 1.43\sqrt{\lambda}$$

$$\frac{1}{0.86} = 1 + 1.43\sqrt{\lambda}$$

$$1.16279 = 1 + 1.43\sqrt{\lambda}$$

$$\sqrt{\lambda} = \frac{0.16279}{1.43} = 0.113839$$

$$\sqrt{\lambda} = 0.012959$$

We know that

(a) $$u_* = \sqrt{\frac{\tau_o}{\rho}} = u_{av}\sqrt{\lambda/8}$$

$$\tau_o = \frac{\rho u_{av}^2 \lambda}{8} = \frac{1000 \times (2.1)^2 \times 0.012959}{8}$$

$= 7.1439$ N/m² **Ans**

(b) $$\frac{\Delta p}{\rho} = \frac{\lambda l u^2}{2d}$$

or, $$\Delta p = \lambda \left(\frac{l}{d}\right) \frac{\rho u^2}{2} = 0.012959 \times \frac{30}{0.15} \times 1000 \times \frac{(2.1)^2}{2}$$

$= 5714.919 = 5.714$ k Pa **Ans**

11.12. In a fully turbulent flow in a 150 mm diameter pipe the centre line velocity is 2.5 m/s and the local velocity at mid radius is 2.28 m/s. Find the discharge and the height of the roughness projections.

Solution

For both smooth and rough boundaries, the maximum velocity u_{max} is at y = R and is given by

$$\frac{u - u_{max}}{u_*} = 5.75 \log_{10} \frac{y}{R}$$

or, $$\frac{u_{max} - u}{u_*} = 5.75 \log_{10} \frac{R}{y} \quad (1)$$

and in terms of u_{max}, u_{av} and u, are related as

$$\frac{u_{max} - u_{av}}{u_*} = 3.75 \quad (2)$$

Using equation (1)

$$\frac{2.5 - 2.28}{u_*} = 5.75 \log_{10} \left(\frac{R}{R/2}\right)$$

$$\frac{0.22}{u_*} = 1.73092$$

$$u_* = 0.12709 \; m/s$$

Using equation (2)

$$\frac{2.5 - u_{av}}{0.127} = 3.75$$

$$2.5 - u_{av} = 0.4766$$

$$u_{av} = 2.023375 \; m/s$$

The discharge $Q = \frac{\pi}{4} d^2 u_{av}$

$$= \frac{\pi}{4}(0.17)^2 \times 2.023375$$

$$= 0.03575 \text{ m}^3/\text{s} = 35.75 \text{ liters/s} \qquad \text{Ans}$$

For rough turbulent flow

$$\frac{u_{max}}{u_*} = 5.75 \log_{10}\frac{R}{K_s} + 8.5$$

or, $$\frac{2.5}{0.127} - 8.5 = 5.75 \log_{10}\frac{R}{K_s}$$

or, $$\frac{19.685 - 8.5}{5.75} = \log_{10}\frac{R}{K_s}$$

or, $$\frac{11.185}{5.75} = \log_{10}\frac{R}{K_s}$$

$$\log_{10}\frac{R}{K_s} = 1.945$$

$$\frac{R}{K_s} = 88.15$$

$$\frac{0.075}{K_s} = 88.15$$

$$K_s = 8.508 \times 10^{-4}\ m = 0.8508 \text{ mm} \qquad \text{Ans}$$

11.13. **A 200 mm diameter pipe line has equivalent sand grain roughness magnitude of 0.2 mm. At what velocity of flow will the roughness**

(a) Cause the flow to depart from smooth pipe regime. The transition from the smooth pipe regime begins at Reynold's roughness number $\frac{R_e\sqrt{\lambda}}{R/K_s} = 17$ **and**

(b) Cause the flow to become fully rough turbulent. Fully rough flow begins at Reynold's roughness number

$$\frac{R_e\sqrt{\lambda}}{R/K_s} = 400 \text{ Take } \nu = 1 \text{ cs.}$$

Solution

(a) $$\frac{R_e\sqrt{\lambda}}{R/K_s} = 17$$

$$R_e\sqrt{\lambda} = 17 \times \left(\frac{0.1}{0.0002}\right) = 8500 \qquad (1)$$

We know that for smooth pipe flow

$$\frac{1}{\sqrt{\lambda}} = 2\log_{10} R_e\sqrt{\lambda} - 0.8 = 2\log_{10} 8500 - 0.8$$

$$= 7.8588 - 0.8 = 7.0588$$

Using equation (1)

$$R_e\sqrt{\lambda} \times \frac{1}{\sqrt{\lambda}} = 8500 \times 7.0588$$

$$R_e = 8500 \times 7.0588 = 60000 = 6 \times 10^4$$

$$R_e = \frac{u\,d}{\nu}$$

$$6 \times 10^4 = \frac{u \times 0.2}{1 \times 10^{-6}}$$

$$u = 0.3 \text{ m/s}$$

Upper limit velocity for smooth pipe regime is 0.3 m/s<T>Ans

(b) $$\frac{R_e\sqrt{\lambda}}{R/K_s} = 400$$

$$R_e\sqrt{\lambda} = 400 \times \frac{0.1}{0.0002} = 2 \times 10^5$$

We know that for rough turbulent pipe flow

$$\frac{1}{\sqrt{\lambda}} = 2\log_{10}\frac{R}{K_s} + 1.74$$

$$\frac{1}{\sqrt{\lambda}} = 2\log_{10}\left(\frac{0.1}{0.0002}\right) + 1.74 = 5.3979 + 1.74 = 7.1379$$

Therefore

$$R_e\sqrt{\lambda} \times \frac{1}{\sqrt{\lambda}} = 2 \times 10^5 \times 7.1379 = 1427588 = 1.427 \times 10^6$$

$$R_e = \frac{U\,d}{\upsilon}$$

$$1.427 \times 10^6 = \frac{u \times 0.2}{1 \times 10^{-6}}$$

$$u = 7.1379 \text{ m/s}$$ Ans

The flow will be at fully rough turbulent at 7.1379 m/s

11.14. **A 100 mm rough pipe line with roughness 0.12 mm carries 38 liters/s of water at 20°C (ν = 1 cs) Calculate (a) friction factor, (b) shear velocity, (c) shear stress at the wall and (d) maximum velocity**

Solution

$$Q = \frac{\pi}{4} d^2 u$$

$$38 \times 10^{-3} = \frac{\pi}{4} (0.1)^2 \times u$$

$$u = 4.838 \text{ m/s}$$

$$R_e = \frac{ud}{\nu} = \frac{4.838 \times 0.1}{1 \times 10^{-6}} = 483831.027 = 4.838 \times 10^5$$

The value of R_e comes out to be greater than 10^5 therefore, Blausius formula cannot be used.

(a) For turbulent flow in rough pipe

$$\frac{1}{\sqrt{\lambda}} = 1.74 + 2\log_{10} \frac{R}{K_s}$$

$$\frac{1}{\sqrt{\lambda}} = 1.74 + 2\log_{10} \frac{0.05}{0.12/1000} = 1.74 + 2 \times 2.61978$$

$$= 1.74 + 5.2395 = 6.9795$$

$$\sqrt{\lambda} = 0.143275$$

$$\lambda = 0.0205$$ Ans

(b) $$u_* = u\sqrt{\lambda/8} = 4.838\sqrt{\frac{0.0205}{8}} = 0.243 \text{ m/s}$$

(c) $$\tau_o = \rho u_*^2 = 1000 \times (0.243)^2 = 59.1\ N/m^2$$

(d) $$\frac{u_{max}}{u_{av}} = 1 + 1.43\sqrt{\lambda} = 1 + 1.43\sqrt{0.0205}$$

$$= 1 + 0.2047 = 1.2047$$

$$u_{max} = 4.838 \times 1.2047 = 5.8285 \text{ m/s}$$ Ans

11.15. **Mean point velocities measured with the help of a pitot tube at mid point and quarter point of a 200 mm diameter pipe were found to be 1.5 m/s and 1.35 m/s respectively. If the flow in the pipe is turbulent, determine discharge, friction factor and average height of roughness projections.**

Solution

We know that

$$\frac{u_{max} - u}{u_*} = 5.75 \log_{10}(R/y)$$

$$\frac{1.5 - 1.35}{u_*} = 5.75 \log_{10}\left(\frac{100}{50}\right)$$

$$\frac{0.15}{u_*} = 5.75 \log_{10} 2$$

$$\frac{0.15}{u_*} = 1.7309$$

$$u_* = 0.0866\ m/s$$

We know a general formula

$$\frac{u - u_{av}}{u_*} = 3.75 + 5.75 \log_{10} \frac{y}{R}$$

When $u = u_{max}$ at $y = R$

Therefore,

$$\frac{u_{max} - u_{av}}{u_*} = 3.75$$

$$\frac{1.5 - u_{av}}{0.0866} = 3.75$$

$$u_{av} = 1.175\ m/s$$

$$Q = \frac{\pi}{4} d^2\ u_{av} \qquad = \frac{\pi}{4}(0.2)^2 \times 1.175 = 0.0369\ \text{m}^3/\text{s}$$ Ans

Also, $u_* = u_{av}\sqrt{\lambda/8}$

$$0.0866 = 1.175\ \sqrt{\lambda/8}$$

$$0.0737 = \sqrt{\lambda/8}$$

$$\lambda/8 = 5.432 \times 10^{-3}$$

$$\lambda = 0.04345$$ Ans

Therefore,

$$\frac{1}{\sqrt{\lambda}} = 2 \log_{10}\left(\frac{R}{K_s}\right) + 1.74$$

$$\frac{1}{\sqrt{0.04345}} = 2 \log_{10}\left(\frac{R}{K_s}\right) + 1.75$$

$$\log_{10}\frac{R}{K_s} = \frac{4.79 - 1.75}{2} = 1.5235$$

$$\frac{R}{K_s} = 33.383$$

$$K_s = \frac{0.1}{33.383} = 2.9955 \times 10^{-3}\ m = 2.9955\ mm$$ Ans

11.16. **In a fully turbulent rough pipe flow occur in a 150 mm diameter pipe carrying oil of density 900 kg/m³. If at a distance of 25 mm from the wall the velocity is 1.5 m/s and the velocity gradient is 7 per second; calculate (a) the boundary shear stress (b) equivalent sand grain roughness, (c) discharge (d) friction factor.**

Solution

In a fully turbulent rough pipe flow

$$\frac{u}{u_*} = 5.75\log_{10}\frac{y}{K_s} + 8.5 \qquad (1)$$

$$u = u_*\left[2.5\ ln\frac{y}{k_s} + 8.5\right]$$

Differentiating with respect to y

$$\frac{du}{dy} = u_*\left[\frac{2.5}{y}\right]$$

$$7 = \frac{2.5\ u_*}{0.025}$$

$$u_* = 0.07\ \text{m/s}$$

$$\tau_o = \rho u_*^2, \quad = 900 \times (0.07)^2 = 4.41\ \text{N/m}^2$$ Ans

(b) u = 1.5 m/s, u_* = 0.07, y = 0.025

Substituting in equation (1)

$$\frac{1.5}{0.07} = 5.75\log_{10}\frac{0.025}{K_s} + 8.5$$

$$\log_{10}\frac{0.025}{K_s} = 2.248$$

$$\frac{0.025}{K_s} = 177.193$$

$$K_s = 1.41 \times 10^{-4}\ m = 0.141\ \text{mm}$$ Ans

(c) For a turbulent pipe flow

$$\frac{u_{av}}{u_*} = 5.75 \log_{10} \frac{R}{K_s} + 4.75$$

$$\frac{u_{av}}{0.07} = 5.75 \log_{10} \left(\frac{0.075}{1.41 \times 10^{-4}} \right) + 4.75$$

$$= 15.67 + 4.75 = 20.422$$

$$u_{av} = 1.4295 \text{ m/s}$$

$$Q = \frac{\pi}{4} d^2 \ u_{av} = \frac{\pi}{4} (0.15)^2 \times 1.4295$$

$$= 0.02526 \text{ m}^3\text{/s} = 25.26 \text{ liters/s}$$ Ans

(d) $u_* = u_{av} \sqrt{\lambda/8}$

$$\sqrt{\lambda/8} = 0.048968$$

$$\sqrt{\lambda/8} = 0.048968$$

or, $\lambda = 0.01918$ Ans

11.17. A liquid a relative density 0.9 and dynamic viscosity 6.5×10^{-4} kg/ms flows through a pipe of diameter 150 mm at the rate of 60 liters/s. If the loss of head in 100 m length of pipe is 5 m, determine whether the pipe is rough or smooth.

Solution

Discharge $Q = \frac{\pi}{4} d^2 u$

$$60 \times 10^{-3} = \frac{\pi}{4} (0.15)^3 \ u$$

$$u = 3.395 \text{ m/s}$$

$$R_e = \frac{\rho \, u \, d}{\mu} = \frac{900 \times 3.395 \times 0.15}{6.5 \times 10^{-4}} = 7.05 \times 10^5$$

$$h_f = \frac{\lambda \, l \, u^2}{2g \, d} \qquad 5 = \frac{\lambda \times 100 \times (3.395)^2}{2 \times 9.81 \times 0.15}$$

$$\lambda = 0.01276$$

Let us assume that the pipe is hydrodynamically rough, then

$$\frac{1}{\sqrt{\lambda}} = 2 \log_{10} \left(\frac{R}{K_s} \right) + 1.74$$

$$\frac{1}{\sqrt{0.01276}} - 1.74 = 2 \log_{10} \frac{R}{K_s}$$

$$2 \log_{10} \frac{R}{K_s} = 8.85 - 1.74$$

$$\log_{10} \frac{R}{K_s} = \frac{7.11}{2}$$

$$\log_{10} \frac{R}{K_s} = 3.555$$

$$\frac{R}{K_s} = 3590.598$$

Thus $\dfrac{R_e \sqrt{\lambda}}{R/K_s} = \dfrac{7.05 \times 10^5 \times \sqrt{0.01276}}{3590.598} = 196.346 \times \sqrt{0.01276} = 22.179$

Since the parameter $\left[\dfrac{R_e \sqrt{\lambda}}{R/K_s}\right]$ is more than 17, and less than 400, the boundary acts in transition. Thus the assumption made for the pipe to be rough is not correct. Hence the pipe is in the transition range.

EXERCISES

A. THEORY

11.1. What is turbulence? Define (a) intensity of turbulence, (b) Reynold's stress. Explain Prandtl mixing length hypothesis and derive the relationship between the mixing length and shear stress.

11.2. For turbulent flow explain how momentum may be transferred between adjoining layers in the fluid which are moving with different velocities and obtain an expression for the resulting shear stress in terms of velocity gradient.

11.3. Define "eddy viscosity" as given by Boussinesq. What is its numerical value in comparison with the viscosity of fluid?

11.4. Define shear velocity or friction velocity for turbulent flow. Give its value in terms of shearing stress.

11.5. Describe Prandtl mixing length theory for finding the shear stress in a turbulent flow in circular pipes. What is universal constant used in this theory?

11.6. Define mixing length concept of Prandtl. Show it by means of sketch. How does the mixing length vary with the distance from the wall in turbulent flow?

11.7. Describe von-karman similarly theory. Where it is used?

11.8. Describe experimental work for pipe friction giving Darcys, Blasius and Nikuradse results.

11.9. State the universal velocity distribution laws for turbulent flow. Derive Prandtl universal velocity distribution equation.

11.10. What is the difference between smooth and rough pipes? Describe with the help of a diagram.

11.11. Describe laminar sub layer in case of turbulent flow in circular pipes. How does it change in case of smooth and rough pipes? What is true thickness of laminar sub layer? How does it differ from thickness of normal laminar sub layer?

11.12. Describe universal resistance laws of turbulent flow in smooth and rough pipes. Derive equations in each case. How is the single relation applicable for both smooth and rough pipes calculated?

11.13. Describe how the roughness of the pipe wall affects the frictional resistance to flow in pipes.

11.14. Explain the difference between wall turbulence and free turbulent shear flow.

11.15. Explain the Reynold's concept of momentum exchange, and derive the following equation for Reynold's shear stress,

$$\tau = -\rho \overline{u' v'}$$

11.16. Define eddy and molecular viscosity

11.17. Show that shear velocity is given by the expression

$$u_* = u_{av} \sqrt{\lambda / 8}$$

11.18. Describe Taylor's vorticity transport theory. Where is it used?

11.19. Name the semiempirical theories of turbulence and explain any one.

11.20. Describe Reynolds's Time Average concept.

11.21. Show that shear velocity is given by the expression

$$u_* = \sqrt{\frac{\tau_o}{\rho}}$$

11.22. Derive the Prandtl's universal velocity distribution equation for turbulent flow in a non dimensional form.

$$\frac{u_{max} - u}{u_*} = 5.75 \log_{10} \frac{R}{y}$$

Explain why this relationship does not apply close to the pipe wall.

11.23. Derive the Prandtl-Karaman velocity distribution equation

$$u = 5.75 \log_{10} (y / y')$$

11.24. Show that for turbulent flow in a smooth pipe

$$\frac{u}{u_*} = 5.5 + 5.75 \log_{10} \frac{u_* y}{\nu}$$

where u = velocity at radius y,

u_* = shear velocity

ν = kinematic viscosity

11.25. Assuming that

$$\frac{u}{u_*} = 5.5 + 5.75 \log_{10} \frac{u_* y}{\nu}$$

Show that for turbulent flow in a smooth pipe

$$\frac{1}{\sqrt{\lambda}} = -0.8 + 2 \log_{10} R_e \sqrt{\lambda}$$

11.26. Show that for turbulent flow in a rough pipe

$$\frac{u}{u_*} = 8.5 + 5.75 \log_{10} \left(\frac{y}{K_s}\right)$$

Where K_s = average height of protrusions

11.27. Assuming that

$$\frac{u}{u_*} = 8.5 + 5.75 \log_{10} (y/K_s)$$

Show that for turbulent flow in a rough pipe

$$\frac{1}{\sqrt{\lambda}} = 1.74 + 2 \log_{10} \frac{R}{K_s}$$

11.28. Show that the velocity distribution for turbulent flow in terms of average velocity for smooth pipes

$$\frac{u_{av}}{u_*} = 1.75 + 5.75 \log_{10} \frac{u_* R}{\upsilon}$$

and hence deduce the formula

$$\frac{u - u_{av}}{u_*} = 3.75 + 5.75 \log_{10} y/R$$

11.29. Show that the velocity distribution for turbulent flow in terms of average velocity for rough pipes

$$\frac{u_{av}}{u_*} = 4.75 + 5.75 \log_{10} \frac{R}{K_s}$$

and hence deduce the formula

$$\frac{u - u_{av}}{u_*} = 3.75 + 5.75 \log_{10} \frac{y}{R}$$

11.30. Show that the average velocity in pipe flow can be related to the centre line (maximum) velocity and the pipe friction factor by

$$\frac{u_{max}}{u_{av}} = 1 + 1.43\sqrt{\lambda}$$

11.31. Derive an expression for the turbulent pipe coefficient as

$$\frac{u}{u_{av}} = 1 + \sqrt{\lambda}\left(1.43 + 2.15 \log_{10} \frac{y}{R}\right)$$

11.32. Plot a curve of the ratio of the average to centre line velocity $\left(\frac{u_{av}}{u_{max}}\right)$ for fully developed turbulent flow in a smooth pipe as a function of the Reynold's number (R_e) of the flow. Use linear scale for $\frac{u_{av}}{u_{max}}$ as ordinate and a log scale for R_e as abscissa.

11.33. What is Colebrook-White formula? where is it used?

11.34. Describe some empirical relationships for finding the friction factor.

11.35. Describe Nikurdse experiments and their significance.

11.36. Describe the results of Nikurdse's experiments on the effect of the relative roughness R/K_s on the value of the resistance coefficient λ for turbulent flow in circular pipes.

11.37. Describe Moody's chart. What are the basis to draw such a diagram? State its uses.

11.38. Show how the smooth pipe and rough pipe formula may be combined to give an approximate relation for λ in terms of the relative roughness K_s/d and the pipe Reynold's number.

11.39. What do you mean by R* = Reynold's roughness number. Give the values of R* for smooth, departure from smooth pipe law and rough pipe law.

11.40. Draw a curve between K_s/δ' and $R_e\sqrt{\lambda}/R\,K_s$ and show on the curve the nature of the boundary.

11.41. Draw Colebrook-White transition curve for commercial pipes and explain its significance.

11.42. Draw a curve between $\frac{1}{\sqrt{\lambda}} - 2\log_{10}\frac{3.7\,d}{K_s}$ and Reynold's roughness number (R_*) and explain it.

11.43. Draw a simplified sketch of Moody's diagram and show the various zones.

11.44. Draw a hydraulic research station charts (HRS) for any value of K_s and discuss it.

11.45. Write down the various explicit equation for friction factor λ for flow in commercial pipes.

(a) Moody's formula; (b) Barr's explicit formula; (c) Murdock formula; (d) Swamee and Jain's explicit formula; (e) SE Haaland's formula

11.46. How is the head loss due to friction in commercial pipe determined? In what way the old pipes will be different from the new pipes.

11.47. What do you understand by roughening of pipes with age? How is roughness of an old pipes determined?

11.48. "The friction factor λ becomes independent of Reynold's number at very high Reynold's number" state the significance attached to this statement.

11.49. Draw Stanton and Pannell curve and discuss it.

11.50. What is Blasius equation and explain its significance.

B. UNSOLVED PROBLEMS

11.1. Fully developed flow in a 300 mm C.I. pipe is at an average velocity of 1.8 m/s. What is the centre line velocity for (a) fuel oil at 10°C (υ = 0.002 m^2/s) and (b) crude oil at 20°C ($\nu = 5.6 \times 10^{-6}$ m^2/s) ? [(a) 3.6 m/s, (b) 2.16 m/s]

11.2. If the velocities in a 600 mm pipe carrying oil are 4.5 m/s and 4.2 m/s on the centre line and at a radial distance of 100 mm from the axis, calculate the discharge in the pipe. [0.958 m^3/s]

11.3. A pipe line carrying water has average height of irregularities projecting from the surface of the boundary of the pipe as 0.15 mm. What type of boundary is it ? The shear stress developed is 4.9 Pa. The kinematic viscosity of water is 1 cs.

$$\left[R_e = \frac{u_* K_S}{\upsilon} = 10.5, \text{ transition state}\right]$$

11.4. A pipe of 250 mm diameter carries air (ρ = 1.22 kg/m^3, $\upsilon = 1.5 \times 10^{-5}$ m^2/s) at an average velocity of 8 m/s. The equivalent sand grain roughness of the pipe is 0.5 mm. Calculate the friction factor λ by assuming the flow to be fully rough-turbulent. What is the shear stress at the boundary ? [λ = 0.0234, τ_0 = 0.2285 Pa]

11.5. Determine the wall shearing stress in a pipe diameter 100 mm, which carries water. The velocities at the pipe centre and 30 mm, from the pipe centre are 2 m/s and 1.5 m/s respectively. [47.676 N/m^2]

11.6. Water (at 20°C) flows in a 75 mm diameter smooth brass pipeline. Wall shear stress is 3.68 N/m^2. Calculate the thickness of the viscous sub layer, the friction factor, the mean velocity and flow rate, the centreline velocity, the shear stress and velocity 25 mm from the pipe centreline, and the head lost in 1000 m of this pipe. [u_* = 0.061 m/s, δ' = 0.19 mm, u_{av} = 1.29 m/s, u_{max} = 1.54 m/s u_{25} = 1.35 m/s, h_f= 20.4 m]

11.7. For the conditions of the problem 12.6, viz, u_{av} = 1.29 m/s, d = 75 mm, ρ = 998 kg/m^3 and ν = 1 cs. Calculate the friction factor, wall shear stress, centreline velocity, and the velocity 25 mm from the pipe centre line, using the seventh root law. [R_e = 96750 λ = 0.018, u_{max} = 1.58 m/s, u_{25} = 1.35 m/s, τ_0 = 3.7 Pa]

11.8. The mean velocity in a 300 mm pipe line is 3 m/s. The relative roughness of the pipe is 0.002 and the kinematic viscosity of water 9 × 10^{-7} m^2/s. Determine the friction factor, centre line velocity, velocity 50 mm from the pipe wall, and the head lost in 300 m of this pipe under the assumption that the pipe is rough. [λ = 0.0234, u_* = 0.162 m/s, u_{max} = 3.62 m/s, u_{50} = 3.17 m/s, h_l = 10.7 m of water]

11.9. What is the pressure drop in 15 m of 6 mm tubing for a horizontal flow of hydraulic fluid (μ = 0.014 kg/ms and S = 0.848) at 50°C at a velocity of 2 m/s? What is the wall shear stress? [37.3 Pa]

11.10. What is the flow rate for water at 15°C in a commercial steel pipe, 250 mm diameter, when the head loss in 300 mm of pipe is 5 m ? Take K_s = 0.000045 m, and ν = 1.40 × 10^{-6} m^2/s [144 liters/s]

11.11. Water flows at the rate of 91 liters/s over a distance of 500 m in a horizontal commercial steel pipe with a pressure drop not to exceed 825 kPa. What is the minimum size pipe which can be used? Take ν = 1 cs. [0.152 m]

11.12. Water at 40°C flows in a 75 mm pipe at Reynold's number 8 × 10^4. If the pipe contains a uniform sand roughness of grain size 0.15 mm diameter, how much head loss is to be expected in 300 m of the pipe ? How much head loss would be expected if this pipe were smooth? [2.1 m, 1.88 m]

11.13. Calculate the loss of head and the pressure drop when air at an absolute pressure of 101.3 kPa and 15°C flows through 600 m of a 450 mm by 300 mm smooth rectangular duct with a mean velocity of 3 m/s. [R_n = 0.09 m, R_e = 73950, λ = 0.019, h_l = 14.5 m of air, δp = 174 Pa]

11.14. A smooth pipe of diameter 80 mm and 800 m long carries water at the rate of 480 liters/min. Calculate the loss of head, wall shearing stress, centreline velocity, velocity and shear stress at 30 mm form the pipe wall. Also calculate the thickness of laminar sub layer. Take ν = 0.015 stokes. [23.92 m, 5.860 Pa, 1.88 m/s, 1.825 m/s, 0.2274 mm]

11.15. For turbulent flow in a pipe of diameter 30 mm, find the discharge when the centreline velocity is 2 m/s and the velocity at a point 100 mm from the centre as measured by pitot tube is 1.6 m/s. [102.7 liters/s].

12

Dimensional Analysis and Simlitude

12.1 INTRODUCTION

Dimensional analysis is one of the most important mathematical tools in the study of fluid mechanics. It is a mathematical technique which makes use of the study of dimensions as an aid to the solution of many engineering problems. The main advantage of a dimensional analysis of a problem is that it reduces the number of variables in the problem by combining dimensional variables to form nondimensional parameters. Dimensional analysis is particularly helpful in experimental work because it provides a guide to those things that significantly influence the phenomena; thus it indicates the direction in which experimental work should go.

Although many practical engineering problems involving fluid mechanics can be solved by using the equations and analytical procedures described in the preceeding chapters, there remain a large number of problems that rely on experimentally obtained data for their solution. In fact, it is probably fair to say that very few problems involving real fluids can be solved by analysis alone. But in the majority of more, complex fluid flow problems, especially those of turbulent flow, there are so many variables involved in the differential equation of fluid motion that a direct mathematical solution is simply out of the question. In these problems dimensional analysis can be used to great advantage in obtaining a functional relationship among the various variables involved in terms of nondimensional parameters.

When experimental testing of a full size prototype is either impossible or prohibitively expensive, the only feasible way of attacking the problem is through model testing in the laboratory. If we are to predict the prototype behaviour from measurements on the model, it is obvious that we cannot run just any test on any model. The model flow and the prototype flow must be related by known scaling laws. The laws of simlitude make it possible to predict the performance of the prototype, which means the full size device, from tests made with the model. It is not necessary that the same fluid be used for the model and its prototype. Neither is the model necessarily small than its prototype. Thus the flow in a Carburiettor might be studied in a very large model. And the flow of water at the entrance to a small centrifugal pump runner might be investigated by the flow of air at the entrance to a large model of the runner.

A few examples where models may be used are ships in towing basins, aeroplane in wind tunnels, hydraulic turbines, centrifugal pumps, spillways of dams, rivel channels, and the study of such phenomena as the action of waves and tides on beaches, soil erosion, and the transportation of sediment.

It should be emphasized that the model need not necessarily be different in size from is prototype. In fact, it may be the same device, the fact, it may be the same device, the variables in the case being the velocity and the physical properties of the fluid.

12.2 DIMENSIONS, DIMENSIONAL HOMOGENEITY AND UNITS

In fluid mechanics we will be dealing with a velocity of fluid characteristics. It is necessary to develop a system for describing these characteristics both qualitatively and quantitatively. The qualitative aspect serves to identify the nature, or type, of the characteristics (such as, length, time, velocity and stress). Whereas quantitative description requires both a number and a standard by which various quantities can be compared such standards are called units, and several systems of units are in common use. The qualitative description of physical quantities can be given in terms of basic dimensions such as mass, M, length, L and time T. Alternatively we could use force, F, L, and T as basic dimensions, since from Newton's second law

$$F = M L T^{-2} \text{ or } \frac{FT^2}{ML} = 1 \tag{12.1}$$

where the symbol = is used to indicate dimensional equality.

Thus, in fluid mechanics, we can select [L] and [T] as fundamental dimensions to gether with either force [F] or mass [M], the remaining quality being regarded as a derived dimensions and expressed in terms of the three fundamental dimensions. In this book, we shall normally select the dimensions of mass [M] as a third fundamental dimension and treat force as a derived dimension, having the dimensions $[MLT^{-2}]$.

The dimensions of a number of common physical quantities are listed in Table 13.1.

The distinction between units and dimensions is that dimensions are properties that can be measured and units are the standard elements in terms of which these dimensions can be described quantitatively and assigned numerical values.

In dimensional analysis, we are concerned only with the nature of the factors involved in the situation and not with their numerical values. The notation adopted to indicate this is to enclose the name or symbol of the quantity in square brackets. Whenever a property in words or symbols, is enclosed in a square bracket it indicates that we are concerned with it only dimensionally and qualitatively, not quantitatively.

Table 12.1 Dimensions associated with common physical quantities.

Quantity	MLT System
Length	L
Area	L^2
Volume	L^3
Time	T
Velocity	LT^{-1}
Acceleration	LT^{-2}
Force	MLT^{-2}
Mass	M
Density	ML^{-3}
Pressure and Stress	$ML^{-1} T^{-2}$
Energy and Work	$ML^{-1} T^{-2}$
Momentum	MLT^{-1}
Power	$ML^2 T^{-3}$
Torque	$ML^{-2} T^{-2}$
Moment of a force	$ML^2 T^{-2}$
RPM	T^{-1}
Angular Velocity	T^{-1}
Dynamic Viscosity	$ML^{-1} T^{-1}$
Kinematic Viscosity	$L^2 T^{-1}$
Surface tension	MT^{-2}
Modulus of elasticity	$ML^{-1} T^{-2}$
Angle	$M^o L^o T^o$
Efficiency	$M^o L^o T^o$
Strain	$M^o L^o T^o$

All theoretically derived equations are dimensionally homogeneous - that is, the dime-nsions of the left side of the equation must be the same as those on the right side, and all additive separate terms must have the same dimensions. We accept as a fundamental permise that all equations describing physical phenomena must be dimensionally homogeneous.

The principle of dimensional homogeneity requires that any equation which fully or correctly represents a possible physical situation must be dimensionally homogenous, so that each term has the same dimensional formula. It is therefore, possible to make a rapid check of any algebraic analysis of such a situation by substituting the dimensions of the quantities in each term and checking that, in the resultant dimensional equation, each fundamental dimension appears to the same power in each term.

Many engineering formulae in day-to-day use are committed to memory and, sometimes, there may be a doubt as to whether they have been recalled correctly. A quick dimensional check will establish the correct form immediately, although, of course, it will not indicate whether the values of any numerical constants are correct.

A dimensionally homogenous equation has the unique characteristics of being indepe-ndent of the units chosen for measurements. For example, equation (12.1) holds true no matter what units of measurements are used.

Alternate forms of equation (12.1) also demonstrate that a dimensionally, homogeneous equation may be transferred to a non dimensional form because of the mutual dependence of fundamental dimensions. Although it is always possible to reduce a dimensionally homogeneous equation to a non dimensional form, the main difficulty in a complicated flow problem is in setting up the correct equation of motion. Therefore, a special mathematical method called dimensional analysis is required to determine the functional relationship among all the variables involved in any complex flow phenomena, in terms of non dimensional parameters.

12.3 DIMENSIONAL ANALYSIS

The fact that a complete physical equation must be dimensionally homogeneous and is, therefore, reducible to a functional equation among nondimensional parameters forms the basis for the theory of dimensional analysis. The following two methods of dimensional analysis are generally used in fluid mechanics. (1) The Rayleigh method and (2) The Buckingham π (pi) method.

Lord Rayleigh in 1899 gives a special form of relationship among the dimensionless groups, and has an inherent drawback that it does not provide any information regarding the number of dimensionless groups to be obtained as a result of dimensional analysis. It is for these reasons that it has become obsolete and is not favoured for use. The second method is the Buckingham method (developed by E. Buckingham in 1914) and is an important over the Rayleigh's method. Buckingham designated the dimensionless groups by the Greek capital letter π (pi). It is, therefore, often called as, Buckingham- π method. It offers an advantages over Rayleigh's method in letting us known, in advance of the analysis as to how many dimensionless groups are to be expected.

12.3.1 The Buckingham π - theorem

The Buckingham π-theorem states that if, in a dimensionally homogenous equation, there are n dimensional variables which are completely described by m fundamental dimensions (such as M, L, T etc.) they may be grouped into (n-m) π terms. Each π- term is a dimensionless parameter.

Buckingham used the symbol π to represent a dimensionless product, and this notation is commonly used. Although the π theorem is a simple one, its proof is not so simple and we will not include it here.

The π-theorem is based on the idea of dimensional homogeneity. Essentially we assume that for any physically meaningful equation involving n variables, such as

$$A_1 = f\left(A_2, A_3\ A_4, \text{----}\ A_n\right)$$

which can be transformed to

$$f_1\left(A_1, A_2, A_3\ A_4 \text{----} A_n\right) = 0 \tag{12.2}$$

The equation (12.2) is a dimensionally homogenous equation. It then follows that we can rearrange. the equation into a set of dimensionless products (pi terms) so that

$$\Phi\left(\pi_1, \pi_2, \text{----}\ \pi_{n-m}\right) = 0 \tag{12.3}$$

or, $$\pi_1 = \Phi_1\left(\pi_2, \pi_3\ , \text{---}\ \pi_{n-m}\right) \tag{12.4}$$

The required number of π-terms is fewer than the number of original variables by m, where m is determined by the minimum number of reference dimensions required to describe the original list of variables. Usually the references dimensions required to describe the variables will be the basic dimensions M, L and T or F, L and T. However, in some instances perhaps only two dimensions, such as L and T, are required, or may be just one, such as L. Also, in a few rare cases the variables may be described by some combination of basic dimensions, such as M/T^2 and L, and in this case m would be equal to two rather than three.

The theorem does not predict the functional form of Φ or Φ_1. The functional relation among the independent, dimensionless π parameters must be determined experimentally.

12.3.2 Selection of repeating variables

Three general rules guide the selection of repeating variables.

(1) In a dimensional analysis of any physical system the repeating variables must include among them all of the m fundamental dimensions. Each repeating variable must be dimensionally independent of the others. This means that the repeating variables cannot themselves be combined to form a dimensionless product. Although this is not an absolute requirement, mass density a characteristic velocity, and characteristic length need to be selected as repeating variables in order to produce common dimensionless groups.

(2) For any given problem we usually are interested in determining how one particular variables is influenced by the other variables. We would consider this variable to be the dependent variable, and we would want this to appear in only one π-term. Thus, do not choose the dependent variable as one of the repeating variables, since the repeating variables will generally appear in more than one π-term.

(3) For a fluid system, the most significant groups will result if the repeating variables are chosen so that one is geometrical characteristics, other is flow characteristics, and still other is fluid properties.

Variables with geometrical properties are, length, diameter, height, breadth, chord span etc.

Variables with flow characteristics are, velocity, angular velocity, rpm, acceleration, mass flow rate, flow rate, pressure, pressure gradient, force (drag, lift, shear) shear stress, torque and moment.

Variables with fluid properties are, Density, dynamic viscosity, kinematic viscosity. Specific weight, Elastic modulus, surface tension.

In ninety percent cases the choice of repeating variables are density, velocity and length (ρ, v and l).

12.3.3 Determining the π groups

We are looking for a method that will allow us to systematically form the π terms so that we are sure that they are dimensionless and independent and that we have the right number. Some experience admittedly is helping in compiling the list. Students, who do not have this experience often are troubled by the need to apply engineering judgment in an apparent massive dose. However, it is difficult to go wrong if a generous selection of parameters is made.

If you suspect that a phenomenon depends on a given parameter, include it. If your suspicion is correct, experiments will show that the parameter must be included to get consistent results. If the parameter is extraneous, an extra π parameter may result, but experiments will show that it may be eliminated from consideration. Therefore, do not be afraid to include all the parameters that you feel are important. With a little practice you will be able to readily complete a dimensional analysis for your problem.

The eight steps listed below are a recommended procedure for determining the π parameters.

Step 1. List all the variables that are involved in the problem. (Let n be the number of parameters).

Step 2. Express each of the variables in terms of basic dimensions. For the fluid mechanics problem the basic dimensions will be either, M, L and T or F, L and T. (Let m be the number of basic dimensional).

Step 3. Determine the required number of π terms.

The number of π terms is equal to n-m, where n is the number of variables in the problem (which is determined from step 1) and m is the reference dimensions required to describe these variables (which is determined from step 2). The reference dimensions usually correspond to the basic dimensions, and can be determined by an inspection of the dimensions of the variables obtained in step 2.

Step 4. Select a number of repeating variables, where the number required is equal to the number of reference dimensions. The repeating parameters chosen may appear in all the dimensions less groups obtained.

Step 5. Set up dimensional equation combining the repeated parameters, selected in step 4 with each of the other parameters in turn to form dimensionless groups.

Essentially each π term will be of the form $A_1^{a_i} A_2^{b_i} A_2^{c_i} A_i$ where A_i is one of the non repeating variables, A_1, A_2 and A_3 are the repeating variables and the exponents a_i, b_i and c_i are determined so that combination is dimensionless.

Step 6. Repeat Step 5 for each of the remaining non repeating variables.

Step 7. Check all the resulting π terms to make sure they are dimensionless.

Step 8. Express the final form as a relationship among the π terms, and think about what it means. Typically the final form can be written as

$$\pi_1 = \Phi\left(\pi_2, \pi_3, \pi_4 \text{ ---} \pi_{n-m}\right)$$

12.3.4 Some additional comments about dimensional analysis

In order to obtain the final expression in the desired manner the following useful suggestions may be noted.

(1) If a quantity is dimensionless (e.g. efficiency, angle etc), it is a π term without going through the above procedure.

For example

$$\pi = \eta, \pi = \theta$$

(2) If any two physical quantities have the same dimensions, their ratio will be one of the π term. For example (l/d) is dimensionless and hence it is a π term dimensionless and hence it is a π term which may be directly written without going through the above procedure.

(3) Any π term may be replaced by another π term obtained by adding or subtracting an absolute numeral from it.

(4) Any π term may be replaced by multiplying it by a numerical constant. For example, π1 may be replaced by $2\pi_1, \pi_1 / \sqrt{2}$ or so.

(5) Any π term may be replaced by multiplying it by another π term. For example π_1 may be replaced by ($\pi_1 \times \pi_2$) or π_1 and π_2 may be replaced by ($\pi_1 \times \pi_2$).

(6) Any π term may be replaced by any power of that term, including negative as well as fractional powers. For example, π_1 may be replaced by $1/\pi_1$ or π_2 may be replaced by π_2^2, π_3 may replaced by $\left(\frac{1}{\pi_3^{3/2}}\right)$ and π_2 can be formed by a combination of say π_3, π_4 and π_5 such as

$$\pi_2 = \frac{\pi_3 \, \pi_4^2}{\sqrt{\pi_5}}$$

then π_2 is not an independent π term.

12.3.5 Uniqueness of π terms

There is not a unique set of π terms which arise from a dimensional analysis. However, the required number of π terms is fixed, and once a correct set is determined all other possible sets can be developed from this set by combinations of products of powers of the original set. For example

$$\pi_1 = \Phi\ (\pi_2, \pi_3) \tag{12.5}$$

We could always form a new set from this one by combining the π terms. For example we could form a new π term, π_3', by letting

$$\pi_3' = \pi^a,\ \pi_3^b$$

Where a and b are arbitrary exponents. Then the relationship could be expressed as

$$\pi_1 = \Phi_1 \left(\pi_2, \pi_3'\right) \tag{12.6(a)}$$

or, $$\pi_3' = \Phi\ (\pi_2, \pi_3) \tag{12.6(b)}$$

All of these would be correct. It should be emphasized, however, that the required number of π terms cannot be reduced by this manipulation; only the form of the π terms is altered.

12.3.6 Limitations of dimensional analysis selection of variables - superfluous and omitted variables

All the fore going procedures deal with straight forward situations. There are some problems which do occur: In order to apply dimensional analysis we must first decide which variables are significant. If the problem is not understood well enough to make a good initial choice of variables, dimensional analysis will seldom provide clarification.

There is no simple procedure whereby the variables can be easily identified. Generally, one must rely on a good understanding of the phenomenon involved and the governing physical laws. If extraneous variables are included, then too many π terms appear in the final solution, and it may be difficult, time consuming, and expensive to eliminate these experimentally. If important variable are omitted, then an incorrect result will be obtain; and again, this may prove to be costly and difficult to ascertain. It is, therefore, imperative that sufficient time and attention be given to this first step in which the variables are determined.

Most engineering problems involve certain simplifying assumptions that have an influ-ence on the variables to be considered. Usually we wish to keep the problem as simple as possible, perhaps even if some accuracy is sacrificed. A suitable balance between simplicity and accuracy is a desired goal. How accurate the solution must be depends on the objective of the study; that is, we may be only concerned with general trends and, therefore, some variables that are thought to have only a minor influence in the problem may be neglected for simplicity.

Since we wish to keep the number of variables to a minimum, it is important that all variables are independent. In a general term, if we have a problem in which the variables are

$$f(p, q, r, --- u, v, w, ---) = 0 \quad (12.7)$$

and it is known that there is an additional relationship among some of the variables, for example

$$P = f_1 \ (u, v, w, ----) \quad (12.8)$$

then p is not required and can be omitted. Conversely, if it is known that the only way the variables u,v,w ------ enter the problem is through the relationship expressed by equation (12.8), then the variables u,v,w, can be replaced by the single variable p, thereby, reducing the number of variables.

It is thus evident that the method of dimensional analysis does not give any clue as to the correctness of the selection of the relevant variables of a particular phenomenon. It merely tells how the variables should be grouped so that from the experimental investigation it may be decided which of the variables are important ones. This is, however, the main limitation of the dimensional analysis.

12.4 MODELING AND SIMILITUDE

Models are widely used in fluid mechanics. A few examples where models may be used are airplanes in wind tunnels, ships in towing basins, spillways of dams, river channels, hydraulic turbines, centrifugal pumps, air and water pollution, harbours, dams, cars and trains, towing test on submarines. A recent application has arisen from the problems of air flow around buildings, which may be studied in wind tunnels and by examining smoke generation and propogation through building models Although the term "model", is used in many different contexts, the "engineering model" generally conforms to the following definition. A model is a representation of a physical system that may be used to predict the behaviour of the system in some desired respect. The physical system for which the predictions are to be made is called the prototype. Although mathematical or computer models may also conform to this definition our interest will be in physical models; that is models that resemble the prototype but are generally of a different size, may involve different fluids, and often operate under different conditions.

Usually a model is smaller than the prototype. Therefore, it is more easily handled in the laboratory and less expensive to constract and operate than a large prototype. Occasionally, if the prototype is very small, it may be advantageous to have a model that is larger than the prototype so that it can be more easily studied. For examples, the flow in a carburettor might be studied in a very large model, the flow of water at the entrance to a small centrifugal runner might be investigated by the flow of air, at the entrance to a large model of the runner and large models have been used to study the motion of red blood cells, which are approximately 8 μm in diameter.

Although studies of models do not always provide ready answers to all questions, the results often guide an engineer toward a better under standing of the problem and aid him in the proper design of the prototype. In fact, a model may be regarded as a mechanical analog for solving differential equations of fluid motion with a given set of rather complicated boundary condition; its main function is to obtain useful quantitative or qualitative information which can be safely transferred to its prototype. A thorough knowledge of principles of model similitude is therefore essential to the proper design, construction, and operation of the model as well as the interpretation of data obtained from model tests.

Silimarity of flow phenomena not only occurs between a prototype and its model but also may exist between various natural phenomena if certain laws of similarity are satisfied. Similarity thus becomes a means of correlating the apparently divergent results obtained from similar fluid phenomena and as such becomes a valuable tool of modern fluid mechanics; the application of the laws of similitude will be found to lead to more compreh-ensive solutions and, therefore, to a better understanding of fluid phenomena in general.

There are, in general three types of similarities to be established for complete similarity to exist between the model and its prototype. These are; geometrically, kinematically and dynamically similarity.

12.4.1 Geometric Similarity

One of the desirable features in model studies is that there be geometric similarity which means that the model and its prototype be identical in shape but differ only in size. The important consideration is that the flow patterns be geometrically similar. If the scale ratio is denoted by l_r, which means the ratio of the linear dimensions of the model to corresponding dimensions in the model.

Mathematically,

$$l_r = \frac{l_m}{l_p} \tag{12.9}$$

Subscript m and p refer to model and prototype respectively.

For true models there will be one scale ratio, and all lengths are fixed in accordance with this scale. The area ratio and volume ratios are written as l_{r2} and l_{r3} respectively. One of the most important factors in maintaining a complete geometric similarity between the two systems is the similarity in surface roughness. Failure to satisfy the geometric similarity in surface roughness results in a model with distorted scale. Therefore, complete geometric similarity is not always easy to attain.

12.4.2 Kinematic similarity

Kinematic similarity is the similarity of motions. To satisfy kinematic similarity, both the model and its prototype must undergo similar time rates of change of motions. Therefore the flow patterns in two systems must be geometrically similar (Fig 12.1). The corresponding velocity and acceleration ratios of two systems are then constant.

$$\text{Velocity ratio} \quad v_r = \frac{v_m}{v_p} \tag{12.10}$$

and its value in terms of lr will be determined by dynamic considerations as explained in the following section.

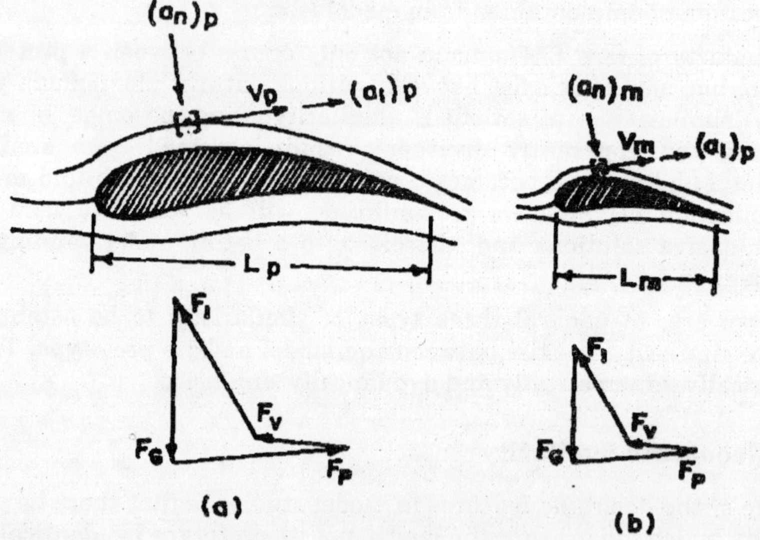

Fig. 12.1. (a) Prototype (b) Model.

As time T is dimensionally L/V, the time scale is

$$T_r = \frac{l_r}{v_r} \tag{12.11}$$

and in similar manner the acceleration scale is

$$a_r = \frac{a_m}{a_p} \tag{12.12}$$

$$\text{or,} \quad a_r = \frac{l_r}{T_r^2} = \frac{v_r^2}{l_r} \tag{12.13}$$

12.4.3 Dynamic similarity

Dynamic similarity means the similarity of forces. The model is said to be dynamically similar to its prototype if the ratios of all forces acting on corresponding fluid particles or corresponding boundary surface in the two systems are constant.

Forces that may act on a fluid element include those due to

(1) Viscous force F_V, (2) Gravity force F_G,

(3) Elastic force F_E, (4) Pressure force F_P, and

(5) Surface tension force F_S.

If the summation of forces on a fluid element does not add up to zero, the element will accelerate in accordance with Newton's law. Such an unbalanced force system can be transferred into a balanced system by adding an inertia force F_I that is equal and opposite to the resultant of the acting forces. Thus generally,

$$\Sigma F = F_V + F_G + F_E + F_P + F_S = \text{Resultant}$$

and $F_I = -\text{Resultant}$

Thus $F_V + F_G + F_E + F_P + F_S + F_I = 0$

For dynamic Similarity, we have

$$\frac{(F_I)_m}{(F_I)_p} = \frac{(F_V)_m}{(F_v)_p} = \frac{(F_G)_m}{(F_G)_p} = \frac{(F_E)_m}{(F_E)_p}$$

$$= \frac{(F_P)_m}{(F_P)_p} = \frac{(F_S)_m}{(F_S)_p} = F_r \tag{12.14}$$

where, F_R is known on force ratio. The subscripts m and p refer to model and prototype.

12.4.4 Standard dimensionless numbers

Over the years, in the history of fluid mechanics a set of dimensionless number of dynamic similarity have come to be well known, and each is associated with the name of a prominent scientist or engineer, usually the one who pioneered its use.

Forces encountered in flowing fluids including those due to inertia, viscosity, gravity, elastic, pressure and surface tension. The ratio of any two forces will be dimensionless.

Inertia forces are important in most fluid mechanics problems. The ratio of the inertia force to each of the other forces listed above leads to five fundamental dimensionless numbers encountered in fluid mechanics.

12.4.4.1 Reynold's number (R_e)

Reynold's number is defined as the ratio of inertia force of a flowing fluid and the viscous force of the fluid. Therefore,

$$R_e = \frac{\text{Inertia force}}{\text{Viscous force}} = \frac{F_I}{F_V} \tag{12.15}$$

F_I = Mass × Acceleration of flowing fluid

= Density × volume × velocity/time

$$= \rho l^3 \left(\frac{u}{t}\right) = \rho l^2 \frac{l}{t} u = \rho l^2 \text{ u u} = \rho l^2 u^2 \tag{12.16}$$

$$F_V = \textit{Shear stress} \times \textit{Area} = \tau A$$

$$= \mu \frac{du}{dy} A \quad = \mu \frac{u}{l} l^2 \quad = \mu \, u \, l$$

Substituting the value of F_I and F_V in equation (12.15)

$$R_e \; \frac{\rho u^2 l^2}{\mu u l} = \frac{\rho u l}{\mu} \qquad [12.17\ (a)]$$

$$= \frac{u l}{\nu} \qquad [12.17\ (b)]$$

where, l and u are a characteristic or typical length and velocity for the system.

For pipe flow,

$$R_e = \frac{ud}{\nu}$$

Where, d is the diameter of the pipe.

For flow between parallel plates

$$R_e = \frac{uh}{\nu}$$

Where, h is the distance between the two plates.

For boundary layer flow

$$R_e = \frac{ul}{\nu}$$

Where l is length from the leading edge.

12.4.4.2 Froude's number (F_R)

Froude's number is defined as the square root of the ratio of inertia force of flowing to the gravity force. Therefore,

$$F_r = \sqrt{\frac{Inertia\ \ force}{Gravity\ \ force}} = \sqrt{\frac{F_1}{F_G}} \quad (12.18)$$

$F_1 = \rho u^2 l^2$ [From equation (12.16)]

F_G = mass × acceleration due to gravity $= \rho\, l^3\, g$

Substituting in equation (12.18)

$$F_r = \sqrt{\frac{\rho u^2 l^2}{\rho l^3 g}} = \frac{u}{\sqrt{lg}} \quad (12.19)$$

12.4.4.3 Mach's number (M)

Mach's number is the square root of the inertia forces to compressibility forces (Elastic forces) in the fluid. Therefore,

$$M = \sqrt{\frac{Inertia\ \ force}{Elastic\ \ force}} = \sqrt{\frac{F_1}{F_E}} \quad (12.20)$$

$F_I = \rho u^2 l^2$ [From equation (12.16)]

F_E = Compressibility force

or Elastic force = Bulk modulus of elasticity × Area $= K\, l^2$

Substituting in equation (12.20)

$$M = \sqrt{\frac{\rho u^2 l^2}{K l^2}} = \frac{u}{\sqrt{K/\rho}} = \frac{u}{c} \quad (12.21)$$

We know that the velocity of sound c is equal to $\sqrt{K/\rho}$.

12.4.4.4 Euler's number (E_u)

Euler's number is defined as the square root of the ratio of inertia force of flowing fluid to the pressure force. Therefore,

$$E_u = \sqrt{\frac{Inertia\ \ force}{Pressure\ \ force}} = \sqrt{\frac{F_I}{F_p}} \quad (12.22)$$

$F_I = \rho\, u^2\, l^2$ [From equation (12.16)]

F_P = Pressure × area = $p\, l^2$

Substituting in equation (12.22)

$$E_u = \sqrt{\frac{\rho u^2 l^2}{p l^2}} = \frac{u}{\sqrt{p/\rho}} \quad (12.23)$$

12.4.4.5 Weber's number (W_b)

Weber number is defined as the squareroot of the ratio of the inertia force of the flowing fluid to the surface tension force. Therefore,

$$W_b = \sqrt{\frac{\text{Inertia force}}{\text{Surface tension force}}} = \sqrt{\frac{F_I}{F_S}} \tag{12.24}$$

$$F_I = \rho u^2 l^2 \qquad \text{[From equation (12.16)]}$$

F_S = Surface tension × length = σl

Substituting in equation (12.24)

$$W_b = \sqrt{\frac{\rho u^2 l^2}{\sigma l}} = \frac{u}{\sqrt{\frac{\sigma}{\rho l}}} \tag{12.25}$$

12.5 MODEL LAWS

The results obtained from the model tests may be transferred to the prototype by the use of model laws which may be developed from the principles of dynamic similarity.

The equation (12.14) can be expressed as

$$\left(\frac{F_I}{F_V}\right)_m = \left(\frac{F_I}{F_V}\right)_p \quad \text{or} \quad R_{e_m} = R_{e_p} \tag{12.26}$$

$$\left(\frac{F_I}{F_G}\right)_m = \left(\frac{F_I}{F_G}\right)_p \quad \text{or} \quad F_{r_m} = F_{r_p} \tag{12.27}$$

$$\left(\frac{F_I}{F_E}\right)_m = \left(\frac{F_I}{F_E}\right)_p \quad \text{or} \quad M_m = M_p \tag{12.28}$$

$$\left(\frac{F_I}{F_P}\right)_m = \left(\frac{F_I}{F_P}\right)_p \quad \text{or} \quad E_{u_m} = E_{u_p} \tag{12.29}$$

$$\left(\frac{F_I}{F_S}\right)_m = \left(\frac{F_I}{F_S}\right)_p \quad \text{or} \quad W_{b_m} = W_{b_p} \tag{12.30}$$

In the foregoing force ratio equations may be written in terms of the dimensionless numbers as indicated above.

It is almost impossible to obtain complete dynamic similarity between the model and its prototype (except models with unit scale ratio) in which all these five conditions are satisfied simultaneously because no two fluids are known to posses all the requisite properties to satisfy all these five conditions. Fortunately, in most engineering problems, only two or three forces may be predominant, whereas the others are relatively insignificant and can be safely neglected. For all practical purposes, therefore, a particular state of fluid motion can usually be simulated in a model by considering the predominant forces only.

12.5.1 Reynold's model law

If two systems, such as model and its prototype, are to be dynamically equivalent so far as inertia and viscous forces are concerned, they must both have the same value of Reynold's number (R_e). Thus for such a case, dynamic similarity is achieved when

$$R_{e_m} = R_{e_p}$$

or, $$\left(\frac{\rho\, u\, l}{\mu}\right)_m = \left(\frac{\rho\, u\, l}{\mu}\right)_p \quad (12.31)$$

or, $$\left(\frac{ul}{\nu}\right)_m = \left(\frac{ul}{\nu}\right)_p \quad (12.32)$$

or, $$\frac{u_r\, l_r}{\nu_r} = 1 \quad (12.33)$$

Where the various quantities with subscript r represent the corresponding scale ratios.

The scale ratios for various physical quantities based on Reynold's model law are obtained as follows

Using equation (12.33), we get

(a) Scale ratio for velocity

$$u_r = \left(\frac{\nu}{l}\right)_r$$

(b) Scale ratio for time

We know that,

$$u_r = \frac{l_r}{T_r}$$

or, $$T_r = \frac{l_r}{u_r}$$

(c) Scale ratio for acceleration

We know that,

$$a_r = \frac{u_r}{T_r}$$

or, $$a_r = \frac{u_r}{l_r / u_r} = \frac{u_r^2}{l_r} = \frac{\nu_r^2}{l_r^3}$$

(d) Scale ratio for discharge

We know that

$$Q_r = \rho_r \, u_r \, l_r^2$$

$$Q_r = \rho_r \frac{v_r}{l_r} l_r^2 = \rho_r \frac{\mu_r}{\rho_r} l_r = \mu_r \, l_r$$

(e) Scale ratio for force

$$F_r = \rho_r \, u_r^2 \, l_r^2 = \rho_r \frac{v_r^2}{l_r^2} l_r^2 = \rho_r \, v_r^2 = \rho_r \frac{\mu_r^2}{\rho_r^2} = \frac{\mu_r^2}{\rho_r}$$

The Reynold's model law is applicable to the following cases

(1) Submarine in the fully submerged condition

(2) Aircraft travelling at a Mach number much smaller than unity (i.e. submerged in air)

(3) Buildings and structures subjected to wind loadings.

(4) Testing of objects in a water tunnel or closed circuit wind tunnel.

(5) Flow through closed conduits.

(6) Flow around immersed bodies.

12.5.2 Froude's model law

If two systems, such as model and its prototype, are to be dynamically equivalent so far as inertia and gravity forces are concerned, they must both have the same value of Froudes number (Fr). Thus for such a case, dynamic similarity is achieved when

$$F_{r_m} = F_{r_p}$$

or, $$\left(\frac{u}{\sqrt{l\,g}}\right)_m = \left(\frac{u}{\sqrt{l\,g}}\right)_p \qquad (12.34)$$

or, $$\frac{u_m}{\sqrt{l_m}} = \frac{u_p}{\sqrt{l_p}}$$

or, $$u_r = \sqrt{l_r} \qquad (12.35)$$

Scale ratio for various physical quantities based on Froude model law are obtained as follows :

(a) Scale ratio for discharge.

Since the velocity varies as $\sqrt{l_r}$ and the cross section are as l_r^2 it follows that

$$Q_r = \frac{Q_m}{Q_p} = v_r\, l_r^2 = \left(\sqrt{l_r}\right) l_r^2 = l_r^{5/2}$$

(b) Scale ratio for time

$$T_r = \frac{T_m}{T_p} = \frac{l_r}{u_r} = \sqrt{l_r}$$

(c) Scale ratio for acceleration

$$a_r = \frac{u_r}{T_r} = \frac{\sqrt{l_r}}{\sqrt{l_r}} = 1$$

(d) Scale ratio for force

$$F_r = \frac{F_m}{F_p} = \frac{\rho_m\, l_m^2\, u_m^2}{\rho_p\, l_p^2\, u_p^2}$$

If the same fluid is used for model and prototype, then

$$F_r = l_r^2\, u_r^2 \quad = l_r^2\, l_r \; = l_r^3$$

(e) Scale ratio for work, energy, torque, moment

Moment = Force × perpendicular distance

$$M_r = \frac{M_m}{M_p} = \frac{(F \times l)_m}{(F \times l)_p} = \frac{l_m^4}{l_p^4} = l_r^4$$

(f) Scale ratio for pressure intensity

$$p = Force/\ Area = \frac{\rho\, u^2\, l^2}{l^2} = \rho\, u^2$$

$$P_r = \frac{p_m}{p_p} = \frac{\rho_m\, u_m^2}{\rho_p\, u_p^2}$$

Using the same fluid for model and prototype, then

$$P_r = u_r^2 = l_r$$

(g) Scale ratio for power

$$\text{Power} = \text{Rate of doing work} = \frac{Work}{Time} = \text{Force} \times \text{Velocity}$$

$$P_r = \frac{P_m}{P_p} = l_r^3\, u_r = l_r^3\, \sqrt{l_r} = l_r^{7/2}$$

The Froude's model law is applicable to the following cases.

(1) Flow with a free surface, e.g. spillways, sluices, channels. weirs, etc.

(2) Motion in river, ocean, etc where waves are likely to be formed.

(3) Hydraulic Jump and stiling basin.

(4) Flow of jets from on orifice or nozzie.

(5) Flow of one fluid past another fluid of different densities.

12.5.3 Mach model law

If two systems, such as model and its prototype, are to be dynamically equivalent so far as inertia and elastic (compressibility) forces are concerned, they must both have the same value of Mach number (M). Thus for such a case, dynamic similarity is achieved when

$$M_m = M_p$$

$$\left(\frac{u}{C}\right)_m = \left(\frac{u}{C}\right)_p \qquad (12.36)$$

Where $C = \sqrt{K/\rho} = \sqrt{\gamma\, R\, T}$ (12.37)

For air $\gamma = 1.4$, R = 287 J/kg K and T in K

Therefore, $C = \sqrt{1.4 \times 287 \times T} = 20.25\sqrt{T}$ (12.38)

Mach model law is applicable to the following cases; e.g.

(1) Commonly used in compressible flow problems, particularly in the fields of gas dynamics and aerodynamics. ($M > 0.3$)

The effect of compressibility is considered only when the Mach number is greater than 0.3.

(2) Unsteady flow-Water hammer problems.

(3) Underwater testing of torpedoes.

(4) Aerodynamic testing.

12.5.4 Euler's model law

If two systems, such as model and its prototype, are to be dynamically equivalent so far as inertia and pressure forces are concerned, they must both have the same value of Euler number (Eu). Thus for such a case, dynamic similarity is achieved when

$$E_{u_m} = E_{u_p}$$

or,

$$\left(\frac{u}{\sqrt{p/\rho}}\right)_m = \left(\frac{u}{\sqrt{p/\rho}}\right)_p \qquad (12.39)$$

Euler number would normally be used in problems in which pressure or the pressure difference between two points is an important variable.

For problems in which cavitation is of concern, the dimensionless group $(P_r - P_v) / \rho u^2/2$ is commonly used, where P_v is the vapour pressure and P_r

is some reference pressure. Although this dimensionless group has the same form as the Euler number, it is generally referred as the cavitation number.

12.5.5 Weber's model low

If two systems, such as model and its prototype, are to be dynamically equivalent so far as inertia and surface tension forces are concerned, they must both have the same value of Weber number (Wb). Thus for such a case, dynamic similarity is achieved when

$$(W_b)_m = (W_b)_p$$

$$\left(\frac{u}{\sqrt{\sigma / \rho l}}\right)_m = \left(\frac{u}{\sqrt{\sigma / \rho l}}\right)_p \qquad (12.40)$$

Weber number may be important in problems in which there is an interface between two fluids. Common examples of problems in which this parameter may be important include the flow of thin films of liquid, or in the formation of droplets or bubbles. Clearly, not all problems involving flows with an interface will require the inclusion of surface tension. More examples are (1) Flow over weirs involving very low heads, (2) Very thin sheet of liquid flowing over a surface, (3) Capillary waves in channels etc.

12.6 UNDISTORTED AND DISTORTED MODELS

A model is said to be perfect or undistorted when it is perfectly similar to its prototype. In undistorted models, the conditions of simlitude are completely satisfied. The model investigation results can be directly used for the prototype.

Models for which one or more of the similarity requirements are not satisfied are called distorted models.

Distorted models can arise for a variety of reasons.

(1) Geometric distortion, (2) Distortion of configuration (3) Material distortion and (4) Distortion of hydraulic quantities such as, discharge, velocity and time.

Distorted models can be successfully used, but the interpretation of results obtained is more difficult than the interpretation of results obtained with true models for which all similarity requirements are met. There are no general rules for handling distorted models, and essentially each problem must be considered on its own merits. The success of using distorted models depends to a large extent on the skill and experience of the investigator responsible for the design of the model, and in the interpretation of experimental data obtained from the model. Distorted models are widely used, especially in the models of rivers and open channels.

(a) River and harbour models

River and horbour engineering projects are costly undertaking and, as analytical techniques only provide approximate prediction of the likely effects of any river widening or harbour improvement, the use of models at an early stage in the design has many advantages. The problems arise when a suitable scale is to be chosen for the model; the adoption of a scale that will give reasonable channel depths will usually result in a model too large to be practical, while choosing a scale on area-cost criteria yields channel depths that are very small.

In order to provide a solution to these problems, distorted scaling is adopted (i.e. geometrical distortion), vertical scales of 1/100 and larger being typical while horizontal scales vary from 1/200 to 1/600. Since the velocity is largely determined by gravity forces, the vertical height of the model is used to calculate its froude number. When comparing Froude numbers of model and prototype, the Froude number of the prototype must also be based on its vertical height. Distortation of this sort is suitable if the overall discharge characteristics of a long length of river are being studied.

(b) Estuary Models

Estuary models may be constructed using the same general principles as outlined for river models; distorted scales are typically 1/50 to 1/150 vertical and 1/300 to 1/2500 horizontal. As the available data on tidal velocity distributions are likely to be sketchy, it is necessary to incorporate in the mode the whole tidal channel system, as well as a substantial portion of adjustment coast line.

The speed of propagation of the tide becomes an important design criteria as the tidal period governs the time available for result recording. The tidal period of the model, thus,

$$\frac{\text{Time in model}}{\text{Time in prototype}} = \frac{(\text{Distance/velocity})_{\text{model}}}{(\text{Distance/velocity})_{\text{prototype}}}$$

Now, as the Froude numbers are equal (as for river models) it follows

$$\frac{(Time)_m}{(Time)_p} = \left(\frac{l_m}{l_p}\right)\left(\frac{u_p}{u_m}\right) = \left(\frac{l_m}{l_p}\right)\sqrt{(Depth)_p/(Depth)_m} \qquad (12.57)$$

The above equation refers to horizontal fluid velocity, the scale factor for vertical silt particle velocity is also of interest and may be derived in the same manner

$$\frac{\text{Rate of fall in model}}{\text{Rate of fall in prototype}} = \frac{\text{Depth Scale}}{\text{Time Scale}} \qquad (12.58)$$

Estuary models are now commonly used to investigate the effects of discharge of power station or industrial cooling water flows and, here, density variations may again have to be modelled, based on temperature. Other uses of estuary models include silting and erosion studies and the spread and deposit of effluent discharged into the sea. Generally, the role of the estuary model should be seen was as a method of comparing the attributes of various design solutions rather than as an accurate method of predicting the effects of one design.

12.7 SCALE EFFECT

It is not possible to predict the exact behaviours of th prototype by model testing. The behaviour of the prototype as predicted by two models with different scale ratios is generally not the same. This difference in the predicted behaviour of the prototype is attributed to difference in scale ratios and is known as scale effect.

Scale effect can be detected by testing a larger number of models with different scale ratios and the exact behaviour of prototype can then be predicted.

12.8 COMMENTS ON MODEL TESTING

While outlining the procedures involved in model we have tried not to that testing is a simple task that automatically gives results which are easy to interpret, accurate and complete. As in all experimental work, careful planning and execution are needed to obtain valid results. Models must be constructed carefully and accurately, and they must include sufficient detail in areas critical to the phenomenon being measured. Aerodynamic balances or other force measuring systems must be aligned carefully and calibrated correctly. Mounting methods must be devised that offer adequate rigidity and model motion, yet do to interfere with the phenomenon being measured.

Experimental facilities must be designated and constructed carefully. The quality of the flow in a wind tunnel must be documented. It should be as nearly uniform as possible, free from angularity, and with little swirl content. If they interfere with measurements boundary layers on tunnel walls must be removed by suction. Pressure gradients in a wind tunnel test section may cause erroneous readings due to horizontal buoyancy in a flow direction.

In the use of models it is essential that the fluid velocity should not be so low that laminar flow exists when the flow in the prototype is turbulent. Also, conditions in the model should not be such that surface tension is importent if such conditions do not exist in the prototype.

While model studies are very important and valuable it is necessary to exercise some judgment in transferring results from the model to the prototype. It is not always necessary or desirable that these various dimensionless ratios be adhered to in every case.

The roughness of a model should be scaled down in the same ratio as the other linear dimensions, which means that a small model should have surfaces that are much smoother than those in its prototype. But this requirement imposes a limit on the scale that can be used if true geometric similarity is to be achieved. However, in the case of a distorted model with a vertical scale larger than the horizontal scale, it may be necessary to make the model surface rough in order to simulate the flow conditions in the prototype.

Special facilities are needed for unusual conditions or for special test requirements, especially to achieved large Reynold's numbers. Many facilities are so large or specialized that they cannot be supported by university laboratories or private industry.

ILLUSTRATIVE EXAMPLES

12.1. Assuming that the resistance to motion, F, of a body moving in a compressible fluid is a function of a characteristic dimension, l, of the body, the velocity u of the body and fluid properties density, ρ, viscosity, μ, and bulk modulus, K; Find the expression for F in terms of these quantities using for F in terms of these quantities using Buckingham's π theorem.

$$F = \frac{1}{2}\left(\rho u^2 l^2\right) \Phi \left[\frac{\rho u l}{\mu}, \frac{u}{\sqrt{K/\rho}}\right]$$

Discuss the significance of the non dimensional groups obtained.

Solution

The relationship between the variables may be expressed as

$$F = \text{function of } \left[u, \rho, l, \mu, K\right]$$

There are six variables in this problem, all of which may be expressed in terms of three primary dimensions M, L and T. Therefore, there will be 6-3 = 3 dimensional parameters or π - groups in the solution.

Applying the π theorem, it is first necessary to select three primary variables which between them include all the primary dimensions (M, L and T) involved. In this problem we may select, ρ, u and *l* as primary variables.

Each of the remaining variables is now taken in turn and used to form a dimensionless π groups as follows

$$\pi_1 = \rho^{a_1} u^{b_1} l^{c_1} F$$

$$\pi_2 = \rho^{a_2} u^{b_2} l^{c_2} \mu$$

$$\pi_3 = \rho^{a_3} u^{b_3} l^{c_3} K$$

Dimensional alaysis for π_1

$$\pi_1 = \rho^{a_1} u^{b_1} l^{c_1} F$$

$$\left[M^o L^o T^o\right] = \left[ML^{-3}\right]^{a_1} \left[LT^{-1}\right]^{b_1} \left[L\right]^{c_1} \left[MLT^{-2}\right]$$

Equating indices of M, L and T, we have

M : 0 = a_1 + 1

L : 0 = – 3 a_1 + b_1 + c_1 + 1

T : 0 = – b_1 – 2

Hence a_1 = –1, b_1 = –2 andcC = –2

Therefore, $\pi_1 = \dfrac{F}{\rho u^2 l^2}$

Dimensional analysis for π_2

$$\pi_2 = \rho^{a_2} u^{b_2} l^{c_2} \mu$$

$$[M^oL^oT^o] = [ML^{-3}]^{a2} [LT^{-1}]^{b2} [L]^{c2} [ML^{-1} T^{-1}]$$

Equating indices

M : 0 = a_2 +1

L : 0 = –3a_2 + b_2 + c_2 –1

T : 0 = – b_2 – 1

Hence a_2 = – 1, b_2 = – 1, c_2 =–1

Therefore, $\pi_2 = \dfrac{\mu}{\rho u l}$

Dimensional analysis for π_3

$$\pi_3 = \rho^{a_3} u^{b_3} l^{c_3} K$$

$$[M^oL^oT^o] = \left[ML^{-3}\right]^{a_3} \left[LT^{-1}\right]^{b_3} [L]^{c_3} \; [ML^{-1} T^{-2}]$$

Equating indices

M : 0 = a_3 +1

L : 0 = –3a_3 + b_3 + c_3 –1

T : 0 = – b_3 – 2

Hence a_3 = – 1, b_3 =–2, c_3 = 0

Therefore, $\pi_3 = \dfrac{K}{\rho u^2}$

Dimensional analysis does not generate any numerical coefficients. In the π_2 term, the numerical constant 1/2, which normally multiplies $(\rho v^2 \; l^2)$ to give it physical significance. It is certainly permissible to multiply any dimensionless group by a dimensionless constant to give it physical significance. Therefore,

$$\pi_1 = 2\,\pi_1 = \frac{F}{\frac{1}{2}\left(\rho u^2 \, l^2\right)}$$

and $$\pi_2 = \frac{1}{\pi_2} = \frac{\rho u l}{\mu}$$

$$\pi_3 = \sqrt{\frac{1}{\pi_3}} = \frac{u}{\sqrt{K/\rho}}$$

The Solution may be stated in the form

$$\pi_1' = \Phi\left(\pi_2' \ \pi_3'\right)$$

or, $$\frac{F}{\frac{1}{2}\rho u^2 l^2} = \Phi\left(\frac{\rho u l}{\mu}, \frac{u}{\sqrt{K/\rho}}\right)$$

or, $$F = \frac{1}{2}\left(\rho u^2 l^2\right)\Phi\left[\frac{\rho u l}{\mu}, \frac{u}{\sqrt{K/\rho}}\right]$$

The group $\frac{F}{\frac{1}{2}\rho u^2 l^2}$ is sometimes referred to as the drag coefficient C_D.

The group $\frac{\rho u l}{\mu}$ is the Reynold's number, R_e, which is a criterion of dynamic similarity when viscous and inertia forces are significant.

The group $\frac{u}{\sqrt{K/\rho}}$ is the Mach number, M, which is a criterion of dynamic similarity when elastic and inertia forces are significant.

12.2. Show by the method of dimensions that for a screw propeller, the relation between the thrust F, torque T, diameter D, speed of travel u, speed of rotation N, mass density ρ and dynamic viscosity μ may be put in the form

$$F = \rho u^2 D^2 \Phi\left[\frac{\rho u^2 D^3}{T}, \frac{DN}{u}, \frac{\rho u D}{\mu}\right]$$

Solution

Let F = function of (ρ, u, D, T, N, μ)

n = number of variables = 7

m = number of primary variables = 3

(here M, L and T)

π groups = n – m = 7 –3 = 4

Selecting ρ, u and D as a repeated variables, then

$$\pi_1 = \rho^{a_1} u^{b_1} D^{c_1} F$$

$$\pi_2 = \rho^{a_2} \; u^{b_2} \; D^{c_2} \; T$$

$$\pi_3 = \rho^{a_3} \; u^{b_3} \; D^{c_3} \; N$$

$$\pi_4 = \rho^{a_4} \; u^{b_4} \; D^{c_4} \; \mu$$

For π1

$$\pi_1 = \rho^{a_1} \; u^{b_1} \; D^{c_1} \; F$$

$$\left[M^o \, L^o \, T^o\right] = \left[ML^{-3}\right]^{a_1} \left[LT^{-1}\right]^{b_1} \left[L\right]^{c_1} \left[MLT^{-2}\right]$$

Equating indices

$$M : 0 = a_1 + 1$$

$$L : 0 = -\,3\,a_1 + b_1 + c_1 + 1$$

$$T : 0 = -\,b_1 - 2$$

Which gives $a_1 = -\,1,\; b_1 = -\,2,\; c_1 = -2$

Therefore, $\pi_1 = \dfrac{F}{\rho\, u^2\, D^2}$

For π2

$$\pi_2 = \rho^{a_2} \; u^{b_2} \; D^{c_2} \; T$$

$$\left[M^o \, L^o \, T^o\right] = \left[ML^{-3}\right]^{a_2} \left[LT^{-1}\right]^{b_2} \left[L\right]^{c_2} \left[ML^2T^{-2}\right]$$

Equating indices

$$M : 0 = a_2 + 1$$

$$L : 0 = -\,3\,a_2 + b_2 + c_2 + 2$$

$$T : 0 = -\,b_2 - 2$$

Which gives $a_2 = -\,1,\; b_2 = -\,2,\; c_2 = -3$

Therefore, $\pi_2 = \dfrac{F}{\rho\, u^2\, D^3}$

or, $\pi_2' = \dfrac{1}{\pi_2} = \dfrac{\rho\, u^2\, D^3}{T}$

For π3

$$\pi_3 = \rho^{a_3} \; u^{b_3} \; D^{c_3} \; N$$

$$\left[M^o \, L^o \, T^o\right] = \left[ML^{-3}\right]^{a_3} \left[LT^{-1}\right]^{b_3} \left[L\right]^{c_3} \left[T^{-1}\right]$$

Equating indices

$$M : 0 = a_3$$

$$L : 0 = -3\,a_3 + b_3 + c_3$$

$$T : 0 = -b_3 - 1$$

Which gives $a_3 = 0, b_3 = -1, c_3 = +1$

Therefore, $\pi_3 = \dfrac{ND}{\mu}$

For π_4

$$\pi_4 = \rho^{a_4}\, u^{b_4}\, D^{c_4}\, \mu$$

$$\left[M^o\, L^o\, T^o\right] = \left[ML^{-3}\right]^{a_4} \left[LT^{-1}\right]^{b_4} \left[L\right]^{c_4} \left[ML^{-1}\, T^{-1}\right]$$

Equating indice

$$M : 0 = a_4 + 1$$

$$L : 0 = -3\,a_4 + b_4 + c_4 - 1$$

$$T : 0 = -b_4 - 1$$

Which gives $a_4 = -1, b_4 = -1, \textit{and}\ c_4 = -1$

Therefore $\pi_4 = \dfrac{\mu}{\rho\, u\, D}$

$$\pi_4' = \frac{1}{\pi_4} = \frac{\rho\, u\, D}{\mu}$$

The functional relation can be expressed as

$$\pi_1 = \Phi\left(\pi_2', \pi_3, \pi_4'\right)$$

$$\frac{F}{\rho\, u^2\, D^2} = \Phi\left[\frac{\rho\, u^2\, D^3}{T}, \frac{DN}{u}, \frac{\rho\, u\, D}{\mu}\right]$$

$$F = \rho\, u^2\, D^2\, \Phi\left[\frac{\rho\, u^2\, D^3}{T}, \frac{DN}{u}, \frac{\rho\, u\, D}{\mu}\right]$$

12.3. **Derive on the basis of Buckingham's π theorem to present the thrust developed by a propeller.**

$$F = \rho\, u^2\, D^2\, \Phi\left[\frac{\omega\, D}{u}, R_e\,, M\right]$$

Where F = thrust developed, ω = angular velocity

u = Speed of advance D = diameter

μ = dynamic viscosity ρ = mass density

and c = speed of sound

Solution

Let F = function of (ρ, u, D, ω, μ, c)

n = number of variables = 7

m = number of primary variables = 3

π groups = 7 – 3 = 4

Choosing ρ, u nd D as a repeated variables, then

$$\pi_1 = \rho^{a_1} u^{b_1} D^{c_1} F$$

$$\pi_2 = \rho^{a_2} u^{b_2} D^{c_2} \omega$$

$$\pi_3 = \rho^{a_3} u^{b_3} D^{c_3} \mu$$

$$\pi_4 = \rho^{a_4} u^{b_4} D^{c_4} c$$

For π_1

$$\pi_1 = \rho^{a_1} u^{b_1} D^{c_1} F$$

$$\left[M^o L^o T^o\right] = \left[ML^{-3}\right]^{a_1} \left[LT^{-1}\right]^{b_1} [L]^{c_1} \left[MLT^{-2}\right]$$

Equating indices

$$M : 0 = a_1 + 1$$

$$L : 0 = -3a_1 + b_1 + c_1 + 1$$

$$T : 0 = -b_1 - 2$$

Which gives $a_1 = -1, b_1 = -2, c_1 = -2$

therefore, $\pi_1 = \dfrac{F}{\rho u^2 D^2}$

For π_2

$$\pi_2 = \rho^{a_2} u^{b_2} D^{c_2} \omega$$

$$\left[M^o L^o T^o\right] = \left[ML^{-3}\right]^{a_2} \left[LT^{-1}\right]^{b_2} [L]^{c_2} \left[T^{-1}\right]$$

Equating indices

$$M : 0 = a_2$$

$$L : 0 = -3a_2 + b_2 + c_2$$

$$T : 0 = -b_2 - 1$$

Which gives $a_2 = -0, b_2 = -1, c_2 = 1$

therefore, $\pi_2 = \dfrac{\omega D}{u}$

For π_3

$$\pi_3 = \rho^{a_3}\ u^{b_3}\ D^{c_3}\ \mu$$

$$\left[M^o\ L^o\ T^o\right] = \left[ML^{-3}\right]^{a_3}\ \left[LT^{-1}\right]^{b_3}\ \left[L\right]^{c_3}\ \left[ML^{-1}T^{-1}\right]$$

Equating indices

$$M : 0 = a_3 + 1$$

$$L : 0 = -\ 3a_3 + b_3 + c_3 - 1$$

$$T : 0 = -\ b_3 - 1$$

Which gives $a_3 = -1, b_3 = -1, c_3 = -1$

therefore, $\pi_3 = \dfrac{\mu}{\rho\, u\, D}$

$$\pi_3' = \frac{1}{\pi_3} = \frac{\rho\, u\, D}{\mu} = \text{the Reynold's number } R_e$$

π_4 can be calculated by inspection

$$\pi_4 = \frac{c}{u}$$

$$\pi_4' = \frac{1}{\pi_4} = \frac{u}{c} = \text{ the Mach number M}$$

Thus the functional relation can be expressed as

$$\pi_1 = \Phi\left(\pi_2, \pi_3', \pi_4'\right)$$

$$\frac{F}{\rho u^2\, D^2} = \Phi\left[\frac{\omega\, D}{u}, \frac{\rho u\, D}{\mu}, \frac{u}{c}\right]$$

$$F = \rho\, u^2\ D^2\ \Phi\left[\frac{\omega D}{u}, R_e, M\right]$$

12.4. **Using Buckingham's π theorem, show that velocity through a circular orifice is given by**

$$V = \sqrt{2gH}\ \Phi\left[\frac{H}{D,}\ \frac{VH}{\nu}\right]$$

Where H is the head causing flow, D is the diameter of the orifice, υ is the kinematic viscosity and g is the acceleration due to gravity.

Solution

Let V = function of (H, D, g, ν)

There are a total number of variables n = 5 Number of primary dimensions (Here L and T), m = 2. Dimensionless π terms = n-m = 5-2 = 3

Assuming H and g as repeating variables, then

$$\pi_1 = H^{a_1}\ g^{b_1}\ V$$

$$\pi_2 = H^{a_2}\ g^{b_2}\ D$$

$$\pi_3 = H^{a_3}\ g^{b_3}\ \upsilon$$

Dimensional analysis for π_1

$$\pi_1 = H^{a_1}\ g^{b_1}\ V$$

$$\left[L^o\, T^o\right] = [L]^{a_1}\left[LT^{-2}\right]^{b_1}\left[LT^{-1}\right]$$

Equating indices

$$L: 0 = a_1 + b_1 + 1$$

$$T: 0 = -2\, b_1 - 1$$

Hence $a_1 = -\frac{1}{2}, b_1 = -\frac{1}{2}$ *and*

$$\pi_1 = \frac{V}{\sqrt{gH}} \text{ or } \pi_1' = \frac{\pi_1}{\sqrt{2}} = \frac{V}{\sqrt{2gH}}$$

Dimensional analysis for π_2

$$\pi_2 = H^{a_2}\ g^{b_2}\ D$$

Applying the above technique or by inspection

$$\pi_2 = \frac{D}{H} \text{ or } \pi_2' = \frac{1}{\pi_2} = \frac{H}{D}$$

Dimensional analysis for π_3

$$\pi_3 = H^{a_3}\ g^{b_3}\ \upsilon$$

$$\left[L^o\, T^o\right] = [L]^{a_3}\left[LT^{-2}\right]^{b_3}\left[L^2\, T^{-1}\right]$$

Equating indices

$$L: 0 = a_3 + b_3 + 2$$

$$T: 0 = -2\, b_3 - 1$$

Hence $a_3 = -\frac{3}{2}, b_3 = -\frac{1}{2}$ *and*

$$\pi_3 = \frac{\nu}{H^{3/2}\, g^{1/2}} = \frac{\nu}{\sqrt{gH}\ H}$$

or, $$\frac{\pi'}{3} = \pi_1 / \pi_3 = \frac{V}{\sqrt{gH}} \times \frac{\sqrt{gH}\ H}{\nu} = \frac{VH}{\nu}$$

Thus the functional relationship can be expressed as

$$\pi_1' = \Phi\left(\pi_2', \pi_3'\right)$$

$$\frac{V}{\sqrt{2g\,H}} = \Phi\left(\frac{H}{D,}\frac{VH}{\upsilon}\right)$$

or, $$V = \sqrt{2gH}\;\Phi\left(\frac{H}{D,}\frac{VH}{\nu}\right)$$

12.5. The efficiency η of a fan depends on the kinematic viscosity υ of the fluid, the angular velocity W, diameter D of the rotor and the discharge Q. Express η in terms of dimensionless parameter using Buckingham's π theorem.

$$\eta = \Phi\left[\frac{\omega\,D^2}{\nu}, \frac{Q}{\omega D^3}\right]$$

Solution

The relationship between the variables may be expressed as

η = function of [ν, ω, D, Q]

There are five variables in this problem, all of which may be expressed in terms of two primary dimensions L and T. Therefore, there will be 5-2 = 3 dimensional parameters or π groups in the solution.

Repeating variables from the expression to be proved, are W and D.

Each of the remaining variables is now taken in turn and used to form a dimensionless π groups as follows

$$\pi_1 = \omega^{a_1}\,D^{b_1}\;\eta$$

$$\pi_2 = \omega^{a_2}\,D^{b_2}\;\upsilon$$

$$\pi_3 = \omega^{a_3}\,D^{b_3}\;Q$$

Dimensional analysis for π_1

$$\pi_1 = \omega^{a_1}\,D^{b_1}\;\eta$$

$$\left[L^o\,T^o\right] = \left[T^1\right]^{a_1}\,[L]^{b_1}\left[L^oT^o\right]$$

Equating indices of L and T, we have

L : 0 = b_1 + 0

T : 0 = $-a_1$ + 0

Hence $a_1 = b_1 = 0$ and

π1 = η

The above value of π_1 can also be written by inspection.

Dimensional analysis for π_2

$$\pi_2 = \omega^{a_2}\,D^{b_2}\;\upsilon$$

$$\left[L^o T^o\right] = \left[T^{-1}\right]^{a_2} \left[L\right]^{b_2} \left[L^2 T^{-1}\right]$$

Equating indices

L : 0 = b_2 + 0

T : 0 = – a_2 – 1

Hence $a_2 = -1, b_2 = -2$ *and*

$$\pi_2 = \frac{v}{\omega D^2}$$

Dimensional analysis for π_3

$$\pi_3 = \omega^{a_3} D^{b_3} Q$$

$$\left[L^o T^o\right] = \left[T^{-1}\right]^{a_3} \left[L\right]^{b_3} \left[L^3 T^{-1}\right]$$

Equating indices

L : 0 = b_3 + 3

T : 0 = –a_3 – 1

Hence $a_3 = -1, b_3 = -3$ *and*

$$\pi_3 = \frac{Q}{\omega D^3}$$

As per given in the question.

From the π_2 term

$$\pi_2^{/} = \frac{1}{\pi_2} = \frac{\omega D^2}{v}$$

Thus the functional relationship can be expressed as

$$\pi_1 = \Phi\left[\pi_2, \pi_3\right]$$

$$\eta = \Phi\left[\frac{\omega D^2}{v}, \frac{Q}{\omega D^3}\right]$$

12.6. Using Buckingham's π theorem show that the discharge, Q, of a centrifugal pump can be expressed as

$$Q = ND^3 \, \Phi\left[\frac{ND}{\sqrt{gh}}, \frac{ND^2}{v}\right]$$

Where N = Speed of the pump in rpm,

D = diameter of the impeller

ρ = Kinematic viscosity

g = acceleration due to gravity

Solution

Let Q = function of (N, D, g, H, ν)

n = number of variables = 6

m = number of primary variables = 2

(Here L and T)

π groups = n – m = 6 – 2 = 4

Selecting N and D as a repeated variables, then

$$\pi_1 = N^{a_1} D^{b_1} Q$$

$$\pi_2 = N^{a_2} D^{b_2} g$$

$$\pi_3 = N^{a_3} D^{b_3} H$$

$$\pi_4 = N^{a_4} D^{b_4} \nu$$

π_3 can be obtained by inspection

$$\pi_3 = \frac{H}{D}$$

For π_1

$$\pi_1 = N^{a_1} D^{b_1} Q$$

$$\left[L^o T^o\right] = \left[T^{-1}\right]^{a_1} [L]^{b_1} \left[L^3 T^{-1}\right]$$

Equating indices

$$L : 0 = b_1 + 3$$

$$T : 0 = -a_1 - 1$$

Which gives $a_1 = -1, b_1 = -3$

Therefore, $\pi_1 = \dfrac{Q}{ND^3}$

For π_2

$$\pi_2 = N^{a_2} D^{b_2} g$$

$$\left[L^o T^o\right] = \left[T^{-1}\right]^{a_2} [L]^{b_2} \left[LT^{-2}\right]$$

Equating indices

$$L : 0 = b_2 + 1$$

$$T : 0 = -a_2 - 2$$

Which gives $a_2 = -2, b_2 = -1$

Therefore, $\pi_2 = \dfrac{g}{N^2 D}$

For π_4

$$\pi_4 = N^{a_4} D^{b_4} \nu$$

$$\left[L^o T^o\right] = \left[T^{-1}\right]^{a_3} [L]^{b_3} \left[L^2 T^{-1}\right]$$

Equating indices

$$L: 0 = b_3 + 2$$

$$T: 0 = -a_3 - 1$$

Which gives $a_3 = -1, b_3 = -2$

Therefore, $\pi_4 = \dfrac{\nu}{ND^2}$

According to question

$$\pi_2' = \frac{1}{\sqrt{\pi_2}} \times \frac{1}{\sqrt{\pi_3}} = \frac{N\sqrt{D}}{\sqrt{g}} \times \frac{\sqrt{D}}{\sqrt{H}} = \frac{ND}{\sqrt{gH}}$$

$$\pi_4' = \frac{1}{\pi_4} = \frac{ND^2}{\nu}$$

Thus the functional relation can be expressed as

$$\pi_1 = \Phi\left(\pi_2', \pi_4'\right)$$

$$Q = ND^3 \, \Phi\left[\frac{ND}{\sqrt{gH}}, \frac{ND^2}{\nu}\right]$$

12.7. The general formula for the discharge Q over a V - notch is

$$Q = H^2 \sqrt{gH} \, \Phi\left[\frac{H\sqrt{gH}}{\upsilon}, \theta\right]$$

Where H = head over apex, υ = Kinematic viscosity, g = acceleration due to gravity and θ = notch angle.

Solution

Q = function of (H, g, υ, θ)

n = number of variables = 5

m = number of primary dimensions = 2

(here, L and T)

n – m = π groups = 5 – 2 = 3

Selecting g and H as repeating variables, then

$$\pi_1 = g^{a_1} H^{b_1} Q$$

$$\pi_2 = g^{a_2} H^{b_2} \, v$$

$$\pi_3 = g^{a_3} H^{b_3} \, \theta$$

For π_1 term

$$\pi_1 = g^{a_1} H^{b_1} \, Q$$

$$\left[L^o T^o\right] = \left[LT^{-2}\right]^{a_1} \left[L\right]^{b_1} \left[L^3 T^{-1}\right]$$

Equating indices

$$L : 0 = a_1 + b_1 + 3$$

$$T : 0 = -2a_1 - 1$$

Hence $a_1 = 1/2,\ b_1 = -5/2$ *and*

$$\pi_1 = \frac{Q}{g^{1/2} H^{5/2}} = \frac{Q}{H^2 \sqrt{gH}}$$

For π_2 term

$$\pi_2 = g^{a_2} H^{b_2} \, v$$

$$\left[L^o T^o\right] = \left[LT^{-2}\right]^{a_2} \left[L\right]^{b_2} \left[L^2 T^{-1}\right]$$

Equating indices

$$L : 0 = a_2 + b_2 + 2$$

$$T : 0 = -2a_1 - 1$$

Hence $a_2 = -\frac{1}{2},\ b_2 = -\frac{3}{2}$ *and*

$$\pi_2 = \frac{\nu}{g^{1/2} H^{3/2}} = \frac{\nu}{H^2 \sqrt{gH}}$$

$$\pi_2' = \frac{1}{\pi_2} = \frac{H\sqrt{gH}}{\nu}$$

For π_3 term

$$\pi_3 = g^{a_3} H^{b_3} \, \theta$$

Applying the above technique or by inspection

$$\pi_3 = \theta$$

Thus the functional relationship can be expressed as

$$\pi_1 = \Phi\left(\pi_2' \ \pi_3\right)$$

$$\frac{Q}{H^2\sqrt{gH}} = \Phi\left(H^2\frac{\sqrt{gH}}{\nu}, \theta\right)$$

or, $$Q = H^2\sqrt{gH}\ \Phi\left(H\frac{\sqrt{gH}}{\nu}, \theta\right)$$

12.8. **A submarine-launched missile, 1 m diameter by 5 m long, is to be studied in a water tunnel to determine the loads acting on it during its underwater launch. The maximum speed during this initial part of the missile's flight is 10 m/s. Calculate the mean water tunnel flow velocity if a 1/20 scale model is to be employed and dynamic similarity is to be achieved. The kinematic viscosity for model and prototype is some.**

Solution

In this problem, Reynold's model law is used.

$$R_{e_m} = R_e$$

$$\frac{u_m l_m}{\nu_m} = \frac{u_p l_p}{\nu_p}$$

$$u_m = u_p \frac{l_p}{l_m} = 10 \times 20 = 200 \text{ m/s}$$ Ans

12.9. **An airship of 3 m diameter and 20 m length is to be studied in a wind tunnel. The air ship speed range to be investigated is at the docking end of its range, a maximum of 2 m/s. Calculate the mean model wind tunnel speed, if the model is made to 1/10 scale. Assume the same air pressure and temperature for model and prototype**

Solution

In the wind tunnel, object is completely submerged hence Reynold's model law is applicable.

$$R_{e_m} = R_{e_p}$$

$$\frac{\rho_m u_m l_m}{\mu_m} = \frac{\rho_p u_p l_p}{\mu_p}$$

For same air pressure and temperature

$$u_m l_m = u_p l_p$$

or, $$u_m = u_p \frac{l_p}{l_m} = 2 \times 10 = 200 \text{ m/s}$$ Ans

12.10. A 1 : 16 model of a submarine is to be tested in a towing tank containing salt water. If the submarine moves at 5 m/s at what velocity should the model be towed for dynamic similarity? Assume kinematic viscosity and density for model and prototype are same. Also determine the drag force ratio between model and its prototype.

Solution

For the submarine, Reynold's number will be used. Therefore,

$$R_{e_m} = R_{e_p}$$

$$\frac{u_m\, l_m}{v_m} = \frac{u_p\, l_p}{v_p}$$

or, $$\frac{u_m}{u_p} = \frac{l_p}{l_m} = \frac{1}{l_r}$$

or, $$u_m = \frac{u_p}{l_r} = 5 \times 16 = 80 = 80 \text{ m/s}$$ Ans

and $$\frac{F_{D_m}}{\rho_m\, u_m^2\, l_m^2} = \frac{F_{D_p}}{\rho_p\, u_p^2\, l_p^2}$$

or, $$\frac{F_{D_m}}{F_{D_p}} = F_{D_r} = \left(\frac{u_m}{u_p}\right)^2 \left(\frac{l_m}{l_p}\right)^2 \left(\frac{\rho_m}{\rho_p}\right)$$

$$= \left(\frac{l_p}{l_m}\right)^2 \times \left(\frac{l_m}{l_p}\right)^2 = 1$$ Ans

12.11. A ship 300 m long moves in sea water, whose relative density is 1.03. A 1 : 100 model of this ship is to be tested in wind tunnel. The velocity of air in the wind tunnel around the model is 30 m/s and the resistance of the model is 600 N. Determine the velocity of ship in sea water and also the resistance of the ship in sea water. Take pressure and temperature as 1.01325 bar and 15°C and kinematic viscosity of air and sea water as 0.018 Stokes and 0.012 stokes.

Solution

$$p = \rho R T$$

$$1.01325 \times 10^5 = \rho \times 287 \times 288$$

$$\rho = 1.2258 \text{ kg/m}^3$$

In the wind tunnel the ship is completely submerged hence Reynold's model law is applicable.

$$R_{e_m} = R_{e_p}$$

$$\frac{u_m l_m}{\nu_m} = \frac{u_p l_p}{\nu_p}$$

$$u_p = u_m \times \frac{l_m}{l_p} \times \frac{\nu_p}{\nu_m} = 30 \times \frac{1}{100} \times \frac{0.012}{0.018} = 0.2\ m/s$$ Ans

Now, $$\frac{F_m}{\rho_m u_m^2 l_m^2} = \frac{F_p}{\rho_p u_p^2 l_p^2}$$

$$F_p = F_m \times \frac{\rho_p}{\rho_m} \times \left(\frac{u_p}{u_m}\right)^2 \times \left(\frac{l_p}{l_m}\right)^2 = 600 \times \frac{1030}{1.2258} \times \left(\frac{0.2}{30}\right)^2 \times (100)^2$$

$= 224071.3548$ N $= 224.071$ kN Ans.

12.12. The performance of a ship was predicted by taking its model and prototype ship was 400 m and its model was 10 m long. The kinematic viscosity of air is 13 times that of water. The velocity of air around the model in the wind tunnel was measured as 39 m/s. The pressure and temperature of air is 1.01325 bar and 293 K. The experiments showed that the resistance of the model was ship 500 N. Find (a) the velocity of actual ship in water, (b) the resistance of the actual ship in water. Take density of sea water as 1030 kg/m³.

Solution

$$p = \rho RT$$

$$1.01325 \times 10^5 = \rho \times 287 \times 293$$

$$\rho = 1.2049\ kg/m^3$$

In the wind tunnel, the ship is completely submerged, hence the Reynold's model law is applicable.

$$R_{e_m} = R_{e_p}$$

$$\frac{u_m l_m}{\nu_m} = \frac{u_p l_p}{\nu_p}$$

$$u_p = u_m \left(\frac{\nu_p}{\nu_m}\right)\left(\frac{l_m}{l_p}\right) = 39\left(\frac{1}{13}\right)\left(\frac{10}{400}\right) = 0.075\ m/s$$

Now, $$\frac{F_p}{\rho_p u_p^2 l_p^2} = \frac{F_m}{\rho_m u_m^2 l_m^2}$$

or, $$F_p = F_m \times \frac{\rho_p}{\rho_m} \times \left(\frac{u_p}{u_m}\right)^2 \times \left(\frac{l_p}{l_m}\right)^2$$

$$= 500 \times \frac{1030}{1.2049} \times \left(\frac{0.075}{39}\right)^2 \times (40)^2$$

$$= 2529.12 \text{ N} = 2.529 \text{ kN}$$ **Ans**

12.13. **A 1/5th scale model of a torpedo is tested in a wind tunnel using air at 23 bars and 20°C, with a speed of 48 m/s and a viscosity of 184 × 10^{-6} poise. The resistance is found to be 76 N. Calculate the corresponding speed of the full scale torpedo, and the power required to drive it in sea water of density 1025 kg/m³ of which the viscosity 164 × 10^{-4} poise. Take for air R = 287 J/kgK.**

Solution

In the wind tunnel object is completely submerged hence Reynold's model law will be used.

$$p_m = \rho_m R T$$

$$23 \times 10^5 = \rho_m \times 287 \times 293$$

$$\rho_m = 27.35 \; kg/m^3$$

Now, $R_{e_m} = R_{e_p}$

$$\frac{\rho_m u_m l_m}{\mu_m} = \frac{\rho_p u_p l_p}{\mu_p}$$

$$\frac{u_p}{u_m} = \frac{\rho_m}{\rho_p} \times \frac{\mu_p}{\mu_m} \times \frac{l_m}{l_p} = \frac{27.55}{1025} \times \frac{164 \times 10^{-4}}{184 \times 10^{-6}} \times \frac{1}{5} = 0.4756$$

$$u_p = u_m \times 0.4756 = 48 \times 0.4756 = 22.832 \text{ m/s}$$

Now, $$\frac{F_p}{\rho_p u_p^2 l_p^2} = \frac{F_m}{\rho_m u_m^2 l_m^2}$$

$$\frac{F_p}{F_m} = \frac{\rho_p}{\rho_m} \times \left(\frac{u_p}{u_m}\right)^2 \times \left(\frac{l_p}{l_m}\right)^2$$

$$= \frac{1025}{27.35} \times (0.4756)^2 \times (5)^2 = 211.9289$$

$$F_p = F_m \times 211.9289 = 76 \times 211.9289$$

$$= 16106.5983 \text{ N} = 16.1 \text{ kN}$$

Power $P = F_p u_p = 16.1 \times 22.832 = 367.745$ kW **Ans**

12.14. **An airplane wing of 1 m chord is to move at 40 m/s in air. A model of 80 mm chord is to be tested in a wind tunnel with air velocity at 48**

m/s. For air temperature of 20°C in each case, what should be the pressure in the wind tunnel? Take $\nu_{20^\circ C} = 14.86 \times 10^{-6}\ m^2/s$

Solution

Airplane is completely submerged in air therefore, Reynold's number is used. Equating Reynold's number of model and prototype.

$$R_{e_m} = R_{e_p}$$

$$\frac{u_m l_m}{\nu_m} = \frac{u_p l_p}{\nu_p}$$

$$\frac{48 \times 0.08}{\nu_m} = \frac{40 \times 1}{14.86 \times 10^{-6}}$$

$$\nu_m = 1.426 \times 10^{-6}\ m^2/s$$

Therefore, $\nu_r = \frac{\nu_m}{\nu_p} = \frac{14.26 \times 10^{-6}}{14.86 \times 10^{-6}} = 0.096 = 1 : 10.416$

We know that $\upsilon_r = \frac{\mu_r}{\rho_r}$

Absolute viscosity is not affected by pressure change but density increase with pressure

Therefore, $\rho_r = \frac{\mu_r}{\nu_p} = 10.416$

$$p_m = \rho_m R T_m$$

$$p_p = \rho_p R T_p$$

or, $\frac{P_m}{P_p} = \frac{\rho_m}{\rho_p} = \rho_r = \frac{1}{1/10.416} = 10.416$

Assuming P_p is one atmosphere ie the standard condition therefore the pressure in the wind tunnel i.e. P_m is 10.416 atm. Ans

12.15. An aircraft is to fly in air at 0°C and 75 percent of sea level standard pressure at a speed of 100 m/s. A 1 : 15 scale model is to be tested in a wind tunnel at a maximum speed of 100 m/s in order to avoid compressibility effects. The air in the wind tunnel is at 25°C, What should be the pressure in the wind tunnel for dynamically similar flow as for the prototype? Give answer in terms of sea level atmospheres. Take $\mu_m = 1.8 \times 10^{-6}$ *Pas and* $\mu_p = 1.7 \times 10^{-6}$ *Pas.*

Solution

$$\frac{l_m}{l_p} = 1:15,\ p_m = \rho_m\ RT_m$$

$$\rho_m = \frac{p_m}{RT_m}$$

and $$\rho_p = \frac{p_p}{RT_p}$$

Equating R_{e_m} *and* R_{e_p} we get

$$R_{e_m} = R_{e_p}$$

$$\frac{\rho_m\, u_m\, l_m}{\mu_m} = \frac{\rho_p\, u_p\, l_p}{\mu_p}$$

or, $$\frac{p_m u_m l_m}{RT_m\, \mu_m} = \frac{p_p u_p l_p}{RT_p\, \mu_p}$$

or, $$\frac{l_m\, p_m}{\mu_m T_m} = \frac{l_p\, p_p}{\mu_p T_p}$$

or, $$p_m = p_p\left(\frac{l_p}{l_m}\right)\left(\frac{T_m}{T_p}\right)\left(\frac{\mu_m}{\mu_p}\right) = p_p(15)\left(\frac{298}{273}\right)\left(\frac{1.8}{1.7}\right)$$

$$= \mathrm{p_p} \times 17.336 = 0.75 \times 17.336 = 13 \text{ sea level atmosphere.}$$

12.16. **The measured resistance in fresh water of a 2.5 m ship model moving at 2 m/s was 45N. (a) What would be the velocity of the 40 m prototype? (b) What force would be required to drive the prototype at this speed in salt water? Take density of sea water as 1030 kg/m³.**

Solution

In this problem gravity forces are prodominantly used; hence Froude model law is applicable.

(a) $F_{rm} = F_{rp}$

$$\frac{u_m}{\sqrt{l_m\, g}} = \frac{u_p}{\sqrt{l_p\, g}}$$

or, $$\frac{u_m}{u_p} = \sqrt{l_r} = \sqrt{\frac{2.5}{40}} = \sqrt{1/16}$$

or, $$u_p = 2 \times \sqrt{16} = 8\ m/s$$ Ans

(b) $$\frac{F_m}{\rho_m\, u_m^2\, l_m^2} = \frac{F_p}{\rho_p\, u_p^2\, l_p^2}$$

or, $$F_p = F_m \left(\frac{\rho_p}{\rho_m}\right)\left(\frac{u_p}{u_m}\right)^2 \left(\frac{l_p}{l_m}\right)^2 = 45\left(\frac{1030}{1000}\right)(16)\,(16)^2$$

$$= 189849.6 \text{ N} = 189.8496 \text{ kN}$$

$\therefore$ Fp = 189.8496 kN **Ans**

12.17. Oil of kinematic viscosity 4.645×10^{-5} m²/s is to be used in a prototype in which both viscous and gravity forces dominate. A model scale of 1 : 5 is also desired. What viscosity of model liquid is necessary to make both the Froudes number and the Reynolds number same in model and prototype?

Solution

Using the scale ratios of Froude and Reynold's number

According to Froudes law

$$F_{r_m} = F_{r_p}$$

$$\frac{u_m}{\sqrt{l_m\, g}} = \frac{u_p}{\sqrt{l_p\, g}}$$

or, $$\frac{u_m}{u_p} = \sqrt{l_r}$$

According to Reynold's law

$$R_{e_m} = R_{e_p}$$

$$\frac{U_m\, l_m}{\upsilon_m} = \frac{u_p\, l_p}{\upsilon_p}$$

or, $$\frac{u_m}{u_p} = \frac{l_p}{l_m} \times \frac{\upsilon_m}{\upsilon_p}$$

Equating (1) and (2)

$$\sqrt{l_r} = \frac{1}{l_r} \times \upsilon_r$$

or, $$\upsilon_p = l_r^{3/2}$$

or, $$\upsilon_m = \upsilon_p \times l_r^{3/2} = 4.645 \times 10^{-5} \times \left(\frac{1}{5}\right)^{3/2}$$

$0.41546 \times 10^{-5}\ m^2/s$ **Ans**

12.18. A Ship model, scale 1 : 50, has a wave resistance of 30 N at its design speed, calculate the prototype wave resistance.

Solution

In this problem Froude number will be used. Equating Froude number of model and prototype.

$$F_{r_m} = F_{r_p}$$

$$\frac{u_m}{\sqrt{l_m g}} = \frac{u_p}{\sqrt{l_p g}}$$

or, $$\frac{u_m}{u_p} = u_r = \sqrt{\frac{l_m}{l_p}} = \sqrt{l_r} = \sqrt{1/50} \qquad (1)$$

and $$\frac{F_{\omega_p}}{\rho_p\, u_p^2\, l_p^2} = \frac{F_{\omega_m}}{\rho_m\, u_m^2\, l_m^2}$$

or, $$F\omega_p = F_{\omega_m}\left(\frac{\rho_p}{\rho_m}\right)\left(\frac{u_p}{u_m}\right)^2\left(\frac{l_p}{l_m}\right)^2 = \frac{F_{\omega_m}}{\left(\sqrt{l_r}\right)^2 l_r^2} = \frac{F_{\omega_m}}{l_r^3}$$

$$= \frac{30}{(1/50)^3} = 30\,(50)^3 = 3750 \times 10^3 \text{ N} = 3750 \text{ kN} \qquad \text{Ans}$$

12.19. If the scale ratio between a model spillway and its prototype is 1 : 49 what velocity, discharge and power ratio should apply between model and prototype? If the model discharge is 1 m³/s, find the prototype discharge.

Solution

For the Spillway, Froude number is used. Equating Froude number of model and prototype

$$F_{r_m} = F_{r_p}$$

$$\frac{u_m}{\sqrt{l_m g}} = \frac{u_p}{\sqrt{l_p g}}$$

or, $$\frac{u_m}{u_p} = u_r = \sqrt{\frac{l_m}{l_p}} = \sqrt{l_r} \qquad \text{Ans}$$

The discharge Q = area × velocity

therefore, $$Q_r = A_r\, u_r = l_r^2\, \sqrt{l_r} = l_r^{5/2} \qquad \text{Ans}$$

The power $$P = \rho\, Q\, g\, h_f$$

therefore, $P_r = Q_r\, h_r = l_r^{5/2}\, l_r \;= l_r^{7/2}$ Ans

Now, $$\frac{Q_m}{Q_p} = l_r^{5/2} = \left(\frac{1}{49}\right)^{5/2} = 5.9499 \times 10^{-5}$$

therefore, $Q_p = 1 \times 10^5 / 5.9499 = 16807\ m^3/s$ Ans

12.20. Water is flowing in an open channel at a depth of 1.5 m with a velocity of 7.5 m/s. At a particular location a hydraulic jump is formed and the depth increases to 2.2 m. Another channel is built where a similar jump is formed. If the flow depth in the new dynamically similar channel is 6 m, estimate the flow velocity and the height of jump.

Solution

For the open channel and Hydraulic jump, Froude number will be used. Equating Froude number for hydraulic jump for both conditions.

$$\left(\frac{u}{\sqrt{l\,g}}\right)_{at\ 1.5\ m} = \left(\frac{u}{\sqrt{l\,g}}\right)_{at\ 6m}$$

Here, $$l_r = \frac{1.5}{6} = \frac{1}{4}$$

therefore, $$u_r = \sqrt{l_r} = \sqrt{\frac{1}{4}} = \frac{1}{2}$$

or, $$\frac{u\ at\ 1.5\ m\ depth}{u\ at\ 6m\ depth} = \frac{1}{2}$$

$\therefore$ $u\ at\ 6\ m\ depth = 2 \times 7.5 = 15\ m/s$ Ans

and $$\frac{h_m}{h_p} = l_r = \frac{1}{4}$$

$h_p = 2 \times 2.2 = 8.8\ m$ Ans

12.21. A model weir built 1/16th full scale is found to discharge 0.5 m³/min when the head over it is 75 mm. Assuming dynamically similar conditions, find (a) the corresponding head over the full scale weir (b) the corresponding discharge over the latter.

Solution

In the weir problem, Froude model law will be used.

$$F_{r_m} = F_{r_p}$$

$$\frac{u_m}{\sqrt{l_m\,g}} = \frac{u_p}{\sqrt{l_p\,g}}$$

$$\frac{u_m}{u_p} = \sqrt{l_r} = \sqrt{1/16} = \frac{1}{4}$$

Therefore, $$\frac{Q_m}{Q_p} = A_r\, u_r \quad = l_r^2 \sqrt{l_r} = l_r^{5/2}$$

$$\therefore \quad Q_p = \frac{Q_m}{l_r^{5/2}} = 0.5 \times (16)^{5/2} \quad = 0.5 \times 1024 \quad = 512 \text{ m}^3\text{/min}$$ **Ans**

and

$$\frac{H_m}{H_p} = l_r$$

or, $$H_p = \frac{H_m}{l_r} = \frac{75}{1/16} = 1200\ mm$$ **Ans**

12.22. **A 1 : 50 scaled model of a proposed power station tail race is to be used to predict prototype flow. If the design load rejection by pass flow is 1200 m³/s what water flow rate should be used on the model?**

Solution

For the power station tail race level the Froude model law will be used. Therefore,

$$F_{r_m} = F_{r_p}$$

$$\frac{u_m}{\sqrt{l_m\, g}} = \frac{u_p}{\sqrt{l_p\, g}}$$

$$u_r = \sqrt{l_r}$$

Therefore, $$Q_r = A_r\, u_r \quad = l_r^2 \sqrt{l_r} = l_r^{5/2}$$

$$Q_m = Q_p\, (l_r)^{5/2} \quad = 1200 \left(\frac{1}{50}\right)^{5/2}$$

$$= 0.06788\ m^3/s \;= 244.376 \text{ m}^3\text{/hr}$$ **Ans**

12.23. **1 in 20 model of a stiling basin the height of the hydraulic jump in the model is observed to be 200 mm. What is the height of the hydraulic jump in the prototype? If the energy dissipated in the model is 75 Watt, What is the corresponding value in prototype?**

Solution

For the hydraulic jump, Froude model law will be used.

$$\frac{H_m}{H_p} = l_r \;= \frac{1}{20}$$

Therefore, $H_p = H_m \times 200 = 200 \times 20 = 4000\ mm = 4\ meters$ Ans

$$F_{r_m} = F_{r_p}$$

or, $$u_r = \sqrt{l_r}$$

and $$Q_r = l_r^{5/2}$$

Therefore, $$P_r = Q_r\, l_r = l_r^{7/2}$$

$$\frac{P_m}{P_p} = \left(\frac{1}{20}\right)^{7/2}$$

or, $$P_p = P_m \times (20)^{7/2} = 75\,(20)^{3.5}$$

$$= 2683281.573 \text{ Watt} = 2.68 \text{ MW}$$ Ans

EXERCISES

A. THEORY

12.1. What is dimensional analysis? state its uses.

12.2. Distinguish between dimensions and units.

12.3. What are fundamental and derived units?

12.4. Explain the principle of dimensional homogeneity.

12.5. What is dimensional homogeneity? What is meant by homogeneous equation? What is a rational equation?

12.6. Name the two methods of dimensional analysis and explain anyone.

12.7. What are the drawbacks of Rayleigh method of dimensional analysis.

12.8. Distinguish between Rayleigh and Buckingham's dimensional analysis methods.

12.9. What are the advantages of Buckingham's π method over the Rayleigh method.

12.10. Describe Buckingham's π theorem to formulate a dimensional homogenous equation between various physical quantities effecting a certain phenomenon.

12.11. How is the choice of repeating variables made in the case of Buckingham's π method.

12.12. What do you mean by repeating variables? How are the repeating variables selected for dimensional analysis?

12.13. Write a recommended procedure for determining the π parameters.

12.14. Write the useful comments about dimensional analysis.

12.15. To obtain the final expression in the desired manner, what are the useful suggestions.

12.16. What is meant by uniqueness of π terms?

12.17. What are limitations of dimensional analysis?

12.18. What do you mean by superfluous and omitted variables?

12.19. What is model analysis and its uses?

12.20. Describe the advantages of model testing.

12.21. What are the three types of similarities used in model analysis? explain.

12.22. What are standard dimensionless numbers? Explain all the numbers.

12.23. Define and explain Reynold's number, Froude's number and Mach number. What are their significances for fluid flow problems?

12.24. What is meant by geometric, kinematic and dynamic similarities? Are these similarities truly attainable? If not why?

12.25. Reynold's and Froude numbers represent ratio of forces influencing fluid motions. Define these force ratios and, hence, determine the conventional expression for each ratio. Illustrate the distinctive applications of these dimensionless numbers to engineering practice.

12.26. What are model laws? Explain, Reynold's model law and its applications.

12.27. Explain Mach's model law and it's applications.

12.28. Explain Euler's model law and its applications.

12.29. Explain Weber model law and its applications.

12.30. Explain model tests of system dependent on Reynold's and Froude numbers.

12.31. Write the sequential steps on system dependent on Reynold's and Froude, numbers.

12.32. Explain model tests of system dependent on Reynold's and Mach numbers.

12.33. Explain the model analysis of turbomachines.

12.34. What do you mean by undistorted and distorted models?

12.35. Explain river and harbour models.

12.36. Explain estuary models.

12.37. Explain scale effect in model testing. How is it found?

12.38. How is scale effect in a model testing minimised?

12.39. What are scale ratios of a distorted model?

12.40. What are distorted models? What are the reasons of constructing such models for river and harbour engineering projects?

12.41. What are the advantages and disadvantages of distorted models?

12.42. What are the limitations under which a hydraulic model could be tested?

12.43. Prove that the scale ratio for a distorted model is given by

(a) Times ratio, $T_r = \frac{l_x}{\sqrt{l_y}}$

(b) Discharge ratio, $Q_r = l_x\, l_y^{3/2}$

(c) Manning's rugovisty constant ratio, $n_r = l_y^{2/3} / \sqrt{l_x}$

(d) Slope ratio, $S_r = {}^{l_y}/_{l_x}$

12.44. River models are often distorted and have their surfaces roughened artificially. Account for these practices.

12.45. A tests on ski-jump spillway are to be made on a geometrically similar model with a scale ratio of l_r. Determine thė corresponding scales of velocity, discharge pressure, force, power, acceleration and time. In what respects, if any, may the flow pattern of the full size spillway be expected to differ from that of a model.

12.46. How will you determine the total drag of a partially submerged object (ship)?

12.47. Derive scale ratios for distorted model for its vertical cross sectional area, velocity of flow, discharge, time of flow, bed slope and the rugosity coefficient in Manning's formula if l_y is the vertical scale and l_r is the horizontal scale.

12.48. Describe how is the total resistance of a cargo vessel, influenced simultaneously by Reynold's and Froude number, determined.

12.49. Describe how the drag force of a supersonic aircraft influenced simultaneously by Reynold's and Froude number, determined.

12.50. Define dimensionless head and power coefficient.

B. UNSOLVED PROBLEMS

(a) Buckingam's π theorem

12.1. The friction torque T of a disc of diameter D rotating at a speed N, fluid of viscosity μ and density ρ in a turbulent flow is given by using π theorem.

$$T = \rho D^5 N^2 \Phi\left[\frac{\mu}{\rho N D^2}\right]$$

12.2. The efficiency η of a fan depends on the density ρ, the dynamic viscosity μ of the liquid, the angular velocity ω, diameter D of the rotar and the discharge Q. Express η in terms of dimensionless parameters by using π theorem.

$$\eta = \Phi\left[\frac{\mu}{\rho\,\omega\,D^2}, \frac{Q}{D^3\,\omega}\right]$$

12.3. The resisting force F of a supersonic plane during flight can be considered as dependent upon the length of the aircraft l, velocity V, air viscosity μ, air density ρ and bulk modulus of air K. Express the functional relationship between these variables and the resisting force by using π theorem.

$$\frac{F}{\rho V^2 l^2} = \Phi\left[\frac{\mu}{\rho V l}, \frac{K}{\rho V^2}\right]$$

12.4. Using Buckingham's π theorem, show that velocity through a circular orifice is given by

$$V = \sqrt{2gH}\,\Phi\left[\frac{D}{H}, \frac{\mu}{\rho VH}\right]$$

Where H is the head causing flow, D is the diameter of the orifice, μ is the coefficient of viscosity, ρ is the mass density and g is the acceleration due to gravity.

12.5. The discharge Q flowing over a V (vee) notch weir with notch of angle θ is some function of the elevation h, of upstream liquid surface above the bottom of the notch, the velocity of approach V, to the weir and gravity g. Using Buckingham's π theorem and adopting g and h as the repeated variables, determine the form of the discharge equation.

$$Q = g^{1/2}\,h^{5/2}\,\Phi\left[\frac{V}{\sqrt{gh}}, \theta\right]$$

12.6. Using Buckingham's π theorem show that the discharge of a centrifugal pump can be expressed as

$$Q = N\,D^3\,\Phi\left[\frac{gH}{N^2\,D^2}, \frac{\mu}{\rho N D^2}\right]$$

Where N = speed of the pump in rpm, D = diameter of the impeller, g = acceleration due to gravity, H = manometric head; μ = dynamic viscosity; ρ = fluid density.

12.7. Show that the power P developed by a hydraulic turbine is given by using π theorem.

$$P = \rho\,D^5\,N^3\,\Phi\left[\frac{N^2\,D^2}{gH}\right]$$

Where ρ is the mass density of the fluid, N the speed of the rotation, D the diameter of the rotor and H is the variable head.

12.8. Prove the formula using Buckingham's π theorem.

$$\frac{Q}{g^{1/2}\,H^{5/2}} = \Phi\left[\frac{\rho D\sqrt{gH}}{\mu}, \frac{D}{H}, \frac{\sigma}{\rho g H^2}, \frac{K_s}{H}\right]$$

Where Q = Volume rate of flow

D = Diameter H = Head

σ = Surface tension K_s = Surface roughness

g = Acceleration due to gravity

ρ = Mass density and μ = Dynamic viscosity

12.9. Assuming that for turbulent flow in rough pipe the resistance per unit area of solid boundary τ is dependent upon the viscosity μ, the density ρ, the velocity V of the fluid, the diameter D of the pipe and size of roughness K, show by using Buckingham's π theorem.

$$\frac{\tau}{\rho V^2} = \Phi\left[\frac{V D \rho}{\mu}, \frac{K}{D}\right]$$

12.10. Show by using Buckingham's π theorem, that one form of the law govering the flow of liquids in pipes is

$$\frac{S}{\rho V^2} = \Phi\left[\frac{\rho V D}{\mu}\right]$$

Where S is the frictional force per unit wetted area $\left[S = \frac{F}{\pi D l}\right]$ and other variables have their usual significance.

12.11. Show by using Buckingham's π theorem, that the formula for flow over a V notch is given by

$$Q = g^{1/2}\,H^{5/2}\,\Phi\left[\frac{g^{1/2}\,H^{3/2}}{\nu}\right]$$

Where

Q is discharge

H is head

g acceleration due to gravity

ν is the kinematic viscosity.

(b) Model Analysis

12.1. The drag of an airfoil at zero angle of attack is a function of the flow density, viscosity, and velocity in addition to a length parameter. A 1 : 10 scale model of an airfoil was tested in a wind tunnel at a Reynold's number of 5.5×10^6 based on chord length. Test conditions in the wind tunnel air stream were 10 atm absolute pressure and 15°C. The prototype airfoil has a chord length 2m, and it is to be flown in air at standard conditions. Determine the velocity at which the wind tunnel model was tested, and the corresponding prototype velocity. [V_m = 39.2 m/s, = V_p]

12.2. An airship to operate at 20 m/s in air at standard conditions. A model is constructed to 1 : 20 scale and tested in wind tunnel at the same air temperature to determine drag. (a) What criterion should be considered to obtain dynamic similarity? (b) If the model is tested at 20 m/s, what pressure should be used in the wind tunnel ? (c) If the model drag force is 250 N, what will be the drag of the prototype? [2.02 MPa, 5 kN]

12.3. A 1 : 5 scale model of a torpedo is to be tested in a wind tunnel to determine the drag force. The prototype operates in water, has a diameter of 533 mm, and is 6.7 m long. The desired operating speed of the prototype is 28 m/s. To avoid compressibility effects in the wind tunnel the maximum speed is limited to 110 m/s. However, the pressure in the wind tunnel can be varied while holding the temperature constant at 20°C. At what minimum pressure should the wind tunnel be operated to achieve a dynamically similar test? At dynamically similar test conditions the drag force on the model is measured as 618 N. Evaluate the drag force to be expected on the full scale torpedo. [1.93 MPa abs, 43.7 kN]

12.4. The moment exerted on a submarine by its rudder is to be studied with a 1 : 20 scale model in a water tunnel. If the torque measured on the water model is 5 Nm for a tunnel velocity of 15 m/s, What are the coresponding torque and speed for the prototype? [T_p = 100 Nm, V_p = 0.75 m/s,]

12.5. A torpedo shaped object 900 mm diameter is to move in air at 60 m/s and its drag is to be estimated from tests in water on a half scale model. Determine the necessary speed of the model and the drag of the full scale object if that of the model is 1140 N. Take densities : air 1.2 kg/m³ & water 1000 kg/m³. Viscoities of air 1.86×10^{-5} Pas and water 1.01×10^{-3} Pas [7.82 m/s, 322 N]

12.6. An aircraft is to fly at a height of 9 km (where the temperature and pressure are -45°C and 30.2 kPa respectively) at 400 m/s. A 1 : 20 scale model is tested in a pressurized wind tunnel in which the air is at 15°C. For complete dynamic similarity what pressure and velocity should be used in the wind tunnel? For air at T Kelvin, $\mu \times T^{3/2} / (T+117)$ [821 kPa, 450 m/s]

12.7. For model and Prototype, show that, when gravity and inertia are the only influences, the ratio of flows Q is equal to the ratio of the length dimension to the five halves power.

12.8. A 1 : 50 scale model of a proposed dam is to be used in a design study. If the design value of the discharge over the spillway at flood stage is

15000 m³/s, what water flow rate should be established in the model to simulate this flow? Also, if a velocity of 1.2 m/s is measured at a point in the model, what is the velocity at a corresponding point in the model, what is the velocity at a corresponding point in the prototype flow? [0.8485 m³/s, 8.485 m/s].

12.9. If the scale ratio between a model spillway and its prototype is 1 : 25 what velocity and discharge ratio should apply between model and prototype? If the prototype discharge is 3000 m³/s What is the model discharge [1/5, 1/3125, 0.96 m³/s].

12.10. A model of a proposed dam spillway was constructed to a scale of 1 : 15. The design flood discharge over the spillway is 1000 m³/s. What discharge should be provided in the model? What is the velocity in the prototype corresponding with a velocity of 1.5 m/s in the model at the corresponding point? [V_p = 7.5 m/s, Q_m = 0.32 m³/s]

13

Boundary Layer Theory

13.1 INTRODUCTION

In 1904 the German physicist and aerodynamicist Ludwig Prandtl (1875-1953) developed a practical theory based on the previous ideas. Prandtl's boundary layer theory helped resolve important contradictions between experimentally measured fluid forces on bodies and those calculated using Euler's momentum equation for inviscid flow. It provides an important link between ideal and real fluid flow. Previously, except for fully developed flows, like pipe flow, the nonlinear mathematics of viscous fluid motions was so formidable that solution was impossible. Even though some brilliant mathematics was developed and used, many important practical flow problems could not be solved. Furthermore, the viscosity of two of the most common fluids, water and air, is so low that it was difficult to imagine that shear forces could be that important.

Most external flows with which we are familiar are associated with moderately sized objects with a characteristic length of the order of 0.01 m < l < 10 m. In addition typical upstream velocities are on the order of 0.01 m/s < U < 100 m/s and the fluids involved are typically water or air. The resulting Reynold's number range for such flows is approximately $10 < R_e < 10^9$. As a rule of thumb, flows with $R_e > 100$ are dominated by inertial effects, whereas flows with $R_e < 1$, are dominated by viscous effects. Hence, most familiar external flows are dominated by inertia.

The discrepancies between theory and practice were resolved in 1904 by Ludwig Prandtl, with his boundary layer theory. The theory concentrates on a thin fluid layer near solid surfaces where the viscous effects provide the missing link in explaining the earlier shortcomings of both theory and practice.

According to the theory, the flow past solid surface, and an "outer" region away from the solid surface (Fig. 13.1).

Inside the thin layer, the velocity changes from zero at the solid surface (no slip condition) to the free stream velocity u = U. Because of this large velocity gradient, fluid deformation, and associated forces are large and cannot be neglected.

Boundary layer phenomenon may be most easily visualized on a smooth flat plate parallel to the oncoming flow as shown in Fig. 13.2.

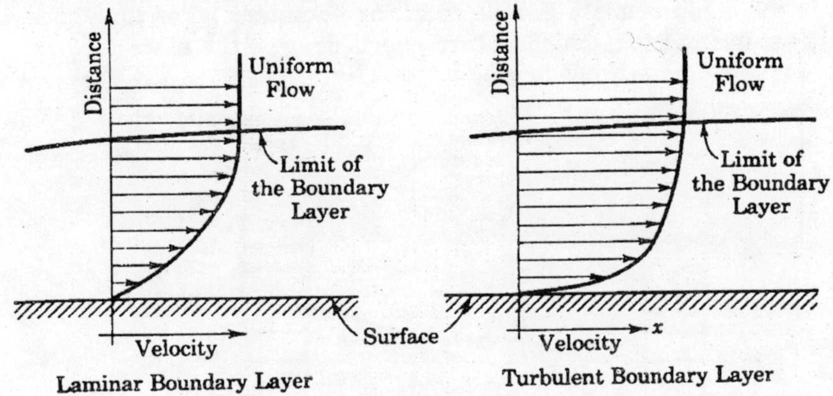

Fig. 13.1. Boundary layer concept

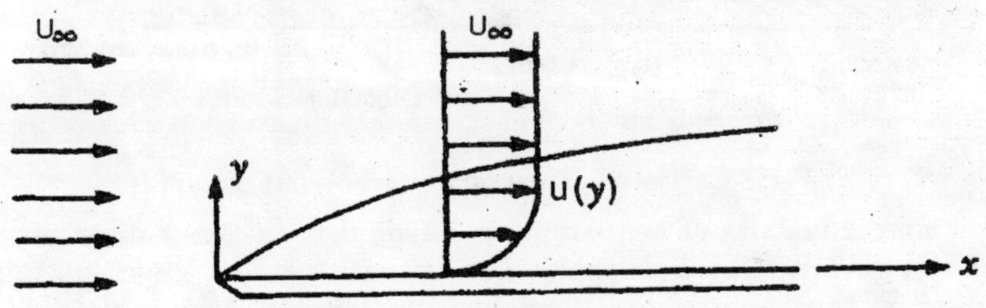

Fig. 13.2. Flow over a flat plate

13.2 DESCRIPTION OF THE BOUNDARY LAYER

A fluid with a free stream velocity U flows along the plate. Immediately before the leading edge of the plate (i.e., x = 0), the fluid velocity is uniform everywhere, but a significant change in the velocity profile occurs as fluid moves along the plate. The viscous forces tend to retard the flow in the regions near the surface. The fluid particles in contact with the surface assume zero velocity, whereas in the region sufficiently far away from the surface, the flow velocity U remains essentially unaffected such a situation implies that the flow velocity U parallel to the surface should increase from a value u = 0 at y = 0 to u = U for y sufficiently large. Then, at each location x, one envisions a point y = δ in the flow field where the axial velocity u is equal to 99 percent of the free-stream velocity U. The locus of the points where u = 0.99 U is called the boundary layer (Fig. 13.3). For some longitudinal distance the flow within the boundary layer is laminar. Downstream from this point the boundary layer flow becomes unstable and eventually becomes turbulent.

The flow field can be assumed to be separated into two distinct regions:

(1) **The boundary layer region,** where the flow velocity u increases rapidly with the distance from the wall surface, and

(2) **The potential flow region** outside the boundary layer, where the velocity gradient are negligible.

It is the steep velocity gradients in the boundary layer that give rise to shear stress in the fluid, which in turn exerts drag on the plate surface.

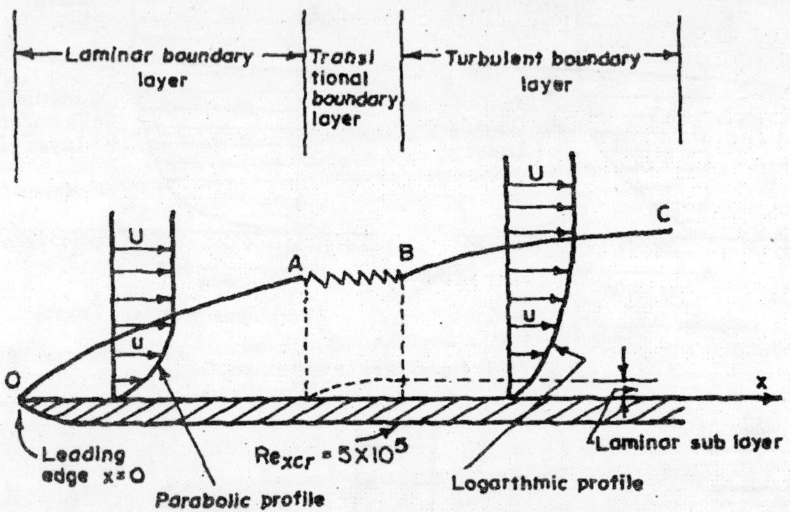

Fig. 13.3. Development of flow over a flat plate

The characteristics of the boundary layer flow in the x direction along the plate is governed by a dimensionless quantity called Reynolds number. Depending on the value of the Reynolds number, the flow in the boundary layer can be laminar, transitional or turbulent. When the flow is laminar, the fluid particles move along streamlines in an orderly manner. The term turbulent is used to indicate that the motion of the fluid is chaotic in nature and involves crosswise mixing or eddying superimposed on the motion of the mainstram. Transitional, as the name implies, indicates the region where the boundary layer flow changes from laminar to turbulent behaviour.

Fig. 13.3 schematically illustrates the laminar, transition and turbulent boundary layers for flow along a flat plate.

In the turbulent boundary layer, next to the wall there is a very thin layer, called the viscous sublayer or laminar sublayer, where the flow retains its viscous flow character.

The boundary layer starts from the leading edge (i.e., x = 0) of the plate as laminar, and the flow stays laminar up to a critical distance $x = x_{cr}$, where the transition from laminar to turbulent flow taken place.

There is no unique value of the Reynold's number at which transition from laminar to turbulent flow occurs in a boundary layer.

On a flat plate with sharp leading edge in a typical air stream, the transition takes place at a distance x from the leading edges given by Re_{cr} =

2×10^5 to 3×10^6. Unless otherwise stated, we will use $Re_{cr} = 5 \times 10^5$ in our calculations.

The actual transition from laminar to turbulent boundary layer flow may occur over a region of the plate, not a specific single location. This occurs, in part, because of the spottiness of the transition. Typically, the transition begins at random locations on the plate in the vicinity of $R_{e_x} = R_{e_{xcr}}$. These spots grow rapidly as they are convected downstream until the entire width of the plate is covered with turbulent flow.

The complex process of transition from laminar to turbulent flow involves the instability of the flow field. Small disturbances imposed on the boundary layer flow will either grow (instability) or decay (stability), depending on where the disturbances is introduced into the flow. If these disturbances occur at a location with $R_{e_x} < R_{e_{xcr}}$ they will die out, and the boundary layer will return to laminar flow at that location. Disturbances imposed at a location with $R_{e_x} > R_{e_{xcr}}$ will grow and transform the boundary layer flow downstream of this location into turbulence.

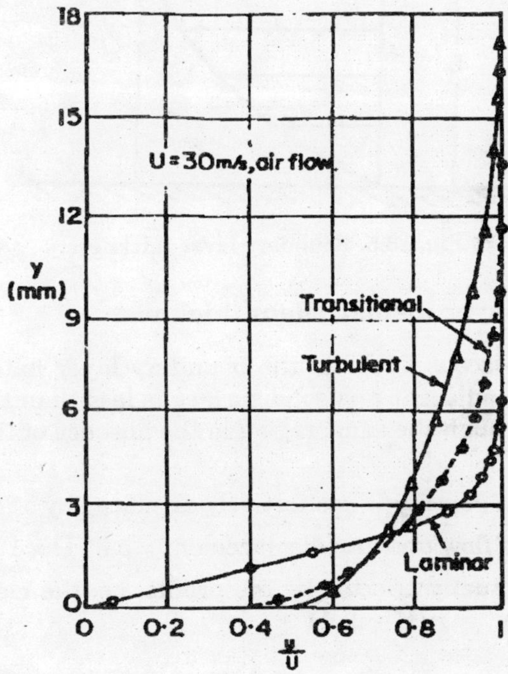

Fig. 13.4. Typical boundary layer profiles on a flat plate for laminar and turbulent flow.

Transition from laminar to turbulent flow also involves a noticeable change in the shape of the boundary layer velocity profile. Typical profiles obtained in the neighborhood of the transition location are indicated in Fig. 13.4. The turbulent profile flatter, have a large velocity gradient at the wall, and produce a large boundary layer than do the laminar profiles.

13.3 BOUNDARY LAYER THICKNESSES

13.3.1 Boundary layer thicknesses δ

The definition of the boundary layer thickness is to a certain extent arbitrary because transition from the velocity in the boundary to that outside it takes place asymptotically. This is, however, of no practical importance, because the velocity in the boundary layer attains a value which is very close to the external velocity already at a small distance from the wall.

The boundary layer thickness δ, is defined as the distance from the surface to the point where the velocity is within 1 percent of the free stream velocity i.e. u = 0.99 U (Fig. 13.5).

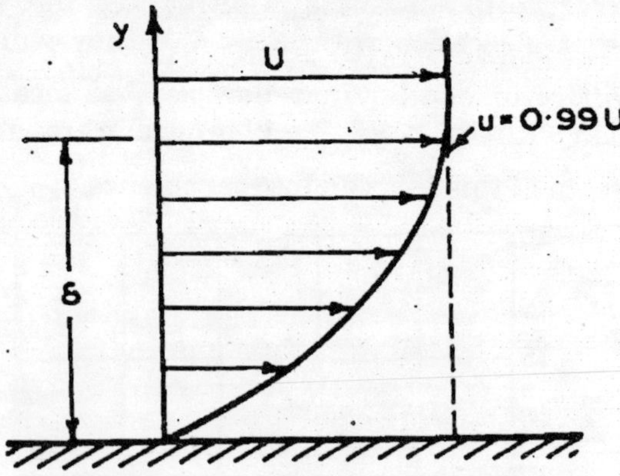

Fig. 13.5. Boundary layer thickness.

13.3.2 Boundary layer displacement thickness

The effect of viscous forces in the boundary layer is to retard the flow. The mass flow rate adjacent to a solid surface is less than the mass flow rate that would pass through the same region in the absence of the boundary layer (Fig. 13.6)

If a free stream velocity U is effectively displaced by δ*

then the loss of flow over the displacement = $\rho\, \delta^*\, U \times 1$ (13.1)

Loss of flow actually affected by the profile, i.e. the elementary hatched area at a distance y, $= \rho\,(U - u)\,dy$

$$\text{Total loss of flow} = \int_0^{\delta} \rho\,(U - u)\,dy) \tag{13.2}$$

Equating (13.1) and (13.2)

$$\rho\, U\, \delta^* = \int_0^{\delta} \rho\,(U - u)\,dy$$

$$\delta^* = \int_0^{\delta} \left(1 - \frac{u}{U}\right) dy \tag{13.3}$$

The displacement thickness is defined as the distance the actual boundary would have to be displaced in order that the actual flow rate would be the same as that of an ideal fluid past the displacement boundary (Fig. 13.6).

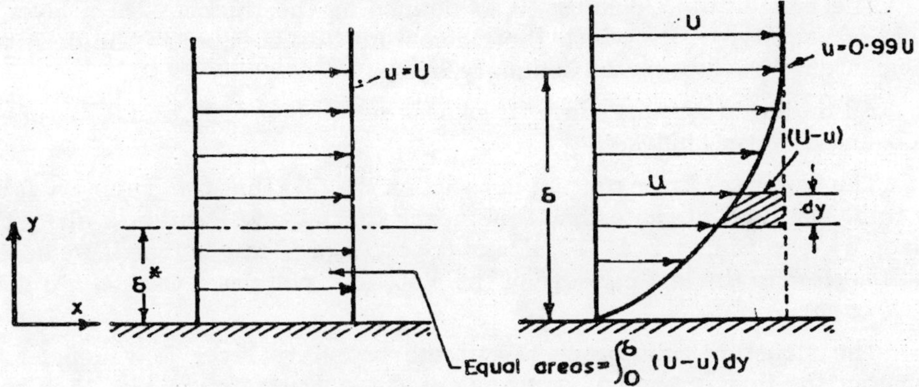

Fig. 13.6. Boundary layer displacement thickness.

13.3.3 Momentum thickness θ

Flow retardation within the boundary layer also results in a reduction in momentum flux at a section compared to inviscid flow (Fig. 13.7).

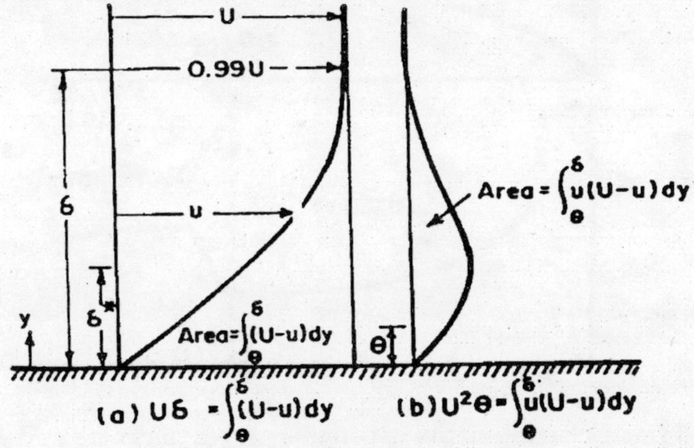

Fig. 13.7. Momentum thickness.

Loss of momentum flux of the free stream

$$= (\rho\, U\, \theta)U = \rho\, U^2\, \theta \tag{13.4}$$

Total loss of momentum flux actually affected by the profile

$$\int_0^\delta \rho\, u\, (U-u)\, dy \tag{13.5}$$

Equating (13.4) and (13.5)

$$\rho\, U^2\, \theta = \int_0^\delta \rho\, u\, (U-u)\, dy$$

or, $$\theta = \int_0^{\delta} \frac{u}{U}\left(1-\frac{u}{U}\right) dy \qquad (13.6)$$

The momentum thickness, θ, is defined as the thickness of a layer of fluid of velocity, U, for which the momentum flux is equal to the deficit of momentum flux through the boundary layer.

All the three boundary layer thickness definitions, δ, δ*, and θ, are of use in boundary layer analysis.

The boundary layer concept is based on the fact that the boundary layer is thin. For the flat plate flow this means that at any location x along the plate, δ << x. Similarly δ* << x and θ << x. Again, this is true if we do not get too close to the leading edge of the plate (i.e., not closer that $R_e = u\,x/\nu$ = 1000 or so)

The structure and properties of the boundary layer flow depend on whether the flow is laminar or turbulent. As is illustrated in Fig. 13.8, both the boundary layer thickness and the wall shear stress are different in these two regimes.

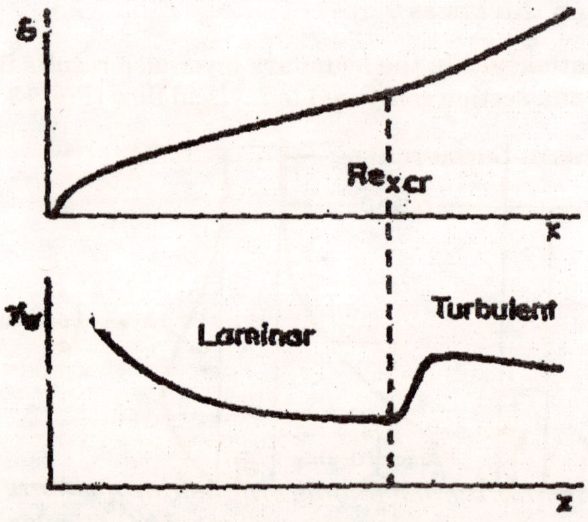

Fig. 13.8. Typical characteristics of boundary layer thickness and wall shear for laminar and turbulent boundary layers.

13.3.4 Energy thickness δ**

It is defined as the distance, measured perpendicular to the boundary of the solid body, by which the boundary should be displaced to compensate for the reduction in kinetic energy of the flowing fluid on account of boundary layer formation.

Kinetic energy loss of the free stream

$$= \frac{1}{2}\rho U^3 \delta^{**} \qquad (13.7)$$

Total loss of K.E. actually affected by the profile

$$= \frac{1}{2} \int_0^{\delta} \rho u \left(U^2 - u^2\right) dy \tag{13.8}$$

Equating (13.7) and (13.8)

$$\frac{1}{2} \rho U^3 \delta^{**} = \frac{1}{2} \int_0^{\delta} \rho u \left(U^2 - u^2\right) dy$$

$$\delta^{**} = \int_0^{\delta} \frac{u}{U} \left(1 - \frac{u^2}{U^2}\right) dy \tag{13.9}$$

13.4 LOCAL SKIN FRICTION AND AVERAGE SKIN FRICTION DRAG COEFFICIENT

13.4.1 Local skin friction drag coefficient C_f

It is defined as the ratio of local wall shear stress to the dynamic pressure of the free stream. In mathematical form,

$$C_f = \frac{\tau_0}{\rho U^2/2} \tag{13.10}$$

13.4.2 Average skin friction drag coefficient C_D

It is defined as the ratio of the total friction drag to the product of dynamic pressure of the free stream and total area being sheared,

$$C_D = \frac{F_D}{\left(\rho U^2/2\right)(\text{Area sheared})} \tag{13.11}$$

C_D is also written as,

$$C_D = \frac{1}{L} \int_0^{L} C_f \, d_x \tag{13.12}$$

The average skin friction drag quite often referred to as the drag coefficient C_D.

These will be determined for a flat plate in a uniform free stream (with no pressure gradients) from both the Prandtl boundary layer equations for a laminar boundary layer and from a momentum analysis for both the laminar and a turbulent boundary layer.

13.5 THE PRANDTL BOUNDARY LAYER EQUATIONS

A solution of the Navier-Stokes equations can be simplified for the boundary layer by using the following assumptions

(a) The flow is steady

(b) The motion is two dimensional

(c) The gravitational effects are neglected

(d) The flow is laminar within the boundary layer

(e) The boundary layer thickness is much smaller than the length of the plate L, i.e. $\delta << L$.

Only those terms in the Navier - Stokes equations of the largest order of magnitude in each equation were retained, those of a lesser order of magnitude were neglected.

The simplified Navier-Stokes equations and the continuity equation for two dimensional flow are known as the Prandtl boundary layer equations. With x and u taken parallel to the boundary and y and v normal to it, they may be written as

$$u\frac{\partial u}{\partial x} + \nu\frac{\partial u}{\partial y} = \nu\frac{\partial^2 u}{\partial y^2} \qquad (13.13)$$

and $$\frac{\partial u}{\partial x} + \frac{\partial v}{\partial y} = 0 \qquad (13.14)$$

With boundary conditions

At $y = 0,\ u = 0\ \nu = 0$

At $y = \infty, u = U = \frac{\partial u}{\partial y} = 0$

13.6 BLASIUS SOLUTION

The boundary layer equations (13.13) and (13.14) represent a set of nonlinear parabolic partial differential equations for which there is no known exact analytic solution. For laminar flow, however, they can be solved using numerical methods.

The closest to an ``exact" solution to equations (13.13) (13.14) is the Blasius solution for steady flow past a flat plate. The solution is based on a special transformation of the variables in the boundary layer equations. The basis of the transformation is the order of magnitude result.

$$\frac{\delta}{x} \sim \sqrt{\frac{\nu}{Ux}} \quad \text{or} \quad \delta \sim \sqrt{\frac{\nu x}{U}}$$

Blausius used this fact to introduce a new nondimensional variable η defined as

$$\eta = \frac{y}{\delta} = y\sqrt{\frac{U}{\nu x}} \qquad (13.15)$$

and a nondimensional stream function $f(\eta)$ given by

$$f(\eta) = \frac{\Psi}{\sqrt{\nu x U}}$$

or $\psi = f(\eta)\sqrt{\nu x U}$ (13.16)

Using the dc. on of ψ

$$u = \frac{\partial \psi}{\partial y} \quad \text{and} \quad v = -\frac{\partial \psi}{\partial x} \tag{13.17}$$

From Equations (13.15), (13.16) and (13.17)

$$u = \frac{\partial \psi}{\partial y} = \frac{\partial \psi}{\partial \eta} \times \frac{\partial \eta}{\partial y}$$

$$u = \frac{\partial f}{\partial \eta}\sqrt{\upsilon x U} \times \sqrt{\frac{U}{\nu x}}$$

$$= U\frac{\partial f}{\partial \eta} \tag{13.18}$$

and $\nu = -\dfrac{\partial \psi}{\partial x}$

$$= \left[\sqrt{\nu x U}\,\frac{\partial f}{\partial x} + \frac{f}{2}\sqrt{\frac{\nu U}{x}}\right]$$

$$= \left[\sqrt{\nu\, x U}\,\frac{\partial f}{\partial x}\frac{\partial \eta}{\partial x} + \frac{f}{2}\sqrt{\frac{\nu\, U}{x}}\right]$$

$$= \frac{1}{2}\sqrt{\frac{\nu U}{x}}\left(\eta\,\frac{\partial f}{\partial \eta} - f\right) \tag{13.19}$$

By differentiating the velocity components, it also can be shown that

$$\frac{\partial u}{\partial x} = -\frac{U}{2x}\eta\eta\frac{d^2 f}{d\eta^2}$$

$$\frac{\partial u}{\partial y} = U\sqrt{U/\nu x}\,\frac{d^2 f}{d\eta^2}$$

$$\frac{\partial^2 u}{\partial y^2} = \frac{U^2}{\nu x}\,\frac{d^3 f}{d\eta^3}$$

Substituting these expressions into equation (13.13), we obtain,

$$2\frac{d^3 f}{d\eta^3} + f\frac{d^2 f}{d\eta^2} = 0$$

$$2f''' + f f'' = 0 \tag{13.20}$$

With boundary conditions :

At $\eta = 0, f = \dfrac{df}{d\eta} = 0$

At $\quad \eta \to \infty, \dfrac{df}{d\eta} = 1$

It is not possible to solve equation (13.20) in closed form. Blasius solved it using a power series expansion about $\eta = 0$ matched to an asymptotic expansion for $\eta \to \infty$. The same equa-tion later was solved more precisely using numerical methods by Howarth. The numerical value of f, df/dη and $\dfrac{d^2f}{d\eta^2}$ in Table 13.1 were calculated by using fourth order Runge Kutta numerical integration.

Table 13.1 The function f(η) for the boundary layer along a flat plat at zero incidence, after L. Howarth [16]

$\eta = y\sqrt{\frac{U}{v}x}$	f	$f' = \frac{u}{U}$	f''
0	0	0	0.33206
1.0	0.16557 ·	0.32979	0.32301
2.0	0.65003	0.62977	0.26675
3.0	1.39682	0.84605	0.16136
4.0	2.30576	0.95552	0.06424
5.0	3.28329	0.99155	0.01591
6.0	4.27964	0.99898	0.00240
7.0	5.27926	0.99992	0.00022
8.0	6.27923	1.00000	0.00001

The velocity profile is obtained in dimensionless form by plotting u/U versus η, using values from Table 13.1. The resulting profile is plotted in Fig. 13.9.

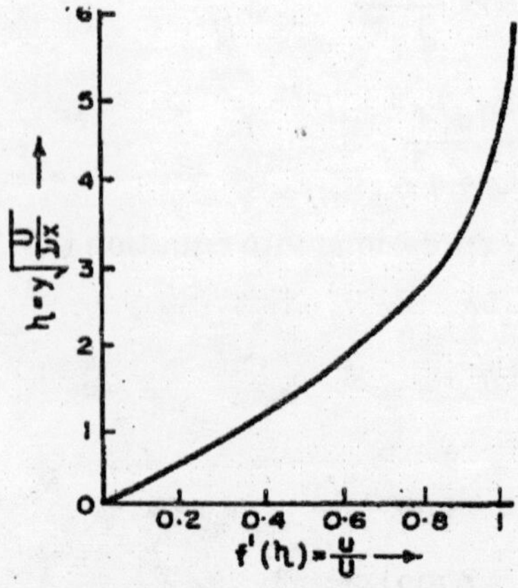

Fig. 13.9. Blasius dimensionless velocity profile for laminar flow past a flat plate.

From Table 13.1, we see that at η = 5, $\frac{u}{U} = 0.992$. With the boundary layer thickness δ, defined as the value of y for which $\frac{u}{U} = 0.99$, equation (13.15) gives

$$\delta \approx \frac{5}{\sqrt{\frac{U}{\nu x}}} = \frac{5\,x}{\sqrt{R_{e_x}}} \tag{13.21}$$

The displacement thickness is given by

$$\delta^* = \int_0^{\delta} \left(1 - \frac{u}{U}\right) dy$$

From Equation (13.15)

$$\eta = y\sqrt{\frac{U}{\nu x}}$$

$$\therefore y = \eta\sqrt{\frac{\nu x}{U}}$$

$$dy = d\eta\sqrt{\frac{\nu x}{U}}$$

Thus,

$$\delta^* = \int_0^{\eta_{max}} \left(1 - \frac{u}{U}\right)\sqrt{\frac{\nu x}{U}}\, d\eta$$

$$= \sqrt{\frac{\nu x}{U}} \int_0^{\eta_{max}} \left(1 - \frac{u}{U}\right) d\eta$$

From equation (13.21)

$$\delta^* = \frac{\delta}{5} \int_0^{\eta_{max}} \left(1 - \frac{u}{v}\right) d\eta$$

From equation (13.18)

$$\delta^* = \frac{\delta}{5} \int_0^{\eta_{max}} \left(1 - \frac{df}{d\eta}\right) d\eta$$

$$= \frac{\delta}{5}\left[\eta - f(\eta)\right]_0^{\eta_{max}}$$

Evaluating at $\eta_{max} = 8$, we obtain

$$\delta^* = \frac{\delta}{5}(8 - 6.2792) = 0.34416\,\delta$$

$$\delta^* = \frac{1.7208\, x}{\sqrt{R_{e_x}}} \tag{13.22}$$

[Closure to 1.73 as $\eta_{max} \to \infty$]

Similarly, the momentum thickness

$$\theta = \int_0^{\delta} \frac{u}{\nu}\left(1 - \frac{u}{U}\right) dy = \frac{0.664\, x}{\sqrt{R_{ex}}} \tag{13.23}$$

The numerically exact solution can now be used to determine other characteristic flow quantities. By definition, the local wall shear stress τ_0 is given as

$$\tau_0 = \mu \left(\frac{\partial u}{\partial y}\right)_{y=0} = \mu \sqrt{\frac{u}{\nu x}}\, f'' \eta = 0$$

Using the numerical solution

$$f'' = (0) = 0.332$$

From Table 14.1

$$\therefore\ \tau_0 = 0.332 \frac{\rho U^2}{\sqrt{R_{e_x}}}$$

$$\frac{\tau_0}{\rho U^2} = \frac{0.332}{\sqrt{R_{e_x}}} \tag{13.24}$$

From equation (13.10)

$$C_f = \frac{\tau_0}{\rho U^2/2} = \frac{0.664}{\sqrt{R_{e_x}}} \tag{13.25}$$

The total resistance force due to viscosity, or skin friction, along the plate surface A is obtained by integration

$$F_D = \int_0^L \tau_0\, B\, dx$$

$$= \int_0^L \frac{0.332\, \rho U^2}{\sqrt{R_{e_x}}} B\, dx$$

$$= 0.332\, \rho U^2 \sqrt{\frac{\nu}{U}}\, W \int_0^L \frac{dx}{\sqrt{x}}$$

$$= 0.664\, \rho U^2 \sqrt{\frac{\nu L}{U}}$$

Where L and B are the length and width of the flat plate, respectively.

$$\therefore\ F_D = 0.664\,\frac{\rho U^2 BL}{\sqrt{R_{e_x}}}$$

or,
$$\frac{F_D}{\rho U^2 (BL)} = \frac{0.664}{\sqrt{R_{e_L}}}$$

Where, $R_{e_L} = \dfrac{UL}{\nu}$

From equation (13.11)

$$C_D = \frac{F_D}{\frac{\rho U^2}{2}} BL = \frac{1.328}{\sqrt{R_{e_L}}} \tag{13.26}$$

13.7 MOMENTUM INTEGRAL BOUNDARY LAYER EQUATION FOR FLAT PLATE OR VON KARMAN INTEGRAL EQUATION

The Blasius solution is obviously quite restrictive in application, applying only to a case of a laminar boundary layer over a flate surface.

Consider a short segment dx of an element boundary layer, ABCDA as shown in Fig. 13.10, having a thickness δ at a distance x from the leading edge of a flat plate of width B.

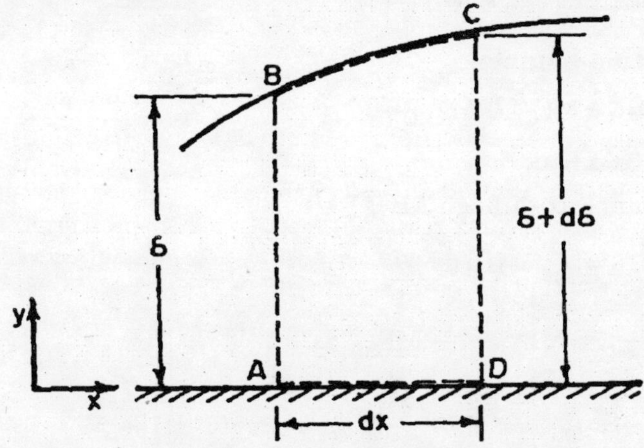

Fig. 13.10. Differential control volume in a boundary layer

It is assumed that the velocity of the boundary layer flow at the outer edge of the boundary layer is the same as that of the free fluid flow, that is, u = U, at y = δ, for the purpose of integration.

It is also assumed that the free stream velocity U outside the boundary layer remains constant in the x direction if the pressure gradient $\partial p / \partial x$ in the x direction is assumed to be zero.

We wish to determine the boundary layer thickness, δ, as a function of x. There will be mass flow across surfaces AB and CD of differential control volume, ABCD. The edge of the boundary layer is not a stream line. Thus there will be mass flow across surface BC. Since control surface AD is adjacent to a solid boundary, there will not be flow across AD. Before considering the forces on the control volume and the momentum fluxes through the control surface, let us apply the continuity equation to determine the mass flux through each portion of the control surface.

Mass balance

$$\dot{M}_{AB} = \dot{M}_{CD} + \dot{M}_{BC}$$

or, $$\dot{M}_{BC} = \dot{M}_{AB} + \dot{M}_{CD} \qquad (13.27)$$

Mass flux

$$\dot{M}_{AB} = \int_0^{\delta} \rho u \, dy$$

and $$\dot{M}_{CD} = \dot{M}_{AB} + \frac{\partial}{\partial x} \dot{M}_{AB} \, dx$$

$$= \int_0^{\delta} \rho u \, dy + \frac{\partial}{\partial x}\left[\int_0^{\delta} \rho u \, dy\right] dx$$

Since there is no flow across the surface of the plate AD, the rate of mass flow out of the element across BC according to continuity equation (13.27).

$$\dot{M}_{BC} = \frac{\partial}{\partial x}\left[\int_0^{\delta} \rho u \, dy\right] dx \qquad (13.28)$$

Momentum balance

$$\dot{MF}_{AB} = \dot{MF}_{CD} + \dot{MF}_{BC}$$

Momentum flux

Momentum flux through AB is

$$\dot{MF}_{AB} = \int_0^{\delta} u \rho u \, dy$$

$$\dot{MF}_{CD} = \dot{MF}_{AB} + \frac{\partial}{\partial x} \dot{MF}_{AB} \, dx$$

$$= \int_0^{\delta} \rho u^2 \, dy + \frac{\partial}{\partial x}\left[\int_0^{\delta} \rho u^2 \, dy\right] dx$$

Since the mass crossing BC has velocity component U in the x direction, the x momentum flux across BC is given by

$$\dot{MF}_{BC} = U \, M_{BC}$$

$$= -U \frac{\partial}{\partial x}\left[\int_0^{\delta} \rho u \, dy\right] dx \qquad (13.29)$$

From the above we can evaluate the net x momentum flux through the control surface as

$$\frac{\partial}{\partial x}\left[\int_0^\delta \rho u^2\, dy\right] dx - U\frac{\partial}{\partial x}\left[\int_0^\delta \rho u\, dy\right] dx \tag{13.30}$$

Let us consider the surface forces acting on the control volume in the *x* direction (Fig. 13.11). A shear force acts on surface AD and shear force acting along surface BC is negligible because velocity gradient goes to zero at the edge of the boundary layer. Normal forces having non zero components in the *x* direction act on three surfaces of the control surface.

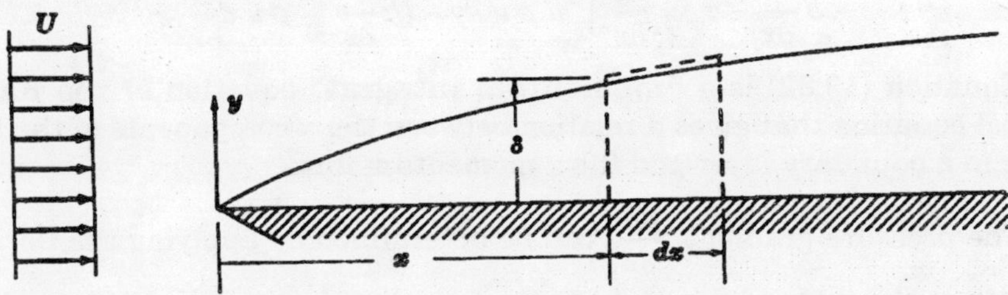

Fig. 13.11. Control volume for integral analysis of the boundary layer.

Force

$$F_{AB} = p\,\delta \times \text{Unit width} = p\,\delta$$

$$F_{CD} = \left(p + \frac{\partial p}{\partial x}\right)\delta + d\,\delta$$

The average pressure acting over surface BC is

$$p + \frac{1}{2}\left(\frac{\partial p}{\partial x}\right) dx$$

Then the normal force acting over BC is given by

$$F_{BC} = \left(P + \frac{1}{2}\left(\frac{\partial p}{\partial x}\right) dx\right) d\,\delta$$

The shear force acting on AD is given by

$$F_{AD} = -\,\tau_0\, dx$$

Summing the x components of all forces acting on the control volume, we obtain

$$F_{AB} + F_{CD} + F_{BC} + F_{AD}$$

$$= p\,\delta - p\,\delta - pd\,\delta - \frac{\partial p}{\partial x} dx\,\delta$$

$$- \frac{\partial p}{\partial x} dx\, d\,\delta + pd\delta + \frac{1}{2}\frac{\partial p}{dx} dx\, d\delta - \tau_o\, dx$$

Neglecting small terms

$$\Sigma F = -\delta\frac{\partial p}{\partial x} dx - \tau_o\, dx \tag{13.31}$$

According to momentum theorem.

$$\sum F = \text{Rate of change of momentum}$$

$$-\delta \frac{\partial p}{\partial x} dx - \tau_o \, dx = \frac{\partial}{\partial x}\left[\int_0^\delta \rho u^2 \, dy\right] dx - U \frac{\partial}{\partial x}\left[\int_0^\delta \rho u \, dy\right] dx$$

[From equations (13.30) and (13.31)]

$$-\delta \frac{\partial p}{\partial x} - \tau_o = \frac{\partial}{\partial x} \int_0^\delta u \, \rho \, u \, dy - U \frac{\partial}{\partial x} \int_0^\delta \rho \, u \, dy \qquad (13.32)$$

Equation (13.32) is a "**momentum integral**" equation or von Karman integral equation that gives a relation between the x components of the forces acting in a boundary layer and the x momentum flux.

The pressure gradient, $\frac{dp}{dx}$, can be determined by applying the Bernoulli equation to the inviscid flow outside the boundary layer

$$\frac{dp}{dx} = -\rho U \frac{dU}{dx}$$

Boundary layer thickness δ can be written as

$$\delta = \int_0^\delta dy$$

Substituting in equation (13.32)

$$+\rho U \frac{dU}{dx} \int_0^\delta dy - \tau_o = \frac{\partial}{\partial x} \int_0^\delta u \, \rho \, u \, dy - U \frac{\partial}{\partial x} \int_0^\delta \rho \, u \, dy$$

$$\tau_o = -\frac{\partial}{\partial x} \int_0^\delta u \, \rho \, u \, dy + U \frac{\partial}{\partial x} \int_0^\delta \rho \, u \, dy + \rho U \frac{dU}{dx} \int_0^\delta dy$$

Since (using integration of parts)

$$U \frac{\partial}{\partial x} \int_0^\delta \rho \, u \, dy = \frac{\partial}{\partial x} \int_0^\delta \rho \, u \, dy$$

We have

$$\tau_o = -\frac{\partial}{\partial x} \int_0^\delta u \, \rho \, u \, dy + \frac{\partial}{\partial x} \int_0^\delta \rho \, u \, U \, dy - \frac{dU}{dx} \int_0^\delta \rho \, u \, dy + \rho U \frac{dU}{dx} \int_0^\delta dy$$

$$= \frac{\partial}{\partial x} \int_0^\delta u \, \rho (U - u) \, dy + \frac{dU}{dx} \int_0^\delta \rho (U - u) \, dy$$

or, $$\tau_o = \frac{\partial}{\partial x} U^2 \int_0^\delta \rho \frac{u}{U}\left(1 - \frac{u}{U}\right) dy + U \frac{dU}{dx} \int_0^\delta \rho \left(1 - \frac{u}{U}\right) dy$$

Using the definitions of displacement and momentum thickness.

$$\frac{\tau_o}{\rho} = \frac{d}{dx}\left(U^2 \theta\right) + \delta^* U \frac{dU}{dx}$$

$$\frac{\tau_o}{\rho} = U^2 \frac{d\theta}{dx} + \theta\,(2U)\frac{dU}{dx} + \delta^*\, U\,\frac{dU}{dx}$$

$$\frac{\tau_0}{\rho U^2} = \frac{d\,\theta}{dx} + (2\,\theta + \delta^*)\frac{1}{U}\frac{dU}{dx} \tag{13.33}$$

Equation (13.33) **is the momentum integral equation or von Karman integral equation**; this equation will yield an ordinary differential equation for boundary layer thickness, provided that a suitable form is assumed for the velocity profile and the wall shear stress can be related to other variables. Once, the boundary layer thickness is determined, momentum thickness displacement thickness, and wall shear stress can be calculated.

13.7.1 Momentum integral equation for zero pressure gradient

The momentum integral equation for zero pressure gradient flow can be obtained by putting U = constant and from Bernoulli's equation p = constant and thus $\frac{dp}{dx} = 0$

The momentum integral equation (13.33) then reduces to

$$\frac{\tau_o}{\rho U^2} = \frac{d\theta}{dx} \tag{13.34}$$

The velocity distribution, $\frac{u}{U}$ in the boundary layer normally is specified as a function of y/δ. (Note that u/U is dimensionless and δ is a function of x only). Consequently, it is convenient to change the variable of integration from y to y/δ. Defining η = y/δ then dy = δ dη and the momentum integral equation (13.34) for zero pressure gradient is written

$$\tau_o = \rho U^2 \frac{d\theta}{dx}$$

$$= \rho U^2 \frac{d\,\delta}{dx} \int_0^1 \frac{u}{U}\left(1 - \frac{u}{U}\right) d\,\eta \tag{13.35}$$

To solve the above equation (13.35) for the boundary layer thickness as a function of x, To do this, we must :

(1) Assume a velocity distribution in the boundary layer–a functional relationship of the form

$$\frac{u}{U} = f\left(\frac{y}{\delta}\right)$$

(a) The assumed velocity distribution should satisfy certain physical boundary conditions

At y = 0, u = 0

At y = δ, u = U

At $y = \delta, \dfrac{\partial u}{\partial y} = 0$

(b) Note that once the velocity distribution has been assumed, then the numerical value of the integral in equation (13.35) is simplify

$$\int_0^1 \frac{u}{U}\left(1 - \frac{u}{U}\right) d\eta = \frac{\theta}{\delta} = \text{constant} = \beta$$

and the momentum integral equation becomes

$$\tau_o = \rho U^2 \frac{d\delta}{dx} \beta$$

13.8 MOMENTUM INTEGRAL METHOD FOR LAMINAR FLOW OVER A FLAT PLATE

For laminar flow over a flat plate, a reasonable assumption for the velocity profile is a polynomial in y.

$$\frac{u}{U} = a + b\left(\frac{y}{\delta}\right) + c\left(\frac{y}{\delta}\right)^2 + d\left(\frac{y}{\delta}\right)^3 + ---- \tag{13.36}$$

The physical boundary conditions are

At $y = 0, \; u = 0$ (Condition 1)

At $y = \delta, \; u = U$ (Condition 2)

At $y = \delta, \; \dfrac{\partial u}{\partial y} = 0$ (Condition 3)

At $y = 0, \; \dfrac{\partial^2 u}{dy^2} = 0$ (Condition 4)

Using boundary conditions 1 and 2, then all constants from c onwards are made zero. Moreover, a becomes zero from condition 1, and b = 1 from condition 2. Thus the velocity distribution is

$$\frac{u}{U} = \frac{y}{\delta}$$

which is linear (Fig. 13.12). Suppose next that we include a third boundary condition. This ensures that the velocity profile has a vertical slope at the edge of the boundary layer and that the velocity is no longer changing with y. Combining boundary conditions 1, 2 and 3 with equation (13.36), we get

$$\frac{u}{U} = \frac{2y}{\delta} - \frac{y^2}{\delta^2}$$

Combining of all four boundary conditions with equation (13.36) yields

$$\frac{u}{U} = \frac{3y}{2\,\delta} - \frac{y^3}{2\delta^3}$$

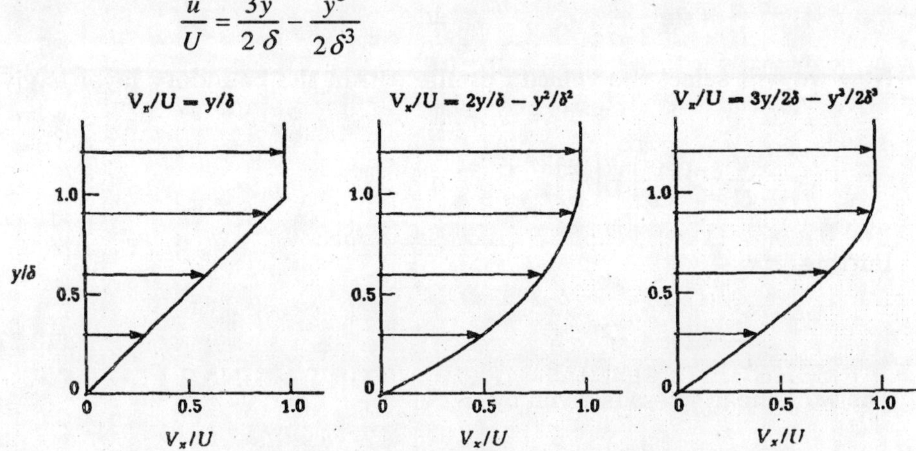

Fig. 13.12. Three possible velocity profiles

The effect of including more boundary conditions is illustrated graphically in Fig. 13.12, where all three distributions are plotted. Fig. 13.13 shows the Blausius solution with polynomial profiles plotted for comparison. It is to be noted that we can select other forms for an assumed profile. One frequently selected profile is

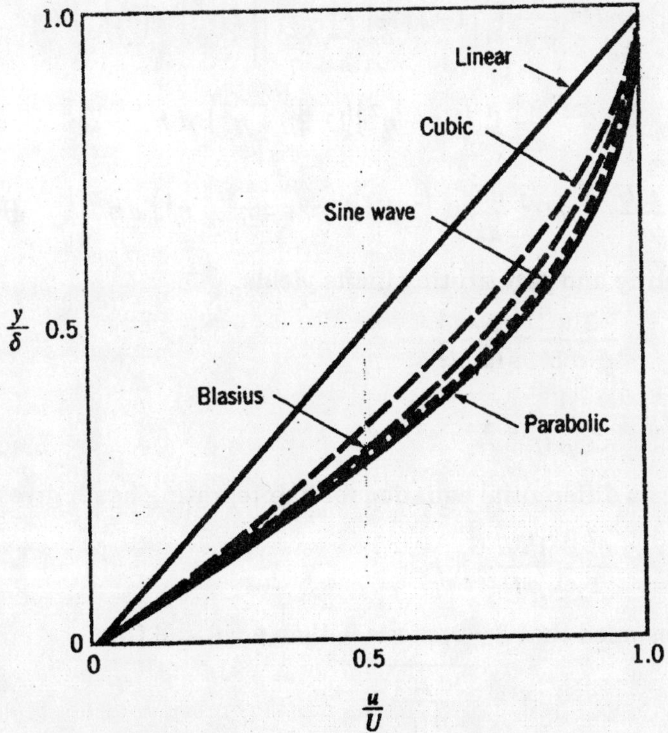

Fig. 13.13. Laminar boundary layer velocity profiles

$$\frac{u}{U} = \sin\frac{\pi}{2}\frac{y}{\delta}$$

Now we will select the velocity profile within the boundary layer is give as

$$\frac{u}{U} = 2\left(\frac{y}{\delta}\right) - \left(\frac{y}{\delta}\right)^2$$

Putting $\eta = \dfrac{y}{\delta}$

$$\frac{u}{U} = 2\eta - \eta^2 \qquad (13.37)$$

The wall shear stress is given by

$$\tau_o = \mu\left(\frac{\partial u}{\partial y}\right)_{y=0}$$

$$= \mu\frac{\partial}{\partial y}\left[2\,U\left(\frac{y}{\delta}\right) - U\,(y/\delta)^2\right]_{y=0} = \frac{2\,\mu\,U}{\delta} \qquad (13.38)$$

We are now in a position to apply the momentum integral equations

$$\tau_o = \rho\,U^2\frac{\partial}{\partial x}\int_0^\delta\left[1 - 2\left(\frac{y}{\delta}\right) + \left(\frac{y}{\delta}\right)^2\right]\left[2\left(\frac{y}{\delta}\right) - \left(\frac{y}{\delta}\right)^2\right]dy$$

or, $$\tau_o = \rho\,U^2\frac{d\delta}{dx}\int_0^1\left(2\eta - \eta^2\right)\left(1 - 2\eta + \eta^2\right)d\eta$$

$$\frac{2\,\mu\,U}{\delta} = \rho\,U^2\frac{d\delta}{dx}\int_0^1\left(2\eta - 5\eta^2 + 4\eta^3 - \eta^4\right)d\eta \qquad \text{[From (13.38)]}$$

Integrating and substituting limits yields

$$\frac{2\,\mu}{\delta\,\rho\,U} = \frac{2}{15}\frac{d\delta}{dx}$$

or, $$\delta\,d\,\delta = \frac{15\,\mu}{\rho\,U}\,d\,x$$

Which is a differential equation for δ, Integrating again gives

$$\frac{\delta^2}{2} = \frac{15\,\mu}{\rho\,U}\,x + e$$

If we assumed that δ = 0, at x = 0, then e = 0 and thus

$$\delta = \sqrt{\frac{30\,\mu\,x}{\rho\,U}}$$

$$\text{or,} \quad \frac{\delta}{x} = \sqrt{\frac{30\,\mu}{\rho\,U\,x}}$$

$$\text{or} \quad \frac{\delta}{x} = \frac{5.48}{\sqrt{R_{e_x}}} \tag{13.39}$$

Once we know the boundary layer thickness, all details of the flow may be determined.

The local skin friction coefficient or wall shear stress is calculated as

$$C_f = \frac{\tau_o}{\frac{1}{2}\rho U^2} = \frac{2\mu(U/\delta)}{\frac{1}{2}\rho\,U^2} = \frac{4\,\mu}{\rho\,U\,\delta}$$

$$= \frac{4\,\mu\,x}{\rho\,U\,x\,\delta} = \frac{4\sqrt{R_e}}{R_{e_x}(5.48)} = \frac{0.73}{\sqrt{R_{e_x}}} \tag{13.40}$$

The average skin friction coefficient over any length of plate l and unit width is given by

$$C_D = \frac{\int_0^l \tau_o\,d\,x}{(\rho U^2/2)l} = \int_0^l \frac{c_f\left(\frac{1}{2}\rho U^2\right)d\,x}{(\rho U^2/2)l} = \frac{1}{l}\int_0^l c_f\,d\,x$$

$$= \frac{0.73}{l\sqrt{\frac{\rho U}{\mu}}}\int_0^l \frac{dx}{\sqrt{x}} = \frac{0.73}{l\sqrt{\frac{\rho U}{\mu}}}\left(2\sqrt{l}\right) = \frac{1.46}{\sqrt{R_{e_l}}} \tag{13.41}$$

Where $R_e = \dfrac{\rho\,u\,l}{\mu}$ is based on total length l of the plate.

Results for linear, parabolic, cubic, sine function velocity profile and the results obtained by solving the Prandtl boundary-layer equations (called Blasius results) are tabulated in Table 13.2 and also given in Fig. 13.13. The more closely the assumed shape approximates the actual (i.e. Blasius) profile, the more accurate the final results.

The Table 13.2 shows the following

(1) The laminar boundary layer thickness, however defined, increases as the square root of the distance x from the leading edge and inversely as the square root of the free stream velocity U.

(2) The local and average skin drag coefficient vary inversely with the square root of both x and u.

(3) The total drag ($F = C_D\,\rho\,U^2\,x/2$ per unit width) varies as the 3/2 power of the free stream velocity and the square root of the length x.

Table 13.2. Results for various assumed Laminar flow velocity profiles

Velocity profile	$\frac{\delta\sqrt{R_{ex}}}{x}$	$\frac{\delta^*\sqrt{R_{ex}}}{x}$	$C_f\sqrt{R_{ex}}$	$C_D\sqrt{R_{ex}}$
(a) Blassius solution	5	1.73	0.664	1.328
(b) Linear $\frac{u}{U} = y/\delta$	3.46	1.73	0.578	1.156
(c) Parabolic $\frac{u}{U} = 2(y/\delta) - (y/\delta)^2$	5.48	1.83	0.730	1.46
(d) Cubic $\frac{u}{U} = \frac{3}{2}\left(\frac{y}{\delta}\right) - \frac{1}{2}\left(\frac{y}{\delta}\right)^3$	4.64	1.74	0.646	1.292
(e) Sine wave $\frac{u}{U} = \sin\frac{\pi}{2}(y/\delta)$	4.79	1.74	0.655	1.31

Experiments indicated that the laminar boundary layer flow is found to remain laminar up to a value of R_{e_x} of between 3×10^5 and 5×10^5. With this range the laminar flow in the boundary layer becomes somewhat unstable, although the exact lower limit of the critical R_{e_x} depends largely on;

(1) the initial turbulence of the fluid stream,

(2) the plate surface roughness and

(3) the geometric configuration at the leading edge of the plate. For large values of R_{e_x} the boundary layer flow becomes turbulet.

13.9 TURBULENT BOUNDARY LAYER

The conditions within the turbulent velocity profile for boundary layers at zero pressure gradient are very similar to those for turbulent flow in pipes and channels. **The momentum integral equation is independent of whether the flow is laminar or turbulent.**

According to Prandtl we can assume that the velocity distribution within a turbulent boundary layer of a flat plate is essentially the same as that in a large diameter pipe. The free stream velocity U for the external boundary layer is comparable with the maximum or centre line velocity u_{max} in the pipe, and the boundary layer thickness is equivalent to the pipe radius.

The universal velocity distribution laws provide the best basis for theoretical studies of the turbulent pipe flow. When von Karman momentum method is used in conjunction with the universal velocity distribution equations to analyze the turbulent boundary layer flow, the calculations are rather complicated and lengthy. However, an approximation, Prandtl/ Blasius one seventh power law for the velocity distribution of turbulent flow in smooth pipes is used.

Prandtl/Blasius suggested that the velocity varies as the seventh root of the distance from the wall for a smooth boundary.

$$\frac{u}{U} = \left(\frac{y}{\delta}\right)^{1/7} \quad \eta^{1/7} \tag{13.42}$$

The shearing stress at the plate surface can no longer be calculated from the expression $\tau_o = \mu\left(\frac{\partial u}{\partial y}\right)_{y=0}$ since τ_o becomes infinitely large at y = 0. It should be recalled that there is a thin laminar sublayer in the immediate vicinity of the plate surface and that the power equation (13.42) applies only to the region outside the laminar sublayer where the flow is turbulent. The shearing stress of the turbulent flow is therefore transmitted to the plane surface through the viscous action in the laminar sublayer (Fig. 13.14).

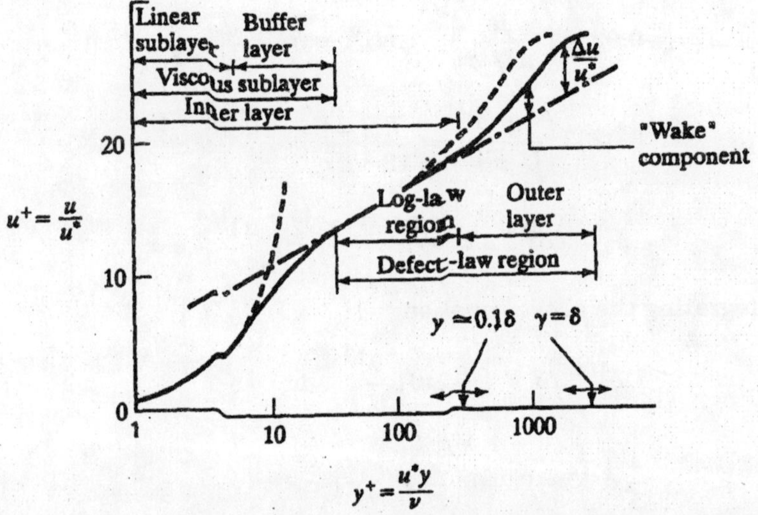

Fig. 13.14. Turbulent boundary layer illustrating wall-layer nomenclature.

The shearing stress at the pipe wall for the turbulent flow in a smooth pipe is written as

$$\tau_o = 0.03325\,\rho\, u_{av}^2 \left(\frac{\mu}{\rho\, u_{av}\, R}\right)^{1/4} \tag{13.43}$$

Where u_{av} is the mean velocity of pipe flow and R is the pipe radius. The relationship between the average velocity u_{av} of pipe flow and the maximum velocity u_{max} at the centre line of the pipe is $u_{av} = 0.817.\, u_{max}$ when the velocity distribution for the turbulent flow in smooth pipes is represented by the one-seventh power equation.

$$\frac{u}{u_{max}} = \left(\frac{y}{R}\right)^{1/7}$$

Substituting $u = u_{max}$ and $R = \delta$ and $u_{av} = 0.817\ u_{max} = 0.817\ U$ into equation (13.43) we obtain

$$\tau_o = 0.0233\, \rho U^2 \left(\frac{\mu}{\rho U \delta}\right)^{1/4} \tag{13.44}$$

The above equation was suggested by Blasius and the applicable from 5×10^5 to 10^7.

We are now in a position to apply the momentum integral equation

$$\tau_o = \rho U^2 \frac{d\delta}{dx} \int_0^1 \frac{u}{U}\left(1 - \frac{u}{U}\right) d\eta \tag{13.45}$$

Substituting τo from equation (13.44) and $\frac{u}{U}$ from equation (13.42)

$$0.0233 \left(\frac{\mu}{\rho U \delta}\right)^{1/4} \rho U^2 = \rho U^2 \frac{d\delta}{dx} \int_0^1 \eta^{1/7} \left(1 - \eta^{1/7}\right) d\eta$$

$$0.0233 \left(\frac{\nu}{U \delta}\right)^{1/4} = \frac{7}{12} \left(\frac{d\delta}{dx}\right)$$

or,
$$\delta^{1/4}\, d\delta = \frac{0.0233 \times 72}{7} \left(\frac{\upsilon}{U}\right)^{1/4} dx$$

Integrating the above equation.

$$\int \delta^{1/4}\, d\delta = \int 0.240 \left(\frac{\nu}{U}\right)^{1/4} dx$$

or,
$$\frac{4}{5} \delta^{5/4} = 0.240 \left(\frac{\nu}{U}\right)^{1/4} x + c$$

If it is assumed that δ ≈ 0 at x = 0 (this is equivalent to assuming turbulent flow from the leading edge), then c = 0 and

$$\delta = 0.382 \left(\frac{\nu}{U}\right)^{1/5} x^{4/5}$$

or,
$$\frac{\delta}{x} = 0.382 \left(\frac{\nu}{u_x}\right)^{1/5}$$

or,
$$\frac{\delta}{x} = \frac{0.382}{\left(R_{e_x}\right)^{1/5}} \tag{13.46}$$

Substituting the value of δ in equation (13.44), we obtain the skin friction coefficient

$$C_f = \frac{0.0594}{\left(R_{e_x}\right)^{1/5}} \tag{13.47}$$

The drag coefficient or average skin friction coefficient can be obtained by integrating the local skin friction coefficient C_f from x = 0 to x = l. and then dividing the integrated result by the plate length. Thus

$$C_D = \frac{1}{l}\int_0^l C_f\, dx = \frac{1}{l}\int_0^l \frac{0.0594}{\left(R_{e_x}\right)^{1/5}}\, dx = \frac{0.0594}{l\left(\frac{U}{\upsilon}\right)^{1/5}}\int_0^l x^{1/5}\, dx$$

$$= \frac{0.0594}{l\left(\frac{U}{\nu}\right)^{1.5}}\left[\frac{5}{4}\, l^{4/5}\right] = \frac{0.074}{\left(\frac{U\, l}{\nu}\right)^{1/5}} = \frac{0.074}{\left(R_{e_x}\right)^{1/5}}$$

$$C_D = \frac{0.074}{\left(R_{e_l}\right)^{1/5}} \tag{13.48}$$

which is valid over the range

$$5 \times 10^5 \leq R_e \leq 10^7$$

Experiments indicate that for Reynold's number between 10^7 and 10^9 the velocity distribution deviates from the one seventh power law and a close fit with experimental data is obtained by the following empirical relationship suggested by Prandtl/schlichting by using logarthmic law of velocity distribution.

$$C_D = \frac{0.455}{\left[\log_{10} R_{e_l}\right]^{2.58}} \tag{13.49}$$

It is seen that for a turbulent boundary layer on a flat plate :

(1) The boundary layer thickness increases as the 4/5 power of the distance from the leading edge as compared with $x^{1/2}$ for a laminar boundary layer.

(2) The local and average skin friction coefficient vary inversely as the fifth root of both x and U, as compared with the square root for a laminar boundary layer.

(3) The total drag varies as the 9/5 power of the free stream velocity ($U^{9/5}$) and the 4/5 power of the plate length ($x^{4/5}$) as compared with $U^{1/5}$ and $x^{1/2}$ for laminar boundary layer.

13.10 COMBINED LAMINAR & TURBULENT BOUNDARY LAYERS

If the boundary layer along a flat plate begin with laminar flow over the forward portion of the plate and becomes turbulent beyond, the total frictional force on the plate decreases.

The total drag, as well as the drag coefficient, may also be estimated for flow on a flat plate with combined laminar and turbulent boundary layers. The methodology recommended by Schlichting is to add the drag owing to the laminar boundary layer to the drag owing to the turbulent portion of the boundary layer, assuming that the turbulent portion beyond the transition point acts as though it were turbulent from the leading edge of the plate. Incorporating this procedure with equation (13.49) yields the Prandtl-Schlichting skin friction equation for a smooth plate

$$C_D = \frac{0.455}{\left[\log_{10} R_{e_x}\right]^{2.58}} - \frac{1700}{R_{e_l}} \qquad (13.50)$$

Where, the valve of A depends on the critical value of the Reynold's number (U x/ν) at which the laminar boundary layer becomes turbulent.

Taking the transition Reynold's number to be 5 × 10^5 $\left[\left(R_{e_x}\right)_{critical} = 5 \times 10^5\right]$, the coefficient of the reduced skin friction which accounts for the laminar region at the front portion of the plate becomes

$$C_D = \frac{0.074}{\left(R_{e_l}\right)^{1/5}} - \frac{1700}{R_{e_l}} \qquad (13.51)$$

The equation (13.51) is valid for the range $5 \times 10^5 < R_e < 10^7$.

For higher Reynold's number ($10^7 < R_{e_x} < 10^9$), the logarthmic form of velocity distribution in the turbulent boundary layer is more appropriate. The boundary layer shear coefficients are expressed by the equation (13.50), in which the value of A is 1610, therefore,

$$C_D = \frac{0.455}{\left(\log_{10} R_{e_l}\right)^{2.58}} - \frac{1700}{R_{e_l}} \qquad (13.52)$$

The term $\frac{1700}{R_{e_l}}$ is so small that omitting it does not cause any appreciable error.

The variation in drag coefficient for a flat plate parallel to the flow is shown in Fig. 13.15

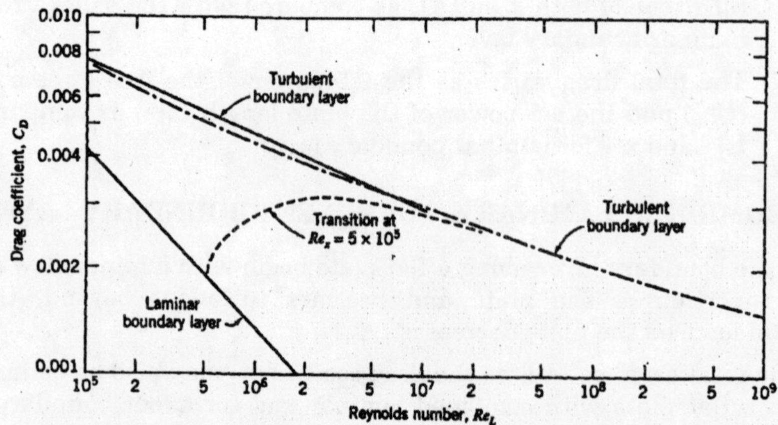

Fig. 13.15. Variation of drag coefficient with Reynolds number for a smooth flat plate parallel to the flow

In the plot at Fig. 13.15, transition was assumed to occur at $R_{e_x} = 5 \times 10^5$ for flows in which the boundary layer was initial laminar. The Reynold's number at which transition occurs depends on a combination of factors, such as surface roughness and free stream disturbance Transition tends to occur earlier (at lower Reynold's number) as surface roughness or free stream turbulence is increased. For transition at other values of Reynold's number, the constant `A' in the second term of equation (13.50) is different. Values of A for various values of U x_c/ υ are shown in Table 13.3.

Table 13.3. Value of A and Reynold's number

$R_e = \dfrac{U x_c}{v}$	10^5	3×10^5	4×10^5	5×10^5	6×10^5	7×10^5	8×10^5
A	300	1050	1400	1700	2080	3300	8700

Although the assumptions involved in this development are not consistent with those employed in the computation of boundary layer thickness, Equation (13.51) is in reasonable agreement with the curves shown in Fig. 13.13 and is more straight forward to apply than other procedures that account for a continuous boundary layer thickness at the transition point.

Fig. 13.15 shows that the drag coefficient is less for a given length of plate, when laminar flow is maintained over the longest possible distance. However, at large R_{e_l} (> 10^7) the contribution of the laminar drag is negligible.

13.11 COEFFICIENT OF DRAG FOR TURBULENT BOUNDARY LAYER FOR ROUGH PLATE

Using the logarithmic universal velocity distribution equation for the turbulent flow in rough pipes, schlichting has also derived an empirical expressions for the local and drag coefficient caused by turbulent boundary layer flow over a rough plate are given by

$$C_f = \frac{1}{\left[2.87 + 1.58 \log_{10}\left(\frac{x}{\in}\right)\right]^{2.5}} \qquad (13.53)$$

and $$C_D = \frac{1}{\left[1.89 + 1.62 \log_{10}\left(\frac{L}{\in}\right)\right]^{2.5}} \qquad (13.54)$$

Where $10^2 < \frac{L}{\in} < 10^6$ in which $\in$ is the plate surface roughness.

Fig. 13.16 shows flat plate drag coefficient for both laminar and turbulent flow conditions. The smooth plate and also with the effect of wall roughness are shown. The effect of wall roughness is quite strong. The proper roughness parameter here is x/∈ or L/∈, by analogy with the pipe parameter ∈/d. In a full rough regime, C_D is independent of Reynold's number, so that the drag varies exactly U^2 and is independent of dynamic viscosity μ. The equation (13.54) is plotted to the right of the dashed line in Fig. 13.16.

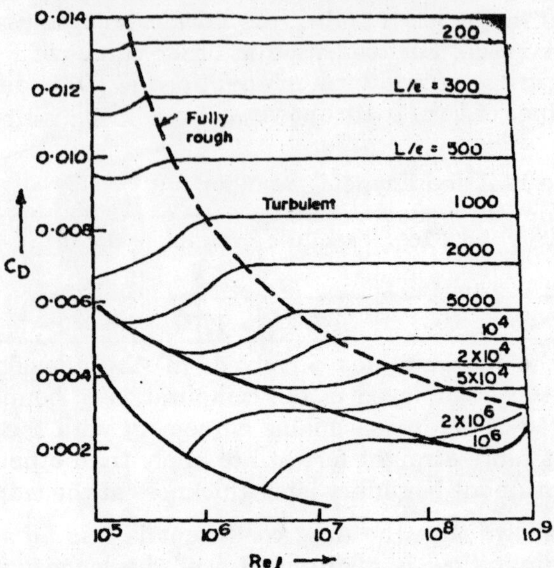

Fig. 13.16. Drag coefficient of laminar and turbulent boundary layer on smooth and rough flat plates. This chart is the flat plate analog of the moody diagram.

13.12 FLOW WITH A PRESSURE GRADIENT

In Blasius's solution for laminar flow over a flat plate, the pressure gradient was zero. The pressure gradient plays a major role in flow separation. Flow 13.17 illustrates the variation in u, $\frac{\partial u}{\partial y}$, and $\frac{\partial^2 u}{\partial y^2}$ across the boundary layer for the case of a zero pressure gradient.

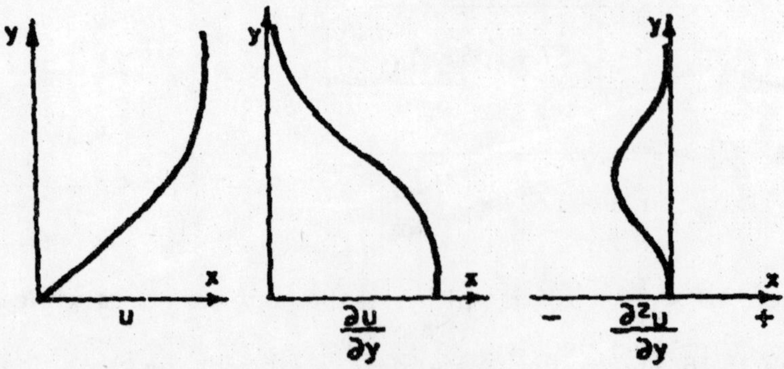

Fig. 13.17. Variation in velocity and velocity derivatives across the laminar boundary layer when dp/dx = 0.

When $\frac{dp}{dx} = 0$, the second derivative of the velocity at the wall must also be zero, hence the velocity profile is linear near the wall. Further out in the boundary layer, the velocity gradient becomes smaller and gradually approaches zero. The decrease in the velocity gradient means that the second derivative of the velocity must be negative.

The derivative $\frac{\partial^2 u}{\partial y^2}$ is shown as being zero at the wall, negative within the boundary layer. It is important to note that the second derivative must approach zero from the negative side as y $\rightarrow$ δ. For values of dp/dx $\times$ 0, the variation in u and its derivative is shown in Fig. 13.18.

A negative pressure gradient is seen to produce a velocity variation somewhat similar to that of the zero pressure gradient case. A positive value of dp/dx, however, requires a positive value of $\frac{\partial^2 u}{\partial y^2}$ at the wall. Since this derivative must approach zero from the negative side, at some point within the boundary layer the second derivative must equal zero. A zero second derivative, it will be recalled, is associated with an inflection point. The inflection point is shown in the velocity profile of Fig. 13.18.

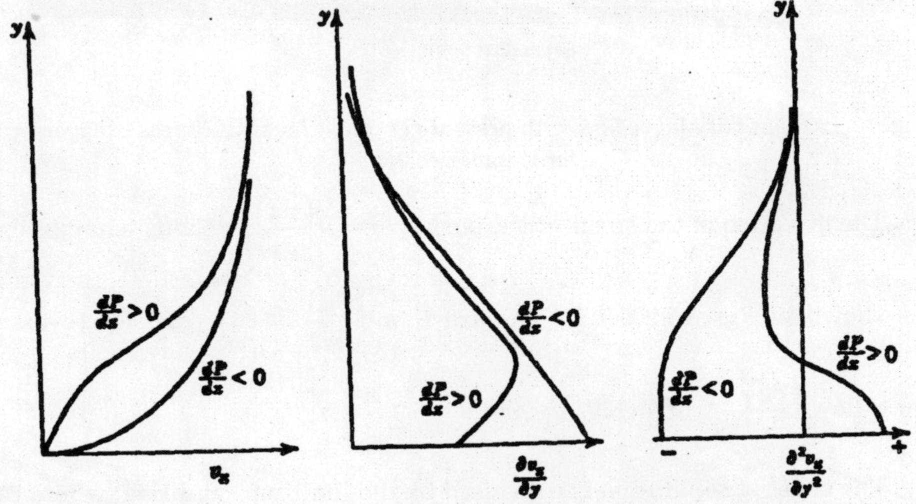

Fig. 13.18. Variation of u and its derivative across the boundary layer for various pressure gradient.

13.13 SEPARATION OF BOUNDARY LAYER FLOW

Prandtl showed that separation is caused by excessive momentum loss near the wall in a boundary layer trying to move downstream against increasing pressure, $\frac{dp}{dx} > 0$, which is called an adverse pressure gradient.

The opposite case of decreasing pressure $\frac{dp}{dx} < 0$, is called a favourable pressure gradient, where flow separation can never occur.

For adverse pressure gradient $\frac{dp}{dx} > 0$, the boundary layer flow thickness vary rapidly and, ultimately, separation from the plate surface. The fluid particles behind the point of separation follow the pressure gradient and move in a direction opposite to the external stream (Fig. 13.19).

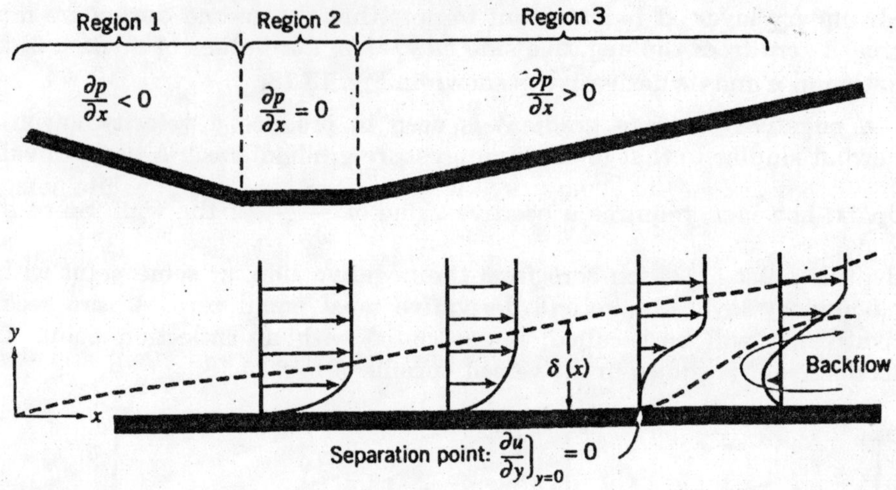

Fig. 13.19. Boundary layer flow with pressure gradient layer thickness exaggerated for boundary clarity.

The direction of the shear at the surface $\tau = \mu \left(\frac{\partial u}{\partial y}\right)_{y=0}$ must change sign : as shown in Fig. 13.20(c) for forward flow $\left(\frac{\partial u}{\partial y}\right)_{y=0} > 0$ and for reversed (back) flow $\left(\frac{\partial u}{\partial y}\right)_{y=0} < 0$ (Fig. 13.20b).

The onset of separation is expressed by the limiting condition, where the shear at the wall vanishes and $\left(\frac{\partial u}{\partial y}\right)_{y=0} = 0$ (Fig. 13.20(b))

The point of separation is defined as the limit between forward and reverse flow in the layer in the immediate neighbourhood of the wall, or

$$\text{Point of separation} : \left(\frac{\partial u}{\partial y}\right)_{y=0} = 0 \qquad (13.55)$$

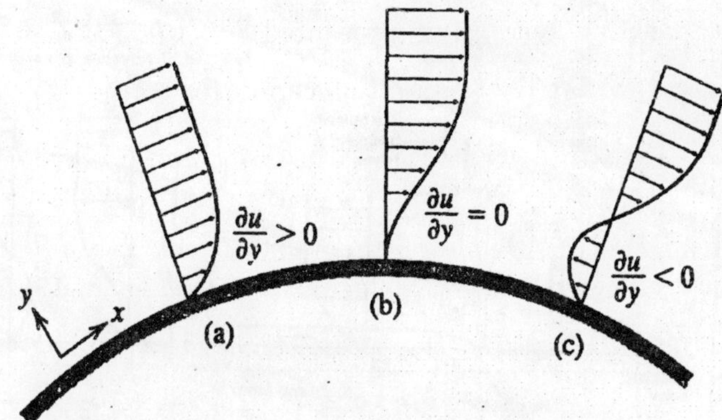

Fig. 13.20. Schematic showing conditions for attached and separated flow

The point on a solid boundary at which $\frac{\partial u}{\partial y} = 0$ is defined as the point of separation. Consequently, we can conclude that for $\frac{\partial p}{\partial x} = 0$, the flow will not separate.

Fig. 13.21 shows boundary layer flow with pressure gradient and sheat at the wall at different points.

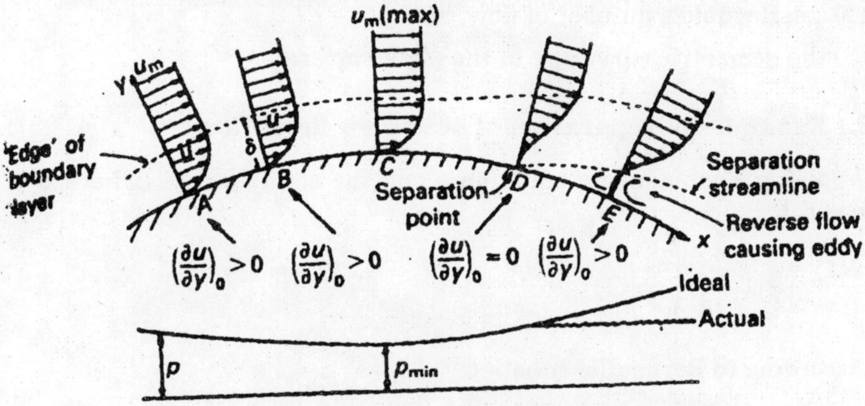

Fig. 13.21. Pressure gradient and shear at the wall.

The phenomenon of boundary layer separation is illustrated in Fig. 13.22

The phenomenon of boundary layer flow and its separation from the plate surface may be explained physically by considering the following three forces which affect the motion of fluid in the boundary layer.

(1) The inertia forces of the fluid in the free stream outside the boundary. These forces are transmitted through the laminar boundary layer by viscous shear and through the boundary layer by means of momentum transfer.

(2) The retarding forces of viscous shear at the solid surface and in the boundary layer flow,

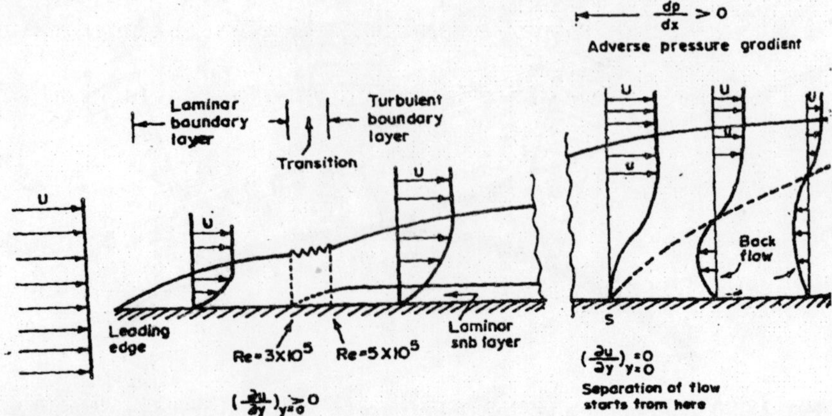

Fig. 13.22. Development of boundary layer separation from a plate surface.

(3) The forces due to the longitudinal pressure gradient in the free stream,

The physically conditions governing the boundary layer flow separation from a plate surface are extremely complex because the position of the point of separation has been found to depend in general on:

(1) the roughness of the plate surface,

(2) the Reynold's number of flow, and

(3) the geometric curvature of the plate surface.

13.13.1 Examples of separation of boundary layer flow

Applying x momentum equation along the solid surface (where u = v = 0), we get

$$\mu\left(\frac{\partial^2 u}{\partial y^2}\right)_{y=0} = \frac{dp}{dx}$$

According to Bernoullis equation

$$\frac{dp}{dx} = -\rho U \frac{dU}{dx}$$

By combining the x momentum and Bernoullis equations, we get

$$\mu\left(\frac{\partial^2 u}{\partial y^2}\right)_{y=0} = -\rho U \frac{dU}{dx} \qquad (13.56)$$

There are two types of external flow : (a) accelerating flows (b) decelerating flows. Whe the flow accelerates, $\left(\frac{\partial u}{\partial x} > 0\right)$, according to equation (13.56) the curvature of the velocity within the boundary layer is negative

$\left(\frac{\partial^2 u}{\partial y^2} < 0\right)$. The velocity then increases monotonically to the outer free stream velocity. Therefore, since there is no inflection point, the flow will not separate.

However, when the flow decelerates $\left(\frac{\partial u}{\partial x} < 0\right)$, the curvature of the velocity at the solid surface is positive $\left[\left(\frac{\partial^2 u}{\partial y^2}\right)_{y=0} > 0\right]$, while at the edge of the boundary layer, it is negative $\left[\left(\frac{\partial^2 u}{\partial y^2}\right)_{y \to \infty} < 0\right]$. Therefore, in decelerating flow, the velocity profile may develop an inflection point. Consequently, separation may possibly occur only in decelerating flows.

The flow around a solid surface remains attached to the surface as long as the flow accelerates, or equivalently, when the pressure gradient is "**favorable**" $\left(\frac{dp}{dx} < 0\right)$. When the flow decelerates, the resulting "**adverse**" pressure gradient $\left(\frac{dp}{dx} > 0\right)$ causes the flow near the surface to slow down to the extent that the flow may be forced to reverse direction and thus to separate.

(1) Converging - diverging nozzle

Fig. 13.23 shows the converging - diverging nozzle. The incompressible flow passing through the nozzle. Because of the decrease in area in the converging part of the nozzle, the flow accelerates, whereas in the diverging part, the flow decelerates, separation, then, can develop only in the diverging section of the nozzle. In the converging part of the nozzle, the flow remains attached to the wall. It is to be noted that the flow will not always separate when the flow decelerates, but if separation is to take place, the flow must be decelerating.

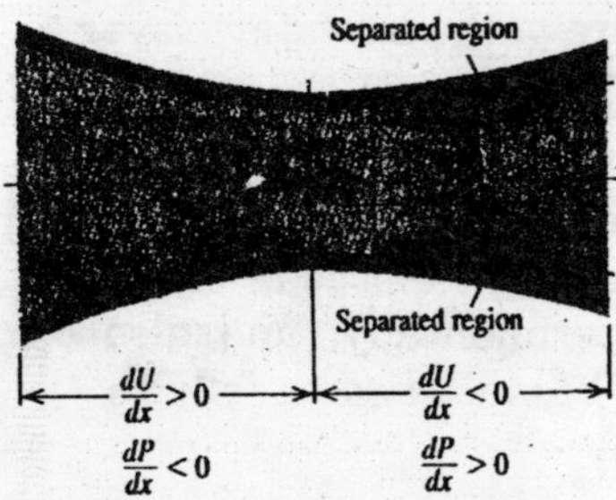

Fig. 13.23. Converging - diverging nozzle.

(2) Diffuser

A practical example is the flow in a duct consisting of a nozzle, throat, and diffuser, as in Fig. 13.24. The nozzle flow is a favorable gradient and never separates nor does the throat flow where the pressure gradient is approximately zero. But the expanding diffuser produces low velocity and increasing pressure, and adverse gradient. If the diffuser angle is too large, the adverse gradient is excessive and the boundary layer will separate at one or both walls, with back flow, increased losses, and poor pressure recovery. In the diffuser literature this condition is called diffuser stall, a term used also in aerofoil aerodynamics to denote aerofoil boundary layer separation. Thus the boundary layer behaviour explains why a large-angle diffuser has heavy flow losses and poor performance.

Presently boundary layer theory can only compute up to the stagnation point, after which it is invalid. New techniques are now being developed for analyzing the strong interaction effects caused by separated flow.

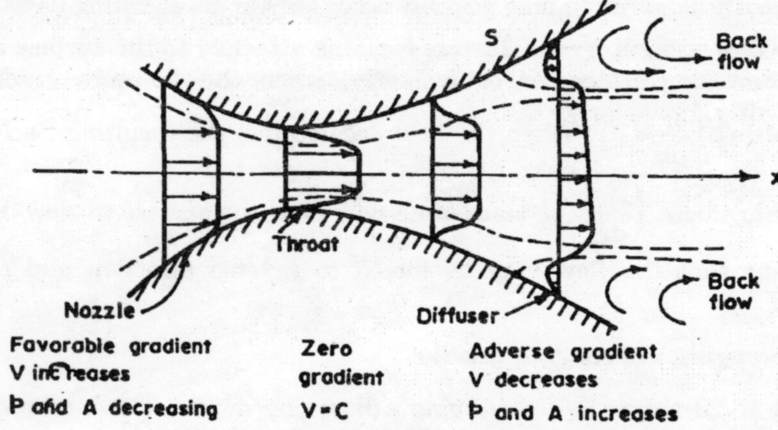

Fig. 13.24. Boundary layer separation in a diffuser.

13.13.2 Separation control

There are in existence several methods which have been developed for the purpose of artificially controlling the behaviour of the boundary layer. The purpose of these methods is to affect the whole flow in a desired direction by influencing the structure of the boundary layer.

The problem of boundary layer control has become very important for a time, in particular in the field of aeronautical engineering; in actual application it is often necessary to prevent separation in order to reduce drag and to attain high lift. Several methods of controlling the boundary layer have been developed experimentally, and also on the basis of theoretical considerations. These can be classified as follows

(1) Cooling of the wall

(2) Suction

(3) Acceleration of the boundary layer (blowing)

(4) Motion of the solid wall

(5) Injection of a different gas (binary boundary layers)

(6) Prevention of transition to turbulent flow by the provision of suitable shapes (Laminar aerofoils).

(1) Cooling of the Wall

In a certain range of supersonic Mach number it is possible completely to stablize the boundary layer by the application of cooling at the wall.

(2) Suction

The effect of suction consists in the removal of decelerated fluid particles from the boun-dary layer before they are given a chance to cause separation Fig. 13.25(a). The application of suction, which was first tried by Prandtl, was later widely used in the design of aircraft wings. By applying suction, considerable greater pressure increases on the upper side of the aerofoil (i.e. Lower absolute pressure) are obtained at larger angles of incidence, and, consequently, much larger maximum lift values.

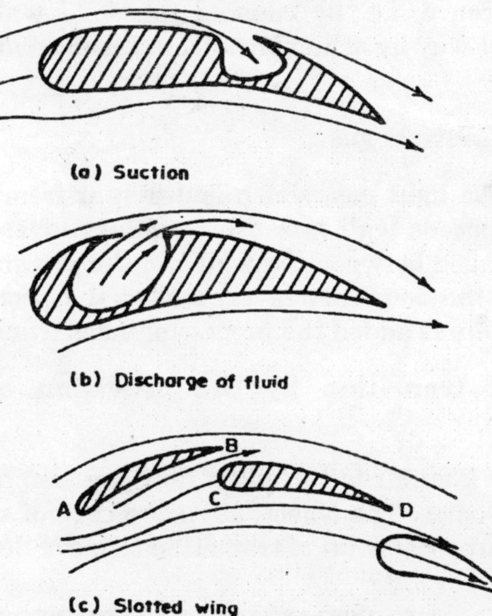

Fig. 13.25. Different arrangement for boundary layer control

In more recent times suction was also applied to reduce drag. By the use of suitable arrangement of suction slits it is possible to shift the point of transition in the boundary layer in the downstream direction; this causes the drag coefficient to decrease because laminar drag is substantially smaller than turbulent drag. The effect of the drag in transition caused by suction is to reduce the boundary layer thickness which then becomes less prone to turning turbulent. Furthermore, the velocity profiles in a boundary layer with suction, being fuller, have forms which are less likely to induce turbulence compared with those in laminar boundary layers without suction and equal thickness.

(3) Acceleration of the boundary layer (blowing)

An alternative method of prevention separation consists in supplying additional energy to the particle of fluid which are being retarded in the boundary layer. This result can be achieved by discharging fluid from the interior of the body with the aid of a special blower (13.25(b)), or by deriving the required energy directly from the main stream. This latter effect can be produced, by connecting the retarded region to a region of higher pressure through a slot in the wing (slotted wing, Fig. 13.25(c)). In either case additional energy is imparted to the particles of the fluid in the boundary layer near the wall.

(4) Motion of the solid wall

The most obvious method of avoiding separation is to attempt to prevent the formation of a boundary layer. Since a boundary layer owes its existence to the difference between the velocity of the fluid and that of the solid wall, it is possible to eliminate the formation of a boundary layer by attempting to suppress that difference, i.e. by causing the solid wall to move with the stream. The simplest way by achieving such a result involves the rotation of a circular cylinder.

(5) Injection of a different gas

The injection of a light gas, which is different from that in the external stream, through a porous wall into the boundary layer reduces the rate at which heat is exchanged between the wall and the stream. Injection creates a gaseous mixture in the boundary layer, and to the processes of momentum and heat transfer there is added the process of mass transfer by diffusion.

(6) Prevention of transition by the provision of suitable shapes Laminar aerofoils

Transition from laminar to turbulent flow can also be delayed by the use of suitably shaped bodies. The object, as in the case of suction, is to reduces friction drag by causing the point of transition to move downstream.

ILLUSTRATIVE EXAMPLES

13.1. The velocity profile in a laminar boundary layer is given by

$$\frac{u}{U} = \frac{y}{\delta}$$

(a) Find the displacement thickness, the momentum thickness, the energy thickness and shape factor $H = \delta^* / \theta$

(b) Find also an expression for boundary layer thickness.

Solution

(a) $$\delta^* = \int_0^\delta \left(1 - \frac{u}{U}\right) dy = \int_0^\delta \left(1 - \frac{y}{\delta}\right) dy = \frac{\delta}{2}$$

$$\theta = \int_0^\delta \frac{u}{U}\left(1 - \frac{u}{U}\right) dy = \int_0^\delta \frac{y}{\delta}\left(1 - \frac{y}{\delta}\right) dy = \frac{\delta}{6}$$

$$\delta^{**} = \int_0^\delta \frac{u}{U}\left[1 - \left(\frac{u}{U}\right)^2\right] dy = \int_0^\delta \frac{y}{\delta}\left[1 - (y/\delta)^2\right] dy = \frac{\delta}{4}$$

Shape factor

$$H = \frac{\delta^*}{\theta} = \frac{(\delta/2)}{(\delta/6)} = 3$$

(b) $$\frac{\tau_o}{\rho U^2} = \frac{\partial \theta}{\partial x} = \frac{\partial}{\partial x}\left[\int_0^\delta \frac{u}{U}\left(1 - \frac{u}{U}\right) dy\right] = \frac{\partial}{\partial x}\left(\frac{\delta}{6}\right)$$

$$\tau_o = \frac{1}{6}\rho U^2 \frac{\partial(\delta)}{\partial x} \qquad (1)$$

The boundary layer thickness δ is obtained by equating the valve of boundary shear stress to its definition given by Newton's.

$$\tau_o = \frac{1}{6}\rho U^2 \frac{\partial}{\partial x}(\delta) = \mu \left(\frac{\partial u}{\partial y}\right)_{y=0} \qquad (2)$$

From the given velocity profile

$$u = U\,(y/\delta)$$

$$\frac{du}{dy} = U/\delta$$

$$\therefore \left(\frac{du}{dy}\right)_{y=0} = \frac{U}{\delta}$$

$$\therefore \; \tau_o = \frac{1}{6}\rho U^2 \frac{\partial}{\partial x}(\delta) = \frac{U}{\delta}\mu$$

$$\delta\,\partial(\delta) = \frac{6\,U\,\mu}{\rho U^2}\partial x$$

$$\therefore \frac{\delta^2}{2} = \frac{6\,\mu}{\rho U}x + c$$

$\delta = 0$, $x = 0$ $\therefore$ $c = 0$

$$\delta = \sqrt{\frac{12\,\mu x}{\rho U}} = \frac{3.464}{\sqrt{R_e}}x$$

$$\therefore \frac{\delta}{x} = \frac{3.464}{\sqrt{R_e}}$$ Ans

13.2. The velocity profile in a laminar boundary layer is given by

$$\frac{u}{U} = \frac{5}{3}\left(\frac{y}{\delta}\right) - \frac{2}{3}\left(\frac{y}{\delta}\right)^3 = \frac{5}{3}\eta - \frac{2}{3}\eta^3$$

What is the boundary layer thickness δ/x in terms of U *x*/ν ?

Solution

$$\tau_o = \mu\left(\frac{du}{dy}\right)_{y=0}$$

$$= \mu\left[\left(\frac{u}{\delta}\right)\frac{\partial(u/U)}{\partial(y/\delta)}\right]_{y=0} = \frac{\mu U}{\delta}\left[\frac{\partial f(\eta)}{\partial \eta}\right]_{y=0}$$

$$= \frac{\mu U}{\delta}\left(\frac{5}{3}\right) = \frac{5\mu U}{3\delta} \quad (1)$$

For a flat plate flow, U = Constant

and $\frac{dp}{dx} = 0$

$$\tau_o = \rho U^2\left(\frac{d\theta}{dx}\right) = \rho U^2 \frac{d\delta}{dx}\int_0^1 \frac{u}{U}\left(1 - \frac{u}{U}\right) d\eta$$

$$= \rho U^2 \frac{d\delta}{dx}\int_0^1 \left[\frac{5}{3}(\eta) - \frac{2}{3}(\eta)^3\right] d\eta$$

$$\tau_o = \rho U^2 \frac{d\delta}{dx}\int_0^1 \left[\frac{5}{3}\eta - \frac{25}{9}\eta^3 + \frac{20}{9}\eta^4 - \frac{4}{9}\eta^6\right] d\eta$$

$$= \rho U^2 \frac{d\delta}{dx}\left[\frac{5}{6}\eta^2 - \frac{25}{27}\eta^3 - \frac{2}{12}\eta^4 + \frac{20}{45}\eta^5 - \frac{4}{63}\eta^7\right]_0^1$$

$$= \rho U^2 \frac{d\delta}{dx}\left[\frac{5}{6} - \frac{25}{27} - \frac{1}{6} + \frac{20}{45} - \frac{4}{63}\right]$$

$$= \rho U^2 \frac{d\delta}{dx}\left[\frac{1575 - 1750 - 315 + 840 - 120}{1890}\right]$$

$$= \rho U^2 \frac{d\delta}{dx}\left[\frac{2415 - 2185}{1890}\right]$$

$$= \rho U^2 \frac{d\delta}{dx}\left[\frac{230}{1890}\right] = \rho U^2 \frac{d\delta}{dx}\left(\frac{23}{189}\right) \quad (2)$$

Equating (1) and (2)

$$\left(\frac{5}{3}\right)\frac{\mu U}{\delta} = \rho U^2 \frac{d\delta}{dx}\left(\frac{23}{189}\right)$$

$$\delta\, d\delta = \left(\frac{945}{69}\right)\frac{\mu x}{\rho U}$$

Integrating

$$\delta^2/2 = \left(\frac{945}{69}\right)\frac{\mu}{\rho U}\, d\, x$$

or, $$\frac{\delta}{x} = \sqrt{\frac{1890}{69}}\sqrt{\frac{\mu}{\rho U x}} = \frac{5.23}{\sqrt{R_{e_x}}}$$ Ans

13.3. The velocity profile in a laminar boundary layer is given by

$$\frac{u}{U} = 2\,(y/\delta) - 2\,(y/\delta)^3 + (y/\delta)^4$$

Find the expressions for the boundary layer thickness, shear intensity and drag force on one side of the plate.

Solution

$$\tau_o = \mu\left(\frac{du}{dy}\right)_{y=0} = \frac{\mu U}{\delta}\left[\frac{\partial\,(u/U)}{\partial\,(y/\delta)}\right]_{y=0}$$

$$= \frac{\mu U}{\delta}\left[\frac{\partial\, f\,(\eta)}{d\,\eta}\right]_{y=0} = \frac{\mu U}{\delta}\,[2] = \frac{2\,\mu U}{\delta} \qquad (1)$$

For a flat plate flow, U = constant and $\frac{dp}{dx} = 0$

$$\tau_o = \rho\, U^2 \frac{d\theta}{dx}$$

$$\tau_o = \rho U^2 \frac{d\delta}{dx}\int_0^1 \frac{u}{U}\left(1 - \frac{u}{U}\right) d\eta$$

$$= \rho\, U^2 \frac{d\delta}{dx}\int_0^1 \left(2\,\eta - 2\,\eta^3 + \eta^4\right)\left(1 - 2\,\eta + 2\eta^3 - \eta^4\right) d\eta$$

$$= \rho\, U^2 \frac{d\delta}{d\,x}\int_0^1$$

$$\left|2\,\eta - 4\,\eta^2 - 2\,\eta^3 + 9\,\eta^4 - 4\,\eta^5 - 4\,\eta^6 + 4\,\eta^7 - \eta^8\right| d\eta$$

$$= \rho\, U^2 \frac{d\delta}{dx}\int_0^1 \left(\frac{37}{315}\right) \qquad (2)$$

Equating (1) and (2)

$$\frac{2\,\mu\,u}{\delta} = \rho\,U^2\,\frac{d\delta}{dx}\left(\frac{37}{315}\right)$$

$$\delta\,d\,\delta = \left(\frac{630}{37}\right)\frac{\mu}{\rho\,U}\,dx$$

Integrating,

$$\frac{\delta^2}{2} = \frac{630}{374}\,\frac{\mu\,x}{\rho\,u}$$

$$\frac{\delta}{x} = \sqrt{\frac{1260}{37}}\sqrt{\frac{\mu}{\rho\,U\,x}} = \frac{5.835}{\sqrt{R_{e_x}}}$$ Ans

(b) Substituting the value of δ in the equation (1) for τo

$$\tau_o = \frac{2\,\mu\,U}{\delta} = \frac{2\,\mu\,U}{5.835\,x}\sqrt{\frac{\rho\,U\,x}{\mu}}$$

$$= 0.34272\,\frac{\mu^{1/2}\,U^{3/2}\,\rho^{1/2}}{x^{1/2}} = 0.6854\,\frac{\rho\,U^2}{2}\left(\frac{\mu}{\rho\,U\,x}\right)^{1/2}$$ Ans

i.e. $$\frac{\tau_o}{\frac{1}{2}\rho\,U^2} = C_f = \frac{0.6854}{\sqrt{R_{e_x}}}$$

(c) $$F_D = \frac{1}{L}\int_0^L \tau_o\,d\,x = \frac{\rho\,U^2}{2}\int_0^L \frac{0.6854}{\sqrt{\frac{\rho U\,x}{\mu}}}\,d\,x$$

$$\frac{F_D}{\frac{1}{2}\rho\,U^2} = C_D = \frac{1.371}{(\rho\,U\;L/\mu)^{1/2}} = \frac{1.371}{\sqrt{R_{e_L}}}$$ Ans

13.4. Velocity profiles in Laminar boundary layers often are approximated by the equation

(i) Linear $\frac{u}{U} = \frac{y}{\delta}$

(ii) Parabolic $\frac{u}{U} = 2\left(\frac{y}{\delta}\right) - \left(\frac{y}{\delta}\right)^2$

(iii) Cubic $\frac{u}{U} = \frac{3}{2}\left(\frac{y}{\delta}\right) - \frac{1}{2}\left(\frac{y}{\delta}\right)^3$

(iv) Sinusoidal $\frac{u}{U} = \sin\left(\frac{\pi}{2}\,\frac{y}{\delta}\right)$

(a) Evaluate of ratio δ^*/δ for each of the laminar boundary layer velocity profiles given above.

(b) Evaluate the ratio θ/δ for each of the laminar boundary layer velocity profiles given above

(c) Evaluate the shape factor H = δ^*/θ for each of the laminar boundary layer P velocity profiles given above.

Solution

(a) We know that

$$\delta^* = \int_0^\delta \left(1 - \frac{u}{U}\right) dy$$

or,

$$\frac{\delta^*}{\delta} = \frac{1}{\delta} = \int_0^\delta \left(1 - \frac{u}{U}\right) dy = \int_0^1 \left(1 - \frac{u}{U}\right) d\left(\frac{y}{\delta}\right) = \int_0^1 \left(1 - \frac{u}{U}\right) d\,\eta$$

(i) Linear

$$\frac{\delta^*}{\delta} = \int_0^1 (1 - \eta)\, d\,\eta = \left[\eta - \frac{1}{2}\eta^2\right]_0^1 = \frac{1}{2}$$

(ii) Parabolic

$$\frac{\delta^*}{\delta} = \int_0^1 (1 - 2\,\eta + \eta)\, d\,\eta = \left[\eta - \eta^2 + \frac{\eta^3}{3}\right]_0^1 = \frac{1}{3}$$

(iii) Cubic

$$\frac{\delta^*}{\delta} = \int_0^1 \left(1 - \frac{3}{2}\eta + \frac{1}{2}\eta^3\right) d\,\eta = \left[\eta - \frac{3}{4}\eta^2 + \frac{1}{8}\eta^4\right]_0^1 = \frac{3}{8}$$

(iv) Sinusoidal

$$\frac{\delta^*}{\delta} = \int_0^1 \left(1 - \sin\frac{\pi}{2}\eta\right) d\,\eta = \left[\eta + \frac{2}{\pi}\cos\frac{\pi}{2}\eta\right]_0^1$$

$$= 1 - \frac{2}{\pi} = 1 - \frac{14}{72} = \frac{22 - 14}{22} = \frac{8}{22}$$

(b) We know that

$$\theta = \int_0^\delta \frac{u}{U}\left(1 - \frac{u}{U}\right) dy$$

$$\frac{\theta}{\delta} = \int_0^1 \frac{u}{U}\left(1 - \frac{u}{U}\right) d\,\eta \qquad \left[\textit{where } \eta = \frac{y}{\delta}\right]$$

(i) Linear

$$\frac{\theta}{\delta} = \int_0^1 \eta\,(1 - \eta)\, d\,\eta = \left[\frac{\eta^2}{2} - \frac{\eta^3}{3}\right]_0^1 = \frac{1}{6}$$

(ii) Parabolic

$$\frac{\theta}{\delta} = \int_0^1 \left(2\eta - \eta^2\right)\left(1 - 2\eta + \eta^2\right) d\,\eta$$

$$= \int_0^1 \left(2\,\eta - 5\,\eta + 4\,\eta^3 - \eta^4\right) d\,\eta$$

$$= \left[\eta^2 - \frac{5}{3}\,\eta^3 + \eta^4 - \frac{\eta^5}{5}\right]_0^1 = \frac{15 - 25 + 15 - 3}{15} = \frac{2}{15}$$

(iii) Cubic

$$\frac{\theta}{8} = \int_0^1 \left(\frac{3}{2}\eta - \frac{1}{2}\eta^3\right)\left(1 - \frac{3}{2}\eta + \frac{\eta^2}{2}\right) d\,\eta$$

$$= \int_0^1 \left(\frac{3}{2}\eta - \frac{9}{4}\eta^2 - \frac{1}{2}\eta^3 + \frac{3}{2}\,\eta^4 - \frac{1}{4}\,\eta^6\right) d\,\eta$$

$$= \left[\frac{3}{4}\,\eta^2 - \frac{9}{12}\,\eta^3 - \frac{1}{8}\,\eta^4 + \frac{3}{10}\,\eta^5 - \frac{1}{28}\,\eta^7\right]_0^1$$

$$= \frac{3}{4} - \frac{9}{12} - \frac{1}{8} + \frac{3}{10} - \frac{1}{28} = \frac{39}{280}$$

(iv) Sinusoidal

$$\frac{\pi}{\delta} = \int_0^1 \left[\sin\left(\frac{\pi}{2}\eta\right)\right]\left[1 - \sin\left(\frac{\pi}{2}\,\eta\right)\right] d\,\eta$$

$$= \int_0^1 \left[\sin\left(\frac{\pi}{2}\eta\right) - \sin\left(\frac{\pi}{2}\eta\right)\right] d\,\eta$$

$$\left[-\frac{2}{\pi}\cos\left(\frac{\pi}{2}\,\eta\right)\right]_0^1 - \frac{2}{\pi}\left[\frac{\pi\,\eta}{4} - \frac{1}{4}\sin\left(\pi\,\eta\right)\right]$$

$$= -0\left(-\frac{2}{\pi}\right) - \frac{2}{\pi}\left(\frac{\pi}{4}\right) - 0 = \frac{2 \times 7}{22} - \frac{1}{2}$$

$$= \frac{14}{22} - \frac{1}{2} = \frac{14 - 11}{22} = \frac{3}{22}$$

(c) Shape factor

$$H = \frac{\delta^*}{\theta}$$

We know that

$$\delta^* = \delta \int_0^1 \left(1 - \frac{u}{U}\right) d\eta$$

and $\quad \theta = \delta \int_0^1 \frac{u}{U}\left(1-\frac{u}{U}\right) d\eta$

Using the answers obtained in the (a) and (b) parts.

(i) Linear $\quad H = \frac{1/2}{1/6} = 3$

(ii) Parabolic $\quad H = \frac{1/3}{2/15} = \frac{1}{3} \times \frac{15}{2} = 2.5$

(iii) Cubic $\quad H = \frac{3/8}{39/280} = \frac{3}{8} \times \frac{280}{39} = 2.692$

(iv) Sinusoidal $\quad H = \frac{8/22}{3/22} = \frac{8}{3} = 2.67$

13.5. Consider two dimensional laminar boundary layer flow along a flat plate. The velocity profile in the boundary layer is assumed to be sinusoidal

$$\frac{u}{U} = \sin\left(\frac{\pi}{2}\, y/\delta\right)$$

Find expressions for

(a) The rate of growth of δ as a function of x.

(b) The displacement thickness, δ,* as a function of x.

(c) The total friction force on a plate of length L, and width B.

Solution

For flat plate flow, U = Constant, $\frac{dp}{dx} = 0$

and $\quad \tau_o = \rho U^2 \frac{d\theta}{dx} = \rho U^2 \frac{d\delta}{dx} \int_0^1 \frac{u}{U}\left(1-\frac{u}{U}\right) d\eta$

or, $\quad \tau_o = \rho U^2 \frac{d\delta}{dx} \int_0^1 \left(\sin\frac{\pi}{2}\eta\right)\left(1-\sin\frac{\pi}{2}\eta\right) d\eta$

$$= \rho U^2 \frac{d\delta}{dx} \int_0^1 \left(\sin\frac{\pi}{2}\eta - \sin^2\frac{\pi}{2}\eta\right) d\eta$$

$$= \rho U^2 \frac{d\delta}{dx}\left(\frac{2}{\pi}\right)\left[-\cos\frac{\pi}{2}\eta - \left(\frac{1}{2}\right)\frac{\pi}{2}\eta + \frac{1}{4}\sin\pi\eta\right]_0^1$$

$$= \rho U^2 \frac{d\delta}{dx}\left(\frac{2}{\pi}\right)\left[0 + 1 - \frac{\pi}{4} + 0 + 0 - 0\right]$$

$$= \rho U^2 \frac{d\delta}{dx}\left[\frac{2}{\pi}\left(1-\frac{\pi}{4}\right)\right]$$

$$= \rho U^2 \frac{d\delta}{dx}[0.1366] = 0.137\ \rho U^2 \frac{d\delta}{dx} \qquad (1)$$

Now, $\tau_o = \mu\left(\frac{\partial u}{\partial y}\right)_{y=0} = \mu\left[\frac{U\,\partial(u/U)}{\delta\,\partial(y/\delta)}\right]_{y=0}$

$$= \mu\frac{u}{\delta}\left(\frac{\pi}{2}\right)\left[\cos\frac{\pi}{2}\eta\right]_{y=0} = \frac{\pi\mu U}{2\delta} \qquad (2)$$

Therefore, τ from (1) and (2)

$$= \frac{\pi\mu U}{2\delta} = 0.137\ \rho U^2 \frac{d\delta}{dx}$$

or, $\delta d\delta = 11.5\frac{\mu}{\rho U}dx$

Integrating

$$\int \delta\, d\delta = 11.5\frac{\mu}{\rho U}\int dx$$

or, $\frac{\delta^2}{2} = 11.5\frac{\mu x}{\rho U} + c$

But c = 0, Since δ = 0 at $x = 0$, therefore,

$$\delta = \sqrt{\frac{23\mu x}{\rho U}}$$

or, $\delta = 4.8\sqrt{\frac{\mu x}{\rho U}}$

or, $\frac{\delta}{x} = 4.8\sqrt{\frac{\mu x}{\rho U}} = \frac{4.8}{\sqrt{R_{e_x}}}$

The displacement thickness δ* is given by

$$\delta^* = \int_0^\delta\left(1-\frac{u}{U}\right)dy = \delta\int_0^1\left(1-\frac{u}{U}\right)d\eta$$

$$= \delta\int_0^1\left(1-\sin\frac{\pi}{2}\eta\right)$$

$$= \delta\left[\eta - \frac{2}{\pi}\cos\frac{\pi}{2}\eta\right]_0^1 = \delta\left[1-0+0-\frac{2}{\pi}\right] = \delta\left[1-\frac{2}{\pi}\right]$$

$$= \frac{4.8}{\sqrt{R_{e_x}}}\left(1 - \frac{2}{\pi}\right)$$ [From equation (3)]

$$= \frac{1.744\ x}{\sqrt{R_{e_x}}} \quad (4)$$

The total friction force on one side of the plate is given by

$$F = \int_A^L \tau_o\ dA\ [dA = B\ d\ x]$$

$$= \int_0^L \tau_o\ Bdx = \int_0^L \rho U^2 \frac{d\theta}{dx} Bdx = \rho U^2\ B \int_0^{\theta_L} d\theta$$

$$= \rho U^2\ B\theta_L \quad (5)$$

Where

$$\theta_L = \int_o^{\delta_L} \frac{u}{U}\left(1 - \frac{u}{U}\right) dy = \delta_L \int_0^1 \frac{u}{U}\left(-\frac{u}{U}\right) d\eta$$

$$= 0.137\ \delta_L$$ [From equation (1)]

$$= 0.137 \times \frac{4.8\ L}{\sqrt{R_{e_L}}}$$ [From equation (3)]

Substituting in equation (5)

$$F = \rho U^2\ B\left(0.658\ L/\sqrt{R_{e_l}}\right) = \frac{0.658\ \rho U^2\ B\ L}{\sqrt{R_{e_L}}}$$ Ans

13.6. Assume that the shear stress varies linearly in a laminar boundary layer such that

$$\tau = \tau_o\left(1 - \frac{y}{\delta}\right)$$

Calculate the displacement and momentum thickness in terms of δ.

Solution

$$\tau = \tau_o\left(1 - \frac{y}{\delta}\right) = \mu\frac{du}{dy}$$

or, $$\frac{du}{dy} = \frac{\tau_o}{u}\left(1 - \frac{y}{\delta}\right)$$

or, $$du = \frac{\tau_o}{\mu}\left(1 - \frac{y}{\delta}\right) dy$$

Integrating

$$u = \frac{\tau_o}{\mu}\left(y - \frac{y^2}{2\,\delta}\right) \quad (1)$$

Boundary condition

u = U at y = δ

Therefore, $U = \frac{\tau_o}{\mu}\left(\delta - \frac{\delta^2}{2\,\delta}\right)$

$$U = \frac{\tau_o\,\delta}{2\,\mu}$$

or, $$\tau_o = \frac{2\,\mu\,U}{\delta}$$

Substituting the valve of τ_o in equation (1)

$$u = \frac{2\,U}{\delta}\left(y - \frac{y^2}{2\,\delta}\right)$$

or, $$\frac{u}{U} = \frac{2y}{\delta} - \frac{2\,y^2}{2\,\delta^2} = \frac{2y}{\delta}\left(1 - y/2\,\delta\right) = 2\,\eta - \eta^2$$

$$\delta^* = \int_0^\delta \left(1 - \frac{u}{U}\right) dy = \delta \int_0^1 \left(1 - \frac{u}{U}\right) d\,\eta$$

$$= \delta \int_0^1 \left(1 - 2\,\eta + \eta^2\right) d\,\eta$$

$$= \delta \left[\eta - \eta^2 + \frac{\eta^3}{3}\right]_0^1 = \frac{\delta}{3}$$ **Ans**

$$\theta = \int_0^\delta \frac{u}{U}\left(1 - \frac{u}{U}\right) dy = \delta \int_0^1 \left(2\,\eta - \eta^2\right)\left(1 - 2\eta + \eta^2\right) d\,\eta$$

$$= \delta \int_0^1 \left(2\,\eta - 5\,\eta^2 + 4\,\eta^3 - \eta^4\right) d\,\eta$$

$$= \delta \left[\eta^2 - \frac{5}{3}\,\eta^3 + \eta^4 - \eta^{5/5}\right]_0^1 = \frac{2\,\delta}{15}$$ **Ans**

13.7. **The velocity profile in a turbulent boundary layer often is approximated by the "power law" equation**

$$\frac{u}{U} = \left(\frac{y}{\delta}\right)^{1/7}$$

(a) Evaluate the ratio δ^*/δ and compare with the value for the cubic, laminar boundary layer velocity profile.

(b) Evaluate the ratio θ/δ and compare with the value for the cubic laminar boundary layer velocity profile.

(c) Evaluate the shape factor H = δ^*/θ and compare with the value for the cubic laminar boundary layer velocity profile

Solution

(a) We know that

$$\delta^* = \delta \int_0^1 \left(1 - \frac{u}{U}\right) d\,\eta$$

$$\frac{\delta^*}{\delta} = \int_0^1 \left(1 - \eta^{1/7}\right) d\,\eta = \left[\eta - \frac{7}{8}\,\eta^{8/7}\right]_0^1 = \frac{1}{8}$$

For the cubic profile

$$\delta^* = \delta \int_0^1 \left(1 - \frac{3}{2}\,\eta + \frac{1}{2}\,\eta^3\right) d\,\eta$$

$$\frac{\delta^*}{\delta} = \left[\eta - \frac{3}{4}\,\eta + \eta^{4/8}\right] = \frac{3}{8}$$

Thus $\delta^*_{turbulent} = \frac{1}{3}\,\delta^*_{laminar}$ Ans

(b) We know that

$$\theta = \int_0^\delta \frac{u}{U}\left(1 - \frac{u}{U}\right) dy$$

$$= \delta \int_0^1 \frac{u}{U}\left(1 - \frac{u}{U}\right) d\,\eta$$

For turbulent boundary layer

or, $$\frac{\theta}{\delta} = \int_0^1 \eta^{1/7}\left(1 - \eta^{1/7}\right) d\,\eta$$

$$= \int_0^1 \left(\eta^{1/7} - \eta^{2/7}\right) d\,\eta$$

$$= \left[\frac{7}{8}\,\eta^{8/7} - \frac{7}{9}\,\eta^{9/7}\right]_0^1 = \frac{7}{8} - \frac{7}{9}$$

$$= 7/72$$

and for Laminar cubic profile

$$\frac{\theta}{\delta} = \int_0^1 \left(\frac{3}{2} - \frac{1}{2}\eta^3\right)\left(1 - \frac{3}{2}\,\eta - \frac{1}{2}\,\eta^3\right) d\,\eta$$

$$= \int_0^1 \left[\frac{3}{2}\eta - \frac{9}{4}\eta^2 - \frac{1}{2}\eta^3 + \frac{3}{2}\eta^4 - \frac{1}{4}\eta^6\right] d\eta$$

$$= \left[\frac{3}{2}\eta^2 - \frac{9}{12}\eta^3 - \frac{1}{8} + \frac{3}{10}\eta^5 - \frac{1}{28}\eta^7\right]_0^1$$

$$= \frac{-35 + 84 - 10}{280} = \frac{39}{280}$$

Thus $$\frac{\theta_{Laminar}}{\theta_{turbulent}} = \frac{39}{280}/7/72 = \frac{39}{280} \times \frac{72}{7}$$

$$= \frac{2808}{1960} = 1.4326$$

or, $$\theta_{Laminar} = \frac{2808}{1960}\theta_{turbulent}$$ **Ans**

(c) $$H = \frac{\delta^*}{\theta}$$

From the answers obtained from the above analysis.

(i) For the turbulent 1/7 power law.

$$\frac{\delta^*}{\delta} = \frac{1}{8} \text{ and } \frac{\theta}{\delta} = \frac{7}{72}$$

Therefore, $$H = \frac{\delta^*}{\theta} = \frac{1/8}{7/72} = \frac{72}{56}$$

(ii) For the cubic laminar profile

$$\frac{\delta^*}{\delta} = \frac{3}{8} \text{ and } \frac{\theta}{\delta} = \frac{39}{280}$$

Therefore, $$H = \frac{\delta^*}{\theta} = \frac{3/8}{39/280} = \frac{3}{8} \times \frac{280}{39} = \frac{840}{312} = \frac{35}{13}$$ **Ans**

13.8. **Workout the following boundary layer parameters for the velocity distribution prescribed by** $\frac{u}{U} = \left(\frac{u}{\delta}\right)^{1/7}$

(a) Displacement and momentum thickness and the shape factor

(b) Energy thickness. Further evaluate the energy loss due to boundary layer if at a particular section the boundary layer thickness is 20 mm and the free stream velocity is 10 m/s. Take ρ = 1.2 kg/m³

Solution

(a) $$\delta^* = \int_0^\delta \left(1 - \frac{u}{U}\right) dy = \int_0^\delta \left[1 - \left(\frac{y}{\delta}\right)^{1/7}\right] dy = \frac{\delta}{8}$$

$$\theta = \int_0^\delta \frac{u}{U}\left(1 - \frac{u}{U}\right) dy$$

$$= \int_0^\delta \left[\left(\frac{y}{\delta}\right)^{1/7}\right]\left[1 - \left(\frac{y}{\delta}\right)^{1/7}\right] dy = \frac{7}{72}\,\delta$$

$$\text{H} = \text{Shape factor} = \frac{\delta^*}{\theta} = \frac{\delta/8}{\frac{7}{72}\delta} = 1.285$$

(b) $$\delta^{**} = \text{Energy thickness} = \int_0^\delta \frac{u}{U}\left[1 - \left(\frac{u}{U}\right)^2\right] dy$$

$$= \int_0^\delta \left[\left(\frac{y}{\delta}\right)^{1/7}\right]\left[1 - \left(\frac{y}{\delta}\right)^{2/7}\right] dy = \frac{7}{40}\,\delta$$

For the given boundary layer thickness

$$\delta^{**} = \frac{7}{40}\,\delta = \frac{7}{40} \times 20 = 3.5\ mm$$

Energy loss, per unit width, due to boundary layer is

$$= \frac{1}{2}\,\rho\,U^3\,\delta^{**}$$

$$= \frac{1}{2} \times (1.2) \times (10)^3 \times \left(\frac{3.5}{1000}\right) = 2.1\ J/s = 2.1\ W$$

13.9. **A roughened thin board 200 mm wide 2.5 m long moves at 4 m/sec through water. The boundary layer is 40 mm thick both sides at rear end of the board and the velocity distribution is prescribed by the relation**

$$\frac{u}{U} = \left(\frac{y}{\delta}\right)^{1/4}$$

Find the drag force in Newtons and express it as a pure number independent of thickness δ.

Solution

$$F_D = \int_0^\delta \rho\,u\,(U - u) \times B\,dy = \rho\,U^2 \int_0^\delta \frac{u}{U}\left(1 - \frac{u}{U}\right) B\,dy$$

$$= \rho U^2 B \int_0^{\delta} \frac{u}{U}\left(1 - \frac{u}{U}\right) dy$$

$$= \rho U^2 B \int_0^{\delta} \left(\frac{y}{\delta}\right)^{1/4} \left[1 - \left(\frac{y}{\delta}\right)^{1/4}\right] dy$$

$$= \rho U^2 B \left[\frac{4}{5}\delta - \frac{2}{3}\delta\right] = \frac{2}{15} \rho U^2 B \delta$$

$$= \frac{2}{15} \times 1000 \times (4)^2 \times 0.2 \times 0.04 = 17.067 \ N$$

Hence drag force on both sides of the plate is 34.13 N

$$\therefore F_D = C_D \frac{1}{2} \rho U^2 \times Area$$

$$17.06 = C_D \times \frac{1}{2} \times 1000 \times (4)^2 \times (0.2 \times 25)$$

$$C_D = 0.00426 = 4.26 \times 10^{-3}$$

13.10. **Air at standard conditions flows over a flat plate. The flow is uniform at the leading edge of the plate. The velocity profile in the boundary layer is of the form**

$$\frac{u}{U} = 2\eta - \eta^2 + c_1 \quad \text{where } \eta = \frac{y}{\delta}$$

At a section, velocity 20 m/s, length 200 mm and boundary layer thickness 5.7 mm. Is the flow at this section laminar or turbulent ? Why? What boundary conditions must the equation for $\frac{u}{U}$ satisfy, and what is the value of C_1 ? Determine δ^* and τ_o at the given section. Take $\nu = 1.45 \times 10^{-5}$ m²/s and $\mu = 1.78 \times 10^{-5} \frac{kg}{ms}$

Solution

$$R_e = \frac{UL}{\nu} = \frac{20 \times 0.2}{1.45 \times 10^{-5}} = 2.76 \times 10^5$$

The value of the R_e number is less than 5×10^5, therefore the flow is laminar.

Using the boundary conditions

(1) At y = 0, u = 0 therefore, $\eta = 0$, $\frac{u}{U} = 0$

hence $c_1 = 0$

(2) At y = δ, u = U therefore, $\eta = 1$, $\frac{u}{U} = 1$

and

(3) $\dfrac{\partial u}{\partial y} = 0 \qquad \dfrac{\partial (u/U)}{\partial (y/\delta)} = 0$

$$\delta^* = \delta \int_0^1 \left(1 - \frac{u}{U}\right) d\eta = \delta \int_0^1 \left(1 - 2\eta + \eta^2\right) d\eta$$

$$= \delta\left(\eta - \eta^2 + \frac{1}{3}\eta^3\right)_0^1 = \frac{\delta}{3} = \frac{1}{3} \times 5.7 = 1.9\ mm$$

Also $\quad \tau_o = \mu\left(\dfrac{\partial u}{\partial y}\right)_{y=0} = \dfrac{\mu U}{\delta}(2 - 2\eta)_{\eta=0} = \dfrac{2\mu U}{\delta}$

$$= \frac{2 \times 1.78 \times 10^{-5} \times 20}{5.7 \times 1000}$$

$$= \frac{2 \times 1.78 \times 20}{5.7 \times 1000} = 0.1249\ N/m^2$$ Ans

13.11. **Calculate the friction drag on a plate 150 mm wide and 450 mm long placed longitudinally in a stream of oil (specific gravity 0.925 and kinematic viscosity 0.9 stokes) flowing with a free stream velocity of 6 m/s. Also find the thickness of the boundary layer and shear stress at the trailing edge.**

Solution

$$R_{e_x} = \frac{Ux}{\nu} = \frac{6 \times 0.45}{0.9 \times 10^{-4}} = 3 \times 10^4$$

The value is less than 5×10^5, flow over the plate is laminar.

By the Blasius equation $C_D = \dfrac{1.328}{\sqrt{R_e}} = \dfrac{1.328}{\sqrt{3 \times 10^4}} = 0.0077$

$$\therefore F_D = \frac{1}{2}\rho A U^2 C_D = \frac{1}{2} \times 925 \times (0.15 \times 0.45) \times 6^2 \times 0.0077$$

= 8.65 N for one surface = 17.30 N for both sides.

The displace thickness is calculated by

$$\frac{\delta}{x} = \frac{5}{\sqrt{R_{e_x}}}$$

$$\delta = \frac{5x}{\sqrt{R_{e_x}}} = \frac{5 \times 0.45}{\sqrt{3 \times 10^4}} = 0.013\ \text{m} = 13\ \text{mm}$$

Shear stress at the trailing edge

$$\tau_o = \frac{0.332\,\rho U^2}{\sqrt{R_e}} = \frac{0.332 \times 925 \times 6^2}{\sqrt{3 \times 10^4}} = 63.83\ N/m^2$$

13.12. **Oil with free stream velocity of 2 m/sec flows over a thin plate 2 m wide and 2 m long. Calculate the boundary layer thickness and shear stress at the trailing end point and determine the total surface resistance of the plate. Take S = 0.86 and ν = 10^{-5} m²/s.**

Solution

$$R_e = \frac{Ux}{\nu} = \frac{2 \times 2}{10^{-5}} = 4 \times 10^5$$

$$\sqrt{R_e} = 632.45$$

Boundary layer is laminar up to 5×10^5

$$\delta = \frac{5\,x}{\sqrt{R_e}}$$

$$= \frac{5 \times 2}{632.45} = 0.0158\ \text{m}$$

$$\tau_o = \frac{0.332\,\rho U^2}{\sqrt{R_e}} = \frac{0.332 \times 860 \times 2^2}{632.45} = 1.8058\ N/m^2$$

$$F_D = \frac{1}{2}\rho U^2 (l \times b) \times C_D$$

Here $C_D = \dfrac{1.328}{\sqrt{R_e}} = 2.099 \times 10^{-3}$

$$\therefore F_D = \frac{1}{2} \times 860 \times 4 \times 4 \times 2.099 \times 10^{-3} = 14.446\ \text{N}$$

Total on both side of the plate

$$F_D = 14.446 \times 2 = 28.892\ \text{N}$$

13.13. **Oil with a free stream velocity of 3 m/s flows over a thin plate 1.25 m wide and 2 m long. Determine the boundary layer thickness and the shear stress at mid length and calculate the total, double sided resistance of the plate. Take ρ = 860 kg/m³, ν = 10^{-5} m²/s.**

Solution

$$R_{e_x} = \frac{Ux}{\nu} = \frac{3 \times 1}{10^{-5}} = 3 \times 10^5$$

$$\sqrt{R_{e_x}} = 547.72$$

R_e is low enough to allow laminar boundary layer to survive over whole plate

$$\tau_o = \frac{0.332\,\rho U^2}{\sqrt{R_{e_x}}} = \frac{0.332 \times 860 \times 3^2}{547.72} = 4.69\,N/m^2$$

$$\tau_o = 0.332\,\frac{\mu U}{x} \times \sqrt{R_e} = 0.332\,\frac{\mu U}{x} \times \frac{R_e}{\sqrt{R_{e_x}}} = \frac{0.332\,\rho U^2}{\sqrt{R_{e_x}}}$$

F_D = skin friction force is given by, double sided.

$$= 2 \times \frac{1}{2}\rho U^2 (l \times b) \times C_D$$

$$= \frac{2}{2} \times 860 \times 3^2 \times 2 \times 1.25 \times 2.42 \times 10^{-3} = 46.91\,N$$

13.14. For the velocity profile in Laminar boundary layer as

$$\frac{u}{U} = 2\left(\frac{y}{\delta}\right) - \left(\frac{y}{\delta}\right)^2$$

Find the thickness of boundary layer at the end of the plate and the drag force on both side of a plate 1.5 m long and 1 m wide when placed in water flowing with a velocity 125 mm per second. Calculate the value of coefficient of drag also. Take ν = 1 cS.

Solution

$$R_e = \frac{Ul}{\nu} = \frac{0.125 \times 1.5}{1 \times 10^{-6}} = 1.875 \times 10^5$$

As the laminar boundary layer exhibits upto 5×10^5

$$\therefore\ \delta = \frac{5.48\,x}{\sqrt{R_{e_x}}}$$

$$= \frac{5.48 \times 1.5}{\sqrt{1.875 \times 10^5}} = \frac{5.48 \times 1.5}{433.012} = 0.01898\text{ m}$$ Ans

(b) $$F_D = C_D \frac{1}{2}\rho A U^2 \times 2$$

$$= 3.37 \times 10^{-3} \times \frac{1}{2} \times 2 \times 1000 \times 1.5 \times 1 \times (0.125)^2 = 0.07898\,N$$ Ans

(c) $$C_D = \frac{1.46}{\sqrt{R_{e_l}}} = \frac{1.46}{\sqrt{1.875 \times 10^5}} = 3.37 \times 10^{-3}$$

$$F_D = 0.73\,B\,\mu\,u\sqrt{\frac{\rho U l}{\mu}}$$

$$= 0.73 \times 1 \times 1 \times 10^{-3} \times .125 \times \sqrt{1.875 \times 10^5}$$

$$= 0.0395 \text{ N For one side}$$

13.15. For the velocity profile in laminar boundary layer as

$$\frac{u}{U} = \frac{3}{2}\left(\frac{y}{\delta}\right) - \frac{1}{2}\left(\frac{y}{\delta}\right)^3$$

Find the thickness of the boundary layer and the shear stress 1.5 m from the leading edge of a plate. The plate is 3 m long and 2 m wide and placed in water which is moving with a velocity of 150 mm/s. Find the total drag force on the plate. Take $\nu = 1$ cs.

Solution

$$R_e = \frac{Ux}{\nu} = \frac{0.15 \times 1.5}{1 \times 10^{-6}} = 0.225 \times 10^6 = 2.25 \times 10^5$$

Flow is laminar

$$\delta = \frac{4.64\, x}{\sqrt{R_{e_x}}} = \frac{4.64 \times 1.5}{\sqrt{2.25 \times 10^5}} = \frac{4.64 \times 1.5}{474.34} = 0.0146 \text{ m} \qquad \text{Ans}$$

$$\sqrt{R_e} = \sqrt{2.25 \times 10^5} = 474.34$$

$$\tau_o = \frac{0.332\, \rho U^2}{\sqrt{R_e}}$$

$$= \frac{0.323 \times 1000 \times (0.15)^2}{\sqrt{2.25 \times 10^5}} = 0.01532\ N/m^2 \qquad \text{Ans}$$

$$C_D = \frac{1.292}{\sqrt{R_{e_l}}} = \frac{1.292}{\sqrt{2.25 \times 10^5}} = 2.7237 \times 10^{-3}$$

$$F_D = C_D \frac{1}{2} \rho A U^2 \times 2$$

$$= 2.7237 \times 10^{-3} \times \frac{1}{2} \times 1000\,(3 \times 2) \times 2 \times (.15)^2$$

$$= 0.7354 \text{ N} \qquad \text{Ans}$$

13.16. Air is flowing over a flat plate 0.75 m long and 1.5 m wide with a velocity of 3 m/s. The velocity profile over the plate as

$$\frac{u}{U} = \sin\left(\frac{\pi}{2} \times \frac{y}{\delta}\right)$$

Find (a) the boundary layer thickness at the end of the plate, (b) shear stress at 500 mm from the leading edge and (c) drag force on both side of the plate. Take $\nu = 1.4 \times 10^{-5}$ m²/s and $\rho = 1.2$ kg/m³.

Solution

$$R_{e_x} = \frac{Ux}{\nu} = \frac{3 \times 0.75}{1.4 \times 10^{-5}} = 160.714 \times 10^3 = 160.714 \times 10^3$$

$$\sqrt{R_{e_x}} = 400.89$$

$$\delta = \frac{4.795\ x}{\sqrt{R_{e_x}}}$$

$$= \frac{4.795 \times 0.75}{400.89} = 8.97 \times 10^{-3}\ m$$

$$\tau_o \ at \ x = 0.5 = \frac{0.327\ \rho U^2}{\sqrt{R_{e_x}}}$$

$$= \frac{0.327 \times 1.2 \times (3)^2}{327.32} = 0.01078\ N/m^2$$ Ans

$$R_{e_x} = \frac{3 \times 0.5}{1.4 \times 10^{-5}} = 107.142 \times 10^{+3}$$

$$\sqrt{R_{e_x}} = 327.32$$

$$C_D = \frac{1.31}{\sqrt{R_{e_x}}} = 3.26 \times 10^{-3}$$

$$F_D = C_D \frac{1}{2} \rho A U^2 \times 2$$

$$= 3.26 \times 10^{-3} \times \frac{1}{2} \times 1.2 \times (.75 \times 1.5) = 0.0397\ N$$ Ans

13.17. **A smooth plate, 3 m × 1.2 m, moves through air (16°C) at a relative velocity of 1.2 m/s parallel to the plate surface and to its length. Calculate the drag force on one side of the plate (a) assuming laminar conditions, and (b) assuming turbulent conditions over the entire plate (c) for laminar conditions, estimate the thickness of the boundary layer at the middle of the plate and at the trailing edge. Take $\nu = 1.47 \times 10^{-5}$ m²/s and $\rho = 1.22$ kg/m³.**

Solution

(a) $$R_e = \frac{Ux}{\nu} = \frac{1.2 \times 3}{1.47 \times 10^{-5}} = 244897.9592$$

$$\sqrt{R_e} = 494.87$$

$$C_D = \frac{1.328}{\sqrt{R_e}} = \frac{1.328}{494.87} = 2.68 \times 10^{-3}$$

$$F_D = C_D \frac{1}{2} \rho A U^2$$

$$= \frac{0.00268 \times 1.22 \times 3 \times 1.2 \times 1.2^2}{2} = 8.4748 \times 10^{-3}\ N$$

$$= 0.0084748 \text{ N}$$

(b) For turbulent flow conditions, with $R_e < 10^7$

$$C_D = \frac{0.074}{(R_e)^{0.2}} = \frac{0.074}{(244897.9592)^{0.2}}$$

$$= \frac{0.074}{11.9618} = 6.186 \times 10^{-3} = 0.006186$$

$$F_D = C_D \frac{1}{2} \rho A U^2$$

$$= 0.006186 \times \frac{1}{2} \times 1.22 \times (3 \times 1.2) \times 1.2^2 = 0.01956 \text{ N}$$

(c) For x = 1.5 m, $R_{e_x} = \dfrac{1.2 \times 1.5}{1.47 \times 10^{-5}} = 122445.9796$

$$\sqrt{R_e} = 349.927$$

$$\delta = \frac{5 \times x}{\sqrt{R_e}} = \frac{5 \times 1.5}{349.927} = 0.02143 \text{ m} \qquad \text{Ans}$$

For x = 3 m

$R_e = 244897.9592$

$\sqrt{R_e} = 494.87$

$$\delta = \frac{5 \times 3}{494.87} = 0.0303 \text{ m} \qquad \text{Ans}$$

13.18. Water flows down a smooth wide concrete apron into a river. Assuming that a turbulent boundary layer forms, estimate the shear stress and the boundary layer thickness 50 m downstream of the entrance to the apron. Use the following data.

U = 7 m/s, $\mu = 1.14 \times 10^{-3}$ kg/ms

$$\frac{u}{U} = \left(\frac{y}{\delta}\right)^{1/7}, \quad \tau_o = \frac{0.0225\, \rho\, u^2}{\left(R_{e_s}\right)^{1/4}}$$

Solution

We know that

$$\tau_o = \rho U^2 \frac{d\theta}{dx}$$

and $$\theta = \int_0^\delta \frac{u}{U}\left(1-\frac{u}{U}\right)dy$$

$$= \int_0^\delta \left(\frac{y}{\delta}\right)^{1/7}\left[1-\left(\frac{y}{\delta}\right)^{1/7}\right]dy = \frac{7}{72}\delta$$

Therefore,

$$\therefore \tau_o = \frac{d}{dx}\left[\rho U^2 \frac{7\,\delta}{72}\right] = \frac{7}{72}\rho U^2 \frac{d\delta}{dx}$$

But $$\tau_o = \frac{0.0225\,\rho U^2}{\left(R_{e_s}\right)^{1/4}}$$

$$\therefore \tau_o = \frac{0.0225\,\rho U^2}{\left(R_{e_s}\right)^{1/4}} = \frac{7}{72}\rho U^2 \frac{d\delta}{dx}$$

$$= \frac{0.2314}{\left(R_{e_s}\right)^{1/4}} = \frac{d\delta}{dx}$$

$$R_{e_s} = \frac{\rho U \delta}{\mu} = \frac{1000 \times 7\,\delta}{1.14\times 10^{-3}}$$

$$(\mathrm{Re}_\delta)^{1/4} = 49.78\,\delta^{1/4}$$

$$\therefore\ dx = \frac{d\delta\left(R_{e_s}\right)^{1/4}}{0.2314}$$

$$x = \int \frac{49.78\,\delta^{1/4}}{0.2314}\,d\delta = \frac{39.82\,\delta^{5/4}}{0.2314}$$

$$\delta = \left(\frac{0.2314\,x}{39.82}\right)^{4/5}$$

$$x = 50 \text{ m}$$

$$\therefore \delta = \left(\frac{0.2314\times 50}{39.82}\right)^{4/5} = 0.372 \text{ m}$$

$$\tau_o = \frac{0.0225\,\rho U^2}{\left(R_{e_\delta}\right)^{1/4}}$$

$$= \frac{0.0225\times 1000\ 7^2}{49.78\times(0.372)^{1/4}} = 28.358\ N/m^2$$ **Ans**

EXERCISES

A. THEORY

13.1. What do you mean by boundary layer? Illustrative with reference to flow over a flat plate.

13.2. Explain the development of boundary layer along a flat plate held parallel to uniform flow. Point out the salient features.

13.3. Flow takes place over a flat plate exposed parallel to free stream. Draw a neat sketch of the boundary layer development showing, (i) Laminar, boundary layer (ii) Turbulent boundary layer, (iii) Transition zone and (iv) Laminar sub layer.

13.4. Write a brief note on Prandtl's boundary layer theory.

13.5. Draw a neat sketch of boundary layer formation on a flat plate and mention the characteristics of flow within the boundary layer along the plate length.

13.6. Explain the characteristics of laminar and turbulent flows that one can obtain within a circular pipe and within boundary layer over a flat plate. What is the significance of critical boundary layer.

13.7. Explain the significance of the following terms: (i) Laminar boundary layer, (ii) Turbulent boundary layer and (iii) Laminar sub layer.

13.8. What is laminar sub layer? Show it in a neat sketch of flow over a flat plate.

13.9. Define the terms boundary layer thickness, displacement thickness and momentum thickness and discuss their practical utility.

13.10. Differentiate between the characteristics of laminar and turbulent boundary layers.

13.11. Differentiate between the characteristics of laminar and Laminar sub layer.

13.12. What is boundary layer? What are the characteristics of boundary layer developing over a flat plate. Define displacement and momentum boundary layer thickness.

13.13. Differentiate between boundary layer thickness and nominal thickness.

13.14. What is the significance of boundary layer separation in fluid flow? Why is it necessary to prevent separation of boundary layer? What are the methods to prevent separation of boundary layer?

13.15. What is boundary layer separation? What is its effect in aircraft propulsion and how can it be controlled?

13.16. Derive the equation for displacement thickness and momentum thickness for boundary layer and compare the both.

13.17. Why boundary layer separation is not desired in general? Discuss the different methods used for its prevention.

13.18. Describe the process of growth of boundary layer on a flat plate for zero pressure gradient. Derive expressions for various thickness parameters used to specify the boundary layer.

13.19. What conditions should be satisfied for separation of boundary layers? Discuss briefly the methods that can be used to prevent separation.

13.20. Describe the development of boundary layer over a flat plate for zero pressure gradient. Explain with a neat sketch the phenomenon of boundary layer separation.

13.21. Derive the equation for momentum thickness for boundary layer. Explain its significance. Also explain the development of boundary layer over a flat plate.

13.22. Starting from Navier stokes equation for two dimensional steady incompressible flow, obtain Prandtl boundary layer equation for a flat plate.

13.23. What is boundary layer control? What are methods to control boundary layer ? Discuss briefly.

13.24. What conditions should be satisfied for separation of boundary layers? Discuss briefly the methods that can be used to prevent separation.

13.25. What are the three forces which affect the motion of fluid in the boundary layer?

13.26. Boundary layer separation depends upon the three conditions. Explain?

13.27. Draw a neat sketch of duct consisting of nozzle, throat and diffuser. Where the adverse pressure gradient is excessive? Explain the diffuser stall.

13.28. Why we are not using the large angle in the diffuser ? Explain the meaning of diffuser stall.

13.29. Draw a neat sketch of boundary layer formation on a Flat plate for adverse pressure gradient.

13.30. How will you determine whether a boundary layer flow is attached flow, detached flow or on the verge of separation?

13.31. What do you mean by separation of boundary layer ? What is the effect of adverse pressure gradient on boundary layer?

13.32. Starting with the Navier-Stokes equations of motion for two dimensional incompressible flow, obtain the Prandtl's boundary layer equations. Give a brief outline of the Blasius solution of laminar boundary layer for flow over a flat plate.

13.33. Explain the essential feature of Blasius method of solving laminar boundary layer equations for a flat plate. Derive expressions for boundary layer thickness and local skin friction coefficient from this solution.

13.34. Derive the Von Karman momentum integral equation for flow past a flat plate for the zero pressure gradient in the form as

$$\tau_o = \rho U^2 \frac{d\theta}{dx}$$

13.35. Derive the von Karman momentum integral equation for the boundary layer on a flat plate in the non dimensional form as,

$$\frac{\tau_o}{\frac{1}{2}\rho U^2} = 2\frac{d\theta}{dx} + \frac{2d}{dx}\int_0^d \frac{u}{U}\left(1-\frac{u}{U}\right)dy$$

13.36. Derive von Karman momentum integral equation for boundary layer flow over a flat plate in the form.

$$\frac{\tau_o}{\rho U^2} = \frac{d\theta}{dx} + (2\theta + \delta^*)\frac{1}{U}\frac{dU}{dx}$$

Write down clearly the assumption made.

13.37. Define boundary layer thickness. Derive Karman's momentum integral equation to determine the wall shear stress.

13.38. Establish a relation for skin coefficient for turbulent boundary layer as

$$C_f = \frac{0.0594}{\left(R_{e_x}\right)^{1/5}}$$

13.39. Establish the following correlation for drag coefficient over a flat plate

$$C_D = \frac{0.074}{\left(R_{e_x}\right)^{0.2}} - \frac{1610}{R_{e_l}}$$

13.40. Draw a curve between C_D and R_e for a smooth flat plate parallel to the flow and discuss it.

13.41. Draw a curve between C_D and R_e for a rough flat plate parallel to the flow and discuss it.

13.42. Derive a relation for turbulent boundary layer over a flat plate for boundary layer thickness for 1/7 power law.

$$\frac{\delta}{x} = \frac{0.371}{\left(R_{e_x}\right)^{0.2}}$$

13.43. Assume $\frac{u}{U} = \left(\frac{y}{\delta}\right)^{1/8}$ in a turbulent boundary layer. Calculate the boundary layer thickness δ / x and the drag coefficient C_D.

13.44. How does the ratio of displacement thickness to boundary layer thickness (δ^* / δ) for a laminar boundary layer compare with that for a turbulent boundary layer? Determine a qualitative answer from an inspection of the velocity profiles within the boundary layers.

13.45. Estimates the ratio of the turbulent boundary layer thickness just after transition at $R_{ex} = 5 \times 10^5$ to the laminar boundary layer thickness just before transition.

13.46. The velocity profile for a laminar boundary layer is to be approximated by the expression.

$$\frac{u}{U} = a + b + \eta + c + \eta^m \text{ where } \eta = \frac{y}{\delta}$$

Evaluate a, b and c for the situation where m = 1.5, Compare the resulting profile with the Blasius.

13.47. Consider a laminar boundary layer on a flat plate with a velocity profiles given by

$$\frac{u}{U} = \frac{3}{2}\eta - \frac{\eta^3}{2}, \ \eta = \frac{y}{\delta}$$

for this profile $\delta x = \dfrac{4.64}{\sqrt{R_{e_x}}}$

Determine the expression for δx

13.48. The velocity profile in a turbulent boundary layer often is approximated by the power law equation

$$\frac{u}{U} = \left(\frac{y}{\delta}\right)^{1/7}$$

Compare the shape of this profile with the parabolic laminar boundary layer profile by plotting $\dfrac{y}{\delta}$ versus $\dfrac{u}{U}$ for both profiles.

13.49. The velocity profile describing laminar boundary layer is given as

$$\frac{u}{U} = \sin\left[\frac{\pi}{2}\left(\frac{y}{\delta}\right)\right]$$

Show that boundary layer thickness δ, wall shear stress τ_o and coefficient of drag C_D are given by

$$\delta = \frac{4.795\,x}{\sqrt{R_{e_x}}} >, \ \tau_o = 0.327\frac{\rho U^2}{\sqrt{R_{e_x}}} \ C_D = \frac{1.31}{\sqrt{R_{e_l}}}$$

Where symbol have their usual meaning.

13.50. If the velocity distribution in a laminar boundary layer on a flat plate is

$$\frac{u}{U} = a + b + \eta + c\,\eta^2 + d\,\eta^3$$

Where $\eta = \dfrac{y}{\delta}$, determine the coefficients. a, b, c and d.

B. UNSOLVED PROBLEMS

13.1. Find the displacement thickness the momentum thickness, energy thickness and the ratio δ*/θ for the velocity distribution in the boundary layer given by

$$\frac{u}{U}=\left(\frac{y}{\delta}\right) \qquad \left[\delta^*=\frac{\delta}{2},\delta^{**}=\frac{\delta}{4},\theta=\frac{\delta}{6},\frac{\delta^*}{\theta}=3\right]$$

13.2. Find the displacement thickness, the momentum thickness energy thickness and δ*/θ for the velocity distribution in the boundary layer given by

$$\frac{u}{U}=2\left(\frac{y}{\delta}\right)-\left(\frac{y}{\delta}\right)^2 \qquad \left[\delta^*=\frac{\delta}{3},\ \delta^{**}=\frac{22}{105}\delta,\theta=\frac{2\delta}{15},\frac{\delta^*}{\theta}=\frac{5}{2}\right]$$

13.3. The velocity distribution in the boundary layer over a high spillway surface was found to be

$$\frac{u}{U}=\left(\frac{y}{\delta}\right)^{0.22}$$

Express the ratio between momentum thickness and displacement thickness.

$$[\delta^*=0.1803\,\delta,\theta=0.1251\,\delta \quad \therefore \delta^*\theta=0.694]$$

13.4. Calculate the friction drag on a plate 15 cm wide and 45 cm long placed longitudinally in a stream of oil (specific gravity 0.925 and kinematic viscosity 0.9 stokes) flowing with a free stream velocity of 6 m/sec. Also find the thickness of the boundary layer and shear stress at the trailing edge.

[$R_e = 3 \times 10^4$ $C_D = 0.0077$ $F_D = 17.30$ N for both sides $\tau_o = 63.85$ N/m^2]

13.5. Oil with a free stream velocity of 3 m/s flows over a thin plate 1.25 m wide and 2 m long. Determine the boundary layer thickness and the shear stress at mid length and calculate the total, double sided resistance of the plate.

Take $\rho=860$ kg/m^3, $\nu=10^{-5}$ m^2/s

$$\left[Re=3\times10^5,\tau_o=4.69\text{ N/m}^2,F_D=46.91\text{ N}\right]$$

13.6. A 1.2 m by 1.2 m thin plate is held to a stream of air moving at 3 m/s (standard conditions). Calculate (a) the surface drag of the plate, (b) the thickness of the boundary layer at the trailing edge and (c) the shear stress at the trailing edge. Take $\nu = 1.486 \times 10^{-5}$ m²/s $\rho = 1.2$ kg/m³.

$$\left[R_e=242261.1036,\ C_D=2.698\times10^{-3},F_D=0.04195\text{ N}\right]$$

$$\delta=0.01219\text{ m},\ \tau=0.00728\text{ N/m}^2$$

14

Flow Over Immersed Bodies

14.1. INTRODUCTION

Whenever a relative motion occurs between an immersed solid body and its surrounding fluid, the former experiences a resistance. Thus a force is required to hold a solid body in its stationary position when it is immersed in a moving fluid. This force counterbalances the fluid resistance. In this chapter we consider various aspects of the flow over bodies that are immersed in a fluid. The knowledge of the a nature of fluid resistance is important in the design of immersed bodies. Examples include the flow of air around aeroplane, automobiles, flow of water around submarines, water droplets in fog, fluid force on the surface vessels (surrounded by two fluids air and water) wind forces on the proper design of building (house or a tall skyscraper), topographic features such as mountains and valleys, bridges etc.

The two approaches (theoretical and experimental) are used to obtain information on the fluid forces developed by the external flows. Theoretical (i.e. numerical and analytical) techniques can provide much of the need information about such flows. However, because of the complexities of the governing equations and the complexities of the geometry of the objects involved, the amount of information obtained from purely theoretical methods is limited. Now a days anticipated advancements in the area of computational fluid mechanics, it is likely that computer prediction of forces and complicated flow patterns will become more readily available.

Much of the information about external flows comes from experiments carried out, for the most part, on scale models of the actual objects. Such testing includes the wind tunnel testing of model aeroplanes, buildings and even entire cities. In some instances the actual device, not a model, is tested in wind tunnels. Better performance of cars, motorcycles, bikes, skiers and numerous other objects has resulted from testing in wind tunnels. The use of water tunnels and towing tanks also provides useful information about the flow around ships and other objects.

Since the resistance results from the relative motion between the immersed solid body and its surrounding fluid, it makes no difference whether the body moves through the fluid or the fluid flows about the immersed body.

In this chapter we consider characteristics of external flows past a variety of objects. We investigate the qualitative aspect of such flows and learn how to determine the various forces on objects surrounded by a moving fluid.

14.2 FORCES ON IMMERSED BODIES-DRAG AND LIFT

When an extensive, incompressible viscous fluid flows past a body submerged in it or when a body moves through this fluid at rest, two types of forces act over the surface of the body. Fig. 14.1 shows a body submerged in a moving fluid. The various forces experienced by the body can be resolved into two components-one along the line of movement of the fluid (or of the body) and one perpendicular to the line of movement of the fluid (or of the body). These are known as drag and lift, respectively, and they are of great importance in many engineering analyses and/or design problems.

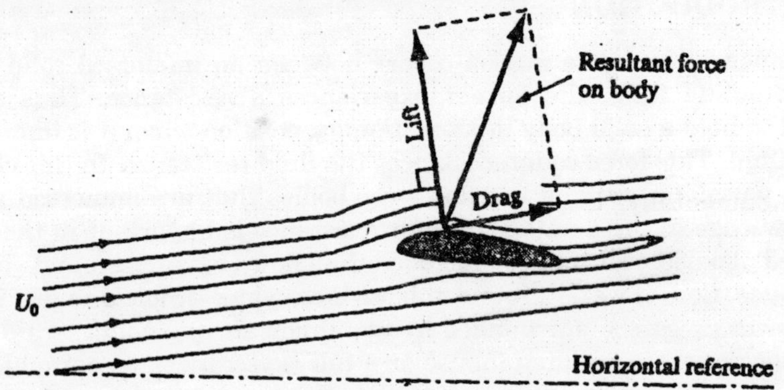

Fig. 14.1. Drag and Lift on body

Drag is, therefore, a force caused by fluid on an submerged (immersed) body along the line of movement of the body or fluid. Drag force resists the movement of the body or fluid. An example of drag is wind resistance on moving automobiles, aeroplanes, trains etc. Engineers usually try to minimize drag.

Lift is a force caused by fluid on submerged (immersed) body in a direction perpendicular to the line of movement of the body or fluid. An example of lift force is the upward force of air on a flying aeroplane's wings (airfoils).

Fig. 14.2 illustrates the direction of the lift force. The arrow V represents the direction of flight, the arrow L represents the lift acting upwards and the arrow W the weight of the aircraft, and shows the downward vertical. Comparison of (a) and (c) shows that this upwards is not fixed relative to the aircraft, while (a), (b) and (d) show that the meaning is not fixed relative to the earth. As a general rule, if it is remembered that the lift is always a component perpendicular to the flight direction, the exact direction in which the lift acts will be obvious, particularly after reference to Fig. 14.2.

In a steady flow of an ideal (inviscid) fluid of infinite extent, only forces from pressure exist and the drag force in all instances except free cavity flows is zero. Lift, however, may be produced in an ideal fluid by superimposing an irrotational vortex, or circulation, around the body. In fact, vortices, or circulation are required in order to produce lift in a viscous fluid as well as in an ideal fluid.

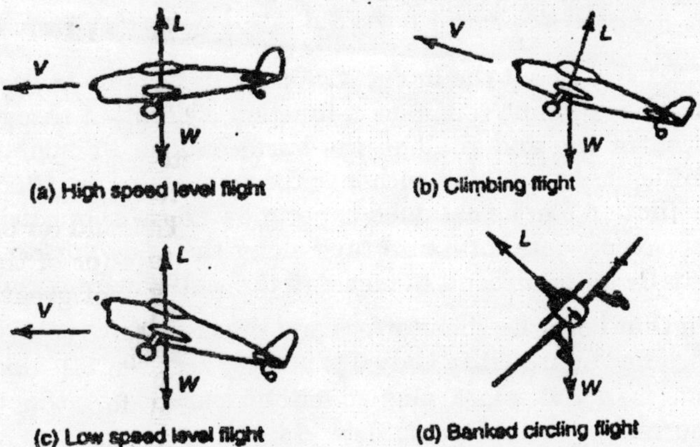

Fig. 14.2. The direction of the lift force

Fundamentally, the total force on each element of the surface can be resolved into normal and a tangential component as shown in Fig. 14.3. This can be described in terms of the stresses (1) wall shear stress τo, due to viscous effects and (2) normal stresses due to pressure, p. Both τo and p vary in magnitude and direction along the surface.

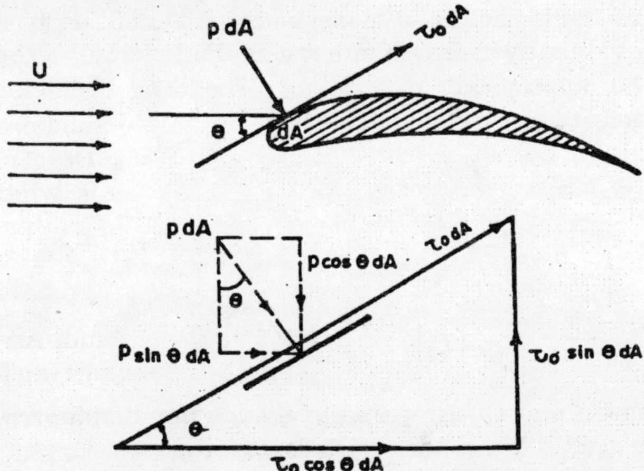

Fig. 14.3. Pressure and shear forces on a small element of the surface of a body

The resultant of the pressure and shear stress distribution can be obtained by integrating the effect of these two quantities on the body surface as is indicated in Fig. 14.3. The x and y components of the fluid force on the small area element dA are.

$$dF_x = pdA\ \sin\theta + \tau_o\ dA\ \cos\theta$$

and $$dF_y = pdA\ \cos\theta + \tau_o\ dA\ \sin\theta$$

Which may be integrated to yield

$$F_D = \int dF_x = \int p \sin\theta \, d\,A + \int \tau_o \cos\theta \, d\,A \tag{14.1}$$

$$F_L = \int dF_y = -\int p \cos\theta \, d\,A + \int \tau_o \sin\theta \, d\,A \tag{14.2}$$

Of course, to carry out the integrations and determine the drag and lift, we must know the body shape (i.e., θ as a function of location along the body) and the distribution of τo and p along the surface. These distribution are often extremely difficult to obtain, either experimentally or theoretically. The pressure distribution can be obtained experimentally without too much difficult by use of a series of static pressure taps along the body surface. On the other hand, it is usually quite difficult to measure the wall shear stress distribution.

It is seen that both the shear stress and pressure force contribute to the lift and drag, since for an arbitrary body θ is neither zero nor 90º along entire body. The exception is a flat plate aligned either parallel to the upstream flow $(\theta = 90°)$ or normal to the upstream flow $(\theta = 0°)$

Although equations (14.1) and (14.2) are valid for any body, the difficulty in their use lies in obtaining the appropriate shear stress and pressure distribution on the body surface. Considerable effort has gone into determining these quantities, but because of the various complexities involved, such information is available only for certain simple situations.

Without detailed information concerning the shear stress and pressure distribution on a body, equations (14.1) and (14.2) can not be used. The widely used alternative is to define dimensionless lift and drag coefficient and determine their values by means of either a simplified analysis, some numerical technique, or an appropriate experiment. The drag coefficient C_L and lift coefficient are defined as

$$C_D = \frac{F_D}{\frac{1}{2}\rho A U^2} \tag{14.3}$$

and

$$C_L = \frac{F_L}{\frac{1}{2}\rho A U^2} \tag{14.4}$$

Where both C_D and C_L are respectively the nondimensional drag and lift coefficients. The area A is a characteristic area, either the largest projected area of the immersed body or the projected area perpendicular to the direction of flow. The term $\rho U^2/2$ is the dynamic pressure of the fluid of density ρ which is moving with a uniform velocity U. Except for a few simple cases, both c_D and c_L are determined experimentally.

The resultant force on the body is, of course, obtained by compounding drag and lift

$$F = \sqrt{F_L^2 + F_D^2} = \frac{1}{2}\rho A U^2 \sqrt{C_L^2 + C_D^2} \tag{14.5}$$

The equations (14.3) and (14.4) which give the mathematical expression for coefficient of drag and lift are derived by the method of dimensional analysis.

14.3 DRAG ON IMMERSED BODIES

Most of the information pertaining to drag on objects is a result of numerous experiments with wind tunnels, water tunnels, towing tanks and other ingenious devices that are used to measure the drag on scale models. These data can be put into dimensionless form and the results can be appropriately rationed for prototype calculations. Typically the result for a given shaped object is a drag coefficient C_D, is given in equation (14.3), and C_D is a function of other dimensionless parameters such as Reynold's number, Mach number, Froude number, and relative roughness of the surface. That is,

$$C_D = \Phi\left(Shape, R_e, M, F_r, \in/l\right)$$

The relative significance of the Reynold's and the Mach number depends on whether the flow is considered incompressible or compressible. For incompressible flow only the Reynold's numbers need be considered. The effect of the Mach number becomes important only when the velocity of flow approaches or exceeds the sonic velocity.

14.4 TYPES OF DRAG

Depending on the nature of fluid flow and the shape of the body drag forces may be classified as follows :

(1) Skin friction drag
(2) Pressure drag
(3) Profile drag
(4) Deformation drag
(5) Wave drag
(6) Induced drag

14.4.1 Skin friction drag

Skin friction drag (shear drag or viscous drag or friction drag), is that part of the drag that is due directly to the shear stress, τ_o, on the object. It is a function of not only the magnitude of the wall shear stress, but also of the orientation of the surface on which it acts. This is indicated by the factor $\tau_o \cos\theta$ in equation (14.1). If the surface is parallel to the upstream velocity the entire shear force contribute directly to the drag (Fig. 14.4). This is true for the flat plate parallel to the flow. If the surface is perpendicular to the upstream velocity, the shear stress contributes nothing to the drag. For highly streamlined bodies or for low Reynold's number flow, however, most of the drag may be due to friction drag.

Fig. 14.4. Thin plate parallel to the flow. (Friction drag)

The friction drag on a flat plate of width b and length l oriented parallel to the upstream flow can be calculated from

$$F_{D_f} = C_{D_f} \frac{1}{2} \rho (bl) U^2$$

Where C_{D_f} is the friction drag coefficient. The value of the drag coefficient depends on the flow-whether it is laminar or turbulent. In laminar flow, the drag coefficient depends only on the Reynold's number and is not affected by the roughness of the plate. In turbulent flow, the drag coefficient is influenced by surface roughness, the location of transition from a laminar to a turbulent boundary layer, and the level of turbulence of free stream in addition to its dependence on the Reynold's number of flow. As with pipe flow, this dependence is a result of the surface roughness elements protruding into or through the laminar sub layer.

Most objects are not flat plates parallel to the flow; instead, they are curved surfaces along which the pressure varies.

The precise determination of the shear stress along the surface of a curved body is quite difficult to obtain. Although approximate results can be obtained by a variety of techniques.

14.4.2 Pressure drag

Pressure drag is that part of the drag that is due directly to the pressure, p, on an object. It is often referred to as form drag because of its strong dependency on the shape or form of the object. Pressure drag is a function of the magnitude of the pressure and the orientation of the surface element on which the pressure force acts.

The pressure force on either side of a flat plate parallel to the flow may be very large, but it does not contribute to the drag because it acts in the direction normal to the upstream velocity. On the other hand, the pressure force on a flat plate normal to the flow provides the entire drag (Fig. 14.5). The pressure drag resulting from difference of pressure over the surface. This force is known as the pressure drag. This is indicated by the factor p sinθ dA in equation (14.1).

As a general rule, skin friction drag predominates with thin plates parallel to the direction of flow (Fig. 14.4) and with well streamlined objects. Pressure drag predominates with spheres, cylinders with their axes perpendicular to the direction of flow, and disks and thin plates perpendicular to the direction of flow (Fig. 14.5).

In almost all cases in which flow takes place round a solid body, the boundary layer separates from the surface at same point. **Downstream of the separation position the flow is greatly disturbed by large scale eddies and this region of eddying motion is known usually as the wake.** As a result of the energy dissipated by the highly turbulent motion in the wake the pressure there is reduced and the pressure drag on the body is thus increased. The magnitude of the pressure drag depends very much on the size of the wake and this, in turn, depends on the position of separation.

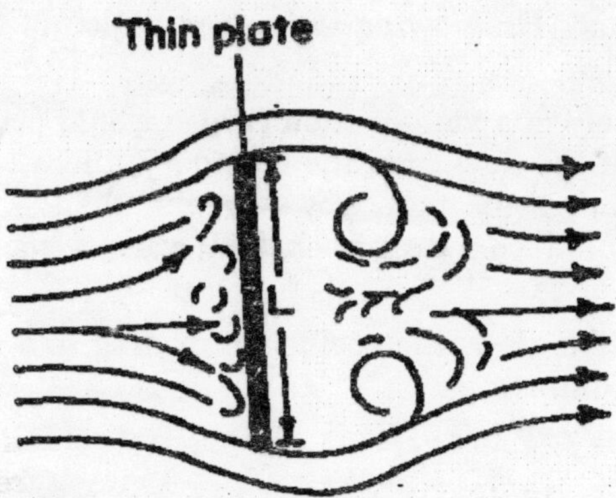

Fig. 14.5. Thin plate perpendicular to the flow (Pressure drag)

14.4.3 Profile drag

Profile drag or total drag on the body, is, therefore, made up of two contributions, namely the pressure (form) drag and the skin friction (viscous) drag.

Thus

Profile drag = Pressure drag + Skin friction drag

The relative contribution of pressure drag and shear drag to the profile drag depends upon the shape of the body and its orientation with respect to the flow. Take, for example, a thin plate. If it is held in a fluid stream as shown in Fig. 14.3, the pressure drag will be negligible. The skin friction drag, however, will be substantial, due to the formation of the boundary layer on both sides of the plate.

If however, the thin plate is held perpendicular to the flow as shown in Fig. 14.5, the drag will be almost entirely due to pressure difference whereas the skin friction drag will be negligible.

14.4.4 Deformation drag

In a laminar flow at very law Reynold's number i.e. R_e less than 0.1, the viscous forces alone needed be considered and the inertia forces may be neglected. The resistance caused by this type of motion is known as deformation drag.

The deformation drag exists in the case of very small object moving at very small velocities through fluids at large viscosity.

According to C.G. Stokes law which is valid upto $R_e = 0.1$, the coefficient of drag $C_D = 24/R_e$.

14.4.5 Wave drag

Waves are generated at a liquid surface as the result of the motion of a body (a ship, a submarine, or a hydrofoil, etc.) on or near that surface. Since energy is required to propagate those waves, a force must be developed between the body and the fluid. Drag owing to the generation of these waves is called wave drag.

With a completely non viscous fluid (which would produce no skin friction and no separation of the flow from the boundary) there would be zero drag in subsonic flow. In supersonic flow, however, the change of pressure across a shock wave, would produce a drag even with a non viscous fluid. This drag is also known as wave drag.

For a compressible flow, the coefficient of drag is a function of Reynold's number and Mach number.

$$C_D = f\left(R_e . M\right)$$

Wave drag is not measured directly, but it is taken as the residual drag remaining after all drag which can be calculated or estimated is subtracted from the total measured drag. These are, of course, sources of uncertainity in these estimates.

14.4.6 Induced drag

Induced drag is caused by a secondary flow along the length of an aerofoil or a lifting vane. The phenomenon of induced drag due to formation of the trailing vortices and the relevant equation for calculation of induced drag coefficient in terms of lift coefficient and aspects ratio of a wing has been given there, will be discussed later on in the article of lift.

14.5 STREAMLINED AND BLUFF BODIES

If the shape of the body is such that the separation occurs only well towards the rear, and the wake is small, the pressure drag is also small, and it is skin friction that makes the major contribution to the total drag. Such a body is termed as streamlined body or well shaped body, e.g. airfoil. Fig. 14.6 (a).

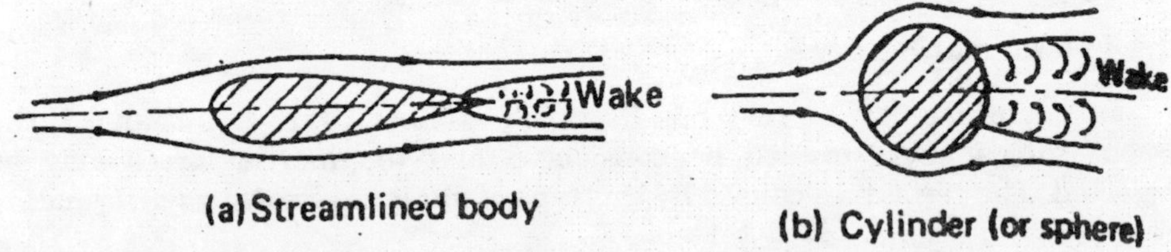

Fig. 14.6. Streamlined and bluff bodies

For a bluff body, on the other hand, the flow is separated over much of the surface, the wake is large and the pressure drag is much greater than the skin friction, [Fig. 14.6 (b)] e.g. Flat plate held normal to the flow, or square body etc. In nutshell, a body on which wake formation are minimum is known as streamlined body, and a body on which wake formation are maximum is known as bluff body.

The body of maximum bluffness is a thin plate normal to the flow (Fig. 14.5).

The object of streamlining a body is to move the point of separation as far back as possible and thus to provide the minimum size of wake; this way the pressure drag is reduced. But the friction drag is increased by making the body longer to promote gradual increase in pressure along the length of the body. Hence, the optimum amount of streamlining is that, for which, the sum of pressure drag and skin friction drag is minimum.

For flows in which viscous effects are large relative to inertial effects (i.e. very small Reynold's number flows), it is found that $C_D = 1/R_e$. This characteristics are similar to the friction factor dependence of $\lambda \sim 1/R_e$ for laminar pipe flow and $\lambda \sim$ constant for large Reynold's number flow.

If the viscosity were zero, the pressure drag on any shaped object (symmetrical or not) in a steady flow would be zero.

14.6 DRAG ON SPHERE

Stokes obtained the solution for drag by expressing the simplified Navier-Stokes equation together with the continuity equation in polar coordinates and using the boundary conditions that all velocity components are zero at the surface of sphere. His solution is well known equation.

$$F_D = 3\pi \, \mu \, U \, D \tag{14.6}$$

This relationship holds true for $R_e < 0.1$ but may be used with negligible error upto $R_e = 0.2$. In this range, often referred to as stokes flow. Stokes further observed that the shear drag is two third of the total drag and pressure drag is one third of the total drag.

Expressing the drag force in the standard form

$$F_D = C_D \frac{1}{2} \rho \, A \, U^2 \tag{14.7}$$

Here, $A = \frac{\pi}{4} D^2$ = Projected area of the sphere.

Equating equations (14.6) and (14.7)

$$C_D \frac{1}{2} \rho U^2 \times \frac{\pi}{4} D^2 = 3\pi \, \mu \, U \, D$$

$$C_D = \frac{24\mu}{\rho U D} = \frac{24}{R_e} \tag{14.8}$$

The equation (14.8) is valid upto $R_e < 0.2$.

At larger values of R_e, separation of the boundary layer occurs and the Navier Stokes equations cannot be used. It is therefore, necessary to rely on empirical expressions.

Corl W. Oseen (1911) made an improvement in the Stokes solution by including in an approximate manner, the inertia terms Stokes omitted. Oseen's solution gives

$$C_D = \frac{24}{R_e}\left(1+\frac{3}{16}R_e\right) \tag{14.9}$$

which is fairly valid upto $R_e = 5$

It is interesting to note that experimental results lie midway between Stokes and Oseen's curve (Fig. 14.7). Thus, since the plot is logarthmic, the experimental data follow the equation (known as Oseen's correction)

$$C_D = \frac{24}{R_e}\left(1+\frac{3}{16}R_e\right)^{1/2} \tag{14.10}$$

quite accurately for Reynold's numbers upto 100.

Beyond $R_e = 100$ it is necessary to use values of c_D as a function of R_e from the graph, as in Fig. 14.7.

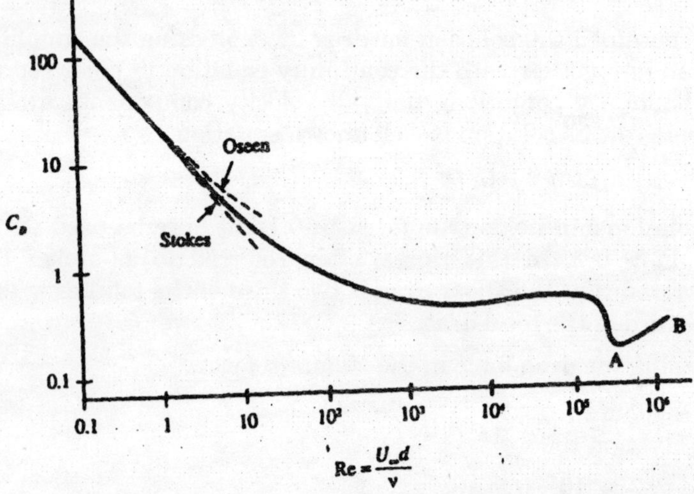

Fig. 14.7. Drag coefficient of a sphere as a function of Reynolds number

At $R_e \approx 1000$ a stable separation position is achieved at about 80° from the front stagnation point. Pressure drag increasingly takes precedence over skin friction, and C_D gradually becomes independent of R_e as shown in Fig. 14.7. The relative contribution of skin friction drag decreases with increasing Reynold's number, at $R_e = 1000$, the friction drag is approximately 5 percent of the total drag. The value of C_D at $R_e = 1000$ is 0.4.

The drag coefficient of the sphere ranges (approximately) between 0.4 and 0.5 from $R_e \approx 1000$ to $R_e \cong 2 \times 10^5$, at which point it suddenly drops more than 50% and then increases gradually with further increase in the Reynold's number (Fig. 14.7 and 14.8). In this range of Reynold's numbers, experiments have shown the separation point to be upstream from the midsection of the sphere, resulting in a wide turbulent wake, the boundary layer on the surface of the sphere from stagnation point to separation point has been found to be laminar up to $R_e = 2 \times 10^5$. In the nutshell the C_D varies approximately 0.4 to 0.5, practically independent of Reynold's number and the flow is laminar.

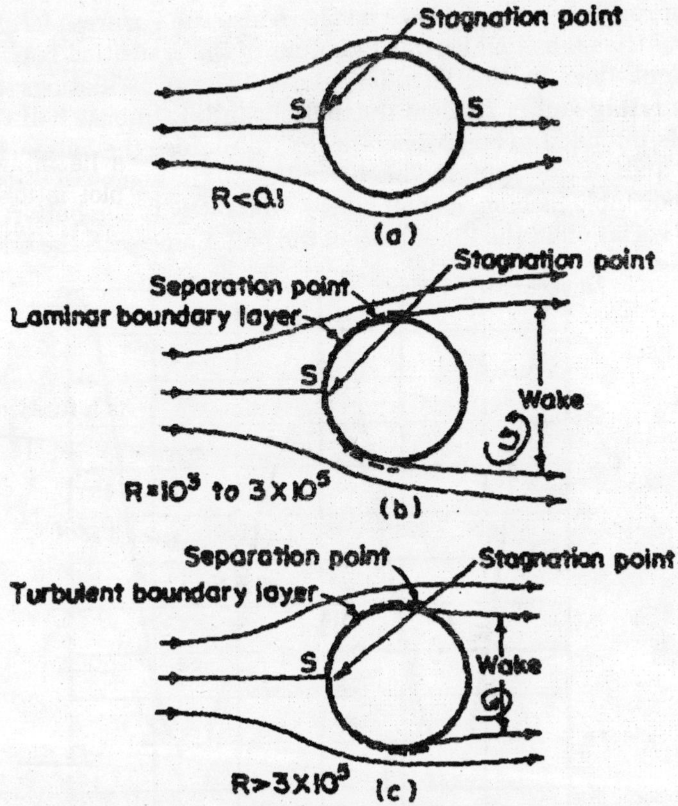

Fig. 14.8. Flow about a sphere at various Reynold's number

For Reynold's number larger than about 2.5×10^5, transition occurs in the boundary layer on the forward portion of the sphere.

At $R_e \cong 3 \times 10^5$, the boundary layer becomes turbulent before separation. A turbulent boundary layer can withstand an adverse pressure gradient better than a laminar boundary layer and therefore separation point shifted from 80° to 110° i.e. moves downstream; the wake becomes smaller and the value of C_D drops sharply from about 0.5 to below 0.2

Turbulent boundary layer flow is desirable on a blunt body because it delays separation and thus reduces the pressure drag.

Transition in the boundary layer is affected by roughness of the sphere surface and turbulence in the flow stream. For rough surfaces and/or highly turbulent free stream flow, transition can occur at a critical Reynold's number as low as 1×10^5.

The drag coefficient with turbulent boundary layer flow is about 5 times less than that for laminar flow near the critical Reynold's number. The corresponding reduction in drag force can affect the range of a sphere (e.g. a golf ball) appreciably. The "dimples" on a golf ball are designed to "trip" the boundary layer and, thus, to guarantee turbulent boundary layer flow and minimum drag.

Adding roughness elements to a sphere also can surpress local oscillations in location of the transition between laminar of the transition between laminar and turbulent flow in the boundary layer. These oscillations can lead to variations in drag and to random fluctuations in lift. In base ball the "Knuckle ball" pitch is intended to behave erratically, to confuse the batter. By throwing the ball with almost no spin, the pitcher relies on the seams to cause transition in an unpredictable fashion, as ball moves an its way to the batter. This causes the desired variation in the flight path of the ball. Same with the cricket ball.

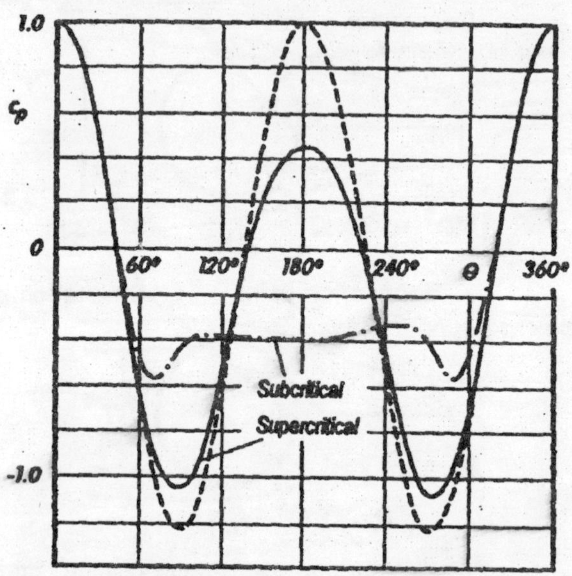

Fig. 14.9. Pressure distribution at a sphere, after measurements by O. Flachshart (1927)

The experimental determination of C_D for a sphere is difficult because, (1), the method of supporting the sphere in the wind tunnel affects the result and (2), because the results depend upon the free stream turbulent level, which is difficult to control, and upon the roughness of the sphere, which is difficult to reproduce.

The subcritical and the supercritical states can be distinguishes at high Reynolds numbers. Fig. 14.9 shows two typical experimentally determined pressure distributions for these two states [E Achenbach (1972)]. At high reynolds numbers there are deviations from the axial symmetry and unsteadly processes occur, as shown by U. Dallmann *et al.* (1993) and B. Schulte - Werning; U. Dallmann (1991), and also E. Achenbach (1974a). The effect of the Mach number on the sphere's drag is shown in Fig. 14.10. E. Achenbach (1974 b) has shown how the flow past a sphere depends on the roughness of the surface.

As seen in Fig. 14.10, the critical Reynolds number in flow past a sphere increases with increasing Mach number, so that for subsonic flows, subcritical can exist up to Re = 10^6. The form drag is then independent of the Reynolds number and agrees well with the limiting solution Re = ∞.

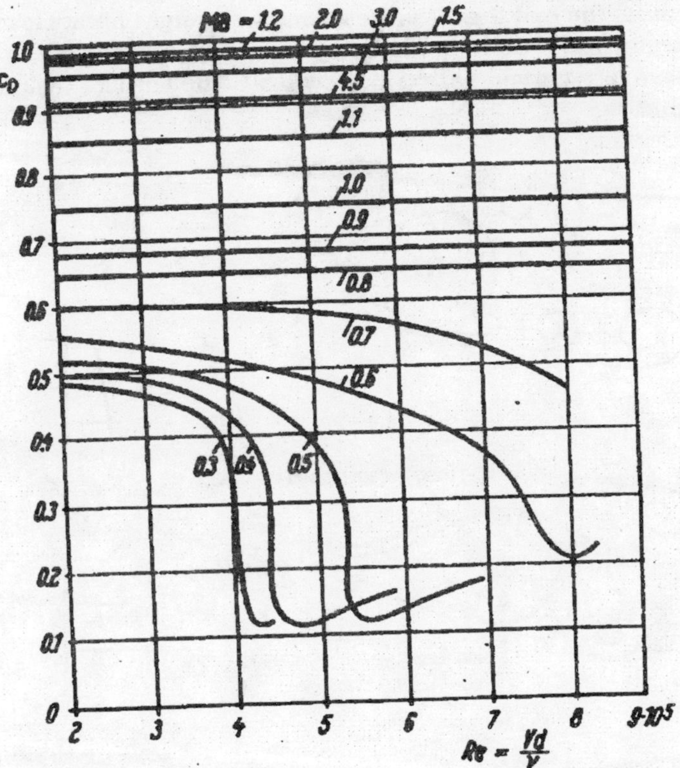

Fig. 14.10. Dependance of the drag coefficient of spheres on the Reynolds number and the mach number, from measurement by A. Naumann (1953)

14.6.1 Dynamics of sports ball

The trajectories of sports ball (such as those involved in cricket ball) bend in the air. The bending is commonly known as swing. The problem has been investigated by wind tunnel tests and by stroboscopic photographs of flight paths in field tests. Evidence indicates that the mechanics of bending is different for spinning and nonspinnning balls.

Cricket balls

The cricket ball has a prominent (1 mm high) seam, and tests show that the orientation of the seam is responsible for bending of the ball's flight path. It is known to bend when thrown at high speed, of around 30 m/s, which is equivalent to a Reynolds number Re = 10^5. The bowling technique is to align the seam at a small angle to the flight path (Fig. 14.11). This is done by spinning the ball about an axis perpendicular to the plane of the seam, and using the gyroscopic inertia to stablize this seam position during the trajectory.

When a ball is thrown (in cricket it is bowled) the air flows around it and at some point it separates (Fig. 14.11). When the separation occurs at the same point all around the ball then it moves along a storaight path. However, when it occurs asymmetrically there is a larger pressure on one side of the ball and so it starts to move in a curved path (i.e. it swings). The bowlers task is to work out how to do this.

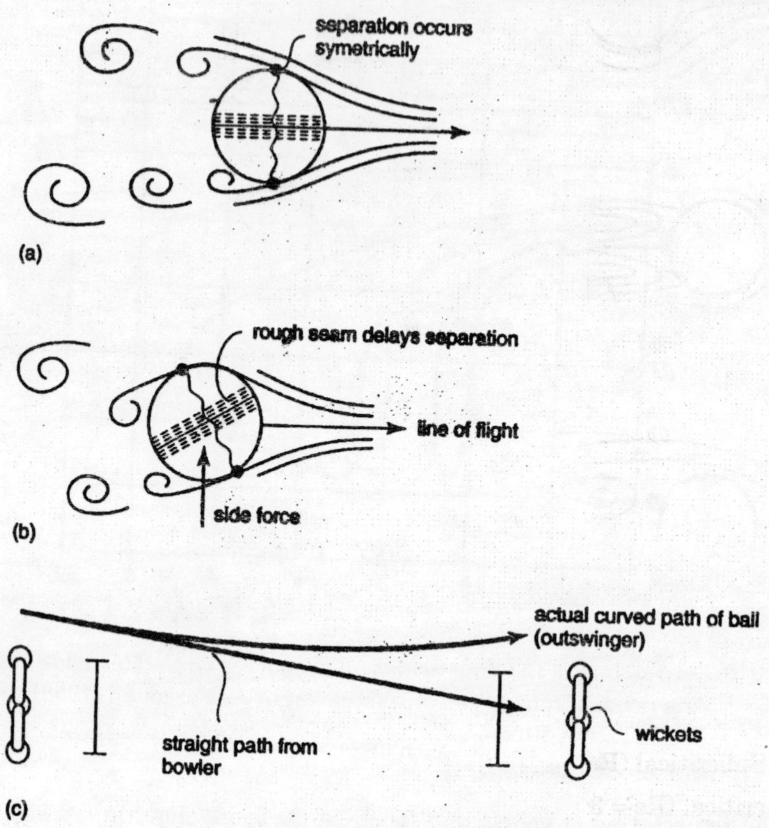

Fig. 14.11. The swing of cricket ball.

14.7 DRAG ON CYLINDER

Fig. 14.12 shows the accumulation of the shed vorticity and the early stages of the wake development. Initially, the flow separates at a single point at the rear stagnation point. This rear stagnation point divides into two separation points, which move forward until they reach a place of near equilibrium at about 78° - 82° from the front stagnation point. At separation, the boundary layer is disrupted and vorticity is introduced into the region directly behind the body. This vorticity accumulates behind the body and forms two large vortices. As the vortices draw circulation from the vorticity shed at separation, they grow stronger. As a result, the vortices induce velocities that cause the flow close to the solid boundaries to decelerate affecting the location of the separation points. The flow remains symmetic until small disturbances destroy the balance between the two vortices.

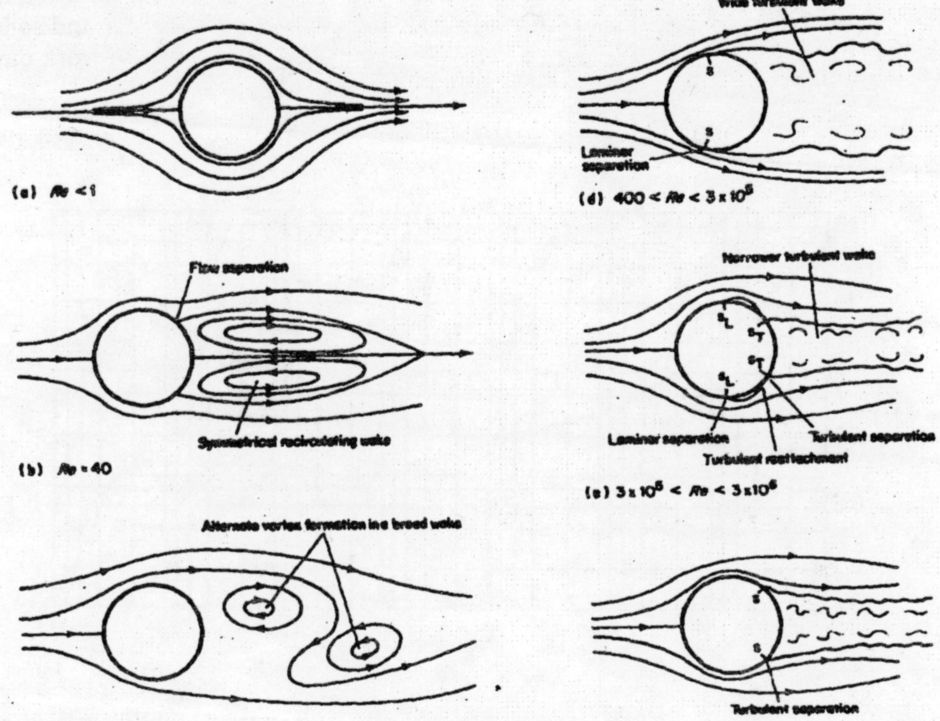

Fig. 14.12. Flow past a cylinder

The wake dynamics are complicated further by inertia effects as expressed by Re. Depending on the range of Re, the flow can be

(a) Subcritical (Re < 10^5), where the flow is laminar

(b) critical (Re = 3 x 10^5), where a rectangular bubble forms,

(c) transcritical (3 x 10^5 < Re < 5 x 10^5), when the reattachment bubble disappears and the flow becomes turbulent,

(d) supercritical (Re > 5 x 10^5), where the flow is fully turbulent.

In the subcritical region, the flow separation, as defined by a vanishing shear stress, developed about 78° - 82° from the forward stagnation point. The drag coefficient in this region is about C_D = 1.14.

Traditionally, the shedding frequently is measured by the non dimensional **Strouhal number**, defined as

$$St = \frac{f_s D}{U} \qquad (14.11)$$

Where, $f_s = \frac{1}{T_s}, T_s$ is the time of shedding between two consecutive vortices from the same side, D is the diameter of the cylinder, and U the free stream velocity.

St may be calculated from an empirical formula

$$St = 0.198\left(1 - \frac{19.7}{R_e}\right)$$

$$= 0.2\left(1 - \frac{20}{R_e}\right) \tag{14.12}$$

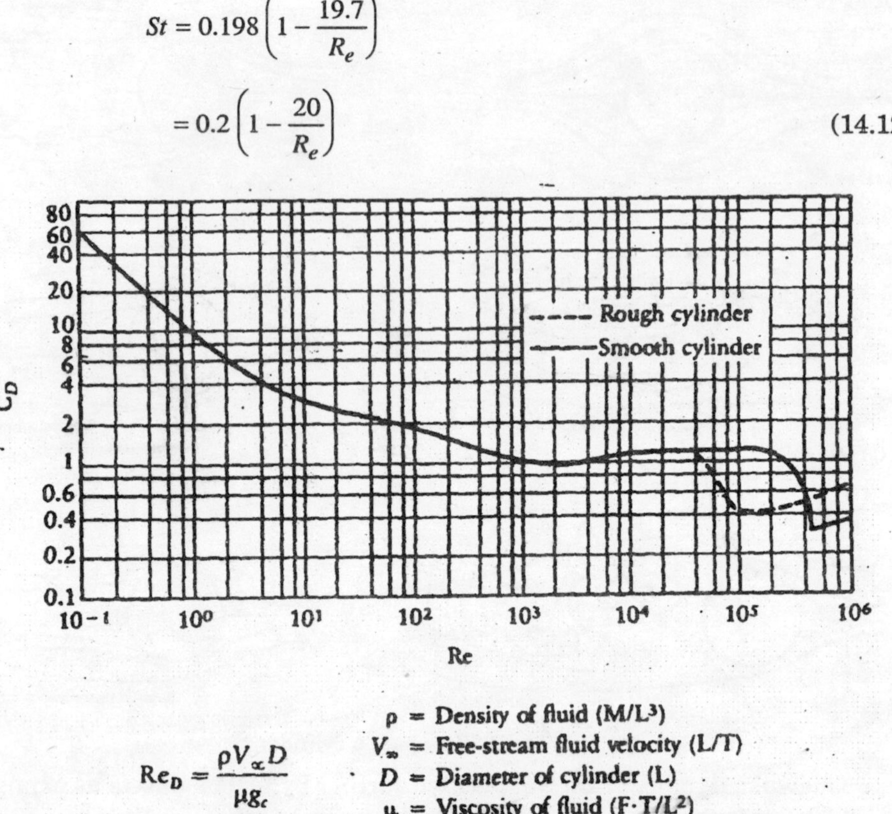

$$Re_D = \frac{\rho V_\infty D}{\mu g_c}$$

$$C_d = \frac{D_f g_c}{\frac{1}{2}\rho V_\infty^2 A}$$

ρ = Density of fluid (M/L³)
V_∞ = Free-stream fluid velocity (L/T)
D = Diameter of cylinder (L)
μ = Viscosity of fluid (F·T/L²)
D_f = Drag force exerted on cylinder (F)
A = Projected frontal area (cylinder diameter × length) (L²)

Fig. 14.13. Drag coefficient of a cylinder as a function of Reynolds number

In the critical region, the flow experiences laminar separation around 94°, but it reattaches downstream as turbulent flow. Experimental measurements downstream from the laminar separation point show an abrupt increase in the skin friction. Final separation is observed about 147°. In the supercritical region, the boundary layer flow is fully turbulent. In this regime the separation points vary between 65° - 145°. This is due to the fact that the turbulent flow penetrates the adverse pressure gradient further downstream than the limit observed in laminar flow. In this regime, the drag coefficient drops abruptly, because the separation points move away from the forward stagnation points, resulting is a smaller wake behind the cylinder. For this reason, golf balls have dimples in order to ensure that the flow around the ball is turbulent so that the drag force is minimized. Fig. 14.13 shows the variation of the drag coefficient as a function of the Reynolds number, which is consistent with the above discussions.

For subcritical Reynolds number the flow past the cylinder is laminar. For large subcritical Reynolds numbers a constant drag coefficient of $C_D = 1.2$. This could possibly hint at an asymptotic state for the hypothetical case if the flow were to remain laminar at arbitrarily high Reynolds numbers. A typical pressure distribution from this Reynolds number regime (Re = 10^5) is shown in Fig. 14.14.

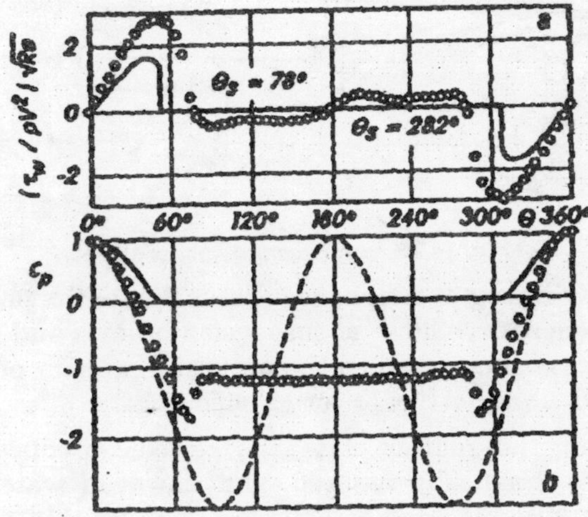

Fig. 14.14. Subcritical flow at a circular cylinder

Fig. 14.15 compares pressure distribution around a cylinder for ideal flow with real flow at low and high values of Reynolds number.

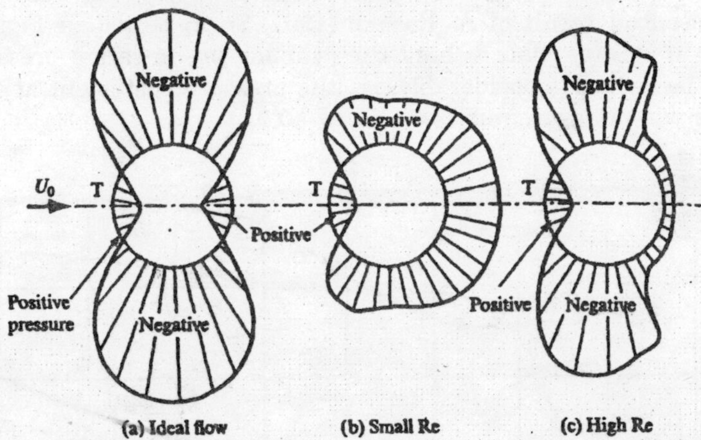

Fig. 14.15. Pressure distribution for ideal and real flow for cylinder

14.7.1 von Karman vortex street

The vortices detach alternatively from the two sides of the cylinder and move downstream forming two rows of vortices in the wake, rotating in opposite directios, called von Karman vortex street. von Karman showed analytically, and confirmed experimentally, that the pattern of vortices in a vortex street follows a mathematical relationship, namely

$$\frac{a}{b} = \frac{1}{\pi} \sin h^{-1} (b) = 0.281 \tag{14.13}$$

where a and b are indicated in Fig. 14.16.

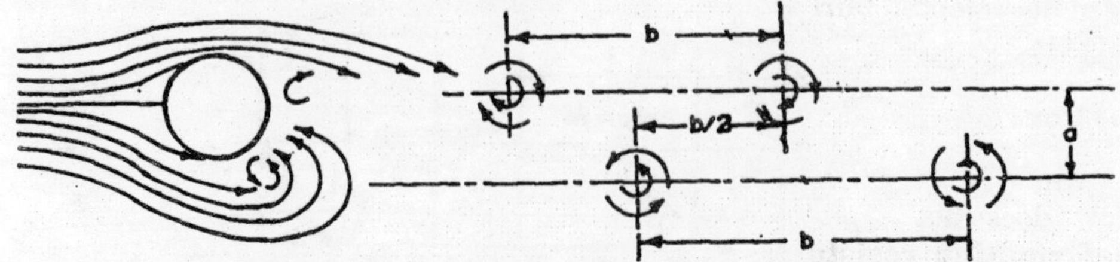

Fig. 14.16. von Karman vortex street

The familiar **Singing** of telephone wires is due to the phenomenon caused by a cross wind, whereas collapse of suspension bridges and flutter of aerofoils is the result of a resonance between the natural frequency of the body and the frequency of forced vibration due to vortex shedding.

By introducing the splitter plate, the periodic production of vortices is halted, or at least drastically reduced, and the dead water pressure which governs the form drag is considerably lowered. In this manner, the drag coefficient of a flow, hypothetically assumed to be steady and laminar, at high reynolds number past a cylinder would be $C_D = 0.5$. According to more recent investigations by B. Fornberg (1987), D.H. Peregrine (1985) and F.T. Smith (1985), the flow seem to be more complicated than this.

The essential result of A. Roshko (1967) is presented in Fig. 14.17. By introducing a splitter plate behind the cylinder the negative pressure at the back could be reduced considerably, i.e. the pressure coefficient at the back of the cylinder was changed from $c_p = -1.1$ to -0.5.

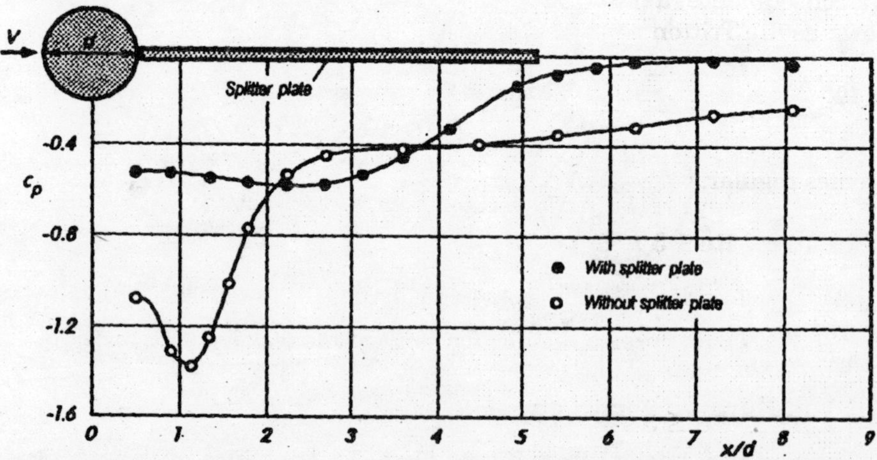

Fig. 14.17. Effect of splitter plate in the wake of a circular cylinder, after A. Roshko (1967)

Re = 14500, the laminar - turbulent transition takes place immediately after laminar separation.

14.7.2 Summary of flow regimes in flow past a circular cylinder

After many years of careful experimentation, the flow behaviour and the wake dynamics for flow past a circular cylinder over the whole range of Re have been documented fairly well [A Roshko, J. of Fluid Mechanics, 10 (1961), p 345] [Fig. 14.12].

(1) For Re $\leq$ 1

In this flow region, the flow maintains its symmetry about both axis as the flow does not separate. The drag force on the cylinder is due to the **deformation resistance,** which is attributed to the disturbance of the flow due to the presence of the body.

(2) 4 $\leq$ Re $\leq$ 40

In this region, the flow separates, resulting in two large vortices, behind the cylinder. These vortices remain attached to the cylinder up to Re = 40. Therefore, the flow maintains its symmetry around the x - axis. The drag force increases with Re, while the lift force is zero.

(3) 40 $\leq$ Re $\leq$ 150

In this range, the regular shedding begins and the symmetry of the flow is entirely lost. The drag and lift forces assume a periodic variation, which is related to the frequency of shedding.

(4) 200 $\leq$ Re $\leq$ 400

The flow inside the vortices undergoes a change from laminar to turbulent. As a conseq-uence, a loss of regularity in the flow is observed. This transition is known as the Tritton transition region.

(5) 400 $\leq$ Re $\leq 3 \times 10^5$

The antisymmetric von Karman vortex street reappears, and the flow becomes regular.

(6) $3 \times 10^5 \leq$ Re $\leq 5 \times 10^5$

The flow undergoes rapid changes. As discussed earlier, the flow could be subcritical transcritical or supercritical. Associated with these changes also are changes in the drag and lift forces.

(7) $5 \times 10^5 \leq$ Re $\leq 3 \times 10^6$

The flow is completely supercritical (i.e. the flow is fully turbulent), and the separation points move towards the rear stagnation point. The width of the wake narrows and the drag coefficient drops abruptly. This drop in the drag force is known as the **drag crisis.**

(8) Re ≤ 3 x 10⁶

The drag force increases from its low during the drag crisis but remains smaller than the drag observed at lower Re. The flow remains fully turbulent.

14.8 LIFT

As is indicated in section 14.2, any object moving through a fluid experience a net force of the fluid on the object. In the section 14.2 we briefly mentioned the lift as a force which acts on an immersed body normal to the relative motion between the fluid and the body.

In the case of both ideal fluid and the real fluid, a circulation is necessary for the production of a lift force.

Lifting vanes are shapes which produce lift and include such things as airfoils, hydrofoils, properller baldes, kites and circular arcs. Because all lifting vanes are fundamentally the same, it is sufficiently informative to study the behaviour of an airfoil.

The lift of an airfoil may be explained in terms of the Kutta-Joukowsky theorem which demonstrates that a circulation.

14.8.1 Magus effect and the circulation theory of lift (Kutta-Joukowski theorem)

If the symmetrical bodies immersed in a fluid stream, with their axis of symmetry directed parallel to the flow, experience a force acting exactly in the direction of flow. This is due to the fact that the flow pattern around the body is symmetrical so that the pressure distribution will also be symmetrical. If however, a symmertrical object, such as a cylinder is placed in a fluid stream and rotated, the rotation causes a distortion of the original symmetrical flow pattern and hence a distortion of the pressure distribution. A transverse force, lift, is observed, the magnitude of which may be changed by altering either the speed of rotation or the stream velocity. The production of lift by a rotating cylinder placed in a fluid stream is known as the Magnus effect.

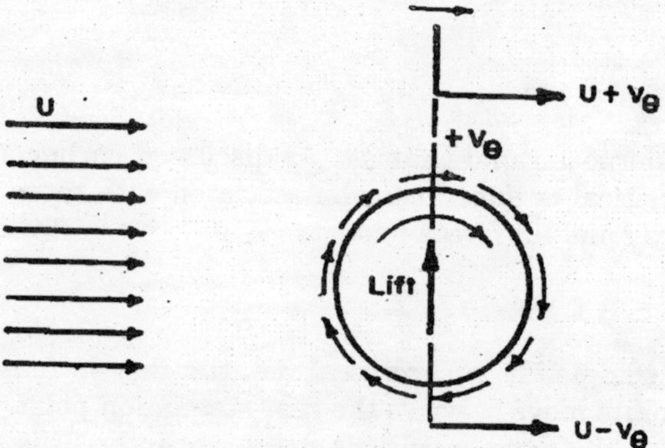

Fig. 14.18. Magnus effect

An explanation of the Magnus effect may be obtained from the following considerations. On the surface of the cylinder the fluid rotates with the speed of the cylinder and viscosity transmits the rotation to adjacent fluid layers. This boosts the velocity of the stream in the direction of rotation and opposes it in the reverse direction. The effect is shown in Fig. 14.18.

The magnus effect was first observed by the German scientist named Magnus in 1852 and is commonly referred to as the Magnus effect. The Magnus effect was used to propel the Flettner rotor ship. Magnus effect may be found in the curved pitch of a base ball; the ball must be "spun" in order to produce the curve.

The circulatory flow, or circulation, has since been closely connected with lift, even with the actual body rotation absent. Indeed, it has been shown that any body placed in a fluid stream may exhibit lift if a circulatory motion or circulation is maintained in same way.

Lift per unit span in an irrotational flow over a two dimensional body of arbitrary cross section is given by

$$F_L = \rho U \tau \tag{14.14}$$

Where U is the free stream velocity and τ is the circulation around the body. Equation (14.14) is called Kutta Joukowski lift theorem.

14.9 LIFT OF AN AEROFOIL

An aerofoil may be defined as a streamlined body designed to produce lift. Aerofoil may be either symmetrical or unsymmetrical. Two aerofoils used to produce lift are shown in Fig. 14.19. Clearly the symmetrical one cannot produce lift unless the angle of attack, α, is non zero. Because of the asymmetry of the non symmetric aerofoil, the pressure distribution on the upper and lower surfaces are different (Fig. 14.20) and a lift is produced even with $\alpha = 0$. Of course there will be a certain value of α (less than zero for this case) for which lift is zero. For this situation, the pressure distribution on the upper and lower surfaces are different, but their resultant (integrated) pressure forces will be equal and opposite.

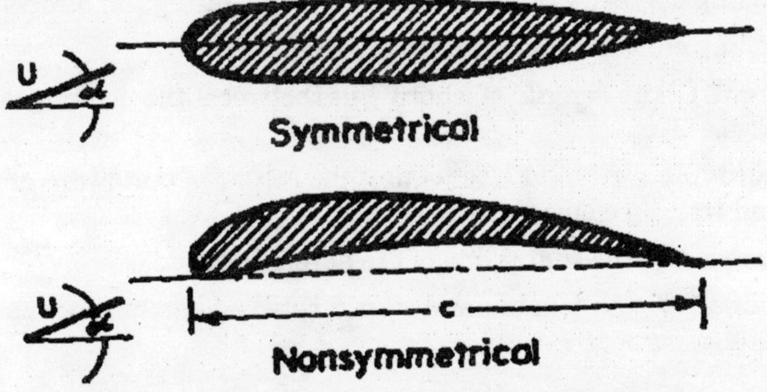

Fig. 14.19. Symmetrical and unsymmetrical aerofoils

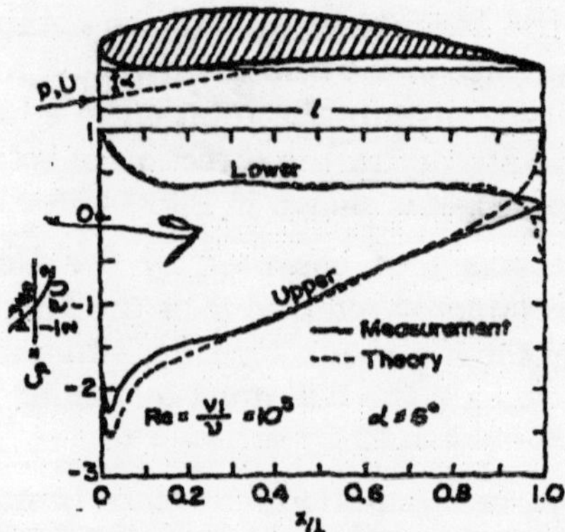

Fig. 14.20. Pressure distribution on an aerofoil. The upper panel shows Cp plotted normal to the surface and lower panel up plotted normal to the chords lin.

14.10 AEROFOIL TERMINOLOGY

There is an accepted terminology concerning aerofoils and familiarization with it is necessary in order to understand the distribution of flow past aerofoils.

Fig. 14.21 shows an aerofoil section and the following are some of the most important terms relating limit.

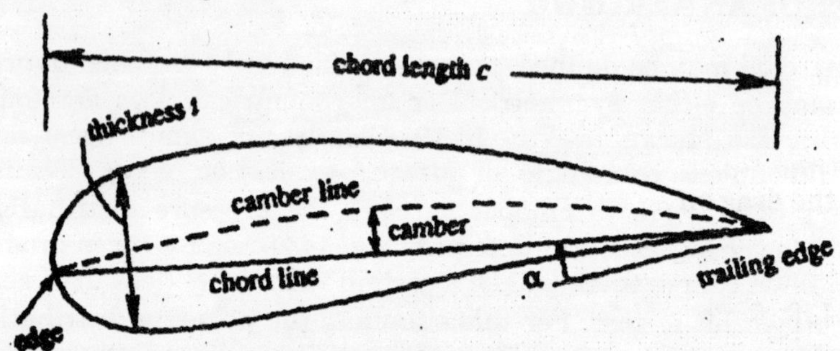

Fig. 14.21. Air foil geometry

(1) Leading edge The front, or upstream edge, facing the direction of flow.

(2) Trailing edge the rear, or downstream edge.

(3) Chord C the length of chord line between the leading and trailing edges.

(4) Chord line a straight line joining the centre of curvature of the leading and trailing edges.

(5) Camber line the centre line of the aerofoil section.

(6) Camber, δ the maximum distance between camber line and the chord line.

(7) Percentage camber = $\frac{100\,\delta}{c}$ percent is a measure of aerofoil curvature.

(8) Span, b the length of the aerofoil in the direction perpendicular to the crosssection.

(9) Plan area A the area of the projection of the aerofoil on the plane containing the chord line. If the aerofoil is of constant cross section A = b × c

(10) Mean chord $\bar{c} = \dfrac{A}{b},$

(11) Aspect ratio $AR = \dfrac{Span}{Mean\ Chord} = \dfrac{b}{c} = \dfrac{b^2}{A}$

(12) Angle of attack α (Incidence) Angle between the chord line and the direction of the fluid stream.

(13) Deviation, Φ Angle between the tangent to camber line at trailing edge and the tangent to camber line at leading edge.

(14) Angle of zero lift, α_o the angle of attack when the lift of an aerofoil is zero.

(15) Centre of pressure the location of the resultant aerodynamic force on an aerofoil.

(16) Stall When an aerofoil is operated at an angle of attack greater than the angle of attack at maximum lift (For clark y aerofoil this is 18.5°), the aerofoil in the stall condition.

(17) Drag force the component of the total force on an aerofoil parallel to the direction of flow

(18) Drag coefficient C_D a dimensionless coefficient that is used to calculate the drag on an aerofoil.

$$C_D = F_D / 1/2\, \rho U^2\, A$$

(19) Lift force the component of the total force on an aerofoil that is perpendicular to the direction of flow.

(20) Lift coefficient C_L a dimensionless coefficient that is used to calculate the lift on an aerofoil.

(21) Dynamic pressure a term that equals $\rho\ U^2/2$.

14.11 INVISCID FLOW PAST A TWO DIMENSIONAL AEROFOIL

The calculation of the inviscid flow past a two dimensional aerofoil gives a flow field as indicated in Fig. 14.22.

The predicted flow field past an aerofoil with no lift [i.e. a symmetrical aerofoil and zero angle of attack, Fig. 14.22(a)] appears to be quite accurate (except for the absence of thin boundary layer regions). However, as is indicated in Fig. 14.22(b), the calculated flow past the same aerfoil at a nonzero angle of attack (but one small enough so that boundary layer separation would not occur) is not proper near the trailing edge. In addition, the calculated lift for a nonzero angle of attack is zero - in conflict with the known fact that such aerofoils produce lift.

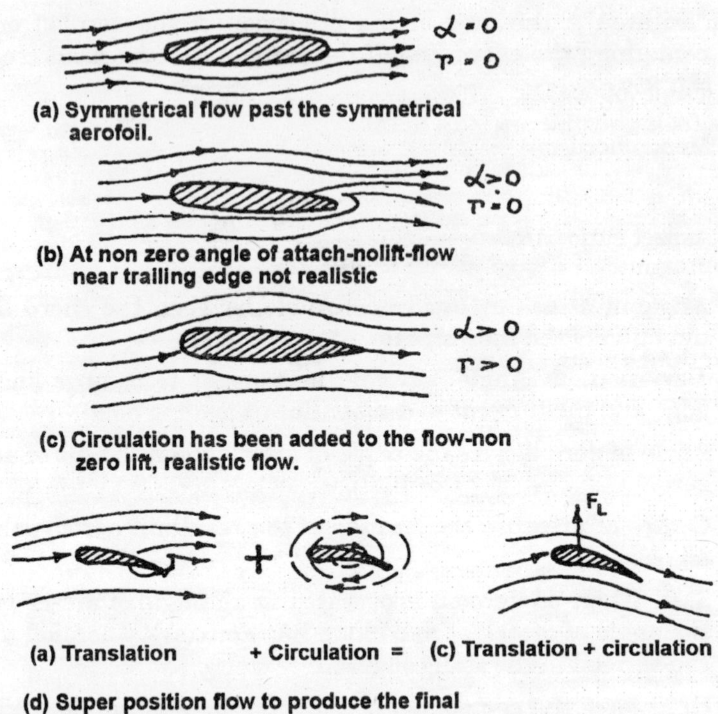

Fig.1 5.22. In viscid flow past an aerofoil

In reality, the flow should pass smoothly over the top surface as is indicated in Fig. 14.22 (c), without the strange behaviour indicated near the trailing edge in Fig. 14.22 (b). As is shown in Fig. 14.22(d), the unrealistic flow situation can be corrected by adding an appropriate clockwise swirling flow around the aerofoil. The results are two fold :

(1) the unrealistic behaviour near the trailing edge is eliminated (i.e. the flow patterm of Fig. 14.22 (b) is changed to that Fig. 14.22 (c), and

(2) the average velocity on the upper surface of the aerofoil is increased while that on the lower surface is decreased. From the Bernoulli's equation, the average pressure on the upper surface is decreased and that on the lower surface is increased. The net effect is to change the original zero lift condition to that of a lift producing aerofoil.

Joukowski discovered a mathematical method for transforming circles into aerofoil shapes and was thus able to transform the streamline pattern around a circular cylinder into the streamline pattern around aerofoils. This method enabled him to calculate the theoretical lift for these "transformed" aerofoils. von Karman and von Mises improved Joukowski mathematical model so that shapes more closely resembling actual wing sections were obtained. Joukowski's thoery indicated that the lift coefficient for an aerofoil in an ideal fluid when the aerofoil thickness and camber approach zero (became a flate plate) is

$$C_L = 2\,\pi \sin \alpha_o = 2\,\pi\,\alpha_o \tag{14.15}$$

for small angle of attack. The angle α_o is the difference between the actual angle of attack and the angle of attack for which the lift is zero. For flat plate and symmetrical aerofoils without camber (curvature), α_o is the same as α, since the lift is zero for zero angle of attack. Equation (14.15) states that the slope of a line relating C_L (ordinate) to α_o (abscissa) is $2\,\pi$ and this is very closely approximated in practice for thin aerofoils at small values of α. Stalled conditions, associated with a decrease in lift coefficient with increased angle of attack due to flow separation along the entire upper surface of the aerofoil, are due to real fluid effects and cannot be described by ideal fluid theory.

Thus ideal flow theory predicts actual lift performance amazingly well but gives zero drag in all instances in an inifinite fluid for steady flow.

14.12 REAL (VISCOUS) FLUID PAST A TWO DIMENSIONAL AEROFOIL

The theory of lift in a real (viscous) fluid is largely credited to Lanchester and its later improvement to Prandtl. An aerofoil in a real fluid must create its own circulation, or vortex field, just as the spining ping pony ball, in order to experience lift.

Results of experiments carried out on aerofoils in wind tunnel show the existence of circulation and furnish strong evidence in favour of the theory. It is found that, except for the thin boundary layer formed on the surface, the main body of the fluid is not sensibly affected by viscosity and may be treated approximately as a perfect fluid, that is, calculations may be based on `potential flow' theory. The relation between lift and circulation applied to real fluids agrees reasonably with the theory as long as aerofoil is placed at small angle of attack. Generally the Kutta-Joukowski law predicts larger lift than is found from observations and the discrepancy increases with increasing angle of attack. The reason for this discrepancy is the omission of viscosity from the theory. The existence of the boundary layer and its associated wake in real fluids and also tends to make the lift less than its theoretical value.

Rotating cylinders placed in flow of real fluids maintain their circulation through viscous action. Aerofoils, however, being non rotational bodies, must have a circulation produced by different means. The process of producing circulation may be described briefly as follows :

When the wing is set in motion from rest, a vortex is formed immediately downstream from the trailing edge, as shown in Fig. 14.23(a). This phenomenon takes place whenever fluid streams of different velocities meet (Fig. 14.23b), The fluid stream separated by the leading edge of the wing meets again at the trailing edge. At the beginning of the motion the fluid velocities are different (higher at the bottom surface than at the top) and upon the meeting of the fluid layers a vortex is produced which is cast off soon after the aerofoil has begun to move forward. After the start the flow rapidly adjusts itself and the two layers attain equal velocities so that no further starting vortices are shed.

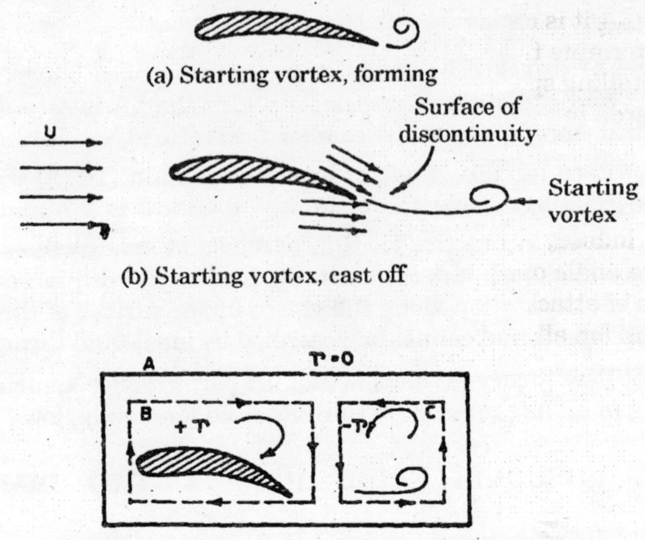

Fig. 14.23. Development of circulation around an aerofoil. (a) Starting vortex, forming; (b) Starting vortex, cast off; (c) Circulation due to cast off starting vortex

In order to counter balance circulation of the starting vortex, another circulation of equal strength to the starting vortex, but of opposite sign, is set up around the aerofoil. The sum of circulations is zero, in accordance with Thomson's (Lord Kelvin) theorem, which states that "the circulation around a closed curve in the fluid does not change with time if one moves with the fluid". Prandtl has been sustantiated with photographs. If we take a contour around the aerofoil in Fig. 14.23 (c) as A, and divide this into two compartments, B and C, one of which contains the starting vortex with + τ and the other, fixed relative to the aerofoil with- τ, we satisfy Thomson's theorem. The starting vortex keeps whirling around Fig. 14.23 (c) until it dies out from viscous effects. The net circulation around a curve including the profile and this vortex is still zero. When their aerofoil comes to a stop or changes its angle of attack, new vortices are formed to effect the necessary change in circulation.

14.13 AEROFOIL CHARACTERISTICS

The typical variation of lift coefficient with angle of attack for an aerofoil is sketched in Fig. 14.24. A low to moderate angles of attack, c_L varies linearly with ∝. In this region, the flow moves smoothly over most of the surface, as shown in the streamline picture at the left of Fig. 14.24. However, as α becomes large, the flow tends to separate from the top surface of the aerofoil, creating a large wake of relatively "dead air" behind the aerofoil as shown at the right of Fig. 14.24. Inside this separated region, the flow is recirculating, the part of the flow is actually moving in a direction to the free stream–so called reversed flow.

The consequences of this separated flow at a high α is a precipitous decrease in lift and a large increase in drag, under such conditions, the aerofoil is said to be stalled. The maximum value of C_L, which occurs just prior to the stall, is denoted by C_{Lmax} it is one of the most important aspects of aerofoil performance, because it determine the stalling speed of an aeroplane. The higher is C_{Lmax}, the lower is the stalling speed. A great deal of moderate aearofoil research has been directed towards increasing C_{Lmax}. Again examining Fig. 14.24 we see that C_L, increases linearly with α until flow separation begins to have an effect. Then the curve becomes non linear, C_L reaches a maximum value, and finally the aerofoils stalls. At the other extreme of the curve, noting Fig. 14.24, the lift at $\alpha = 0$ is finite; indeed, the lift goes to zero only when the aerofoil is pitched to some negative angle of attack. The value of α when lift equals zero is called the zero lift angle of attack and denoted by $\alpha_L = 0$ or α_o. For symmetrical aerofoil, $\alpha_o = 0$, whereas for all aerofoils with positive camber (camber above the chord line), α_o is negative value, usually on the order of -2 or -3°.

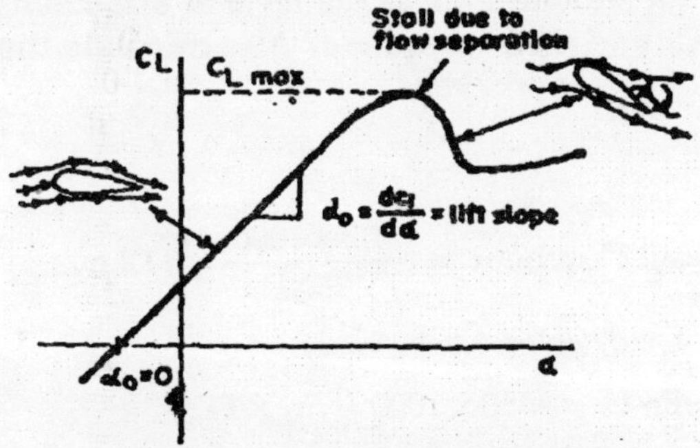

Fig. 14.24. C_L v/s α for an aerofoil

The inviscid flow aerofoil theory allows us to predict the lift slope and α_o for a given aerofoil. It does not allow us to calculate C_{Lmax} (Fig. 14.25) which is a difficult viscous flow problem.

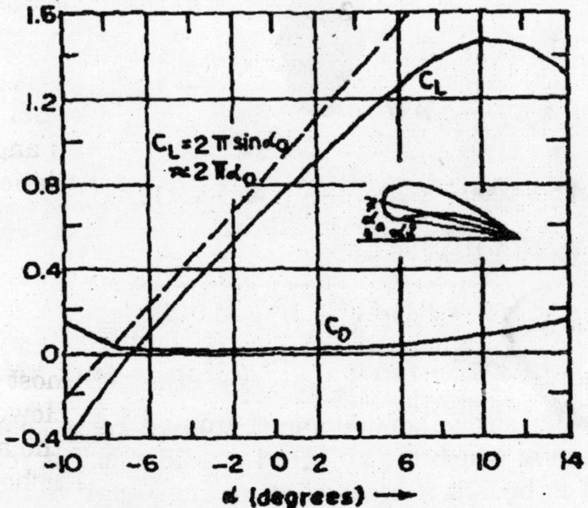

Fig. 14.25. C_L v/s α for infinite length of span.

The angle of attack is α_o and obviously this has considerable effect upon the lift. Therefore, α_o is an important variable in determining the characteristics of an aerofoil. Fig. 14.25 shows actual variation of both C_L and C_D with the angle of attack for a Joukowsky aerofoil. Plotted as a broken line is equation (14.15), the theoretical variation of lift with angle of attack. Note that the experimental curve is not greatly different from the theoretical one for small angles of attack (including negative angles). When the angle reaches above 8°, however, a marked deviation develops.

The coefficient C_L and C_D are determined experimentally, and for a aerofoil of finite length they differ somewhat from the infinite aerofoil length, because of induced drag.

ILLUSTRATIVE EXAMPLES

14.1 A wing of 46.5 m² plan form area is to produce a lift of 44.5 kN in level flight through standard air at sea level at 402 km/hr by expending 450 kW. What C_L and C_D are required? Also calculate the total aerodynamic force.

Solution

$$U = 402 \text{ km/hour} = \frac{402 \times 1000}{3600} = 112 \text{ m/s}$$

$\rho = 1.225$ kg/m² at sea level.

$$P = F_D U = 450 \times 1000 = F_D \times 112$$

$$F_D = 4017.857 \text{ N} = 4.017 \text{ kN}$$

$$F_D = C_D \frac{1}{2} \rho A U^2$$

$$4017.857 = C_D \times \frac{1.225}{2} \times 46.5 \times (112)^2$$

$\therefore C_D = 0.0112$ **Ans**

$$F_L = C_L \frac{1}{2} \rho A U^2$$

$$44.5 \times 1000 = C_L \times \frac{1.225}{2} \times 46.5 \times (112)^2$$

$\therefore C_L = 0.1295$ **Ans**

$$F_R = \sqrt{F_L^2 + F_D^2} = \sqrt{(44.5)^2 + (4.017)^2}$$

$$= \sqrt{1980.25 + 16.136} = \sqrt{1996.3856} = 44.68 \text{ kN}$$ **Ans**

14.2 A rectangular aerofoil of 300 mm chord and 1.5 m span was tested in a wind tunnel in which the air speed was 36 m/s. The resultant force was measured to be 310 N inclined at an angle of 4° to the vertical axis or axis of the lift. Calculate the coefficient of lift and drag. Take density of air as 1.25 kg/m³.

Solution

A = Chord × Span = 0.3 × 1.5 = 0.45 m²

F_R = 310 N

$F_L = F_R \cos\theta$ and $F_D = F_R \sin\theta$

= 310 cos 4° = 310 sin 4°

= 309.24 N = 21.62 N

Therefore,

$$F_D = C_D \frac{1}{2} \rho A U^2$$

$$21.62 = C_D \times \frac{1.25}{2} \times 0.45 \times (36)^2$$

C_D = 0.059 Ans

Therefore,

$$F_L = C_L \frac{1}{2} \rho A U^2$$

$$309.24 = C_L \times \frac{1.25}{2} \times 0.45 \times (36)^2$$

C_L = 0.848 Ans

14.3 An aerofoil of 50 m² area is travelling at 30 m/s with an angle of attack of 6°, If c_D = 0.05 for this angle of attack, what power is required to maintain this velocity in air –23°C and pressure 0.85 bar. The density of air at N.T.P. is to be 1.2931 kg/m³.

Solution

Let subscript 0 to NTP and x for elevation at –23°C and 0.85 bar

$$p_o = \rho_o R T_o$$

and $$p_x = \rho_x R T_x$$

Therefore,

$$\frac{p_o}{p_x} = \frac{\rho_o}{\rho_x} \times \frac{T_o}{T_x}$$

$$\frac{1.01325}{0.85} = \frac{1.2931}{\rho_x} \times \frac{273}{250}$$

or, $\rho_x = 1.18 \ kg/m^3$

Therefore,

$$F_D = C_D \frac{1}{2} \rho A U^2$$

$$= 0.05 \times \frac{1}{2} \times 1.18 \times 50 \times (30)^2 = 1332.63 \text{ N}$$

$$Power\ P = \frac{F_D U}{1000} = \frac{1332.63 \times 30}{1000} = 39.98 \text{ kW} \qquad \text{Ans}$$

14.4 **A jet plane which weighs 50 kN and having a wing area of 25 m² flies at a velocity of 900 km/hr, when the engine delivers 7.5 MW, 60% of the power is used to overcome the drag resistance of the wing. Calculate the coefficient of lift and drag for the wing. The atmospheric pressure and temperatures are 100 kPa and 285 K. Take characteristic gas constant as 0.287 kJ/kg K.**

Solution: 900 km/hr = 900 × 1000/3600 = 250 m/s

Using gas equation

$$P = \rho R T$$

$$100 = \rho \times 0.287 \times 285$$

$$\rho = 1.222\ kg/m^3$$

$$F_L = W = C_L \frac{1}{2} \rho A U^2$$

$$50 \times 1000 = C_L \times \frac{1.222}{2} \times 25 \times (250)^2$$

$$C_L = 0.0523 \qquad \text{Ans}$$

Total power = 7.5 MW

60% of the total power is used to overcome the drag resistance, therefore

Effective power = 60% of 7.5 MW = 4.5 MW

Therefore,

$$P = F_D U$$

$$4.5 \times 10^6 = F_D \times 250 \quad \therefore F_D = 18000 \text{ N}$$

Therefore,

$$F_D = C_D \frac{1}{2} \rho A U^2$$

$$18000 = C_D \times \frac{1.222}{2} \times 25 \times (250)^2 \quad \therefore C_D = 0.018885 \qquad \text{Ans}$$

14.5 **The wing area of a small aeroplane weighs 6.25 kN is 10.25 m² (a) If the crusing speed of the plane is 225 km/hour, determine the lift coefficient of the wing. (b) If the engine delivers 150 kW at this speed, and if 65%**

of this power represents propeller loss and body resistance, what is the drag coefficient of the wing? Take density of air to be 1.225 kg/m³

Solution

(a) $U = 225 \times 1000/3600 = 62.5$ m/s

$$F_L = C_L \frac{\rho}{2} A U^2 = W$$

$$6.25 \times 1000 = C_L \times \frac{1.225}{2} \times 10.25 \times (62.5)^2 \quad \therefore C_L = 0.2548 \qquad \text{Ans}$$

(b) $P = F_D U$

$$(150 \times 1000) \times 0.35 = F_D \times 62.5$$

$F_D = 840$ N

and $F_D = C_D \frac{1}{2} \rho A U^2$

$$840 = C_D \times \frac{1.225}{2} \times 10.25\,(62.5)^2 \quad \therefore C_D = 0.0345 \qquad \text{Ans}$$

14.6 A wing of a small aeroplane is rectangular in plan having a span of 10 m and a chord of 1.2 m. In a straight and level flight at 240 km/hour the total aerodynamic force acting on the wing is 20 kN. If the ratio of lift to drag is 10, calculate the coefficient of lift and the total weight of the aeroplane can carry. Assume air density to be 1.2 kg/m³.

Solution

Total aerodynamic force $F_R = \sqrt{F_L^2 + F_D^2}$

or, $$20 \times 1000 = \sqrt{F_L^2 + F_D^2} \qquad (1)$$

$$\frac{F_L}{F_D} = 10 \qquad (2)$$

Therefore,

$$20000 = \sqrt{(10F_D)^2 + (F_D)^2} \qquad \text{[Using equation (2)]}$$

$$20000 = \sqrt{101\ F_D{}^2}$$

$$4 \times 10^8 = 101\ F_D{}^2$$

$$F_D{}^2 = \frac{4 \times 10^8}{101} = 3960396.04$$

$$F_D = 1990.074 \text{ N}$$

Therefore,

$$F_L = 10 \times 1990.074\ N$$

$$= 19900.74 \text{ N} = \text{Total weight of aeroplane}$$ Ans

$$F_L = C_L \frac{1}{2} \rho A U^2$$

$$19900.74 = C_L \times \frac{1.2}{2} \times (10 \times 1.2) \times (66.67)^2$$

$$19900.74 = C_L \times 32000$$

$$\therefore \quad C_L = 19900.74 / 32000 = 0.62189$$ Ans

14.7 An aeroplane is designed according to the following specifications

Weight = 13.5 kN Wing area = 30 m²

Take off speed = 30 m /s

Model tests show that the lift and drag coefficient vary with the angle of attack of the wing according to the approximate relations

$C_D = 0.008 (1 + \alpha)$ $C_L = 0.35 (1 + 0.2\,\alpha)$

For small α, where α is the angle of attack measured in degrees. The atmospheric density is 1.29 kg/m³. Find the angle of attack that ensures take off at the design speed, and power required for take off.

Solution

$$W = F_L = C_L \frac{1}{2} \rho A U^2$$

$$13.5 \times 1000 = C_L \times \frac{1.29}{2} \times 30 \times (30)^2$$

$$13500 = C_L \times 17415$$

$$C_L = 0.7752$$

But, $C_L = 0.35 (1 + 0.2\,\alpha)$

Therefore,

$$0.7752 = 0.35 (1 + 0.2\,\alpha)$$

$$\alpha = 6.0742^\circ = 6^\circ\ 4'\ 27''$$ Ans

$$C_D = 0.008 (1 + 6.0742) = 0.008 (7.0742) = 0.05659$$

$$F_D = C_D \frac{1}{2} \rho A U^2$$

$$F_D = 0.05659 \times \frac{1.29}{2} \times 30 \times (30)^2 = 985.577 \text{ N}$$

Power, $P = F_D U / 1000 = \dfrac{985.577 \times 30}{1000} = 29.5673 \text{ kW}$ Ans

14.8 **An aeroplane weighs 24 kN and has a wing area of 36 m². If the C_L varies linearly from 0.35 to 0.8, when the α varies from 0° to 6°, calculate the angle of attack, where the aeroplane moves at a constant speed of 144 km/hour. Take density of air 1.2 kg/m³.**

Solution

$$W = F_L = C_L \frac{1}{2} \rho A U^2$$

$$24 \times 1000 = C_L = \frac{1.2}{2} \times 36 (40)^2$$

$$24000 = C_L \times 34560$$

$$C_L = \frac{24000}{34560} = 0.694$$

For a straight line

$$m = \tan \Phi = \frac{0.8 - 0.35}{6 - 0} = \frac{0.45}{6} = 0.075$$

Therefore,

$$y = m x + c$$

$$0.694 = 0.075 x + 0.35$$

$$0.075 x = 0.694 - 0.35 = 0.344$$

$$x = \frac{0.344}{0.075} = 4.586$$

Therefore, $\alpha = 4.586°$ Ans

14.9 **A rectangular plate 1.5 m by 1 m moves at 15 m/s in still air at an angle of 12° with horizontal. If the coefficient of drag and lift are 0.17 and 0.72 respectively compute**

(a) The drag force

(b) The lift force

(c) The resultant force

(d) It's direction,

(e) Frictional force in the direction of plate and

(f) The power required to keep the plate moving

Take ρ = 1.2 kg/m³.

Solution

$$A = 1.5 \times 1 = 1.5 \text{ m}^2$$

(a) Drag Force $F_D = C_D \frac{1}{2} \rho A U^2$

$$= 0.17 \times \frac{1.2}{2} \times 1.5 \times (15)^2 \; N = 34.425$$ **Ans**

(b) Lift Force $F_L = C_L \frac{1}{2} \rho A U^2$

$$= 0.72 \times \frac{1.2}{2} \times 1.5 \times (15)^2 = 145.8 \text{ N}$$ **Ans**

(c) $F_R = \sqrt{F_L^2 + F_D^2} = \sqrt{(145.8)^2 + (34.425)^2} = 149.8 \text{ N}$ **Ans**

(d) $\tan\theta = \frac{F_L}{F_D} = \frac{145.8}{34.425} = 4.23$

$$\theta = \tan^{-1}(4.23) = 76.715^\circ$$ **Ans**

(e) Frictional resistance in the direction of plate

$$F_D = F_R \cos(\theta + \propto) = 149.8 \cos(76.715 + 12)$$

$$= 149.8 \times 0.0224 = 3.359 \text{ N}$$ **Ans**

(f) $P = F_D U = 3.359 \times 15 = 50.385$ Watt. **Ans**

14.10 **A parachute of 10 m diameter when carrying a load W descends at a constant velocity of 5.5 m/s in atmospheric air at a temperature of 18°C and pressure of 750 mm of mercury. Determine the load W if the drag coefficient for the parachute is 1.4.**

Solution

We know that

1 bar = 750 mm of mercury

Therefore, $P = \rho RT$

$$1 \times 10^5 = \rho \, 287 \times 291$$

$$\rho = 1.19736 \; kg/m^3$$

$$F_D = mg = Load = C_D \frac{1}{2} \rho A U^2$$

$$F_D = 1.4 \times \frac{1.19736}{2} \times \frac{\pi}{4} \times (10)^2 \times (5.5)^2 = 1991.3 \text{ N}$$ **Ans**

14.11 **The vertical component of the landing speed of a parachute is to be less than 6 m/s. The parachute may be treated as an open hemisphere. The total mass of chute and jumper is 120 kg. Determine the minimum diameter of the open parachute. Take C_D = 1.45 and ρ = 1.29 kg/m³**

Solution

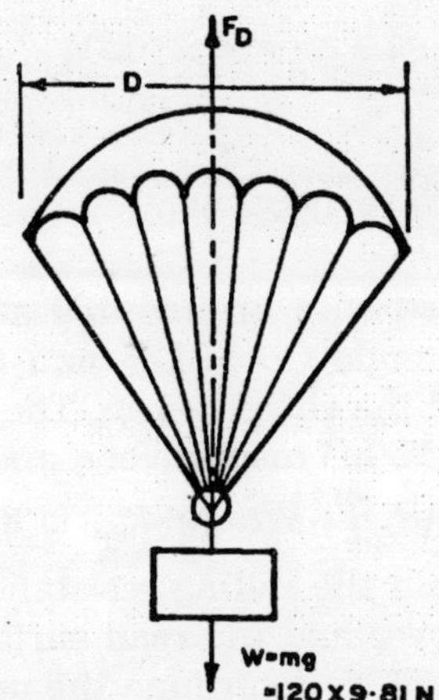

Fig. 14.26. Parachute P. 14.11.

$$F_D = mg = \frac{1}{2} C_D \rho A U^2$$

or, $$mg = \frac{1}{2} C_D \rho \frac{\pi}{4} D^2 U^2$$

or, $$120 \times 9.81 = \frac{1}{2} \times 1.45 \times 1.29 \times \frac{\pi}{4} D^2 \times (6)^2$$

or, $$D^2 = 44.5174$$

$\therefore$ D = 6.672 m Ans

14.12 **A vehicle having a projected area of 6 m² is travelling at 90 km/hr. The total resistance to the motion of the vehicle is 2200 N. If 20 % and 10 % of the total resistance area due to rolling friction and surface friction respectively and balance resistance is the form drag. Determine the form drag coefficient. Take density of air to be 1.2 kg/m³.**

Solution

Total resistance = 2200 N

Rolling resistance = 20 % of 2200 = 440 N

Surface drag = 10 % of 2200 = 220 N

Total resistance = Surface drag + Form drag + Rolling resistance

2200 = 220 + 440 + Form drag

$\therefore$ Form drag = 1540 N

From drag = $C_D \frac{1}{2} \rho A U^2$

$$1540 = C_D \frac{1.2}{2} \times 6 \left(\frac{90 \times 1000}{3600}\right)^2$$

$$C_D = \frac{1540}{2250} = 0.684$$ Ans

14.13 A racing car is fitted with an inverted aerofoil inclined at 10° to the horizontal. At this angle C_D = 0.35 and C_L = 1.4. The length of the aerofoil is 1.5 m and the chord is 1 m. The car length and body surface area are 4.5 m and 12 m² respectively, and the skin friction coefficient is given by $0.0741/(R_{e_l})^{0.2}$ Where R_{e_l} is based on car length. The car weight is 12.5 kN and the rolling resistance is 40 N per kN of normal force between the tyres and the road surface. Assuming that the form drag on the car is 450 N, determine the power required to maintain a constant velocity of 50 m/s. Assume ambient conditions of 1 bar and 12°C. Take R = 287 J/kg K and μ = 17.5 × 10^{-5} kg/ms.

Solution

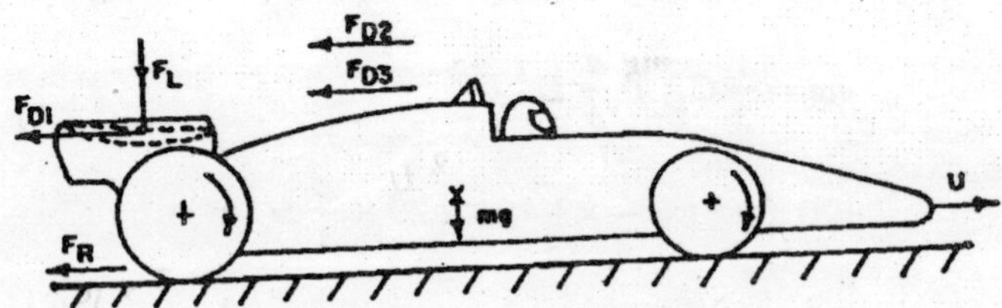

Fig. 14.27. Racing car P. 14.13

$$p = \rho RT$$

$$1 \times 10^5 = \rho \times 287 \times 285$$

$$\rho = 1.222 \ kg/m^3$$

$$R_e = \frac{\rho U l}{\mu} = \frac{1.222 \times 50 \times 4.5}{17.5 \times 10^{-5}} = 15.7 \times 10^5$$

$$R_e^{0.2} = (15.7 \times 10^5)^{0.2} = 17.342$$

$$C_{D_2} = \frac{0.0741}{(R_{e_l})^{0.2}} = \frac{0.0741}{17.342} = 4.27 \times 10^{-3}$$

Now, $$F_{D_1} = C_{D_2} \frac{1}{2} \rho A U^2$$

$$= 0.35 \times \frac{1.222}{2} \times (1.5 \times 1) \times (50)^2 = 801.9375 \text{ N}$$

$$F_{D3} = C_{D3} \frac{1}{2} \rho A U^2$$

$$= \left(4.27 \times 10^{-3}\right) \times \frac{1.222}{2} \times 12 \times (50)^2 = 78.2691 \text{ N}$$

$F_{D3} = 450$ N

Now the total normal force F_N between the car wheels and the road surface is equal to the sum of the car weight and the downward lift force of the aerofoil.

$$F_N = F_L + mg \qquad (1)$$

$$F_L = C_L \frac{1}{2} \rho A U^2$$

$$= 1.4 \times \frac{1.222}{2} \times (1.5 \times 1) \times (50)^2 = 3202.5 \text{ N}$$

$$mg = 12.5 \times 1000 = 12500 \text{ N}$$

Substituting in equation (1)

$$F_N = 3202.5 + 12500 = 15702.5 \text{ N} = 15.7025 \text{ kN}$$

Therefore,

$$F_R = 40 F_N = 40 \times 15.7025 = 628.1 \text{ N}$$

Total resistance to motion

$$F = F_{D1} + F_{D2} + F_{D3} + F_R$$

$$= 801.9375 + 78.2691 + 450 + 628.100 = 1958.3$$

$$P = F \times U / 1000 = \frac{1958.3 \times 50}{1000} = 97.9153 \text{ kW}$$ Ans

14.14 The foils of a hydrofoil type watercraft have an effective area of 0.7 m². Their coefficients of lift and drag are 1.6 and 0.5 respectively. The total mass of the craft in running trim is 1800 kg. Determine the minimum speed which the craft is supported by the hydrofoils. At this speed, find the power required to overcome water resistance. If the craft is fitted with a 110 kW engine, estimate its top speed.

Solution

Minimum spped is when $F_L = mg$

$$= C_L \frac{1}{2} \rho A U_{min}^2$$

or, $$1800 \times 9.81 = 1.6 \times \frac{1000}{2} \times 0.7\, U_{min}^2$$

or, $$U_{min}^2 = 31.532$$

$$U_{min} = 5.615 \text{ m/s}$$

At U_{min},

$$F_D = \frac{C_D}{C_L} F_L = \frac{0.5}{1.6} \times 1800 \times 9.81 = 5518.125 \text{ N}$$

$$p = \frac{F_D U_{min}}{1000} = \frac{5518.125 \times 5.615}{1000} = 30.98 \text{ kW}$$

At increased speeds, craft will ride higher in the water so that the lifting area is decreassed such that $F_L = W$, Thus

$$P_{max} = F_D\, U_{max} = \frac{C_D}{C_L}(mg)\, U_{max}$$

$$110 \times 1000 = \frac{0.5}{1.6} \times 1800 \times 9.81 \times U_{max}$$

$U_{max} =$ 19.934 m/s Ans

EXERCISES

A.THEORY

14.1 Define the following terms. (a) Drag, (b) Lift, (c) Coefficient of drag and (d) Coefficient of lift.

14.2 Define the following terms related to drag force. (a) Skin friction drag (b) Pressure drag, (c) Profile drag, (d) Deformation drag, (e) Induced drag and (f) wave drag.

14.3 Differentiate between streamlined body and bluff body.

14.4 Differentiate between shear drag and form drag. Show how these can be calculated for flow over a flat plate.

14.5 What are coefficients of drag and lift ? Show that these are dependent on Reynold's number and characteristic area of te body immersed.

14.6 Draw C_D v/s R_e diagram for a sphere and discuss all the peculiarities of the curve.

14.7 Draw pressure distribution around a sphere and compare with the ideal flow.

14.8 Draw C_D v/s R_e diagram for a sphere in Reynold's number range $0.01 < R_e < 10^6$ and explain why C_D suddenly drops at $R_e \approx 3 \times 10^5$.

14.9 Write a formula of stokes. Stokes formula is valid upto $R_e \approx 0.2$. Why is it necessary for making correction or improvement by Carl W Oseen and makes it useable up to $R_e = 100$?

14.10 Prove the formula

$$C_D = 24/R_e$$

Write down the limitations of the formula.

14.11 Explain why C_D practically independent of Reynold's number in the range of 0.4 to 0.5?

14.12 Explain why C_D suddenly rises from 0.4 to 0.5 and suddenly drops from 0.5 to 0.2.

14.13 Explain why the drag of a rough sphere is less than the drag of a smooth curve. Explain with the help of examples.

B. UNSOLVED PROBLEMS

14.1 A flat plate 1.5 m × 1.5 m moves at 50 km/hr in a stationary air of density 1.15 kg/m³. If the coefficient of drag and lift are 0.15 and 0.75 respectively, determine (a) lift force, (b) the drag force, (c) the resultant force, and (d) the power required to keep the plate in motion. [187.15 N, 37.425 N, 190.85 N, 0.5189 kW]

14.2 An aeroplane of gross weight 29.43 kN is flying at an angle of attack of 5° at a speed of 30 m/s. What is the area of the wing, required to support this load ? Take $C_L = 0.75$ and $\rho = 1.18$ kg/m³. [73.89 m²]

14.3 An aeroplane is travelling at 88 m/s and its rudder consists of a flat surface of area 0.4 m². Find the force on the rudder normal to the direction of flight when turned through an angle of 4°. Assume C_L for a flat surface of 4° incidence to be 0.7 and the density of air to be 1.2 kg/m³ [1276.189 N]

14.4 A circular disc 3 m in diameter is held normal to a 26.4 m/s using a density $\rho = 1.2$ kg/m³. What force is required to hold it at rest ? Take $C_D = 1.1$ for disc. [3250.75 N]

14.5 An aerofoil is moving at an altitude of 3000 m through the standard atmosphere with a velocity of 100 m/s. If the shape of the aerofoil is Clark Y, having a chord of 2 m and span equal to 12 m with an angle of attack $\alpha = 5.4°$ find the weight which the wing will carry and power required to drive it. Take C_L and C_D for Clark Y at 5.4°, are 0.46 and 0.02 respectively. Take density of air to be 1.25 kg/m³. [3000 N, 300 kW]

14.6 Calculate the diameter of a parachute to be used for dropping an object weighing 981 N so that maximum terminal velocity of dropping is 5 m/s. The drag coefficient for parachute, which may be treated as hemispherical is 1.3. Take $\rho = 1.216$ kg/m³. [7.95 m]

14.7 The drag coefficient of a parachute is 1.33. How large a parachute would be required to limit the terminal velocity of a 1000 N parachutist and parachute combination of a value no greater than the free fall velocity from a 3 m height? [5.15 m]

14.8 A 900 mm × 1200 mm plate moves at 13.5 m/s in still air at an angle of attack 12° with the horizontal using $C_D = 0.17$ and $C_L = 0.72$, determine (a) The resultant force exerted by the air on the plate, (b) The H.P. required to make the plate moving. Take $\rho = 1.2$ kg/m³. [$F_L = 84.75$ N $F_D = 20$ N, $F_R = 87$ N]

14.9 A kite weighing 1 kg and having an area of 1 m² makes an angle of 7° to the horizontal when flying in a wind of 36 km/hr. If pull on the string attached to the kite is 49 N and if it is inclined to horizontal at 45°, calculate the lift and drag coefficient. [0.712, 0.55]

14.10 A kite 0.8 m to 0.8 m weight 0.4 kg assumes an angle of 12° to the horizontal. The string attached to the kite makes an angle of 45° to the horizontal. The pull on the string is 24.525 N when the wind is blowing at a speed of 30 km/hr. Find the corresponding coefficient of drag and lift. Take density of air as 1.25 kg/m³. [$C_D = 0.624$, $C_L = 0.765$]

14.11 Determine the velocity of fall of rain drops of a 30×10^{-2} mm diameter, density 1.2 kg/m³ and kinematic viscosity 0.15 stokes/s. [U = 3.27 mm/s]